T0228971

In Praise of *Digital Design: An Embedded Systems Approach Using VHDL*

"Peter Ashenden is leading the way towards a new curriculum for educating the next generation of digital logic designers. Recognizing that digital design has moved from being gate-centric assembly of custom logic to processor-centric design of embedded systems, Dr. Ashenden has shifted the focus from the gate to the modern design and integration of complex integrated devices that may be physically realized in a variety of forms. Dr. Ashenden does not ignore the fundamentals, but treats them with suitable depth and breadth to provide a basis for the higher-level material. As is the norm in all of Dr. Ashenden's writing, the text is lucid and a pleasure to read. The book is illustrated with copious examples and the companion Web site offers all that one would expect in a text of such high quality."

— GRANT MARTIN,   Chief Scientist, Tensilica Inc.

"Dr. Ashenden has written a textbook that enables students to obtain a much broader and more valuable understanding of modern digital system design. Readers can be sure that the practices described in this book will provide a strong foundation for modern digital system design using hardware description languages."

— GARY SPIVEY,   George Fox University

"The convergence of miniaturized, sophisticated electronics into handheld, low-power embedded systems such as cell phones, PDAs, and MP3 players depends on efficient, digital design flows. Starting with an intuitive exploration of the basic building blocks, Digital Design: An Embedded Systems Approach introduces digital systems design in the context of embedded systems to provide students with broader perspectives. Throughout the text, Peter Ashenden's practical approach engages students in understanding the challenges and complexities involved in implementing embedded systems."

— GREGORY D. PETERSON,   University of Tennessee

"Digital Design: An Embedded Systems Approach *places emphasis on larger systems containing processors, memory, and involving the design and interface of I/O functions and application-specific accelerators. The book's presentation is based on a contemporary view that reflects the real-world digital system design practice. At a time when the university curriculum is generally lagging significantly behind industry development, this book provides much needed information to students in the areas of computer engineering, electrical engineering and computer science."*

— DONALD HUNG, San Jose State University

"Digital Design: An Embedded Systems Approach *presents the design flow of circuits and systems in a way that is both accessible and up-to-date. Because the use of hardware description languages is state-of-the-art, it is necessary that students learn how to use these languages along with an appropriate methodology. This book presents a modern approach for designing embedded systems starting with the fundamentals and progressing up to a complete system—it is application driven and full of many examples. I will recommend this book to my students."*

— GOERAN HERRMANN, TU Chemnitz

"Digital Design: An Embedded Systems Approach *is surprisingly easy to read despite the complexity of the material. It takes the reader in a journey from the basics to a real understanding of digital design by answering the 'whys' and 'hows'—it is persuasive and instructive as it moves deeper and deeper into the material."*

— ANDREY KOPTYUG, Mid Sweden University

"This up-to-date text on digital design is written in a very accessible style using a modern design methodology and the real world of embedded systems as its contexts. Digital Design: An Embedded Systems Approach provides excellent coverage of all aspects of the design of embedded systems, with chapters not just on logic design itself, but also on processors, memories, input/output interfacing and implementation technologies. It's particularly good at emphasizing the need to consider more than just logic design when designing a digital system: the design has to be implemented in the real world of engineering, where a whole variety of constraints, such as circuit area, circuit interconnections, interfacing requirements, power and performance, must be considered. For those who think logic design is mundane, this book brings the subject to life."

— ROLAND IBBETT, University of Edinburgh

Digital Design
An Embedded Systems Approach
Using VHDL

ABOUT THE AUTHOR

Peter J. Ashenden is an Adjunct Associate Professor at Adelaide University and the founder of Ashenden Designs, a consulting business specializing in electronics design automation (EDA).

From 1990 to 2000, Dr. Ashenden was a member of the faculty in the Department of Computer Science at Adelaide. He developed curriculum and taught in a number of areas for both the Computer Science and the Electrical and Electronic Engineering departments. Topics included computer organization, computer architecture, digital logic design, programming and algorithms, at all levels from undergraduate to graduate courses. He was also actively involved in academic administration at many levels within the university.

In 2000, Dr. Ashenden established Ashenden Designs. His services include training development and delivery, advising on design methodology, research in EDA tool technology, development of design languages, and standards writing. His clients include industry and government organization in the United States, Europe and SE Asia.

Since 1992, Dr. Ashenden has been involved in the IEEE VHDL standards committees, and continues to play a major role in ongoing development of the language. From 2003 to 2005 he was Chair of the IEEE Design Automation Standards Committee, which oversees development of all IEEE standards in EDA. He is currently Technical Editor for the VHDL, VHDL-AMS, and Rosetta specification language standards.

In addition to his research publications, Dr. Ashenden is author of *The Designer's Guide to VHDL* and *The Student's Guide to VHDL*, and coauthor of *The System Designer's Guide to VHSL-AMS* and *VHDL-2007: Just the New Stuff*. His VHDL books are highly regarded and are the best-selling references on the subject. From 2000 to 2004, he was Series Coeditor of the Morgan Kaufmann Series on Systems on Silicon, and from 2001 to 2004 he was a member of the Editorial Board of *IEEE Design and Test of Computers* magazine.

Dr. Ashenden is a Senior Member of the IEEE and the IEEE Computer Society. He is also a volunteer Senior Firefighter of 12 years standing with the South Australian Country Fire Service.

Digital Design
An Embedded Systems Approach
Using VHDL

PETER J. ASHENDEN

Adjunct Associate Professor
School of Computer Science
University of Adelaide

ELSEVIER

AMSTERDAM • BOSTON • HEIDELBERG • LONDON
NEW YORK • OXFORD • PARIS • SAN DIEGO
SAN FRANCISCO • SINGAPORE • SYDNEY • TOKYO
Morgan Kaufmann is an imprint of Elsevier

Publishing Director	Joanne Tracy
Publisher	Denise E. M. Penrose
Acquisitions Editor	Charles Glaser
Publishing Services Manager	George Morrison
Senior Production Editor	Dawnmarie Simpson
Developmental Editor	Nate McFadden
Editorial Assistant	Kimberlee Honjo
Production Assistant	Lianne Hong
Cover Design	Eric DeCicco
Cover Image	Getty Images
Composition	diacriTech
Technical Illustration	Peter Ashenden
Copyeditor	JC Publishing
Proofreader	Janet Cocker
Indexer	Joan Green
Interior printer	Sheridan Books, Inc.
Cover printer	Phoenix Color, Inc.

Morgan Kaufmann Publishers is an imprint of Elsevier.
30 Corporate Drive, Suite 400, Burlington, MA 01803, USA

This book is printed on acid-free paper.

© 2008 by Elsevier Inc. All rights reserved.

Designations used by companies to distinguish their products are often claimed as trademarks or registered trademarks. In all instances in which Morgan Kaufmann Publishers is aware of a claim, the product names appear in initial capital or all capital letters. Readers, however, should contact the appropriate companies for more complete information regarding trademarks and registration.

No part of this publication may be reproduced, stored in a retrieval system, or transmitted in any form or by any means—electronic, mechanical, photocopying, scanning, or otherwise—without prior written permission of the publisher.

Permissions may be sought directly from Elsevier's Science & Technology Rights Department in Oxford, UK: phone: (+44) 1865 843830, fax: (+44) 1865 853333, E-mail: permissions@elsevier.com. You may also complete your request online via the Elsevier homepage (http://elsevier.com), by selecting "Support & Contact" then "Copyright and Permission" and then "Obtaining Permissions."

Library of Congress Cataloging-in-Publication Data
Ashenden, Peter J.
  Digital design: an embedded systems approach using VHDL / Peter J. Ashenden.
    p. cm.
  Includes index.
  ISBN 978-0-12-369528-4 (pbk. : alk. paper) 1. Embedded computer systems. 2. VHDL (Computer hardware description language) 3. System design. I. Title.
  TK7895.E42.A69 2007
  621.39'16–dc22

                                                        2007023241

ISBN: 978-0-12-369528-4

For information on all Morgan Kaufmann publications,
visit our Web site at *www.mkp.com* or *www.books.elsevier.com*

Printed and bound by CPI Group (UK) Ltd, Croydon, CR0 4YY

Transferred to digital print 2012

Working together to grow
libraries in developing countries

www.elsevier.com | www.bookaid.org | www.sabre.org

ELSEVIER   BOOK AID International   Sabre Foundation

To my daughter, Eleanor
—PA

CONTENTS

PREFACE

APPROACH

This book provides a foundation in digital design for students in computer engineering, electrical engineering and computer science courses. It deals with digital design as an activity in a larger systems design context. Instead of focusing on gate-level design and aspects of digital design that have diminishing relevance in a real-world design context, the book concentrates on modern and evolving knowledge and design skills.

Most modern digital design practice involves design of embedded systems, using small microcontrollers, larger CPUs/DSPs, or hard or soft processor cores. Designs involve interfacing the processor or processors to memory, I/O devices and communications interfaces, and developing accelerators for operations that are too computationally intensive for processors. Target technologies include ASICs, FPGAs, PLDs and PCBs. This is a significant change from earlier design styles, which involved use of small-scale integrated (SSI) and medium-scale integrated (MSI) circuits. In such systems, the primary design goal was to minimize gate count or IC package count. Since processors had lower performance and memories had limited capacity, a greater proportion of system functionality was implemented in hardware.

While design practices and the design context have evolved, many textbooks have not kept track. They continue to promote practices that are largely obsolete or that have been subsumed into computer-aided design (CAD) tools. They neglect many of the important considerations for modern designers. This book addresses the shortfall by taking an approach that embodies modern design practice. The book presents the view that digital logic is a basic abstraction over analog electronic circuits. Like any abstraction, the digital abstraction relies on assumptions being met and constraints being satisfied. Thus, the book includes discussion of the electrical and timing properties of circuits, leading to an understanding of how they influence design at higher levels of abstraction. Also, the book teaches a methodology based on using abstraction to manage complexity, along with principles and methods for making design trade-offs. These intellectual tools allow students to track evolving design practice after they graduate.

Perhaps the most noticeable difference between this book and its predecessors is the omission of material on Karnaugh maps and related logic optimization techniques. Some reviewers of the manuscript argued that such techniques are still of value and are a necessary foundation for students learning digital design. Certainly, it is important for students to understand that a given function can be implemented by a variety of equivalent circuits, and that different implementations may be more or less optimal under different constraints. This book takes the approach of presenting Boolean algebra as the foundation for gate-level circuit transformation, but leaves the details of algorithms for optimization to CAD tools. The complexity of modern systems makes it more important to raise the level of abstraction at which we work and to introduce embedded systems early in the curriculum. CAD tools perform a much better job of gate-level optimization than we can do manually, using advanced algorithms to satisfy relevant constraints. Techniques such as Karnaugh maps do have a place, for example, in design of specialized hazard-free logic circuits. Thus, students can defer learning about Karnaugh maps until an advanced course in VLSI, or indeed, until they encounter the need in industrial practice. A web search will reveal many sources describing the techniques in detail, including an excellent article in Wikipedia.

The approach taken in this book makes it relevant to Computer Science courses, as well as to Computer Engineering and Electrical Engineering courses. By treating digital design as part of embedded systems design, the book will provide the understanding of hardware needed for computer science students to analyze and design systems comprising both hardware and software components. The principles of abstraction and complexity management using abstraction presented in the book are the same as those underlying much of computer science and software engineering.

Modern digital design practice relies heavily on models expressed in hardware description languages (HDLs), such as Verilog and VHDL. HDL models are used for design entry at the abstract behavioral level and for refinements at the register transfer level. Synthesis tools produce gate-level HDL models for low-level verification. Designers also express verification environments in HDLs. This book emphasizes HDL-based design and verification at all levels of abstraction. The present version uses VHDL for this purpose. A second version, *Digital Design: An Embedded Systems Approach Using Verilog*, substitutes Verilog for the same purpose.

OVERVIEW

For those who are musically inclined, the organization of this book can be likened to a two-act opera, complete with overture, intermezzo, and finale.

Chapter 1 forms the overture, introducing the themes that are to follow in the rest of the work. It starts with a discussion of the basic ideas of the digital abstraction, and introduces the basic digital circuit elements.

It then shows how various non-ideal behaviors of the elements impose constraints on what we can design. The chapter finishes with a discussion of a systematic process of design, based on models expressed in a hardware description language.

Act I of the opera comprises Chapters 2 through 5. In this act, we develop the themes of basic digital design in more detail.

Chapter 2 focuses on combinational circuits, starting with Boolean algebra as the theoretical underpinning and moving on to binary coding of information. The chapter then surveys a range of components that can be used as building blocks in larger combinational circuits, before returning to the design methodology to discuss verification of combinational circuits.

Chapter 3 expands in some detail on combinational circuits used to process numeric information. It examines various binary codes for unsigned integers, signed integers, fixed-point fractions and floating-point real numbers. For each kind of code, the chapter describes how some arithmetic operations can be performed and looks at combinational circuits that implement arithmetic operations.

Chapter 4 introduces a central theme of digital design, sequential circuits. The chapter examines several sequential circuit elements for storing information and for counting events. It then describes the concepts of a datapath and a control section, followed by a description of the clocked synchronous timing methodology.

Chapter 5 completes Act I, describing the use of memories for storing information. It starts by introducing the general concepts that are common to all kinds of semiconductor memory, and then focuses on the particular features of each type, including SRAM, DRAM, ROM and flash memories. The chapter finishes with a discussion of techniques for dealing with errors in the stored data.

The intermezzo, Chapter 6, is a digression away from functional design into physical design and the implementation fabrics used for digital systems. The chapter describes the range of integrated circuits that are used for digital systems, including ASICSs, FPGAs and other PLDs. The chapter also discusses some of the physical and electrical characteristics of implementation fabrics that give rise to constraints on designs.

Act II of the opera, comprising Chapters 7 through 9, develops the embedded systems theme.

Chapter 7 introduces the kinds of processors that are used in embedded systems and gives examples of the instructions that make up embedded software programs. The chapter also describes the way instructions and data are encoded in binary and stored in memory and examines ways of connecting the processor with memory components.

Chapter 8 expands on the notion of input/output (I/O) controllers that connect an embedded computer system with devices that sense and affect real-world physical properties. It describes a range of devices that are used in embedded computers and shows how they are accessed by an embedded processor and by embedded software.

Chapter 9 describes accelerators, that is, components that can be added to embedded systems to perform operations faster than is possible with embedded software running on a processor core. This chapter uses an extended example to illustrate design considerations for accelerators, and shows how an accelerator interacts with an embedded processor.

The finale, Chapter 10, is a coda that returns to the theme of design methodology introduced in Chapter 1. The chapter describes details of the design flow and discusses how aspects of the design can be optimized to better meet constraints. It also introduces the concept of design for test, and outlines some design for test tools and techniques. The opera finishes with a discussion of the larger context in which digital systems are designed.

After a performance of an opera, there is always a lively discussion in the foyer. This book contains a number of appendices that correspond to that aspect of the opera. Appendix A provides sample answers for the Knowledge Test Quiz sections in the main chapters. Appendix B provides a quick refresher on electronic circuits. Appendix C is a summary of the subset of VHDL used for synthesis of digital circuits. Finally, Appendix D is an instruction-set reference for the Gumnut embedded processor core used in examples in Chapters 7 through 9.

For those not inclined toward classical music, I apologize if the preceding is not a helpful analogy. An analogy with the courses of a feast came to mind, but potential confusion among readers in different parts of the world over the terms appetizer, entrée and main course make the analogy problematic. The gastronomically inclined reader should feel free to find the correspondence in accordance with local custom.

COURSE ORGANIZATION

This book covers the topics included in the Digital Logic knowledge area of the Computer Engineering Body of Knowledge described in the IEEE/ACM *Curriculum Guidelines for Undergraduate Degree Programs in Computer Engineering*. The book is appropriate for a course at the sophomore level, assuming only previous introductory courses in electronic circuits and computer programming. It articulates into junior and senior courses in embedded systems, computer organization, VLSI and other advanced topics.

For a full sequence in digital design, the chapters of the book can be covered in order. Alternatively, a shorter sequence could draw on Chapter 1 through Chapter 6 plus Chapter 10. Such a sequence would defer material in Chapters 7 through 9 to a subsequent course on embedded systems design.

For either sequence, the material in this book should be supplemented by a reference book on the VHDL language. The course work should also include laboratory projects, since hands-on design is the best way to reinforce the principles presented in the book.

WEB SUPPLEMENTS

No textbook can be complete nowadays without supplementary material on a website. For this book, resources for students and instructors are available at the website:

textbooks.elsevier.com/9780123695284

For students, the website contains:

▶ Source code for all of the example HDL models in the book

▶ Tutorials on the VHDL and Verilog hardware description languages

▶ An assembler for the Gumnut processor described in Chapter 7 and Appendix D

▶ A link to the ISE WebPack FPGA EDA tool suite from Xilinx

▶ A link to the ModelSim Xilinx Edition III VHDL and Verilog simulator from Mentor Graphics Corporation

▶ A link to an evaluation edition of the Synplify Pro PFGA synthesis tool from Synplicity, Inc. (see inside back cover for more details).

▶ Tutorials on use of the EDA tools for design projects

For instructors, the website contains a protected area with additional resources:

▶ An instructor's manual

▶ Suggested lab projects

▶ Lecture notes

▶ Figures from the text in JPG and PPT formats

Instructors are invited to contribute additional material for the benefit of their colleagues.

Despite the best efforts of all involved, some errors have no doubt crept through the review and editorial process. A list of detected errors will be available accumulated on the website mentioned above. Should you detect such an error, please check whether it has been previously recorded. If not, I would be grateful for notice by email to

peter@ashenden.com.au

I would also be delighted to hear feedback about the book and supplementary material, including suggestions for improvement.

ACKNOWLEDGMENTS

This book arose from my long-standing desire to bring a more modern approach to the teaching of digital design. I am deeply grateful to the folks at Morgan Kaufmann Publishers for supporting me in realizing this goal, and for their guidance and advice in shaping the book. Particular thanks go to Denise Penrose, Publisher; Nate McFadden, Developmental Editor and Kim Honjo, Editorial Assistant. Thanks also to Dawnmarie Simpson at Elsevier for meticulous attention to detail and for making the production process go like clockwork.

The manuscript benefited from comprehensive reviews by Dr. A. Bouridane, Queen's University Belfast; Prof. Goeran Herrmann, Chemnitz University of Technology; Prof. Donald Hung, San Jose State University; Prof. Roland Ibbett, University of Edinburgh; Dr. Andrey Koptyug, Mid Sweden University; Dr. Grant Martin, Tensilica, Inc.; Dr. Gregory D. Peterson, University of Tennessee; Brian R. Prasky, IBM; Dr. Gary Spivey, George Fox University; Dr. Peixin Zhong, Michigan State University; and an anonymous reviewer from Rensselaer Polytechnic Institute. Also, my esteemed colleague Jim Lewis of SynthWorks Design, Inc., provided technical reviews of the VHDL code and related text. To all of these, my sincere thanks for their contributions. The immense improvement from my first draft to the final draft is due to their efforts.

The book and the associated teaching materials also benefited from field testing: in alpha form by myself at the University of Adelaide and by Dr. Monte Tull at The University of Oklahoma; and in beta form by James Sterbenz at The University of Kansas. To them and to their students, thanks for your forbearance with the errors and for your valuable feedback.

INTRODUCTION AND METHODOLOGY

1

This first chapter introduces some of the fundamental ideas underlying design of modern digital systems. We cover quite a lot of ground, but at a fairly basic level. The idea is to set the context for more detailed discussion in subsequent chapters.

We start by looking at the basic circuit elements from which digital systems are built, and seeing some of the ways in which they can be put together to perform a required function. We also consider some of the nonideal effects that we need to keep in mind, since they impose constraints on what we can design. We then focus on a systematic process of design, based on models expressed in a hardware description language. Approaching the design process systematically allows us to develop complex systems that meet modern application requirements.

1.1 DIGITAL SYSTEMS AND EMBEDDED SYSTEMS

This book is about digital design. Let's take a moment to explore those two words. *Digital* refers to electronic circuits that represent information in a special way, using just two voltage levels. The main rationale for doing this is to increase the reliability and accuracy of the circuits, but as we will see, there are numerous other benefits that flow from the digital approach. We also use the term *logic* to refer to digital circuits. We can think of the two voltage levels as representing truth values, leading us to use rules of logic to analyze digital circuits. This gives us a strong mathematical foundation on which to build. The word *design* refers to the systematic process of working out how to construct circuits that meet given requirements while satisfying constraints on cost, performance, power consumption, size, weight and other properties. In this book, we focus on the design aspects and build a methodology for designing complex digital systems.

Digital circuits have quite a long and interesting history. They were preceded by mechanical systems, electromechanical systems, and analog electronic systems. Most of these systems were used for numeric computations in business and military applications, for example, in ledger calculations and in computing ballistics tables. However, they suffered from numerous disadvantages, including inaccuracy, low speed, and high maintenance.

Early digital circuits, built in the mid-twentieth century, were constructed with relays. The contacts of a relay are either open, blocking current flow, or closed, allowing current to flow. Current controlled in this manner by one or more relays could then be used to switch other relays. However, even though relay-based systems were more reliable than their predecessors, they still suffered from reliability and performance problems.

The advent of digital circuits based on vacuum tubes and, subsequently, transistors led to major improvements in reliability and performance. However, it was the invention of the *integrated circuit* (IC), in which multiple transistors were fabricated and connected together, that really enabled the "digital revolution." As manufacturing technology has developed, the size of transistors and the interconnecting wires has shrunk. This, along with other factors, has led to ICs, containing billions of transistors and performing complex functions, becoming commonplace now.

At this point, you may be wondering how such complex circuits can be designed. In your electronic circuits course, you may have learned how transistors operate, and that their operation is determined by numerous parameters. Given the complexity of designing a small circuit containing a few transistors, how could it be possible to design a large system with billions of transistors?

The key is *abstraction*. By abstraction, we mean identifying aspects that are important to a task at hand, and hiding details of other aspects. Of course, the other aspects can't be ignored arbitrarily. Rather, we make assumptions and follow disciplines that allow us to ignore those details while we focus on the aspects of interest. As an example, the *digital abstraction* involves only allowing two voltage levels in a circuit, with transistors being either turned "on" (that is, fully conducting) or turned "off" (that is, not conducting). One of the assumptions we make in supporting this abstraction is that transistors switch on and off virtually instantaneously. One of the design disciplines we follow is to regulate switching to occur within well-defined intervals of time, called "clock periods." We will see many other assumptions and disciplines as we proceed. The benefit of the digital abstraction is that it allows us to apply much simpler analysis and design procedures, and thus to build much more complex systems.

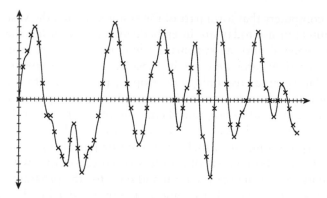

FIGURE 1.1   A pressure waveform of a sound, continuously varying over time, and the discrete representation of the waveform in a digital system.

The circuits that we will deal with in this book all perform functions that involve manipulating information of various kinds over time. The information might be an audio signal, the position of part of a machine, or the temperature of a substance. The information may change over time, and the way in which it is manipulated may vary with time.

Digital systems are electronic circuits that represent information in discrete form. An example of the kind of information that we might represent is an audio sound. In the real world, a sound is a continuously varying pressure waveform, which we might represent mathematically using a continuous function of time. However, representing that function with any significant precision as a continuously varying electrical signal in a circuit is difficult and costly, due to electrical noise and variation in circuit parameters. A digital system, on the other hand, represents the signal as a stream of discrete values sampled at discrete points in time, as shown in Figure 1.1. Each sample represents an approximation to the pressure value at a given instant. The approximations are drawn from a discrete set of values, for example, the set $\{-10.0, -9.9, -9.8, \dots, -0.1, 0.0, 0.1, \dots, 9.9, 10.0\}$. By limiting the set of values that can be represented, we can encode each value with a unique combination of digital values, each of which is either a low or high voltage. We shall see exactly how we might do that in Chapter 2. Furthermore, by sampling the signal at regular intervals, say, every 50µs, the rate and times at which samples arrive and must be processed is predictable.

Discrete representations of information and discrete sequencing in time are fundamental abstractions. Much of this book is about how to choose appropriate representations, how to process information thus represented, how to sequence the processing, and how to ensure that the assumptions supporting the abstractions are maintained.

The majority of digital systems designed and manufactured today are *embedded systems*, in which much of the processing work is done by one or more computers that form part of the system. In fact, the vast majority of computers in use today are in embedded systems, rather than in PCs and other general purpose systems. Early computers were large systems in their own right, and were rarely considered as components of larger systems. However, as technology developed, particularly to the stage of IC technology, it became practical to embed small computers as components of a circuit and to program them to implement part of the circuit's functionality. Embedded computers usually do not take the same form as general purpose computers, such as desktop or portable PCs. Instead, an embedded computer consists of a *processor core*, together with memory components for storing the program and data for the program to run on the processor core, and other components for transferring data between the processor core and the rest of the system.

The programs running on processor cores form the *embedded software* of a system. The way in which embedded software is written bears both similarities and differences with software development for general purpose computers. It is a large topic area in its own right and is beyond the scope of this book. Nonetheless, since we are dealing with embedded systems in this book, we need to address embedded software at least at a basic level. We will return to the topic as part of our discussion of interfacing with embedded processor cores in Chapters 8 and 9.

Since most digital systems in use today are embedded systems, most digital design practice involves developing the interface circuitry around processor cores and the application-specific circuitry to perform tasks not done by the cores. That is why this book deals specifically with digital design in the context of embedded systems.

1.2 BINARY REPRESENTATION AND CIRCUIT ELEMENTS

The simplest discrete representation that we see in a digital system is called a *binary* representation. It is a representation of information that can have only two values. Examples of such information include:

▶ whether a switch is open or closed

▶ whether a light is on or off

▶ whether a microphone is active or muted

We can think of these as logical conditions: each is either true or false. In order to represent them in a digital circuit, we can assign a high voltage level to represent the value true, and a low voltage level to represent the value false. (This is just a convention, called *positive logic*, or *active-high logic*. We could make the reverse assignment, leading to *negative logic*, or *active-low logic*, which we will discuss in Chapter 2.) We often use the values 0 and 1 instead of false and true, respectively.

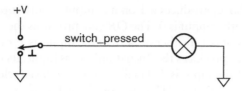

FIGURE 1.2   A circuit in which a switch controls a lamp.

The values 0 and 1 are binary (base 2) digits, or *bits*, hence the term binary representation.

The circuit shown in Figure 1.2 illustrates the idea of binary representation. The signal labeled switch_pressed represents the state of the switch. When the push-button switch is pressed, the signal has a high voltage, representing the truth of the condition, "the switch is pressed." When the switch is not pressed, the signal has a low voltage, representing the falsehood of the condition. Since illumination of the lamp is controlled by the switch, we could equally well have labeled the signal lamp_lit, with a high voltage representing the truth of the condition, "the lamp is lit," and a low voltage representing the falsehood of the condition.

A more complex digital circuit is shown in Figure 1.3. This circuit includes a light sensor with a digital output, dark, that is true (high voltage) when there is no ambient light, or false (low voltage) otherwise. The circuit also includes a switch that determines whether the digital signal lamp_enabled is low or high (that is, false or true, respectively). The symbol in the middle of the figure represents an *AND gate*, a digital circuit element whose output is only true (1) if both of its inputs are true (1). The output is false (0) if either input is false (0). Thus, in the circuit, the signal lamp_lit is true if lamp_enabled is true *and* dark is true, and is false otherwise. Given this behavior, we can apply laws of logic to analyze the circuit. For example, we can determine that if there is ambient light, the lamp will not light, since the logical AND of two conditions yields falsehood when either of the conditions is false.

The AND gate shown in Figure 1.3 is just one of several basic digital logic components. Some others are shown in Figure 1.4. The AND gate, as

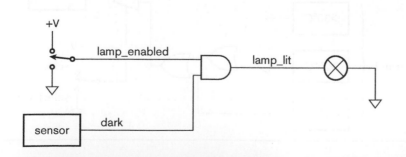

FIGURE 1.3   A digital circuit for a night-light that is only lit when the switch is on and the light sensor shows that it is dark.

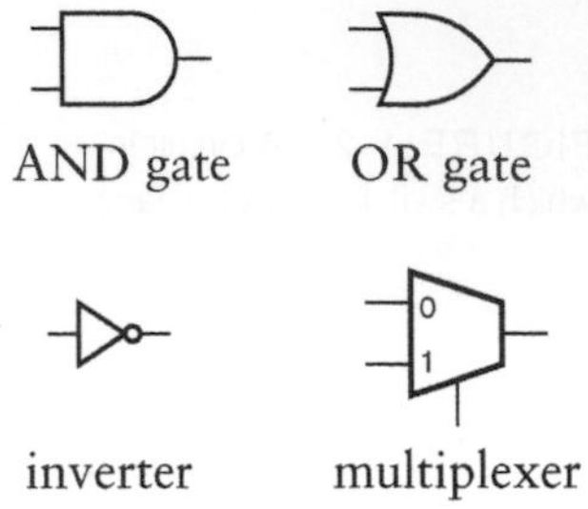

AND gate   OR gate

inverter   multiplexer

FIGURE 1.4   Basic digital logic gates.

we mentioned above, produces a 1 on its output if both inputs are 1, or a 0 on the output if either input is 0. The *OR gate* produces the "inclusive or" of its inputs. Its output is 1 if either or both of the inputs is 1, or 0 if both inputs are 0. The *inverter* produces the "negation" of its input. Its output is 1 if the input is 0, or 0 if the input is 1. Finally, the *multiplexer* selects between the two inputs labeled "0" and "1" based on the value of the "select" input at the bottom of the symbol. If the select input has the value 0, then the output has the same value as that on the "0" input, whereas if the select input has the value 1, then the output has the same value as that on the "1" input.

We can use these logic gates to build digital circuits that implement more complex logic functions.

EXAMPLE 1.1   Suppose a factory has two vats, only one of which is used at a time. The liquid in the vat in use needs to be at the right temperature, between 25°C and 30°C. Each vat has two temperature sensors indicating whether the temperature is above 25°C and above 30°C, respectively. The vats also have low-level sensors. The supervisor needs to be woken up by a buzzer when the temperature is too high or too low or the vat level is too low. He has a switch to select which vat is in use. Design a circuit of gates to activate the buzzer as required.

SOLUTION   For the selected vat, the condition for activating the buzzer is "temperature not above 25°C or temperature above 30°C, or level low." This can be implemented with a gate circuit for each vat. The switch can be used to control the select input of a multiplexer to choose between the circuit outputs for the two vats. The output of the multiplexer then activates the buzzer. The complete circuit is shown in Figure 1.5.

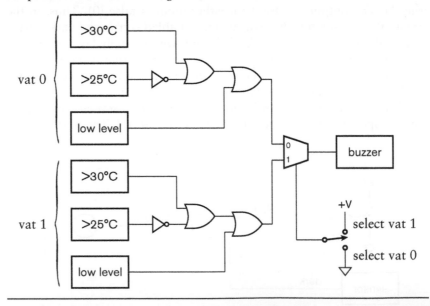

FIGUREV 1.5   The vat buzzer circuit.

Circuits such as those considered above are called *combinational*. This means that the values of the circuit's outputs at any given time are determined purely by combining the values of the inputs at that time. There is no notion of storage of information, that is, dependence on values at previous times. While combinational circuits are important as parts of larger digital systems, nearly all digital systems are *sequential*. This means that they do include some form of storage, allowing the values of outputs to be determined by both the current input values and previous input values.

One of the simplest digital circuit elements for storing information is called, somewhat prosaically, a *flip-flop*. It can "remember" a single bit of information, allowing it to "flip" and "flop" between a stored 0 state and a stored 1 state. The symbol for a *D flip-flop* is shown in Figure 1.6. It is called a "D" flip-flop because it has a single input, D, representing the value of the data to be stored: "D" for "data." It also has another input, clk, called the clock input, that indicates when the value of the D input should be stored. The behavior of the D flip-flop is illustrated in the *timing diagram* in Figure 1.7. A timing diagram is a graph of the values of one or more signals as they change with time. Time extends along the horizontal axis, and the signals of interest are listed on the vertical axis. Figure 1.7 shows the D input of the flip-flop changing at irregular intervals and the clk input changing periodically. A transition of clk from 0 to 1 is called a *rising edge* of the signal. (Similarly, a transition from 1 to 0 is called a *falling edge*.) The small triangular marking next to the clk input specifies that the D value is stored only on a rising edge of the clk input. At that time, the Q output changes to reflect the stored value. Any subsequent changes on the D input are ignored until the next rising edge of clk. A circuit element that behaves in this way is called *edge-triggered*.

While the behavior of a flip-flop does not depend on the clock input being periodic, in nearly all digital systems, there is a single clock signal that synchronizes all of the storage elements in the system. The system is composed of combinational circuits that perform logical functions on the values of signals and flip-flops that store intermediate results. As we

FIGURE 1.6   A D flip-flop.

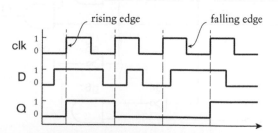

FIGURE 1.7   Timing diagram for a D flip-flop.

shall see, use of a single periodic synchronizing clock greatly simplifies design of the system. The clock operates at a fixed *frequency* and divides time into discrete intervals, called *clock periods*, or *clock cycles*. Modern digital circuits operate with clock frequencies in the range of tens to hundreds of megahertz (MHz, or millions of cycles per second), with high-performance circuits extending up to several gigahertz (GHz, or billions of cycles per second). Division of time into discrete intervals allows us to deal with time in a more abstract form. This is another example of abstraction at work.

EXAMPLE 1.2   Develop a sequential circuit that has a single data input signal, S, and produces an output Y. The output is 1 whenever S has the same value over three successive clock cycles, and 0 otherwise. Assume that the value of S for a given clock cycle is defined at the time of the rising clock edge at the end of the clock cycle.

SOLUTION   In order to compare the values of S in three successive clock cycles, we need to save the values of S for the previous two cycles and compare them with the current value of S. We can use a pair of D flip-flops, connected in a *pipeline* as shown in Figure 1.8, to store the values. When a clock edge occurs, the first flip-flop, ff1, stores the value of S from the preceding clock cycle. That value is passed onto the second flip-flop, ff2, so that at the next clock edge, ff2 stores the value of S from two cycles prior.

The output Y is 1 if and only if three successive value of S are all 1 or are all 0. Gates g1 and g2 jointly determine if the three values are all 1. Inverters g3, g4 and g5 negate the three values, and so gates g6 and g7 determine if the three values are all 0. Gate g8 combines the two alternatives to yield the final output.

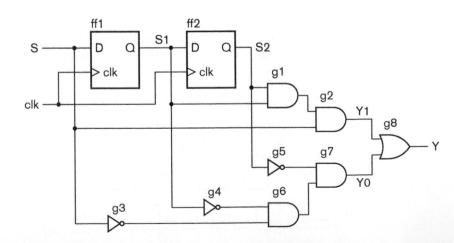

FIGURE 1.8   A sequential circuit for comparing successive bits of an input.

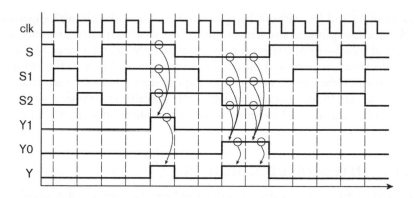

FIGURE 1.9 Timing diagram for the sequential comparison circuit.

Figure 1.9 shows a timing diagram of the circuit for a particular sequence of input values on S over several clock cycles. The outputs of the two flip-flops follow the value of S, but are delayed by one and two clock cycles, respectively. This timing diagram shows the value of S changing at the time of a clock edge. The flip-flop will actually store the value that is on S immediately before the clock edge. The circles and arrows indicate which signals are used to determine the values of other signals, leading to a 1 at the output. When all of S, S1 and S2 are 1, Y1 changes to 1, indicating that S has been 1 for three successive cycles. Similarly, when all of S, S1 and S2 are 0, Y0 changes to 1, indicating that S has been 0 for three successive cycles. When either of Y1 or Y0 is 1, the output Y changes to 1.

---

1. What are the two values used in binary representation?

2. If one input of an AND gate is 0 and the other is 1, what is the output value? What if both are 0, or both are 1?

3. If one input of an OR gate is 0 and the other is 1, what is the output value? What if both are 0, or both are 1?

4. What function is performed by a multiplexer?

5. What is the distinction between combinational and sequential circuits?

6. How much information is stored by a flip-flop?

7. What is meant by the terms *rising edge* and *falling edge*?

KNOWLEDGE
TEST QUIZ

1.3   REAL-WORLD CIRCUITS

In order to analyze and design circuits as we have discussed, we are making a number of assumptions that underlie the digital abstraction. We have assumed that a circuit behaves in an ideal manner, allowing us to think in terms of 1s and 0s, without being concerned about the circuit's electrical behavior and physical implementation. Real-world circuits, however, are made of transistors and wires forming part of a physical device or package. The electrical properties of the circuit elements, together with the physical properties of the device or package, impose a number of constraints on circuit design. In this section, we will briefly describe the physical structure of circuit elements and examine some of the most important properties and constraints.

1.3.1   INTEGRATED CIRCUITS

Modern digital circuits are manufactured on the surface of a small flat piece of pure crystalline silicon, hence the common term "silicon chip." Such circuits are called integrated circuits, since numerous components are integrated together on the chip, instead of being separate components. We will explore the process by which ICs are manufactured in more detail in Chapter 6. At this stage, however, we can summarize by saying that transistors are formed by depositing layers of semiconducting and insulating material in rectangular and polygonal shapes on the chip surface. Wires are formed by depositing metal (typically copper) on top of the transistors, separated by insulating layers. Figure 1.10 is a photomicrograph of a small area of a chip, showing transistors interconnected by wires.

FIGURE 1.10   Photomicrograph of a section of an IC.

The physical properties of the IC determine many important operating characteristics, including speed of switching between low and high voltages. Among the most significant physical properties is the minimum size of each element, the so-called *minimum feature size*. Early chips had minimum feature sizes of tens of microns (1 micron = $1 \mu m = 10^{-6} m$). Improvements in manufacturing technology has led to a steady reduction in feature size, from $10 \mu m$ in the early 1970s, through $1 \mu m$ in the mid 1980s, with today's ICs having feature sizes of 90nm or 65nm. As well as affecting circuit performance, feature size helps determine the number of transistors that can fit on an IC, and hence the overall circuit complexity. Gordon Moore, one of the pioneers of the digital electronics industry, noted the trend in increasing transistor count, and published an article on the topic in 1965. His projection of a continuing trend continues to this day, and is now known as Moore's Law. It states that the number of transistors that can be put on an IC for minimum component cost doubles every 18 months. At the time of publication of Moore's article, it was around 50 transistors; today, a complex IC has well over a billion transistors.

One of the first families of digital logic ICs to gain widespread use was the "transistor-transistor logic" (TTL) family. Components in this family use bipolar junction transistors connected to form logic gates.

The electrical properties of these devices led to widely adopted design standards that still influence current logic design practice. In more recent times, TTL components have been largely supplanted by components using "complementary metal-oxide semiconductor" (CMOS) circuits, which are based on field-effect transistors (FETs). The term "complementary" means that both n-channel and p-channel MOSFETs are used. (See Appendix B for a description of MOSFETS and other circuit components.) Figure 1.11 shows how such transistors are used in a CMOS circuit for an inverter. When the input voltage is low, the n-channel transistor at the bottom is turned off and the p-channel transistor at the top is turned on, pulling the output high. Conversely, when the input voltage is high, the p-channel transistor is turned off and the n-channel transistor is turned on, pulling the output low. Circuits for other logic gates operate similarly, turning combinations of transistors on or off to pull the output low or high, depending on the voltages at the inputs.

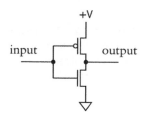

FIGURE 1.11   CMOS circuit for an inverter.

1.3.2 LOGIC LEVELS

The first assumption we have made in the previous discussion is that all signals take on appropriate "low" and "high" voltages, also called *logic levels*, representing our chosen discrete values 0 and 1. But what should those logic levels be? The answer is in part determined by the characteristics of the electronic circuits. It is also, in part, arbitrary, provided circuit designers and component manufacturers agree. As a consequence, there are now a number of "standards" for logic levels. One of the contributing factors to the early success of the TTL family was its adoption of uniform logic levels for all components in the family. These *TTL logic levels* still form the basis for standard logic levels in modern circuits.

Suppose we nominate a particular voltage, 1.4V, as our *threshold voltage*. This means that any voltage lower than 1.4V is treated as a "low" voltage, and any voltage higher than 1.4V is treated as a "high" voltage. In our circuits in preceding figures, we use the ground terminal, 0V, as our low voltage source. For our high voltage source, we used the positive power supply. Provided the supply voltage is above 1.4V, it should be satisfactory. (5V and 3.3V are common power supply voltages for digital systems, with 1.8V and 1.1V also common within ICs.) If components, such as the gates in Figure 1.5, distinguish between low and high voltages based on the 1.4V threshold, the circuit should operate correctly. In the real world, however, this approach would lead to problems. Manufacturing variations make it impossible to ensure that the threshold voltage is exactly the same for all components. So one gate may drive only slightly higher than 1.4V for a high logic level, and a receiving gate with

FIGURE 1.12 Problems due to variation in threshold voltage. The receiver would sense the signal as remaining low.

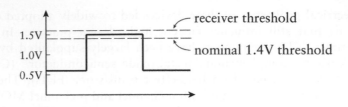

FIGURE 1.13 Problems due to noise on wires.

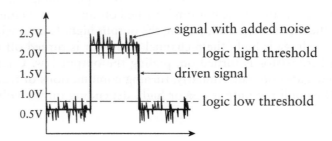

a threshold a little more above 1.4V would interpret the signal as a low logic level. This is shown in Figure 1.12.

As a way of avoiding this problem, we separate the single threshold voltage into two thresholds. We require that a logic high be greater than 2.0V and a logic low be less than 0.8V. The range in between these levels is not interpreted as a valid logic level. We assume that a signal transitions through this range instantaneously, and we leave the behavior of a component with an invalid input level unspecified. However, the signal, being transmitted on an electrical wire, might be subject to external interference and parasitic effects, which would appear as voltage noise. The addition of the noise voltage could cause the signal voltage to enter the illegal range, as shown in Figure 1.13, leading to unspecified behavior.

The final solution is to require components driving digital signals to drive a voltage lower than 0.4V for a "low" logic level and greater than 2.4V for a "high" logic level. That way, there is a *noise margin* for up to 0.4V of noise to be induced on a signal without affecting its interpretation as a valid logic level. This is shown in Figure 1.14. The symbols for the voltage thresholds are

▸ V_{OL}: output low voltage—a component must drive a signal with a voltage below this threshold for a logic low

▸ V_{OH}: output high voltage—a component must drive a signal with a voltage above this threshold for a logic high

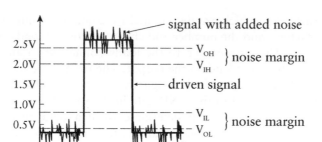

FIGURE 1.14   Logic level thresholds with noise margin.

▸ V_{IL}: input low voltage—a component receiving a signal with a voltage below this threshold will interpret it as a logic low

▸ V_{IH}: input high voltage—a component receiving a signal with a voltage above this threshold will interpret it as a logic high

The behavior of a component receiving a signal in the region between V_{IL} and V_{IH} is unspecified. Depending on the voltage and other factors, such as temperature and previous circuit operation, the component may interpret the signal as a logic low or a logic high, or it may exhibit some other unusual behavior. Provided we ensure that our circuits don't violate the assumptions about voltages for logic levels, we can use the digital abstraction.

1.3.3   STATIC LOAD LEVELS

A second assumption we have made is that the current loads on components are reasonable. For example, in Figure 1.3, the gate output is acting as a source of current to illuminate the lamp. An idealized component should be able to source or sink as much current at the output as its load requires without affecting the logic levels. In reality, component outputs have some internal resistance that limits the current they can source or sink. An idealized view of the internal circuit of a CMOS component's output stage is shown in Figure 1.15. The output can be pulled high by closing switch SW1 or pulled low by closing switch SW0. When one switch is closed, the other is open, and *vice versa*. Each switch has a series resistance. (Each switch and its associated resistance is, in practice, a transistor with its on-state series resistance.) When SW1 is closed, current is sourced from the positive supply and flows through R1 to the load connected to the output. If too much current flows, the voltage drop across R1 causes the output voltage to fall below V_{OH}. Similarly, when SW0 is closed, the output acts as a current sink from the load, with the current flowing through R0 to the ground terminal. If too much current flows in this direction, the voltage drop across R0 causes the output voltage to rise above V_{OL}. The amount of current that flows in each case depends on the

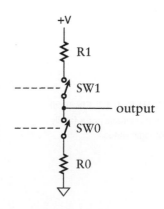

FIGURE 1.15   An idealized view of the output stage of a CMOS component.

output resistances, which are determined by component internal design and manufacture, and the number and characteristics of loads connected to the output. The current due to the loads connected to an output is referred to as the *static load* on the output. The term *static* indicates that we are only considering load when signal values are not changing.

The load connected to the AND gate in Figure 1.3 is a lamp, whose current characteristics we can determine from a data sheet or from measurement. A more common scenario is to connect the output of one gate to the inputs of one or more other gates, as in Figure 1.5. Each input draws a small amount of current when the input voltage is low and sources a small amount of current when the input is high. The amounts, again, are determined by component internal design and manufacture. So, as designers using such components and seeking to ensure that we don't overload outputs, we must ensure that we don't connect too many inputs to a given output. We use the term *fanout* to refer to the number of inputs driven by a given output. Manufacturers usually publish current drive and load characteristics of components in data sheets. As a design discipline when designing digital circuits, we should use that information to ensure that we limit the fanout of outputs to meet the static loading constraints.

EXAMPLE I.3   The data sheet for a family of CMOS logic gates that use the TTL logic levels described earlier lists the characteristics shown in Table 1.1. Currents are specified with a positive value for current flowing into a terminal and a negative value for current flowing out of a terminal. The

TABLE 1.1 Electrical characteristics of a family of logic gates.

PARAMETER	TEST CONDITION	MIN	MAX
V_{IH}		2.0V	
V_{IL}			0.8V
I_{IH}			5µA
I_{IL}			−5µA
V_{OH}	$I_{OH} = -12mA$	2.4V	
	$I_{OH} = -24mA$	2.2V	
V_{OL}	$I_{OL} = 12mA$		0.4V
	$I_{OL} = 24mA$		0.55V
I_{OH}			−24mA
I_{OL}			24mA

parameters I_{IH} and I_{IL} are the input currents when the input is at a logic high or low, respectively, and I_{OH} and I_{OL} are the static load currents at an output driving logic high or low, respectively. What is the maximum fanout for an output driving multiple inputs using this logic family, taking account of static loading only?

SOLUTION   For both high and low logic levels, an output can source or sink up to 24mA of current, and an input load is 5μA. Thus each output can drive up to 24mA/5μA = 4800 inputs. However, in sourcing that much current in the high level, the output voltage may drop to 2.2V, and in the low level, the output voltage may rise to 0.55V. This gives a noise margin of only 0.2V for a high level and 0.15V for a low level. If we want to preserve our 0.4V noise margins, we need to limit the output currents to 12mA, in which case the maximum fanout would be 2400 inputs.

In practice, we cannot connect anywhere near as many inputs to an output as this example might suggest. Static loading is only one factor that determines maximum fanout. In the next part of this section, we will describe another factor that limits fanout more significantly in most designs.

1.3.4 CAPACITIVE LOAD AND PROPAGATION DELAY

A further assumption we've made in the preceding discussion has been that signals change between logic levels instantaneously. In practice, level changes are not instantaneous, but take an amount of time that depends on several factors that we shall explore. The time taken for the signal voltage to rise from a low level to a high level is called the *rise time*, denoted by t_r, and the time for the signal voltage to fall from a high level to a low level is called the *fall time*, denoted by t_f. These are illustrated in Figure 1.16.

One factor that causes signal changes to occur over a nonzero time interval is the fact that the switches in the output stage of a digital component, illustrated in Figure 1.15, do not open or close instantaneously. Rather, their resistance changes between near zero and a very large value over some time interval. However, a more significant factor, especially in CMOS circuits, is the fact that logic gates have a significant amount of capacitance at each input. Thus, if we connect the output of one

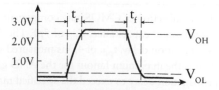

FIGURE 1.16   Rise time and fall time for a signal whose value is changing.

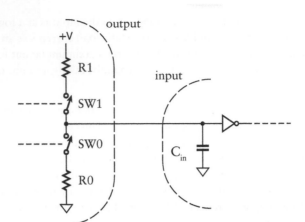

FIGURE 1.17   Connection of
an output stage to a capacitively
loaded input.

component to the input of another, as shown in Figure 1.17, the input capacitance must be charged and discharged through the output stage's switch resistances in order to change the logic level on the connecting signal.

If we connect a given output to more than one input, the capacitive loads of the inputs are connected in parallel. The total capacitive load is thus the sum of the individual capacitive loads. The effect is to make transitions on the connecting signal correspondingly slower. For CMOS components, this effect is much more significant than the static load of component inputs. Since we usually want circuits to operate as fast as possible, we are constrained to minimize the fanout of outputs to reduce capacitive loading.

A similar argument regarding time taken to switch transistors on and off and to charge and discharge capacitance also applies within a digital component. Without going into the details of a component's circuit, we can summarize the argument by saying that, due to the switching time of the internal transistors, it takes some time for a change of logic level at an input to cause a corresponding change at the output. We call that time the *propagation delay*, denoted by t_{pd}, of the component. Since the time for the output to change depends on the capacitive load, component data sheets that specify propagation delay usually note the capacitive load applied in the test circuit for which the propagation delay was measured, as well as the input capacitance.

EXAMPLE 1.4   For a collection of CMOS gate components, the manufacturer's data sheet specifies a typical input capacitance, C_{in}, of 5pF. The AND gate component has a maximum propagation delay, t_{pd}, of 4.3ns measured with a load capacitance, C_L, of 50pF. What is the maximum fanout for the AND gate that can be used without causing the propagation delay to exceed the specified maximum?

SOLUTION   Allowing only for the capacitive loading effect of the inputs, the maximum fanout is

$$C_L / C_{in} = 50pF / 5pF = 10$$

In practice, other stray capacitance between the output and the inputs would limit the maximum fanout to a smaller value.

In many components, the propagation delay differs depending on whether the output is changing from 0 to 1 or from 1 to 0. If it is important to distinguish between the two cases, we can use the symbol t_{pd01} for the propagation delay when the output changes from 0 to 1, and t_{pd10} for the propagation delay when the output changes from 1 to 0. If we don't need to make this distinction, we usually use the largest of the two values, that is,

$$t_{pd} = \max (t_{pd01}, t_{pd10})$$

1.3.5   WIRE DELAY

Yet another assumption we've made about the behavior of digital systems is that a change in the value of a signal at the output of a component is seen instantaneously at the input of other connected components. In other words, we've assumed that wires are perfect conductors that propagate signals with no delay. For very short wires, that is, wires of a few centimeters on a circuit board or up to a millimeter or so within an IC, this assumption is reasonable, depending on the speed of operation of the circuit. For longer wires, however, we need to take care when designing high-speed circuits. Problems can arise from the fact that such wires have parasitic capacitance and inductance that are not negligible and that delay propagation of signal changes. Such wires should be treated as transmission lines and designed carefully to avoid unwanted effects of reflection of wavefronts from stubs and terminations. A detailed treatment of design techniques in these cases is beyond the scope of this book. However, we need to be aware that relatively long wires add to the overall propagation delay of a circuit. Later, we will describe the use of computer-based tools that can help us to understand the effects of wire delays and to design our circuits appropriately.

1.3.6   SEQUENTIAL TIMING

In our discussion of sequential digital systems in Section 1.2, we assumed that a flip-flop stores the value of its data input at the moment the clock input rises from 0 to 1. Moreover, we assumed that the stored value is reflected on the output instantaneously. It should not be surprising now

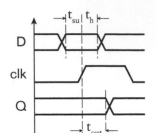

FIGURE 1.18   Setup, hold and clock-to-output times for a flip-flop.

that these assumptions are an abstraction of realistic sequential circuit behavior, and that we must observe design disciplines to support the abstraction. Real flip-flops require that the value to be stored be present on the data input for an interval, called the *setup time*, before the rising clock edge. Also, the value must not change during that interval and for an interval, called the *hold time*, after the clock edge. Finally, the stored value does not appear at the output instantaneously, but is delayed by an interval called the *clock-to-output delay*. These timing characteristics are shown in Figure 1.18. In this timing diagram, we have drawn the rising and falling edges as sloped, rather than vertical, to suggest that the transitions are not instantaneous. We have also drawn both 0 and 1 values for the data input and output, suggesting that it is not the specific values that are relevant, but the times at which values change, shown by the coincident rising and falling values. The diagram illustrates the constraint that changes on the data input must not occur within a time window around the clock rising edge, and that the data output cannot be assumed correct until after some delay after the clock edge.

In most sequential digital circuits, the output of one flip-flop is either connected directly to the data input of another, or passes through some combinational logic whose output is connected to the data input of another flip-flop. In order for the circuit to operate correctly, a data output resulting from one clock edge must arrive at the second flip-flop ahead of a setup interval before the next clock edge. This gives rise to a constraint that we can interpret in two ways. One view is that the delays in the circuit between flip-flops are fixed and place an upper bound on the clock cycle time, and hence on the overall speed at which the circuit operates. The other view is that the clock cycle time is fixed and places an upper bound on the permissible delays in the circuit between flip-flops. According to this view, we must ensure that we design the circuit to meet that constraint. We will examine timing constraints for sequential circuits in much more detail in Chapter 4, and describe a design discipline that ensures that we meet the constraints, thus allowing us to use the timing abstraction of periodic clock cycles.

1.3.7 POWER

Many modern applications of digital circuits require consideration of the power consumed and dissipated. Power consumption arises through current being drawn from a constant-voltage power supply. All gates and other digital electronic components in a circuit draw current to operate the transistors in their internal circuitry, as well as to switch input and output transistors on and off. While the current drawn for each gate is very small, the total current drawn by millions of them in a complete system can be many amperes. When the power supply consists of batteries, for example, in portable appliances such as phones and notebook computers, reducing power consumption allows extended operating time.

The electrical power consumed by the current passing through resistance causes the circuit components to heat up. The heat serves no useful purpose and must be exhausted from the circuit components. Designers of the physical packaging of ICs and complete electronic systems determine the rate at which thermal energy can be transferred to the surroundings. As circuit designers, we must ensure that we do not cause more power dissipation than can be handled by the thermal design, otherwise the circuit will overheat and fail. Puffs of blue smoke are the usual sign of this happening!

There are two main sources of power consumption in modern digital components. The first of these arises from the fact that transistors, when turned off, are not perfect insulators. There are relatively small *leakage currents* between the two terminals, as well as from the terminals to ground. These currents cause *static power* consumption. The second source of power consumption arises from the charging and discharging of load capacitance when outputs switch between logic levels. This is called *dynamic power* consumption. To a first approximation, the static power consumption occurs continuously, independent of circuit operation, whereas dynamic power consumption depends on how frequently signals switch between logic levels.

As designers, we have control over both of these forms of power consumption. We can control static power consumption of the circuit by choosing components with low static power consumption, and, in some cases, by arranging for parts of circuits that are not needed for periods of time to be turned off. We can control dynamic power consumption by reducing the number and frequency of signal transitions that must occur during circuit operation. This is becoming an increasingly important part of the design activity, and computer-based tools for power analysis are gaining increased use. We will discuss the topic of power analysis in more detail throughout this book.

1.3.8 AREA AND PACKAGING

In most applications of digital electronics, cost of the final manufactured product is an important factor, particularly when the product is to be sold in a competitive market. There are numerous factors that influence cost, many of them based on business decisions rather than engineering design decisions. However, one factor over which designers have control and that strongly affects the final product cost is circuit area.

As we mentioned earlier, digital circuits are generally implemented as integrated circuits, in which transistors and wires are chemically formed on the surface of a wafer of crystalline silicon (see Figure 1.19). The more transistors and wires in our circuit, the more surface area it consumes. The manufacturing process for ICs is based on wafers of a fixed size, up to 300mm in diameter, with a fixed cost of manufacture. Multiple ICs

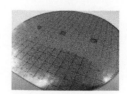

FIGURE 1.19 A silicon wafer, on which multiple ICs are manufactured.

FIGURE 1.20 A packaged IC soldered onto a printed circuit board.

are manufactured on a single wafer in a series of steps. The wafer is then broken into individual ICs, which are encapsulated in packages that can be soldered onto the circuit board of a complete system (see Figure 1.20). Thus, the larger the individual IC, the fewer there are per wafer, and so the greater their cost. Unfortunately, the manufacturing process is not perfect, so defects occur, distributed across the surface of the wafer. Those ICs that have a defect that cause them not to function correctly are discarded. Since the cost of manufacturing a wafer is fixed, the functional ICs must bear the cost of those that are nonfunctional, increasing the final product cost of the IC. The larger an individual IC, the greater the proportion that have defects. So the final cost of an IC is disproportionately dependent on area.

Since each IC is packaged individually, the cost of the package is a direct cost of the final product. The package serves two purposes. One is to provide connection pins, allowing the wires of the IC to be connected to external wires in the larger digital system, as well as providing for power supply and ground connections. An IC with more external connections requires more pins and, thus, a more costly package. Therefore, the pin count of the IC is a factor that constrains our designs. The other purpose served by the IC package is to transfer heat from the IC to the surroundings so that the IC does not overheat. If the IC generates more thermal power than the package can dissipate, additional cooling devices, such as heat sinks, fans or heat pipes, are required, adding to cost. Thus, thermal concerns arising from packaging also constrain our designs.

As we have suggested, a packaged IC may not be the final product of a design. The IC may be one of several components on a printed circuit board, which, in turn, is assembled with other items into a complete packaged product. Similar arguments to those above can be made about the cost of a printed circuit board based on the number of ICs and other components, the number of external connections, the area or size of the board and package, and heat dissipation in the enclosing case or cabinet.

KNOWLEDGE TEST QUIZ

1. What are the TTL output voltage levels, input threshold voltages and noise margins?

2. What is meant by the term *fanout*?

3. How is the propagation delay of a component defined?

4. Why do we try to minimize fanout of components?

5. Do wires contribute to delay in a circuit?

6. What is meant by the terms *setup time*, *hold time* and *clock-to-output time* of a flip-flop?

7. What are the two sources of power consumption in a digital component?

8. Is the cost of an IC proportional to the area of the IC?

1.4 MODELS

As children, many of us will have made or played with models of real-world things, for example, model airplanes. One way of thinking of a model is that it is a representation of an object that incorporates aspects of interest for a particular purpose while omitting aspects that are not relevant. In other words, it is an abstraction of an object. A model airplane may, for example, have the look of a real airplane, but does not have the scale or detailed mechanical aspects of the real thing. The model incorporates just those aspects that satisfy a child's wish to play with an airplane.

Now that we've grown up (mostly!) and turned to digital circuit design, our task is to design circuits that perform certain required functions while meeting various constraints. We could try to build a prototype circuit to check that it performs as required, but that would be costly and time consuming, since we would usually need to work through numerous versions before we get things right. A more effective approach is to develop *models* of our designs and to evaluate them. A model of a digital circuit is an abstract expression in some modeling language that captures those aspects that we are interested in for certain design tasks and omits other aspects. For example, one form of model may express just the function that the circuit is to perform, without including aspects of timing, power consumption or physical construction. Another form of model may express the logical structure of the circuit, that is, the way in which it is composed of interconnected components such as gates and flip-flops. Both of these forms of model may be conveniently expressed in a *hardware description language* (HDL), which is a form of computer language akin to programming languages, specialized for this purpose.

Functional models may also be expressed in mathematical notations, such as Boolean equations and finite state machine notations, that we will introduce in later chapters. Structural models may also be expressed in the form of graphical circuit schematics, such as those in earlier figures in this chapter. We will use all of these forms of models where appropriate, but we will focus on models expressed in an HDL, since that allows us to take advantage of *computer-aided design* (CAD) tools to help us with design tasks. Designing electronic circuits using CAD tools is also called *electronic design automation* (EDA).

In this book, we will introduce and use an HDL called *VHDL*. VHDL is an acronym, standing for VHSIC Hardware Description Language. VHSIC is, in turn, an acronym for Very High Speed Integrated Circuit, which was the name of a U.S. Government sponsored program in the 1980s that promoted development of tools and technology for IC design and manufacture. VHDL was one of the products of that program. Since then, the specification of VHDL has been standardized in the United States by the Institute of Electrical and Electronic Engineers (IEEE) and internationally by the International Electrotechnical Commission (IEC), and the language has been widely adopted by designers and tool developers.

VHDL is not the only HDL used for digital system design. Others include Verilog and its more recent successor, SystemVerilog. Also, SystemC, an extension of the C++ programming language, is gaining increased usage. While these languages have many basic features in common, they vary in their more advanced features. Moreover, they are all evolving, with new features being added in successive revisions to meet evolving design challenges. Choice among them is often dictated by tool availability and organizational culture, as well as the kind of design work to be performed.

EXAMPLE 1.5   Develop a VHDL model that expresses the logical structure of the gate circuit in Figure 1.5. Assume that the sensor signals and the switch signal are inputs to the model, and that the buzzer signal is the output from the model.

SOLUTION   The VHDL model has two parts, one describing the inputs and output of the circuit, and the other describing the interconnection of gate components that realizes the circuit. The first of these takes the form of an *entity declaration*:

```
library ieee; use ieee.std_logic_1164.all;

entity vat_buzzer is
  port ( buzzer                                : out std_logic;
         above_25_0, above_30_0, low_level_0 : in  std_logic;
         above_25_1, above_30_1, low_level_1 : in  std_logic;
         select_vat_1                         : in std_logic );
end entity vat_buzzer;
```

The circuit is represented in VHDL as an entity, in this case named vat_buzzer. The first line identifies a standard library, ieee, and a package, std_logic_1164, containing definitions that we want to reference in our model. (Not all of the modeling features we need are built into VHDL. Instead, the language provides library and package mechanisms to allow definition of extensions. We include the first line shown to specify that we need to use the extensions defined in the std_logc_1164 package in the ieee library in our model. This is a little like referencing a header file using an include directive in the C programming language.) The entity has *ports*, described in the port list of the entity declaration. Each port is given a name and is either an output (out) or an input (in). The port description also specifies the *type* of each port, that is, the set of values that the port can take on. In this example, each port is of type std_logic, defined by the std_logic_1164 package. The std_logic type includes the values '0' and '1', corresponding to the binary values 0 and 1 introduced earlier. The difference is that VHDL requires the quotation marks to distinguish the std_logic values from the numbers 0 and 1.

The second part of the VHDL model takes the form of an *architecture*:

```
library dld; use dld.gates.all;

architecture struct of vat_buzzer is

  signal below_25_0, temp_bad_0, wake_up_0 : std_logic;
  signal below_25_1, temp_bad_1, wake_up_1 : std_logic;

begin

  -- components for vat 0
  inv_0 : inv port map (above_25_0, below_25_0);
  or_0a : or2 port map (above_30_0, below_25_0, temp_bad_0);
  or_0b : or2 port map (temp_bad_0, low_level_0, wake_up_0);

  -- components for vat 1
  inv_1 : inv port map (above_25_1, below_25_1);
  or_1a : or2 port map (above_30_1, below_25_1, temp_bad_1);
  or_1b : or2 port map (temp_bad_1, low_level_1, wake_up_1);

  select_mux : mux2 port map (wake_up_0, wake_up_1,
                              select_vat_1, buzzer);

end architecture struct;
```

The architecture contains the details of the circuit model. In this case, the circuit is modeled as a collection of interconnected components. We use the term *structural architecture* to refer to a model in this form.

The first line of the architecture identifies a design library, dld, and a package, gates, containing definitions of components that we want to include in our model. (The source code of this library, along with documentation about its content and how to use it, are available on the companion website for this book.) The architecture also references the definitions in the std_logic_1164 package, but since we identified that package and its library in the entity declaration, we don't need to duplicate the library and use clauses in the architecture.

The architecture is named struct and is associated with the vat_buzzer entity. Within the architecture, a number of signals are declared for connecting the components together. Each signal is given a name and has a specified type, in this case, std_logic. Between the begin and end lines, the architecture contains a number of *component instances*. Each has a label to distinguish it, and specifies which kind of component is instantiated. For example, inv_0 is an instance of the inv component, representing the inverter for vat 0 in the circuit. Within parentheses, the signals and circuit inputs and outputs connected to the ports of the component are listed. For example, the inverter inv_0 has the input above_25_0 connected to its first port and the signal below_25_0 connected to its second port. We have also included some comments in this architecture to provide documentation. A comment in VHDL starts with two hyphen characters and extends to the end of the line. Figure 1.21 shows the vat buzzer circuit again with the signal names and component labels included for reference.

FIGURE 1.21 The vat buzzer circuit, showing signal names and component labels.

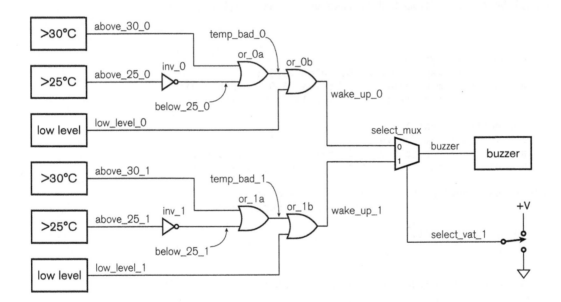

EXAMPLE 1.6 Develop a VHDL model that expresses the function performed by the gate circuit in Figure 1.5.

SOLUTION We can use the same entity declaration as that for the structural model, since the inputs and outputs are the same. We simply need an alternative architecture:

```
architecture behavior of vat_buzzer is
begin
  buzzer <= low_level_1 or (above_30_1 or not above_25_1)
              when select_vat_1 = '1' else
              low_level_0 or (above_30_0 or not above_25_0);
end architecture behavior;
```

An architecture of this form, in which we describe the function performed by the circuit, is called a *behavioral architecture*. (Some people also use this term for more abstract models of functionality, but in the absence of a better term, we will use it as described here.) Since we don't need to use definitions of any components, we omit the library and package references that we included in the structural architecture. The architecture is named behavior and is associated with the vat_buzzer entity. Within the architecture, we have a single *assignment statement* that determines the value of the output port using the values of the input ports. The operators or and not correspond in function to the OR gates and inverters, respectively, in the circuit. The choice of value to assign to the output port depends on the condition in the when ... else construct, corresponding to the multiplexer in the circuit.

The examples above illustrate the general organization of structural and behavioral VHDL models. A circuit is modeled by an entity declaration and one or more corresponding architectures, each representing a different aspect or implementation of the circuit. The VHDL tutorial and reference material included on the companion website for this book provide more detailed information on the specifics of writing VHDL models. The source code for all of the example models in this book is also provided on the companion website. We follow the convention of naming each source code file after the entity or architecture it contains. This makes it easier to locate the file containing a given model.

There are three principal design tasks that benefit from the use of VHDL. The first of these is *design entry*, that is, expressing a model in a form that can be input to a CAD tool. For simple circuits, design entry can also be done using graphical schematics, and many CAD tools provide for that form of input. However, schematics for larger and more complex circuits can be cumbersome, particularly when they are annotated with information such as the types of signals. Moreover, a textual form, such as VHDL, allows for richer forms of expression, as we shall see in later examples. It also works better with other computer tools, such as scripting tools and source-code control tools. For these reasons, we shall focus on VHDL descriptions of circuits and use circuit schematics for illustrative purposes in this book.

The second design task that benefits from use of VHDL is *verification*, that is, ensuring that the design meets its requirements and constraints. There are several aspects to verification. *Functional verification* involves ensuring that the design performs the required function. *Timing verification* involves ensuring that the design meets its timing constraints. Timing constraints are ultimately derived from the performance requirements of the circuit. For example, in a circuit that processes a digital representation of an audio signal, the processing steps must be performed at the sampling rate of the signal. Other verification tasks include *power verification* (ensuring that the circuit meets power consumption and dissipation constraints); *manufacturability verification* (ensuring that the circuit will operate correctly for all variations that might arise in the manufacturing processes); and *test verification* (ensuring that the manufactured circuit can be tested to identify defective parts). All of these forms of verification involve *analysis* of models of the circuit to determine the relevant properties and checking that the property values are acceptable.

Functional verification is often the most time consuming part of the entire design process. One approach to functional verification is *simulation*, in which the model is interpreted as an executable computer program by a CAD tool called a *simulator*. This involves applying different combinations and sequences of values to the input ports, executing the model code, and ensuring that it produces the required values on the output ports. For behavioral models, the expressions in assignment statements can be executed directly. For structural models, each component instance must have a corresponding behavioral model that can be invoked by the simulator. The simulator passes values produced at the outputs of components along the interconnecting signals to the inputs of other components.

The problem with simulation, particularly for large and complex models, is that covering all possible combinations and sequences of values that might arise in the real circuit is time consuming and resource intensive, and is generally not feasible. An alternative to simulation is *formal verification*, which involves mathematical proof of properties of the design. The properties take the form of logical statements relating values of inputs and outputs, or sequences of values of inputs and outputs, that express the functional requirements, usually in a more abstract form than that of the model being verified. The analysis of the model and proof of the properties is performed by a CAD tool called a *model checker*. Formal verification is a relatively new technology, and can require significant computational resources. In practice, functional verification of real-world circuits is most effectively done using a combination of simulation and formal verification. We will return to this issue in more detail in our discussions of design methodology in Section 1.5 and Chapter 10.

The third design task that benefits from use of VHDL is *synthesis*. This involves automatic refinement and optimization of a model at a higher level of abstraction into a structural model at a lower level of abstraction. For example, the *register transfer level* (RTL) of abstraction in VHDL involves expressing behavior in terms of assignment statements and expressions, such as those in Example 1.6, as well as *processes*, that we shall come to later. An RTL synthesis CAD tool automatically refines an RTL model into an optimized gate-level model, that is, a structural model using gate components such as that of Example 1.5. Since a synthesis tool automates a task that we would otherwise have to perform manually, it greatly improves our productivity. In particular, it makes more complex designs tractable.

Since hardware description languages such as VHDL help so significantly with these design tasks, they have become central to modern design methods. Throughout this book, as we introduce digital components and circuits, we will also show VHDL models that describe them. As we introduce design methods, we will show how CAD tools that process models help us perform these methods.

1. What are the two parts of a VHDL model of a circuit?

2. What information is specified for each port in a VHDL entity declaration?

3. What is meant by the terms *structural architecture* and *behavioral architecture*?

4. What are functional verification and timing verification?

5. Identify two approaches to functional verification.

6. What is meant by synthesis?

KNOWLEDGE
TEST QUIZ

1.5 DESIGN METHODOLOGY

Designing a digital system of any significant complexity is a large undertaking, requiring a systematic approach. This is especially important when many people are collaborating on a design, as is usually the case. Depending on the complexity of the product, design teams can range in size from a handful of engineers for a relatively simple product, to several hundred people for a complex IC or packaged system. We use the term *design methodology* to refer to the systematic process of design, verification and preparation for manufacture of a product. A design methodology specifies the tasks undertaken, the information required and produced by each task, the relationships between the tasks, including dependencies and sequencing, and the CAD tools used. A mature design methodology will also be reflective, specifying measurements that will be made of the design process, such as adherence to schedule and budget, and numbers of design errors detected and missed. Accumulated data from previous projects can be used to improve the design methodology for subsequent projects. The benefit of a good methodology is that it makes the design process more reliable and predictable, thus reducing risk and cost. Even a small design project benefits from a design methodology, though perhaps on a reduced scale.

Given its importance, we will focus on design methodology throughout this book. We will start by outlining a relatively simple methodology, since we are in the early stages of learning digital design. In Chapter 10, we will see what's involved in a more complete methodology for real-world systems.

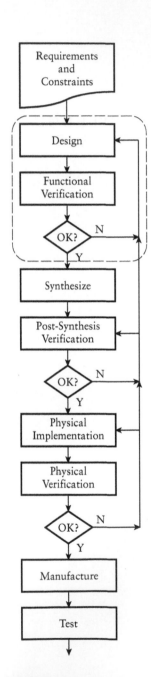

FIGURE 1.22 A simple design methodology.

Figure 1.22 illustrates a simple design methodology. The starting point is a collection of requirements and constraints. These are usually generated externally to the design team, for example, by the marketing group of a company or by a customer for whom a product development is undertaken. They usually include function requirements (what the product is to do), performance requirements (how fast it is to do it), and constraints on power consumption, cost and packaging. The design methodology specifies three tasks—design, synthesis and physical implementation—each of which is followed by a verification task. (The design and functional verification tasks are outlined to indicate that they are actually a bit more involved than is shown on the diagram. We will come back to this shortly.) If verification fails at any stage, we must revisit a previous task to correct the error. Ideally, we would like to revisit only the immediately preceding task and make a minor correction. However, if the error is severe enough, we may need to backtrack further to make more significant changes. Hence, when performing a given design task, it is worth keeping in mind the constraints applying in subsequent tasks, so as to avoid introducing errors that will be detected later. Once the tasks and associated verification activities have been performed, the product can be manufactured, and each unit tested to ensure that it is functional. We will now spend a little time examining the stages in this methodology in more detail.

The design task involves understanding the requirements and constraints and developing a specification of a digital circuit that meets the requirements and constraints. The information produced by this task is a collection of models that describe the design. The methodology then specifies that we verify the function of the design, using techniques such as simulation and formal verification. In preparation for the verification task, we should prepare a *verification plan* that identifies what input and output cases should be verified, and what CAD tools should be used. We will illustrate development of verification plans in parallel with design tasks throughout this book.

We've already discussed use of abstractions to make the design task more manageable, and the need to adhere to design disciplines to ensure that we don't violate assumptions underlying the abstractions. However, for a system of any significant complexity, that is still not sufficient to make the task tractable. Another form of abstraction that allows us to manage design complexity is *hierarchical composition*. This involves developing a subcircuit that performs some relatively simple function, then treating it as a "black box." Provided we adhere to assumptions made in designing the subcircuit, we can then use it in a larger circuit that performs a more complex function. As an example, the subcircuit might be a small liquid-crystal display (LCD) controller, which is used as part of the user interface of a cordless phone. We can repeat the step of using one subcircuit as part of a more complex circuit. For example, the user-interface subcircuit might be used as a black box within the cordless phone handset. At each level of the hierarchical design, we can focus on the aspects that are relevant and hide the details of the lower-level components. Using abstraction in this way makes the task of designing complex systems tractable. It also allows us to re-use subcircuits, either from previous projects or from third-party providers. Design re-use can potentially save significant design effort and cost.

Hierarchical composition in a design also makes functional verification more tractable. We can first verify each of the most primitive subcircuits as independent units. Next, we can verify a subsystem that uses the subcircuits by treating the subsystem as a collection of black boxes. In particular, we can use more abstract models of the black box subcircuits instead of detailed models that describe their internal composition. This approach means that the verification tools have less work to do, allowing them to verify more input/output cases for the subsystem. We can repeat this process until we have verified the entire system.

Returning to the design and verification tasks shown in Figure 1.22, we can expand the design and functional verification tasks to illustrate use of hierarchical composition, as shown in Figure 1.23. This approach is often called *top-down design*. Architectural design involves analyzing the requirements and developing the overall organization of a digital system to meet them. One of the main tools used for this level of design is a white board, on which system architects draw (and redraw) block diagrams describing the main subsystems and their interconnections. The next step is unit design, in which the subsystems and sub-subsystems are designed. Each unit can then be verified, possibly requiring some redesign if any of the units fail verification. Finally, the units can be integrated and the subsystems and entire system verified, as we described above. Again, if verification fails, units may need to be redesigned. If the failure is severe enough, it may be necessary to revise the architectural organization of the system, and then to reflect the changes in the unit designs.

The tasks immediately after functional verification in the design methodology of Figure 1.22 are *synthesis* and *post-synthesis verification*. We described synthesis in the previous section as automatic refinement and optimization of a model at a higher level of abstraction to a structural model at a lower level of abstraction. Currently, synthesis is usually performed from register-transfer level to gate level, as CAD tool technology for this level of refinement is quite mature. Behavioral synthesis (also called high-level synthesis), from higher levels of abstraction to RTL, is much less mature, though the subject of much active development work.

In order to perform RTL synthesis, we specify information about the implementation fabric that we intend to use for our design. We might also annotate the RTL models with additional information to guide the synthesis CAD tool in its optimization task. The tool then selects primitive

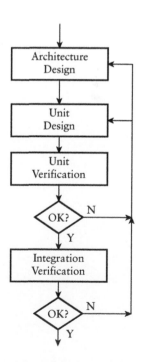

FIGURE 1.23 Hierarchical design and verification.

components from a library of components available in the chosen implementation fabric and constructs a circuit with the same function as that of the RTL design. The library may contain further information about the properties of each component, such as timing parameters, power dissipation, and so on. Our design methodology shows that we use this information, together with the refined design produced by the synthesis tool, to further verify the design. Using the timing parameters and the information about the way components are interconnected, a *static timing analysis* tool can estimate propagation delays in the circuit and verify that timing constraints are met. Similarly, using information about the number or transistors and amount of wiring required for components, a *floor planning* tool can estimate the area of the design and verify that area and packaging constraints are met. Note that the properties used at this stage are estimates of final property values for the manufactured circuit, and need to be refined later in the design process. As a further step in verification, an *equivalence checker* can compare the function of the refined design with that of the original RTL design to verify that the synthesis tool has done its job correctly and that the functional requirements are still met.

The next task in the design methodology is physical implementation. This involves using the refined design, expressed as an interconnection of primitive circuit elements, and generating the information required to manufacture the circuit. The precise steps to be performed depend on the implementation fabric chosen for the circuit. By implementation fabric, we mean the kind of IC used to implement a design. The two main implementation fabrics in common use today are *field-programmable gate arrays* (FPGAs) and *application-specific integrated circuits* (ASICs). An FPGA consists of a large number of gates and flip-flops whose interconnection can be determined, or *programmed*, after the IC is manufactured. In this book, we will focus on FPGAs, especially in lab projects, since they are widely used for a range of circuits of varying size and complexity, can be reprogrammed, and are cost-effective for nearly all but large-volume applications. ASICs, as their name suggests, are ICs that are customized for a particular application, and cannot be programmed. We will describe these implementation fabrics in more detail in Chapter 6. However, for now, we can identify some general steps that are common to physical implementation on both of these fabrics, as well as on printed circuit boards.

The first of these steps is *mapping*, which involves determining the particular circuit resources to be used for each of the components in the refined design. Next, *placement and routing* determines where each mapped component is to be positioned in the physical circuit and where the interconnecting wires run. Once mapping, placement and routing are done, refined estimates of circuit properties can be extracted. In particular, since the physical wiring details have been determined, propagation delays through wires can be included in the timing estimates. These refined estimates are used to perform final physical verification. Finally, one or more files of information are generated for manufacturing the circuit. When that step is passed, we reach a golden milestone, called *tape out* for ASIC design, referring historically to production of a magnetic tape containing the manufacturing data to be shipped out to the manufacturer. These days, the data is more likely to be transferred by file transfer over the Internet. Nonetheless, reaching the milestone is usually reason for the design team to hold a party!

The final tasks shown in the design methodology are the manufacturing and test tasks. For ASICs, manufacturing is done by a foundry that uses the design information to control the chemical processes that form ICs on silicon wafers. For FPGAs, prefabricated parts are programmed using the design information. In your lab work, you will encounter the CAD tools and equipment needed to program FPGAs. The test task for ASICs involves exercising each manufactured circuit to ensure that it operates correctly. Some parts, as we've mentioned, will fail to operate due to defects in their manufacture and must be discarded. Alternatively, all of the manufactured parts may fail due to design errors that escaped the various verification steps we performed. Identifying the errors that cause such failures is very difficult and costly, involving use of measuring instruments to probe wires within the circuit to trace actual operation. It is much better to avoid bug escapes by verifying the design more thoroughly earlier in the design process. Testing of FPGA ICs also occurs once they are manufactured, but before they are delivered to customers for programming. Once an FPGA has been programmed, the programming device will often read back the program to verify that it has been correctly installed.

1.5.1 EMBEDDED SYSTEMS DESIGN

In Section 1.1 we introduced the idea of an embedded system, a digital system in which one or more computers are used as components. Each embedded computer comprises a processor core, memory and interfaces with other parts of the system. Since the computers must be programmed to implement part of the system's functionality, we must augment our design methodology to include embedded software design.

Recall that the initial inputs to the design methodology are the functional requirements and constraints for the system. As part of our architectural design considerations, we can choose which aspects of functionality can be implemented by embedded software on a processor core, and which parts can be implemented as digital subcircuits, that is, by hardware. Designing the hardware and software for a system together is called *hardware/software codesign*. Deciding which parts to put in hardware and which in software is called *partitioning*. There are numerous trade-offs to consider. Functionality that involves testing many conditions and taking alternative actions can be hard to implement in hardware, but is relatively straightforward in software. On the other hand, functionality that involves performing rapid computations on large amounts of data or data that arrives at a high rate may need a very high performance (and hence costly and power hungry) processor core, and so may more readily be performed by customized hardware. A further consideration is that embedded software may be stored in memory circuits that may be reprogrammed after the system is manufactured or deployed. Thus, the software may be upgraded to correct design errors or add functionality without revising the hardware design or replacing deployed systems.

Once functionality has been partitioned between hardware and software, development of the two can proceed concurrently, as shown in Figure 1.24. For those aspects of the embedded software that depend

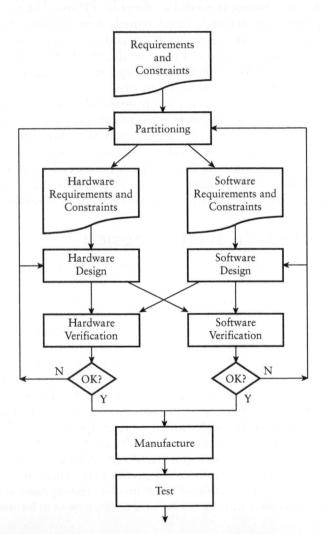

FIGURE 1.24   A design methodology for hardware/ software codesign.

on hardware, the abstract behavioral models from the hardware design task can be used to verify the software design. This can be done using an instruction-set simulator for the processor core working in tandem with a simulator for the hardware model. A similar approach can be used to verify parts of the hardware that interface directly with a processor core. Test programs can be run using the processor simulator running in tandem with the hardware simulator. The benefits of developing hardware and software concurrently include avoiding the extra time involved in developing one after the other, and early detection of errors in the interplay of software and hardware.

KNOWLEDGE TEST QUIZ

1. What is meant by the term *design methodology*?

2. Why is a design methodology beneficial?

3. If verification fails during some stage of a design methodology, what action is taken?

4. What is meant by *top-down design*?

5. Name two implementation fabrics for digital circuits.

6. What is an embedded system?

7. What is meant by the term *hardware/software codesign*?

1.6 CHAPTER SUMMARY

▸ Abstraction means identifying aspects that are important to a task at hand and hiding details of other aspects. Using abstractions requires following design disciplines to avoid violating assumptions inherent in the abstractions.

▸ The digital abstraction considers voltages to be high or low logic levels, and time to be a sequence of intervals called clock periods.

▸ Binary representation uses bits (0 and 1) to represent logical conditions. These can be implemented in a circuit using low and high logic levels.

▸ Logic gates are circuit elements that implement logical operations on binary-represented information. Logic gates can be interconnected in a circuit to perform more complex logical functions.

▸ Combinational circuits are those whose outputs depend only on the current values of inputs. They do not include any storage of information. Sequential circuits are those whose outputs depend on current and past input values. They include storage elements.

▸ A flip-flop is a storage element that stores one bit of information. An edge-triggered flip-flop stores the value of its input when a clock input changes, that is, when a clock edge occurs.

▸ The output low voltage of a driver is lower than the input low threshold of a receiver, and the output high voltage of a driver is higher than the input high threshold of a receiver. The differences are called the noise margins.

▸ Static and capacitive loading limits the fanout of a driver, that is, the number of inputs that can be connected to the output.

▸ Propagation delay depends on delay within components, capacitive loading and wire delays. Flip-flops have setup and hold time windows and clock-to-output delays.

▸ Circuits consume and dissipate static power, due to current leakage, and dynamic power, due to switching between logic levels.

▸ Circuit area and packaging have significant effects on cost.

▸ A model written in a hardware description language allows us to enter a design description into CAD tools, to verify it (using simulation and formal verification), and to synthesize it.

▸ A behavioral model describes the function performed by a circuit. A structural model describes the circuit as an interconnection of components.

▸ A design methodology specifies the tasks to be performed, the information required and produced by each task, the dependencies and sequencing of tasks, and the CAD tools used.

▸ Verification involves analyzing a model to ensure that requirements and constraints are met.

▸ Embedded systems are digital systems that contain one or more processor cores, each running embedded software.

1.7 FURTHER READING

"Cramming more components onto integrated circuits," Gordon E. Moore, *Electronics*, Volume 38, Number 8, April 19, 1965. ftp://download.intel.com/museum/Moores_Law/Articles-Press_Releases/Gordon_Moore_1965_Article.pdf. The article describing trends in IC manufacture, from which Moore's Law originated.

Foundations of Analog and Digital Electronic Circuits, Anant Agarwal and Jeffrey H. Lang, Morgan Kaufmann Publishers, 2005. As well as providing a thorough grounding in analog circuit analysis, this book introduces the basics of digital gate circuits and their analog behavior. Topics covered include static and dynamic loading, propagation delays, power dissipation, binary representation and gate-level circuits.

LVC and LV Low Voltage CMOS Logic Data Book, Texas Instruments, 1998. A comprehensive listing of the manufacturer's component products, with detailed data on electrical and timing parameters. The book also contains application reports covering details of electrical design on digital circuits. Available from www.ti.com.

The Designer's Guide to VHDL, 2nd Edition, Peter J. Ashenden, Morgan Kaufmann Publishers, 2002. A comprehensive reference on VHDL.

The Student's Guide to VHDL, Peter J. Ashenden, Morgan Kaufmann Publishers, 1998. A condensed version of *The Designer's Guide to VHDL*.

The Verilog® Hardware Description Language, 5th Edition, Donald E. Thomas and Philip R. Moorby, Springer, 2002. A comprehensive reference on Verilog.

A Verilog HDL Primer, 3rd Edition, J. Bhasker, Star Galaxy Publishing, 2005. A tutorial-style introduction to Verilog.

System Verilog for Design: A Guide to Using System Verilog for Hardware Design and Modeling, 2nd Edition, Stuart Sutherland, Simon Davidmann, Peter Flake, and P. Moorby, Springer, 2006. Describes how SystemVerilog extends Verilog, and shows how the extensions can be used to model digital systems.

SystemC: From the Ground Up, David C. Black, Jack Donovan, Bill Bunton, and Anna Keist, Springer, 2004. Describes the language, presents examples of its use, and shows how it fits within a system design methodology.

The Electronic Design Automation Handbook, Dirk Jansen (Editor), Springer, 2003. Provides information on EDA tools, methodologies, and systems, and a tutorial guideline on how to apply these concepts to high-performance ASIC design.

Reuse Methodology Manual for System-On-A-Chip Designs, 3rd Edition, Michael Keating, Russell John Rickford, and Pierre Bricaud, Springer, 2006. Describes a design methodology for creating reusable ASIC designs.

Comprehensive Functional Verification: The Complete Industry Cycle, Bruce Wile, John C. Goss, and Wolfgang Roesner, Morgan Kaufmann Publishers, 2005. Describes the place of verification in a design methodology, simulation-based verification and formal verification.

ASIC and FPGA Verification, Richard Munden, Morgan Kaufmann Publishers, 2005. A reference on writing VHDL models of digital components for use in verifying function and timing of digital systems.

Surviving the SOC Revolution: A Guide to Platform-Based Design, Henry Chang *et al.*, Springer, 1999. Describes a design methodology based on reuse of programmable hardware/software platforms.

Computers as Components: Principles of Embedded Computing System Design, Wayne Wolf, Morgan Kaufmann Publishers, 2001. Includes descriptions of software and hardware components, design and analysis techniques, and design methodology.

EXERCISES

EXERCISE 1.1   Suppose a digital system samples a sinusoidal waveform every 10μs, with each sample in the discrete set {−10.0, −9.0, −8.0, ... , −1.0, 0.0, 1.0, ... , 0.0, 10.0}. Draw graphs similar to that in Figure 1.1 showing the sample values over a 100μs interval if the waveform has:

a)   a period of 100μs and peak-to-peak amplitude of 10.0

b)   a period of 30μs and a peak-to-peak amplitude of 4.0

c)   a period of 100μs and a peak-to-peak amplitude of 0.4

EXERCISE I.2 Devise a circuit for a simple burglar alarm that activates a siren if either a motion sensor detects motion or a sensor on a window detects that the window is open.

EXERCISE I.3 Revise the night-light circuit of Figure 1.3 by adding an override switch that turns the lamp on, regardless of any other conditions.

EXERCISE I.4 Revise the night-light circuit of Figure 1.3 to include a switch that selects between activating the light when it is dark and activating the light during night-time hours. Assume there is a timer that produces a 1 output at night.

EXERCISE I.5 Suppose a factory has a vat with a sensor that outputs 1 when the vat is empty, a 0 otherwise. The vat also has a pump to empty it, and a control switch to activate the pump. Devise a circuit that turns the pump on when the switch is set to activate the pump and the vat is not empty.

EXERCISE I.6 Complete the timing diagram in Figure 1.25, showing the operation of a rising-edge-triggered D flip-flop.

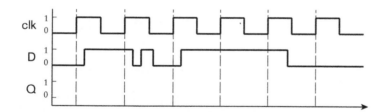

FIGURE 1.25

EXERCISE I.7 Develop a sequential circuit with a single data input S and a single data output Y. The output is 1 when the input value in the current clock cycle is different from the input value in the previous clock cycle, as shown in the timing diagram in Figure 1.26.

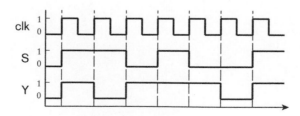

FIGURE 1.26

EXERCISE I.8 Suppose, for a family of logic components, V_{IL} is 0.6V and V_{IH} is 1.2V. What voltages are required for V_{OL} and V_{OH} to provide a noise margin of 0.2V?

EXERCISE 1.9   Suppose the gate components described in Example 1.4 were used in a circuit that added 5pF of stray capacitance to each input. What would the maximum fanout be reduced to?

EXERCISE 1.10   Use graph paper to estimate how many whole 15mm × 15mm ICs would fit on a 300mm diameter wafer. Bear in mind that the ICs must be aligned in rows and columns so that they can be separated by cutting the wafer in straight lines. What percentage of the wafer area is wasted?

COMBINATIONAL BASICS

2

In this chapter, we look at combinational circuits in some detail. We start with some of the theory underpinning combinational circuits, and show how circuits of gates correspond to formulas in the theory. Next, we show how information can be represented in binary form for processing by digital circuits. We then survey a range of components that can be used as building blocks in larger combinational circuits. Finally, we return to our design methodology and discuss verification of combinational circuits.

2.1 BOOLEAN FUNCTIONS AND BOOLEAN ALGEBRA

In Chapter 1, we showed how a digital signal can be used to represent information with two possible values, such as the truth or falsehood of a logical condition. We will now expand on that discussion and show how the laws of logic can be used to analyze and design digital systems that use binary representation. The theoretical foundation that we will use is called *Boolean algebra*, named after the nineteenth century British mathematician, George Boole, who invented the mathematical theory that deals with logical propositions.

2.1.1 BOOLEAN FUNCTIONS

According to our abstract view, a digital logic circuit has inputs and outputs, each of which has a low or high voltage at any given time. We think of these two voltage levels as electrical implementations of two *Boolean values*, 0 and 1, respectively. We could choose other names for the Boolean values, such as F and T, corresponding to falsehood and truth of logical conditions. However, that would make them harder to distinguish from the names of variables that we also introduce. Use of 0 and 1 is equally valid, less confusing, and closer to the way we express Boolean values in hardware description languages.

The combinational circuits that we mentioned in Chapter 1 have outputs that depend only on the current input values. In such circuits, each output value is a *Boolean function* of one or more inputs. This means that, for each possible combination of Boolean input values, the output takes on a specified Boolean value. This is analogous to functions on other sets of values, such as addition on numbers, where for each possible combination of operand numbers, a function yields a result number.

The most direct way of defining a Boolean function is simply to list the result values for each combination of input values. We call a table containing such a list a *truth table*. Table 2.1 shows truth tables for three basic Boolean functions that we will denote with the symbols "+", "·" and the overbar notation ("‾"). The "+" function is the *logical OR* of its two operands, and the "·" function is the *logical AND* of its operands. We use these operator symbols because the functions have many properties in common with arithmetic addition and multiplication. However, there are some differences, as we will see. The function denoted by the overbar notation is the logical negation (logical "NOT") of its single operand.

Another way of defining a Boolean function is to use a *Boolean expression*, in which we combine the literal values 0 and 1 and Boolean variables with Boolean operators. We will use alphanumeric names such as x, y and z for variables. Each variable represents a Boolean value, such as the value of a signal in a digital circuit or the value of a logical condition. Note that the column headings in Table 2.1 are simple Boolean expressions. More generally, we can include an arbitrary number of literals, variables and operators, and can use parentheses to specify an order of evaluation. We adopt the convention of giving "·" higher precedence than "+", allowing us to omit parentheses in expressions such as $(a \cdot b) + c$, giving the equivalent expression $a \cdot b + c$.

In practical terms, the literal values 0 and 1 are usually implemented as low-voltage and high-voltage digital signal values, respectively. The operator "+" is implemented as an *OR gate*, "·" as an *AND gate*, and "‾" as an *inverter*. (We introduced these basic gates in Chapter 1.) Named variables in Boolean expressions are implemented by digital signals of the same name. A complete Boolean expression is implemented by a circuit of interconnected gates, in which there is one gate corresponding to each operator in the expression. We can also write a *Boolean equation* in which one

TABLE 2.1 Truth tables for the logical OR, AND and negation functions.

x	y	$x + y$	x	y	$x \cdot y$	x	$\bar{x}$
0	0	0	0	0	0	0	0
0	1	1	0	1	0	1	0
1	0	1	1	0	0		
1	1	1	1	1	1		

Boolean expression is defined to be equal to another. A Boolean equation in which a single variable of a given name is defined to be equal to a Boolean expression is implemented by the circuit for the expression yielding an output with the given name. For example, the Boolean equation

$$f = (x + y) \cdot \overline{z}$$

is implemented by the digital logic circuit shown in Figure 2.1.

We can show that truth tables and Boolean expressions are equally valid ways of specifying Boolean functions. For any Boolean expression, we can write a truth table with a column for each variable mentioned in the expression and a column for the expression value. We systematically fill in a row for each combination of variable values. For an expression with n distinct variables, there are 2^n combinations, so we need 2^n rows. For each combination, we substitute the variable values into the expression and evaluate the result. We write the result in the same row as the variable values, under the expression column.

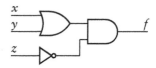

FIGURE 2.1   Circuit implementing a Boolean equation.

EXAMPLE 2.1   Derive the truth table corresponding to the Boolean expression $(x + y) \cdot \overline{z}$.

SOLUTION   There are three distinct variables in the expression, namely, x, y and z, so we will need $2^3 = 8$ rows in our truth table, as shown in Table 2.2. The easiest way to systematically fill in the variable values is to start with the value 0 for x in the first half of the table and 1 in the second half. Then, in each half, fill in the value 0 for y in the first half of that half and 1 in the second half of that half. In general, keep on filling in columns to the right, reducing the number of successive 0s and 1s by half each time, until single 0s and 1s alternate in the column for the last variable. Now evaluate the expression for the first row, substituting 0 values for x, y and z, to get the result 0. For the second row, substitute 0 for x and y and 1 for z, also giving the result 0. Continue in this way until all rows are filled in.

---

x	y	z	$(x + y) \cdot \overline{z}$
0	0	0	0
0	0	1	0
0	1	0	1
0	1	1	0
1	0	0	1
1	0	1	0
1	1	0	1
1	1	1	0

TABLE 2.2   Truth table for a Boolean expression.

We can also work in the reverse direction and derive a Boolean expression for a function represented by a truth table. We do this by examining the rows for which the expression has the value 1. For each such row, we form the logical AND of those variables for which the input value is 1, together with the negation of those variables for which the input value is 0. Such a conjunction is called a *minterm* of the function. For example, the third row of Table 2.2 gives us the minterm $\overline{x} \cdot y \cdot \overline{z}$. The complete expression for the function is then the logical OR of all the minterms for which the function value is 1. Thus, for the function of Table 2.2, the expression is

$$\overline{x} \cdot y \cdot \overline{z} + x \cdot \overline{y} \cdot \overline{z} + x \cdot y \cdot \overline{z}$$

Note that this is not the same expression as $(x + y) \cdot \overline{z}$, but it does have the same value for all combinations of input values. We say that the two expressions are *equivalent*, denoting the same function, and write the Boolean equation

$$(x + y) \cdot \overline{z} = \overline{x} \cdot y \cdot \overline{z} + x \cdot \overline{y} \cdot \overline{z} + x \cdot y \cdot \overline{z}$$

The right-hand expression in this equation is in *sum-of-products* form, meaning that it is the "sum" (logical OR) of a number of "product" (logical AND) terms, or *p-terms*, of variables. Note that each term in a sum-of-products expression need not be a minterm; that is, it need not include every variable that is mentioned in the expression. For example, another sum-of-products expression that is equivalent to the above expression is

$$\overline{x} \cdot y \cdot \overline{z} + x \cdot \overline{z}$$

An implication of equivalence of Boolean expressions is that digital circuits corresponding to equivalent expressions also implement the same function. For example, the two circuits shown in Figure 2.2, corresponding to the equivalent expressions $(x + y) \cdot \overline{z}$ and $\overline{x} \cdot y \cdot \overline{z} + x \cdot \overline{z}$, are functionally equivalent. This is a very important idea, as it means we can choose among the various equivalent circuits to implement a given function in order to satisfy nonfunctional constraints. Making such choices is a form of *optimization*, and is central to digital logic design. Note that a circuit with the minimal number of logic gates may not be the best choice in all circumstances. It depends on the particular constraints that apply. For example, if we are constrained to implement the function in certain kinds of programmable logic device, the circuit on the left may actually have more delay than the circuit on the right. We will return to the idea of constraint-dependent optimization many times throughout this book. In particular, in Section 2.1.2, we will look at some ways in which we can determine equivalent circuits for a given Boolean function.

An interesting thing about the logical OR, AND and negation operators is that any Boolean function can be written as an expression involving just these operators. One way to see the truth of this statement is to recognize that any function can be written as a truth table, and from there as a sum of products of minterms. Such an expression only involves the basic operators. A corollary is that any Boolean function can

FIGURE 2.2   Two equivalent digital circuits.

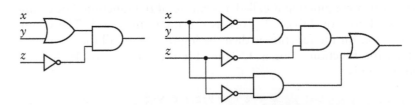

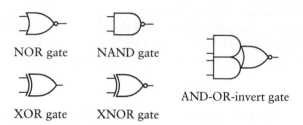

NOR gate   NAND gate

XOR gate   XNOR gate

AND-OR-invert gate

FIGURE 2.3 Complex logic gates.

be implemented using only OR gates, AND gates and inverters. However, such an implementation may not be optimal or even meet constraints. In fact, in most implementation fabrics, these gates are not the most simple that we can use. Figure 2.3 shows a number of other gates. They are often called *complex gates*, as their functions are combinations of the basic logical operations. The NOR, NAND and AND-OR-invert gates are of particular interest, since their internal circuitry in many implementation fabrics is very simple, and hence fast. Use of those gates can often lead to smaller and faster circuits for a given Boolean function than circuits involving OR and AND gates.

The function implemented by the *NOR* gate is the negation of the OR operation. Similarly, the function implemented by the *NAND* gate is the negation of the AND operation. The term *XOR* is short for *exclusive OR*, denoted by the operator "⊕" in Boolean expressions. The result of the exclusive OR operator is 1 if either, but not both, of the inputs is 1; and is 0 if both inputs are 0 or both inputs are 1. This is closer to what we usually mean when we say "or" informally in English. For example, when we're asked if we'd like ice cream or cake for dessert, we usually don't expect both! The function implemented by the *XNOR* gate is the negation of the exclusive OR operation. It is 1 when both inputs are the same and 0 when the inputs differ. For this reason, it is also called an *equivalence* gate. Finally, the *AND-OR-invert* gate performs the logical AND on each of two pairs of inputs, then performs a NOR operation on the two results. While it may look overly complicated to be called a single gate, the electrical implementation as a transistor circuit is surprisingly simple, which is why we include it here. The truth table for the functions implemented by the two-input gates are shown in Table 2.3. The truth table for the AND-OR-invert gate is left as an exercise.

a	b	$\overline{a + b}$	$\overline{a \cdot b}$	$a \oplus b$	$\overline{a \oplus b}$
0	0	1	1	0	1
0	1	0	1	1	0
1	0	0	1	1	0
1	1	0	0	0	1

TABLE 2.3 Truth table for functions implemented by complex gates.

EXAMPLE 2.2 Use truth tables to show that the following two Boolean functions are equivalent. Design a circuit using NOR and NAND gates for the first function, and a circuit using OR and AND gates and inverters for the second.

$$f_1 = \overline{\overline{a \cdot b} + c} \quad \text{and} \quad f_2 = (a \cdot b) \cdot \overline{c}$$

SOLUTION The truth table for f_1 is shown in Table 2.4, and that for f_2 is shown in Table 2.5. For each combination of input values, both functions have the same result value, so they are equivalent.

TABLE 2.4 Truth table for the first function.

a	b	c	$\overline{a \cdot b}$	$\overline{a \cdot b} + c$	f_1
0	0	0	1	1	0
0	0	1	1	1	0
0	1	0	1	1	0
0	1	1	1	1	0
1	0	0	1	1	0
1	0	1	1	1	0
1	1	0	0	0	1
1	1	1	0	1	0

TABLE 2.5 Truth table for the second function.

a	b	c	$a \cdot b$	$\overline{c}$	f_2
0	0	0	0	1	0
0	0	1	0	0	0
0	1	0	0	1	0
0	1	1	0	0	0
1	0	0	0	1	0
1	0	1	0	0	0
1	1	0	1	1	1
1	1	1	1	0	0

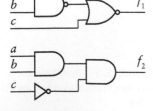

FIGURE 2.4 Two equivalent gate circuits.

The function f_1 involves the NAND operation applied to a and b, followed by the NOR operation applied to the result and c. The circuit implementing this function is shown at the top of Figure 2.4. The function f_2 involves the AND operation applied to a and b, followed by the AND operation applied to the result and the negation of c. The circuit for this function is shown at the bottom of Figure 2.4. Note that, since NAND and NOR gates are considerably simpler and faster in most implementation fabrics, the circuit at the top would be the preferred implementation.

___

There is one further Boolean function that we need to consider, namely, the *identity* function. This function has one input, and the function's value is just the value of the input. The simplest implementation of the identity function is a piece of wire. However, there is also a gate component, called a *buffer*, that implements the identity function. The symbol for a buffer is shown in Figure 2.5.

FIGURE 2.5   Symbol for a buffer.

It might seem strange to waste precious circuit area and power on a component that doesn't do anything. However, if we recall our discussion in Chapter 1 of static and capacitive loading of component outputs, we realize that buffer components are useful when we need to connect a given output to many inputs. If we just connect the output directly to the inputs, the output may be overloaded, affecting its ability to drive proper logic levels or to change between logic levels with acceptable rise and fall times. By inserting buffers between the output and the inputs, as shown in Figure 2.6, we can reduce the loading on the output to just that of the buffer inputs. Furthermore, each buffer output is now driving a fraction of the original inputs. When the number of inputs to be driven is very large, we can buffer the outputs of the buffers, and so on, forming a *buffer tree*, as shown in Figure 2.7. This is a two-level buffer tree, meaning that the original output drives each of the original inputs through two intervening buffers. If we extrapolate this arrangement, we can see that the number of inputs that can be driven from an output increases exponentially with the number of levels in the buffer tree.

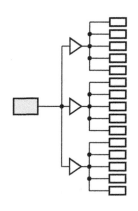

FIGURE 2.6   Using buffers to reduce loading on a component.

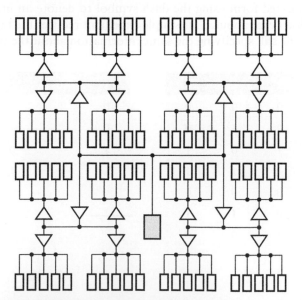

FIGURE 2.7   A two-level buffer tree.

As we shall see later, one important use for buffer trees is for connecting a clock signal from a clock-generator circuit to all of the flip-flops in a system. Meanwhile, however, we just need to be aware that buffers and buffer trees can be used in combinational circuits where many inputs are to be driven from a single output.

Don't Care Notation

While truth tables provide a systematic way to completely define a Boolean function, they can be cumbersome, particularly when the function has a lot of inputs. In many such cases, we can write the truth table in more compact form using the *don't care* notation for function inputs. In this book, we use the notation "−" for don't care, but "X" is another commonly used notation. Use of the don't care notation takes advantage of the property of many Boolean functions that, if some inputs have given values, the values of other inputs don't affect the result value. This is illustrated in Table 2.6, which shows the complete truth table and the compacted truth table for the function

$$z = \bar{s} \cdot a + s \cdot b$$

This is a Boolean equation for the multiplexer component that we introduced in Chapter 1. The input s represents the select input, and a and b represent the two data inputs: a is selected when $s = 0$ and b is selected when $s = 1$.

Note that, for this function, when $s = 0$, we don't care what value b has, and the output is the same as a. Similarly, when $s = 1$, we don't care what value a has, and the output is the same as b. This is shown in the compacted form using the dash symbol to denote an input whose value we don't care about. This simple expedient reduces the table to half the size, while still specifying the same information about the function.

TABLE 2.6 Complete and compacted truth tables for the multiplexer function.

s	a	b	z
0	0	0	0
0	0	1	0
0	1	0	1
0	1	1	1
1	0	0	0
1	0	1	1
1	1	0	0
1	1	1	1

s	a	b	z
0	0	−	0
0	1	−	1
1	−	0	0
1	−	1	1

In some designs, we can also use the don't care notation for the result of a function. We can do this if the design only requires a *partial function*, that is, if the function result need only be specified for some combinations of inputs and not for others. Usually, the input combinations for which we don't care about the result are those combinations that cannot arise during operation of the circuit; the combinations are logically impossible, given the functionality of the system of which the circuit is a part. However, any real circuit that we design will yield some value, either 0 or 1, for all possible input combinations. The benefit of specifying "don't care" for the impossible combinations, rather than arbitrarily choosing 0 or 1 as the function result, is that it gives us more scope for optimizing the circuit. We might be able to identify two candidate circuits that both produce the required outputs for the combinations we do care about, but that differ in their output for the "don't care" combinations. If one of the candidates better meets constraints than the other, we would choose it, accepting whatever result it yields for the "don't care" combinations.

EXAMPLE 2.3   The truth table in Table 2.7 has two don't care entries for the function f, since a result of 0 or 1 is equally acceptable for those two "impossible" input combinations. Compare the circuits that result from choosing 0 or 1 as the actual function result for both of the don't care combinations.

a	b	c	f	f_1	f_2
0	0	0	–	0	1
0	0	1	0	0	0
0	1	0	1	1	1
0	1	1	0	0	0
1	0	0	–	0	1
1	0	1	1	1	1
1	1	0	0	0	0
1	1	1	0	0	0

TABLE 2.7   Truth table for a function with "don't care" results, and two realizations of the function.

SOLUTION   If a value of 0 is chosen for both of the input combinations, the resulting function can be expressed as the sum of two minterms, $f_1 = \overline{a} \cdot b \cdot \overline{c} + a \cdot \overline{b} \cdot c$, and can be implemented by the circuit shown at the top of Figure 2.8. If a value of 1 is chosen for the combinations, the resulting function has more minterms, but can be reduced to the sum of products $f_2 = a \cdot \overline{b} + \overline{a} \cdot \overline{c}$,

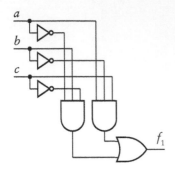

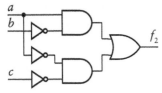

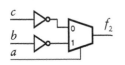

FIGURE 2.8　Realizations of a partial function.

implemented by either of the middle or bottom circuits in Figure 2.8. Our choice among these circuits may depend on the implementation fabric to be used. If we are simply concerned with minimizing the number of gate inputs, we would choose the middle circuit, yielding a result of 1 for the impossible input combinations. If our implementation fabric is based on sum-of-product circuit, and the minterms can also be shared as part of other functions in the system, we would choose the first, yielding a result of 0 for the impossible input combinations. Some implementation fabrics are based on multiplexers, introduced in Chapter 1, as the primitive circuit elements. If we were using such a fabric, we would choose the bottom circuit.

2.1.2　BOOLEAN ALGEBRA

The mathematical abstraction that we use as the foundation for digital design is *Boolean algebra*. It deals with Boolean expressions containing symbols that denote Boolean values, variables and operations. We can interpret the symbols as representing digital signals and gates.

Boolean algebra is based on a number of *axioms*. These are just Boolean equations that we take as given without requiring proof. The axioms of Boolean algebra are:

▶ Commutative laws:

$$x + y = y + x \tag{2.1}$$

$$x \cdot y = y \cdot x \tag{2.2}$$

▶ Associative laws:

$$(x + y) + z = x + (y + z) \tag{2.3}$$

$$(x \cdot y) \cdot z = x \cdot (y \cdot z) \tag{2.4}$$

▶ Distributive laws:

$$x + (y \cdot z) = (x + y) \cdot (x + z) \tag{2.5}$$

$$x \cdot (y + z) = (x \cdot y) + (x \cdot z) \tag{2.6}$$

▶ Identity laws:

$$x + 0 = x \tag{2.7}$$

$$x \cdot 1 = x \tag{2.8}$$

▶ Complement laws:

$$x + \overline{x} = 1 \tag{2.9}$$

$$x \cdot \overline{x} = 0 \tag{2.10}$$

Although we don't have to prove these laws, we can see that they make sense, since any consistent substitution of 0 and 1 values for variables in

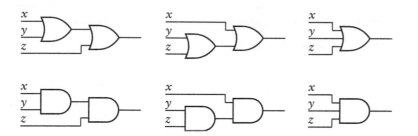

FIGURE 2.9 Circuits whose equivalence follows from the associative laws.

each law demonstrates the equality. The laws also suggest ways in which we can transform digital circuits while maintaining functional equivalence. For example, the commutative laws tell us that it doesn't matter which way around we connect the two inputs of an OR gate or an AND gate; we will get the same result either way. Similarly, the associative laws tell us that we don't need the parentheses when forming the logical OR or logical AND of three values, and that the circuits in each row in Figure 2.9 are equivalent. The distributive laws suggest how we can transform a circuit into sum-of-products form. This can be very useful, since many implementation fabrics allow efficient implementation of sum-of-product circuits.

Notice that we have presented the axioms in pairs, with each axiom being similar in form to the other in the pair. Each axiom is called the *dual* of the other in the pair. The *duality principle* of Boolean algebra states that we can take any Boolean equation and form its dual by interchanging the "+" and "·" operators and interchanging occurrences of 0 and 1; the dual is then a valid Boolean equation.

Given the axioms of Boolean algebra listed above, we can derive a number of further useful theorems:

▸ Idempotence laws:

$$x + x = x \tag{2.11}$$

$$x \cdot x = x \tag{2.12}$$

▸ Further identity laws:

$$x + 1 = 1 \tag{2.13}$$

$$x \cdot 0 = 0 \tag{2.14}$$

▸ Absorption laws:

$$x + (x \cdot y) = x \tag{2.15}$$

$$x \cdot (x + y) = x \tag{2.16}$$

▸ DeMorgan laws:

$$\overline{(x + y)} = \overline{x} \cdot \overline{y} \tag{2.17}$$

$$\overline{(x \cdot y)} = \overline{x} + \overline{y} \tag{2.18}$$

EXAMPLE 2.4   Prove the idempotence laws using just the axioms.

SOLUTION   To prove law 2.11:

$$
\begin{aligned}
x + x &= (x + x) \cdot 1 && \text{by identity law 2.8} \\
&= (x + x) \cdot (x + \overline{x}) && \text{by complement law 2.9} \\
&= x + (x \cdot \overline{x}) && \text{by distributive law 2.5} \\
&= x + 0 && \text{by complement law 2.10} \\
&= x && \text{by identity law 2.7}
\end{aligned}
$$

Law 2.12 immediately follows, since it is the dual of law 2.11.

EXAMPLE 2.5   Suppose we are to implement the following Boolean function using AND and OR gates and inverters:

$$f = (x + y \cdot \overline{z}) \cdot \overline{(y \cdot z)}$$

If we were to implement it directly, as shown in Figure 2.10, the longest path through the circuit is four gates. Show how the Boolean equation for f can be transformed into sum-of-products form, thus reducing the length of the longest path.

FIGURE 2.10   A circuit that directly implements a Boolean function.

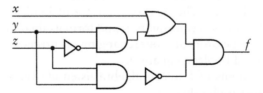

SOLUTION   We can transform the Boolean equation as follows:

$$
\begin{aligned}
f &= (x + y \cdot \overline{z}) \cdot \overline{(y \cdot z)} \\
&= (x + y \cdot \overline{z}) \cdot (\overline{y} + \overline{z}) && \text{DeMorgan law 2.18} \\
&= x \cdot (\overline{y} + \overline{z}) + (y \cdot \overline{z}) \cdot (\overline{y} + \overline{z}) && \text{distributive law 2.6} \\
&= x \cdot \overline{y} + x \cdot \overline{z} + y \cdot \overline{z} \cdot \overline{y} + y \cdot \overline{z} \cdot \overline{z} && \text{distributive law 2.6 twice} \\
&= x \cdot \overline{y} + x \cdot \overline{z} + 0 \cdot \overline{z} + y \cdot \overline{z} \cdot \overline{z} && \text{complement law 2.10} \\
&= x \cdot \overline{y} + x \cdot \overline{z} + 0 + y \cdot \overline{z} \cdot \overline{z} && \text{identity law 2.14} \\
&= x \cdot \overline{y} + x \cdot \overline{z} + 0 + y \cdot \overline{z} && \text{idempotence law 2.12} \\
&= x \cdot \overline{y} + x \cdot \overline{z} + y \cdot \overline{z} && \text{identity law 2.7}
\end{aligned}
$$

This reduced sum-of-products form of the Boolean equation can be implemented by the circuit of Figure 2.11, in which the longest path is reduced to three gates. Moreover, the circuit may be more efficiently implemented in this form in many fabrics.

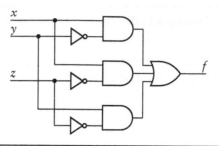

FIGURE 2.11   A circuit that implements the sum-of-products form.

The laws of Boolean algebra can be used to transform Boolean equations and their corresponding circuits, and to verify equivalence of Boolean expressions and circuits. However, they don't provide a recipe for finding an optimal circuit. That's mainly because the criteria for optimization depend on many different factors, including the implementation fabric to be used, power consumption constraints, physical packaging requirements, design resources available, and others. Optimization procedures, such as use of Karnaugh maps and the Quine-McClusky procedure, are described in many textbooks on digital logic design. They and other more involved procedures are founded on the laws of Boolean algebra. Given the complexity of the Boolean equations in real-world systems and the fact that computer aided design tools are needed to make optimization tractable, we won't go into the detail of the procedures in this book. Rather, we will focus on identifying the constraints that apply so that we can bring appropriate tools to bear on design problems.

2.1.3 VHDL MODELS OF BOOLEAN EQUATIONS

In the design methodology described in Chapter 1, we focused on the use of models expressed in an HDL such as VHDL. Modern CAD tools are very good at analyzing, verifying and synthesizing Boolean functions expressed in an HDL. In this section, we will see how to express Boolean equations in VHDL. Later, as we introduce more complex combinational components and circuits, we will also show how they can be expressed in VHDL.

As we mentioned earlier, a Boolean equation in which a name is defined to be equal to a Boolean expression can be implemented by the circuit for the expression yielding an output with the given name. We can write a Boolean equation directly in VHDL using an *assignment statement* within an architecture. We write the name of a signal or port on the left hand side of the assignment symbol, "<=", and a VHDL expression corresponding to the Boolean expression on the right hand side.

EXAMPLE 2.6   Develop a VHDL model for a circuit that implements the Boolean equation of Example 2.5.

SOLUTION   The equation refers to three inputs, x, y and z, and one output, f. We represent them as input and output ports in the entity declaration, as follows:

```
library ieee; use ieee.std_logic_1164.all;

entity circuit is
  port ( x, y, z : in  std_logic;
         f       : out std_logic );
end entity circuit;
```

The architecture contains an assignment statement that represents the Boolean equation, as follows:

```
architecture boolean_eqn of circuit is
begin
  f <= (x or (y and not z)) and not (y and z);
end architecture boolean_eqn;
```

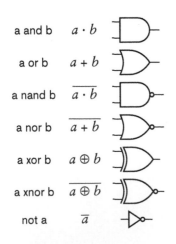

a and b	$a \cdot b$	
a or b	$a + b$	
a nand b	$\overline{a \cdot b}$	
a nor b	$\overline{a + b}$	
a xor b	$a \oplus b$	
a xnor b	$\overline{a \oplus b}$	
not a	$\overline{a}$	

FIGURE 2.12   VHDL operators and their corresponding Boolean operations and gates.

In order to write arbitrary Boolean equations in VHDL, we need to know how to form VHDL expressions that mean the same as Boolean expressions. The example above uses the VHDL operators and, or and not, corresponding to the Boolean operators ".", "+" and the overbar notation, respectively. VHDL also provides nand, nor, xor and xnor operators, corresponding to the operations and complex gates we introduced in Section 2.1.1. The VHDL operators, the Boolean expressions they represent and the corresponding gates are summarized in Figure 2.12. Note that VHDL does not make the same assumptions about precedence of logical operations that we have made for Boolean expressions. The not operator is evaluated first, but all of the remaining operators have equal precedence. Hence we need to include parentheses in VHDL expressions, as we did in the assignment in Example 2.6, to ensure that and operators are evaluated before or operators. Furthermore, since nand and nor are not associative, we need to parenthesize combinations of each of these operators applied to more than two operands, such as (a nand b) nand c.

When we write VHDL models for combinational circuits, we should generally not try to rearrange the Boolean expressions to imply any particular circuit of gates or other components. Rather, we should express the Boolean equations in the way that makes them most readily understood, then let our CAD tools synthesize and optimize a circuit based on constraints and our chosen implementation fabric. CAD tools can usually do a much better job at this than we could do manually. Where a CAD tool requires us to rearrange an expression to enable an optimization, we should clearly document the change and the reason for it using comments in the model code.

EXAMPLE 2.7   Develop a VHDL model for a combinational circuit that implements the following three Boolean equations, representing part of the control logic for an air conditioner:

$$\text{heater\_on} = \text{temp\_low} \cdot \text{auto\_temp} + \text{manual\_heat}$$

$$\text{cooler\_on} = \text{temp\_high} \cdot \text{auto\_temp} + \text{manual\_cool}$$

$$\text{fan\_on} = \text{heater\_on} + \text{cooler\_on} + \text{manual\_fan}$$

SOLUTION   The entity declaration of the VHDL model defines the input and output ports as follows:

```
library ieee; use ieee.std_logic_1164.all;

entity aircon is
  port ( temp_low, temp_high, auto_temp        : in  std_logic;
         manual_heat, manual_cool, manual_fan  : in  std_logic;
         heater_on, cooler_on, fan_on          : out std_logic );
end entity aircon;
```

The architecture body contains assignment statements for the Boolean equations:

```
architecture eqns of aircon is
  signal heater_on_tmp, cooler_on_tmp : std_logic;
begin
  heater_on_tmp <= (temp_low  and auto_temp) or manual_heat;
  cooler_on_tmp <= (temp_high and auto_temp) or manual_cool;
  fan_on        <= heater_on_tmp or cooler_on_tmp or manual_fan;
  heater_on     <= heater_on_tmp;
  cooler_on     <= cooler_on_tmp;
end architecture eqns;
```

One point to note is that VHDL does not allow us to use output ports, such as heater_on and cooler_on, as inputs in assignment statements. Hence, we have declared two internal signals, heater_on_tmp and cooler_on_tmp, for use within the architecture. The last two assignments simply make the output ports mirror

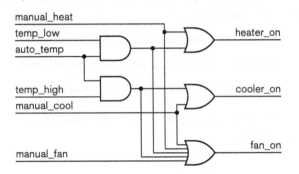

FIGURE 2.13 Circuits corresponding to the assignment statements for the air conditioner control logic.

the values of the internal signals. The hardware realization of these assignments is directly connecting wires. Apart from this difference, the VHDL assignments directly express the Boolean equations.

A straightforward synthesis of a digital circuit from this model is shown at the top of Figure 2.13. There are two subcircuits, one each for heater_on and cooler_on. The outputs of these circuits then drive the third subcircuit for fan_on. For some implementation fabrics, however, CAD tools might transform the circuit as shown at the bottom of Figure 2.13. The logical OR operations that produce the heater_on and cooler_on outputs are replicated and merged with the logical OR operation for fan_on. This circuit would fit well in a sum-of-products implementation fabric, and would have reduced propagation delay in that fabric.

KNOWLEDGE TEST QUIZ

1. Write a truth table for the Boolean function $f = a \cdot \overline{b} + \overline{c}$.

2. Use truth tables to show that the Boolean expression $\overline{a \cdot b}$ is equivalent to $\overline{a} + \overline{b}$.

3. What is meant by a Boolean expression being in *sum-of-products* form?

4. Write the truth table for the AND-OR-invert gate shown in Figure 2.3.

5. Why are buffers used in digital circuits?

6. Use the "don't care" notation for inputs to compact the truth table for the function f_1 shown in Table 2.4.

7. What is the benefit of using the "don't care" notation for outputs in a truth table?

8. What is the dual of the following Boolean equation?

$$\overline{a + b \cdot c} = \overline{a} \cdot \overline{b} + \overline{a} \cdot \overline{c}$$

9. Write a VHDL assignment statement to model the Boolean equation $f = a \cdot \overline{b} + \overline{c}$.

10. Why should we generally not try to optimize Boolean equations manually when modeling them in VHDL?

2.2 BINARY CODING

Thus far, we have looked at digital representation of information that has two possible values and shown how we can use Boolean algebra as the formal basis for circuits that deal with such information. We now extend our discussion to dealing with information involving more than two values. An obvious example is numeric information. However, since representation and computation of numeric information is such an important and extensive topic, it deserves a chapter of its own (Chapter 3). First, we will look at more general principles that underlie digital representation of all forms of information.

We saw in Chapter 1 that we can represent two-valued information with two distinct voltage levels in a circuit. Using our digital abstraction, we called the levels "low" and "high," but then refined them to ranges of voltages for pragmatic reasons. If we need to represent information that can take on N possible values, we could choose N distinct voltage levels (or voltage ranges, with intermediate thresholds). However, designing electronic circuits that can distinguish between more than two levels is extremely complex, and we would lose many of the benefits of binary digital circuits.

A better approach is to use multiple binary signals to represent a multivalued piece of information. Since each individual signal is binary, we can continue to use binary logic gates in our circuits with all of the advantages that they afford. We will use the values 0 and 1, as we did when discussing Boolean algebra, as the abstract values for each binary signal. We will continue to use the term *bit* to refer to these values.

Suppose that we have two signals, a_1 and a_0, available for representing some information. There are four possible combinations of binary values for the pair (a_1, a_0), namely, $(0, 0)$, $(0, 1)$, $(1, 0)$ and $(1, 1)$. Each possible combination is called a *code word*, and the set of all of the code words is called a *binary code*. Since a two-bit code has four possible code words, we can use a two-bit code to represent information with any number of values up to and including four. We just need to specify which code word corresponds to which value of the information. We say that a code word *encodes* the corresponding value.

EXAMPLE 2.8   Devise a binary code for the state of a road traffic light. The possible states are red, yellow and green.

SOLUTION   Since there are three possible values to represent, we can use a two-bit binary code with one code word unused. One possible code is

$$\text{red: } (0, 0) \qquad \text{yellow: } (0, 1) \qquad \text{green: } (1, 0)$$

In this case, the code word $(1, 1)$ is unused.

If two bits, with four possible code words, are not sufficient for the information we need to represent, we can just use more bits. In general an n-bit code has 2^n possible code words, so an n-bit code can represent information with up to 2^n values. Conversely, if we need to represent information with N values, we need at least $\lceil \log_2 N \rceil$ bits in our code. (The notation $\lceil x \rceil$ is called the *ceiling* of x, and denotes the smallest integer that is greater than or equal to x.) We might choose a longer code, for a variety of reasons that we will explore, in which case there will be more unused code words.

EXAMPLE 2.9   Many ink-jet printers have six cartridges for different colored ink: black, cyan, magenta, yellow, light cyan and light magenta. A multibit signal in such a printer indicates selection of one of the colors. Devise a minimal length code for the signal.

SOLUTION   Since there are six values to encode, the minimal length code is $\lceil \log_2 6 \rceil = 3$ bits long. There are $2^3 = 8$ possible code words, so two will remain unused. One possible code is:

black: $(0, 0, 1)$	cyan: $(0, 1, 0)$	magenta: $(0, 1, 1)$
yellow: $(1, 0, 0)$	light cyan: $(1, 0, 1)$	light magenta: $(1, 1, 0)$

While it might make sense in some cases to use the shortest code, in other cases a longer code is better. A particular case of a non–minimal-length code is a *one-hot* code, in which the code length is the number of values to be encoded. Each code word has exactly one 1 bit with the remaining bits 0. The advantage of a one-hot code becomes clear when we want to test whether the encoded multibit signal represents a given value; we just test the single-bit signal corresponding to the 1 bit in the code word for that value.

EXAMPLE 2.10   Devise a one-hot code for the state of the traffic light described in a preceding example.

SOLUTION   Since there are three values to encode, we need a 3-bit one-hot code. A possible code is:

red: (1, 0, 0)   yellow: (0, 1, 0)   green: (0, 0, 1)

With this code, the left-most bit can be used to activate the red light, the middle bit to activate the yellow light, and the right-most bit to activate the green light. No additional circuitry is needed to decode the encoded signals to determine which light to activate.

2.2.1   USING VECTORS FOR BINARY CODES

Since a collection of binary coded bits conceptually represents a single piece of information, it would be convenient to be able to represent it as a single signal in VHDL. We can do so using a signal of type std_logic_vector, instead of using several individual signals of type std_logic. For example, if we need a signal s to carry a 5-bit binary coded value, we could declare it as

```
signal s : std_logic_vector(4 downto 0);
```

This defines s to be a collection of five signals, s(4), s(3), s(2), s(1) and s(0), each of which is a std_logic element. Apart from condensing the declaration of the signals quite considerably, using vectors for encoded values gives us many other benefits, as we shall see throughout this book.

When we declare a vector signal or port, the part in parentheses (4 downto 0 in the above example) specifies the index range for the elements of the vector. The first value is the index of the left-most element, and the second value is the index of the right-most element. If we want to number elements in descending order, we use the word downto, as in the above example. We can also number elements in ascending order by using the word to, as in the following:

```
signal a : std_logic_vector(1 to 3);
```

Here, the elements from left to right are a(1), a(2) and a(3). The choice between ascending and descending order is often a question of style, and may be addressed by coding guidelines used in an organization. This example also shows that we don't have to use 0 for the least index value; it can be any number.

EXAMPLE 2.11   Assume that the one-hot code for the traffic lights in Example 2.10 is represented using a 3-element vector with element 1 corresponding to red, 2 to yellow and 3 to green. Develop a VHDL model for a light controller that has an encoded input, an encoded output, and a single-bit input that enables the lights. When the enable input is 1, the encoded output is the same as the encoded input. When the enable input is 0, all bits of the output are 0.

SOLUTION   The entity declaration is

```
library ieee; use ieee.std_logic_1164.all;

entity light_controller is
  port ( lights_in  : in  std_logic_vector(1 to 3);
         enable     : in  std_logic;
         lights_out : out std_logic_vector(1 to 3) );
end entity light_controller;
```

We can control each bit of the output by "AND-ing" the corresponding input with the enable bit. An architecture that does this is

```
architecture and_enable of light_controller is
begin
  lights_out(1) <= lights_in(1) and enable;
  lights_out(2) <= lights_in(2) and enable;
  lights_out(3) <= lights_in(3) and enable;
end architecture and_enable;
```

An alternative architecture is

```
architecture conditional_enable of light_controller is
begin
  lights_out <= lights_in when enable = '1' else
                "000";
end architecture conditional_enable;
```

The assignment in this architecture is called a *conditional assignment*. It uses the enable input to determine whether to copy the input to the output (when enable = '1') or to set the output to all 0 bits otherwise. Recall that, in VHDL, we use single quotation marks around 0 and 1 values for std_logic values. Also note that we can use a string of 0s and 1s, enclosed in double quotation marks, to form a std_logic_vector value.

2.2.2 BIT ERRORS

While digital circuits are much more immune to noise than analog electrical circuits, they are not completely immune from interference. The effect of interference is occasionally to change the value of a signal from 0 to 1 or from 1 to 0. We sometimes prosaically call this a *bit flip*. If the signal is a single bit representing a logical condition, the rest of the circuit continues operating on the incorrect value, possibly causing erroneous outputs. If the signal is one of several bits in a binary-coded representation of some information, there are two possibilities. The flipped bit results in the code word being changed either to another valid code word or to a bit combination that is not a valid code word. If the result is a valid code word, the rest of the circuit operates on the incorrect value, as in the single-bit case, possibly producing erroneous outputs. If the result is an invalid code word, operation of the circuit depends on how we deal with invalid codes in the design.

One design approach is to consider invalid code words as "impossible" inputs, and not to specify the behavior of circuits that operate on invalid inputs. If we adopt this approach, the actual behavior of the circuits will depend on the implementation for the valid-code-word cases and on optimizations performed by CAD tools. It may be acceptable not to care about the circuit output values for invalid code words, particularly if cost reduction is a driving constraint. For example, in a mass-produced consumer toy, no one really cares about a once-a-year glitch, particularly if fixing it would increase the cost from \$1.00 to \$1.05.

If, on the other hand, the application demands more deterministic outputs, we can adopt a "fail safe" design approach. We can design our circuit to produce correct outputs for valid code words, and to produce known safe outputs should an invalid code word arise due to interference. For example, in our ink-jet printer of Example 2.9, if interference caused the signal for selecting the color to take on the code word (1, 1, 1), we could deliberately select no color, rather than spoiling a printout with incorrect colors or damaging the mechanism by trying to select more than one color at once.

EXAMPLE 2.12   In Example 2.10, we suggested that the bits of the one-hot-coded signal could be used to activate the red, yellow and green lights, respectively. However, an error in the three-bit signal could cause multiple lights to activate, or no light to activate. Design a circuit that causes the three lights to activate normally for valid one-hot code words, and for the red light to be activated alone for invalid code words.

SOLUTION   Let us represent the three-bit signal with the bits *s_red*, *s_yellow* and *s_green*. The green light should be activated only when *s_green* is 1 and *s_yellow* and *s_red* are both 0. The Boolean equation is

$$green = \overline{s\_red} \cdot \overline{s\_yellow} \cdot s\_green$$

Similarly, the yellow light should be activated when s_yellow is 1 and s_green and s_red are both 0, giving the Boolean equation

$$yellow = \overline{s\_red} \cdot s\_yellow \cdot \overline{s\_green}$$

The red light should be activated when s_red is 1 and s_yellow and s_green are both 0, but it should also be activated in all other cases when neither the green nor yellow light is activated. The Boolean equation is

$$red = s\_red \cdot \overline{s\_yellow} \cdot \overline{s\_green} + \overline{(green + yellow)}$$

There are many other ways we could write this last Boolean equation, for example, by substituting for *green* and *yellow* and using the laws of Boolean algebra to rearrange it. However, we can leave that to a CAD tool, and simply enter the equations in the form above as part of a VHDL model.

A third design approach to dealing with errors introduced by interference is to have the circuit detect when they occur and then to take exceptional action. This is, in a sense, an extension of the "fail safe" approach. However, rather than producing a safe "normal" output, the circuit produces an "exceptional" output that indicates the circuit's function has not been performed correctly. An example of this approach is seen in modern cars that include digital circuits to manage the engine. If an error arises, the circuit detects the error and illuminates a warning light in the instrument panel as its exceptional output. Detecting that interference has flipped a bit in a code word requires that the code include unused code words, and that the bit flip change a valid code word to one of the invalid code words. Circuits that use the encoded information can check for invalid code words and take action, such as suppressing outputs or activating an error signal. Of course, if interference causes a valid code word to change to a different valid code word, the error would not be detected.

One technique that is often used for error detection is *parity*, which refers to the number of bits that are 1 in a code word. Parity error checking involves increasing the code length by one bit, called the *parity bit*. In the *even parity* scheme, the parity bit in each augmented code word is set to 0 or 1 to ensure that the total number of 1 bits is even. For example, if the original code word is 1011, the augmented code word is 10111. (The converse *odd parity* scheme sets the parity bit to ensure that the total number of 1 bits is odd.) In an even parity scheme, valid augmented code words have even parity, and invalid augmented code words have odd parity. If interference causes a 0 bit to change to 1, the number of 1 bits is increased by one, making the parity odd. Similarly, if interference changes a 1 bit to 0, the number of 1 bits is decreased by one, again making the parity odd. So to check whether a bit has flipped, we simply count the number of 1 bits, including the parity bit. If the count is odd, parity has been reversed, so an error has occurred. If the count is even, either no error has occurred, or an even number of bits have been flipped, which we can't detect. In many applications, the probability of two or more bits flipping is much lower than the probability of one bit flipping, so it is acceptable not to be able to detect an even number of bit flips.

Counting the number of bits in a code word might, at first, seem a rather complicated function to perform. However, since we're only interested in whether the total is even, the task is much simpler. For a code of original length 2, the function p to generate the parity bit so that the augmented code has even parity is shown in the truth table in Table 2.8. As we can see, this function is equivalent to the exclusive-OR function. So we can use an exclusive-OR gate to generate the parity bit to augment a 2-bit code. We can extend this to augment a 3-bit code by taking the exclusive OR of the parity of two bits with the third bit. In general, for a code of any length, we can just take the exclusive OR of all of the bits. Since the exclusive-OR function is commutative and associative, the order in which we apply the exclusive OR to the bits of the code doesn't matter. A common approach is to use a parity tree, as shown in Figure 2.14, since it keeps the overall propagation delay small and avoids using gates with large numbers of inputs. The tree at the left of the figure generates the parity bit to augment an 8-bit code, creating a code of nine bits with even parity. The tree at the right checks the augmented code and yields a 1 if there is a parity error.

There are two problems with parity schemes. First, if interference flips two bits, parity is preserved, so we miss that error. The same applies if four, six, or any even number of bits are flipped. In many applications, however, the probability of multiple bits being flipped is extremely low, so the cost of a more elaborate error detection scheme is not warranted. The second problem is that for any given invalid code word, there are several possible bit flips from a valid code word that could yield the invalid code word. So while we can detect occurrence of a single-bit error, we can't tell which bit is in error. If detection of errors and taking some exceptional action is sufficient for the application, parity is a good choice. However, if corrective action is needed, the approach can be extended by including sufficient invalid code words in the code that a flip of any given bit yields a distinct invalid code word. When that invalid code word is detected, it indicates that the given bit has been flipped. So correcting the error is simply a matter of flipping it back, that is, using the negation of that bit's value. This kind of code is called an *error correcting code* (ECC).

a_1	a_0	p
0	0	0
0	1	1
1	0	1
1	1	0

TABLE 2.8 Truth table for the parity bit of a code of original length 2, giving even parity for the augmented code.

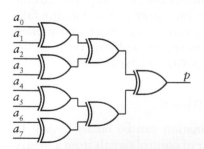

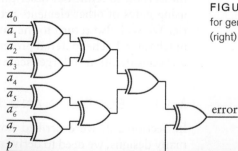

FIGURE 2.14 Parity trees for generating (left) and checking (right) even parity.

The design of codes to provide for error detection and correction is a very broad topic area. We will return to it as part of our discussion of storage in Chapter 5, since that is one place where errors can arise. Meanwhile, when we design circuits that operate on binary coded information, we should think about how they should behave when interference produces bit errors.

KNOWLEDGE TEST QUIZ

1. How many code words are possible with a code of 5 bits?

2. What is the minimum number of bits needed to encode information with 12 possible values?

3. Devise a one-hot code to represent the days of the week (Monday through Sunday).

4. Write a VHDL declaration for a signal, named s, representing an 8-bit binary coded value.

5. Write a VHDL assignment that drives each bit of s with a 0 value.

6. Does a single bit flip in a one-hot code word produce an invalid code word always, never, or sometimes?

7. How does extending a code with a parity bit to ensure odd parity enable detection of single-bit errors?

8. Can parity checking be used to correct the effect of a bit flip? If so, how? If not, why not?

2.3 COMBINATIONAL COMPONENTS AND CIRCUITS

In this section, we will introduce a number of combinational circuit components that are used as building blocks in larger digital systems. While these components can, themselves, be constructed from gates, it is generally not useful to do so. Instead, we will work at a higher level of abstraction. We will think of these components as basic blocks that, together with gates, are used to construct complex combinational circuits. We will rely on synthesis tools to refine our descriptions of such circuits into implementations using gates or other elements provided by the target implementation fabric. We will also return to the notion of negative logic, briefly mentioned in Chapter 1. The material presented in this section will form the basis for our consideration of larger-scale digital systems in later chapters.

2.3.1 DECODERS AND ENCODERS

In Section 2.2, we described how information can be binary coded. In many designs, we need to derive a number of control signals from a binary coded signal, with one control signal corresponding to each valid code word. When the encoded signal takes on a given code value, the corresponding control signal is activated. We call a circuit that derives the control signals in this way a *decoder*. For an *n*-bit code, if every code word is valid, the decoder will have 2^n outputs. As we shall see in Chapter 5, decoders are an important building block in memory designs.

We can derive the Boolean equation for each output of a decoder by looking at the corresponding code word. To illustrate, suppose we have an encoded 4-bit input signal (a_3, a_2, a_1, a_0), and we need to determine the Boolean equation for the output corresponding to the code word 1011. The output is 1 only when $a_3 = 1$, $a_2 = 0$, $a_1 = 1$ and $a_0 = 1$. Thus, the output is the value of the expression

$$a_3 \cdot \bar{a}_2 \cdot a_1 \cdot a_0.$$

A similar argument applies for other outputs. Each is the logical AND of the input bits, either directly (for bits that are 1 in the corresponding code word) or negated (for bits that are 1 in the corresponding code word).

EXAMPLE 2.13   Develop a VHDL model for a decoder for use in the ink-jet printer described in Example 2.9. The decoder has three input bits representing the choice of color cartridge and six output bits, one to select each cartridge.

SOLUTION   The entity declaration is

```
library ieee; use ieee.std_logic_1164.all;

entity ink_jet_decoder is
  port ( color2, color1, color0      : in  std_logic;
         black, cyan, magenta, yellow,
         light_cyan, light_magenta   : out std_logic );
end entity ink_jet_decoder;
```

The architecture body contains assignments representing the Boolean equations for the outputs, as follows:

```
architecture eqn of ink_jet_decoder is
begin
  black         <= not color2 and not color1 and     color0;
  cyan          <= not color2 and     color1 and not color0;
  magenta       <= not color2 and     color1 and     color0;
  yellow        <=     color2 and not color1 and not color0;
  light_cyan    <=     color2 and not color1 and     color0;
  light_magenta <=     color2 and     color1 and not color0;
end architecture eqn;
```

If an invalid code occurs on the input bits, none of the outputs is activated. This can be considered a "fail safe" design.

The inverse of a decoder is called an *encoder*. It has, as inputs, a number of single-bit signals, and as outputs, a collection of signals representing the bits of an encoded value. We will assume for the moment that at most one of the inputs is 1 at any time, and the others are all 0. The code word at the output corresponds to the particular input that is 1.

We can derive the Boolean equation for each bit of the output by identifying those inputs for which the output bit is 1. The output bit is then the logical OR of those inputs. However, we need to take account of the possibility that none of the inputs is 1, since that would cause our encoder to output a code word of all 0 bits. If that code word is invalid, we can use it to imply that no inputs are 1, essentially extending the code. Alternatively, if the all-0s code word is valid and corresponds to one of the inputs being 1, we need to have a separate output that indicates when any of the inputs is 1. When this output is 0, we ignore the code word produced by the encoder.

EXAMPLE 2.14   Design an encoder for use in a domestic burglar alarm that has sensors for each of eight zones. Each sensor signal is 1 when an intrusion is detected in that zone, and 0 otherwise. The encoder has three bits of output, encoding the zone as follows:

Zone 1: 000   Zone 2: 001   Zone 3: 010   Zone 4: 011
Zone 5: 100   Zone 6: 101   Zone 7: 110   Zone 8: 111

SOLUTION   Since all code words are used, we need a separate output to indicate when there is a valid code-word output. The entity declaration is

```
library ieee; use ieee.std_logic_1164.all;

entity alarm is
  port ( zone           : in  std_logic_vector(1 to 8);
         intruder_zone  : out std_logic_vector(2 downto 0);
         valid          : out std_logic );
end entity alarm;
```

The left-most bit of the output code is 1 when any of the zone 5 through zone 8 inputs is 1, so the equation for that output is the logical OR of those zone inputs. The equations for the other two output code bits are derived similarly. The valid output is the logical OR of all of the zone inputs. The architecture is

```
architecture eqn of alarm is
begin
  intruder_zone(2) <= zone(5) or zone(6) or zone(7) or zone(8);
  intruder_zone(1) <= zone(3) or zone(4) or zone(7) or zone(8);
  intruder_zone(0) <= zone(2) or zone(4) or zone(6) or zone(8);
  valid <= zone(1) or zone(2) or zone(3) or zone(4) or
           zone(5) or zone(6) or zone(7) or zone(8);
end architecture eqn;
```

Now let's consider the possibility of more than one input to an encoder being 1 at a time. The design we described above would produce an incorrect output, possibly an invalid code word. The solution is to assign priorities to the inputs, so that if multiple inputs are 1, the encoder outputs the code word corresponding to the input with highest priority. Such an encoder is called, not surprisingly, a *priority encoder*. One application of priority encoders is to prioritize interrupts in embedded systems. (We describe interrupts in Chapter 8.)

EXAMPLE 2.15   Revise the encoder for the burglar alarm to be a priority encoder, with zone 1 having highest priority, down to zone 8 having lowest priority.

SOLUTION   The entity declaration is unchanged, since we need the same inputs and outputs for the encoder. The truth table for the priority encoder is shown in Table 2.9. From this, we can derive the Boolean equations for each bit of the output. A revised VHDL architecture is shown on the next page.

| zone | | | | | | | | intruder_zone | | | |
(1)	(2)	(3)	(4)	(5)	(6)	(7)	(8)	(2)	(1)	(0)	valid
1	–	–	–	–	–	–	–	0	0	0	1
0	1	–	–	–	–	–	–	0	0	1	1
0	0	1	–	–	–	–	–	0	1	0	1
0	0	0	1	–	–	–	–	0	1	1	1
0	0	0	0	1	–	–	–	1	0	0	1
0	0	0	0	0	1	–	–	1	0	1	1
0	0	0	0	0	0	1	–	1	1	0	1
0	0	0	0	0	0	0	1	1	1	1	1
0	0	0	0	0	0	0	0	–	–	–	0

TABLE 2.9   Truth table for a priority encoder for a burglar alarm.

```
architecture priority of alarm is
  signal winner : std_logic_vector(1 to 8);
begin
  winner(1) <= zone(1);
  winner(2) <= zone(2) and not zone(1);
  winner(3) <= zone(3) and not (zone(2) or zone(1));
  winner(4) <= zone(4) and not (zone(3) or zone(2) or zone(1));
  winner(5) <= zone(5) and not (zone(4) or zone(3) or zone(2) or
                                zone(1));
  winner(6) <= zone(6) and not (zone(5) or zone(4) or zone(3) or
                                zone(2) or zone(1));
  winner(7) <= zone(7) and not (zone(6) or zone(5) or zone(4) or
                                zone(3) or zone(2) or zone(1));
  winner(8) <= zone(8) and not (zone(7) or zone(6) or zone(5) or
                                zone(4) or zone(3) or zone(2) or
                                zone(1));

  intruder_zone(2) <= winner(5) or winner(6) or
                      winner(7) or winner(8);
  intruder_zone(1) <= winner(3) or winner(4) or
                      winner(7) or winner(8);
  intruder_zone(0) <= winner(2) or winner(4) or
                      winner(6) or winner(8);

  valid <= zone(1) or zone(2) or zone(3) or zone(4) or
           zone(5) or zone(6) or zone(7) or zone(8);
end architecture priority;
```

In this architecture, each element of the internal signal winner indicates when the corresponding zone is 1 and has not lost to a higher priority zone. The encoder then uses the elements of the internal signal instead of the zone inputs directly to generate the output code word. Another way of expressing this in VHDL is shown in the following architecture:

```
architecture priority_1 of alarm is
begin
  intruder_zone <= "000" when zone(1) = '1' else
                   "001" when zone(2) = '1' else
                   "010" when zone(3) = '1' else
                   "011" when zone(4) = '1' else
                   "100" when zone(5) = '1' else
                   "101" when zone(6) = '1' else
                   "110" when zone(7) = '1' else
                   "111" when zone(8) = '1' else
                   "000";
```

(continued)

```
    valid <= zone(1) or zone(2) or zone(3) or zone(4) or
             zone(5) or zone(6) or zone(7) or zone(8);
end architecture priority_1;
```

The conditional assignment in this architecture tests a series of conditions to determine the value to assign to the signal intruder_zone. First the zone 1 input is tested, and the result assigned 000 if the zone 1 input is 1. Otherwise, the zone 2 input is tested, and the result assigned 001 if the zone 2 input is 1. Testing continues in this way, with priority implied by the order of testing the conditions. This form of assignment for priority encoding is much easier to understand, and leaves the hard work of determining and optimizing the Boolean equations to the synthesis CAD tool.

BCD Code and 7-Segment Decoders

One form of information that we might wish to encode is numeric information. As we mentioned earlier, we will look at this topic in detail in Chapter 3. However, in this section, we will look at a particular form of numeric coding called *binary coded decimal* (BCD). If we consider just a single decimal digit, the ten possible values are 0, 1, 2, 3, 4, 5, 6, 7, 8 and 9. We need at least 4 bits in a binary code for these values. There are a large number of possible codes, but BCD is the most common, having the following code words:

$$0: 0000 \quad 1: 0001 \quad 2: 0010 \quad 3: 0011 \quad 4: 0100$$
$$5: 0101 \quad 6: 0110 \quad 7: 0111 \quad 8: 1000 \quad 9: 1001$$

If we have more than one decimal digit of information to represent, we simply use groups of four bits, with each group corresponding to one decimal digit. For example, a system that deals with three-digit numbers would use a 12-bit code. The number 493 would be encoded as 0100 1001 0011.

Many digital systems display decimal numbers using 7-segment displays. Each display digit consists of seven separate lights, arranged as shown in Figure 2.15. If we have a digit encoded using BCD and we need to display the digit on a 7-segment display, we need a *7-segment decoder*. Strictly speaking, we should call it a "7-segment code converter," since it converts from a BCD code input to a 7-segment code output. However, the term "7-segment decoder" is widely used. Assuming a segment is lit if its input is 1, we need a 7-bit code for representing the digits 0 through 9. The code word for each digit has a 1 bit corresponding to each segment that is lit and a 0 bit corresponding to each segment that is not lit. A 7-segment decoder then converts between BCD and this 7-bit code. One possible code is shown in Figure 2.16, with the bits corresponding left to right with segments g through a.

FIGURE 2.15   A 7-segment display digit. The segments are named "a" through "g," as shown.

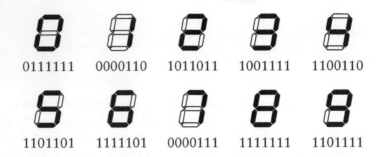

FIGURE 2.16 A 7-segment code for decimal digits. In each code word, the bits correspond to segments g through a in left-to-right order.

EXAMPLE 2.16 Develop a VHDL model for a 7-segment decoder. Include an additional input, blank, that overrides the BCD input and causes all segments not to be lit.

SOLUTION   The entity declaration is

```
library ieee; use ieee.std_logic_1164.all;

entity seven_seg_decoder is
  port ( bcd   : in  std_logic_vector(3 downto 0);
         blank : in  std_logic;
         seg   : out std_logic_vector(7 downto 1) );
end entity seven_seg_decoder;
```

We could determine the BCD code words that result in each segment being lit, and so derive Boolean equations for each segment output. However, that would make the model hard to understand. A better approach is to list the 7-bit code word corresponding to each BCD code word, as we did in Figure 2.16. An architecture that does this is

```
architecture behavior of seven_seg_decoder is
  signal seg_tmp : std_logic_vector(7 downto 1);
begin
  with bcd select
    seg_tmp <= "0111111" when "0000",    -- 0
               "0000110" when "0001",    -- 1
               "1011011" when "0010",    -- 2
               "1001111" when "0011",    -- 3
               "1100110" when "0100",    -- 4
               "1101101" when "0101",    -- 5
               "1111101" when "0110",    -- 6
               "0000111" when "0111",    -- 7
               "1111111" when "1000",    -- 8
```

(continued)

```
                "1101111" when "1001",  -- 9
                "1000000" when others;  -- "-" for invalid code

    seg <= "0000000" when blank = '1' else
           seg_tmp;
end architecture behavior;
```

The first assignment uses the BCD encoded input to select which 7-bit code to assign to the internal signal seg_tmp. This is an example of a form of assignment in VHDL called a *selected assignment*. The last line of that assignment includes the clause "when others" to deal with invalid codes. VHDL requires us to handle all possible values of the signal controlling selection, so a "when others" clause is a convenient way to handle cases not listed explicitly. The second assignment uses the blank input to determine whether to drive the encoded output with all 0s, causing all segments not to be lit, or to copy the value decoded from the BCD input to the output.

2.3.2 MULTIPLEXERS

Multiplexers are an important building block in many digital systems. We introduced a simple multiplexer in Section 1.2. It has two data inputs, one data output, and a select input that determines which input value is used for the output value. We can expand on this simple multiplexer along two dimensions. First, we can add more data inputs, which also requires adding further select inputs to encode the choice of input to drive the output. Second, we can use multiplexers in parallel to select between two sources of multibit encoded data. Let's look at the alternatives in more detail.

Suppose that, instead of selecting between two input bits, we need to select between four input bits. Since there are four input sources, we need to have four values for the select input. We can encode the select input using two bits. Figure 2.17 shows a 4-to-1 multiplexer. The select input is drawn as a thicker line to indicate that it is a multibit encoded input. In this book, we will mostly use line thickness to distinguish between single-bit and multibit signals. Occasionally, where we want to emphasis that a signal is multibit, we will add a stroke across the line and show the number of bits, as in Figure 2.17. The code for the select input is

00: input 0   01: input1   10: input 2   11: input 3

We could describe a gate circuit to implement the multiplexer, but there is little point, for two reasons. First, a synthesis tool would probably optimize the circuit, changing it from what we specify. Second, in a number of implementation fabrics, multiplexers can be constructed from individual transistors more efficiently than as a circuit of gates. Multiplexers

FIGURE 2.17 A 4-to-1 multiplexer.

would be considered primitive elements in those fabrics. So instead of a gate-level circuit, we will just consider how to express a multiplexer function in VHDL.

EXAMPLE 2.17   Develop a VHDL model for a 4-to-1 multiplexer.

SOLUTION   The entity declaration and architecture are

```
library ieee; use ieee.std_logic_1164.all;

entity multiplexer_4_to_1 is
  port ( a   : in  std_logic_vector(3 downto 0);
         sel : in  std_logic_vector(1 downto 0);
         z   : out std_logic );
end entity multiplexer_4_to_1;

architecture eqn of multiplexer_4_to_1 is
begin
  with sel select
    z <= a(0) when "00",
         a(1) when "01",
         a(2) when "10",
         a(3) when others;
end architecture eqn;
```

The selected assignment in the architecture uses the value of the sel input to determine which input bit to copy to the output. Note that we might try to write the when others alternative as when "11", to make it symmetric with the other alternatives. However, this would cause a VHDL tool to object. The reason is that the std_logic type includes additional values beyond '0' and '1' that are used for modeling purposes that we will describe later. Since we need to write alternatives for all possible values for the selection expression when writing a selected assignment, we use a when others alternative to include values beyond just '0' and '1' values. A synthesis CAD tool will interpret this as we intend and generate a circuit for a multiplexer.

We can further expand this multiplexer to have eight data inputs, which would require a 3-bit select input. The number of data inputs need not be a power of 2. If it is not, then the select input code will have unused code words. We must then ensure that an invalid code word is never presented to the select input. In general, a multiplexer having N input bits needs $\lceil \log_2 N \rceil$ bits for the select input, since the select input carries a binary code requiring N values.

Now let's consider using multiplexers to select between two sources of encoded data. If the code length is m (that is, each code word has

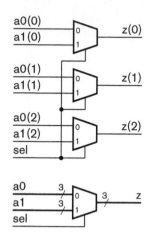

m bits), we can use m two-input multiplexers, one for each bit of the two data sources. This is illustrated in Figure 2.18 for selecting between two sources each of three bits. The circuit at the top of the figure shows the three separate 2-to-1 multiplexers. At the bottom of the figure is a symbol that represents a 2-to-1 multiplexer operating on the 3-bit encoded data inputs and output.

EXAMPLE 2.18   Develop a VHDL model for the 3-bit 2-to-1 multiplexer.

SOLUTION   The entity declaration and architecture are

```
library ieee; use ieee.std_logic_1164.all;

entity multiplexer_3bit_2_to_1 is
  port ( a0, a1 : in  std_logic_vector(2 downto 0);
         sel    : in  std_logic;
         z      : out std_logic_vector(2 downto 0) );
end entity multiplexer_3bit_2_to_1;

architecture eqn of multiplexer_3bit_2_to_1 is
begin
  z <= a0 when sel = '0' else
       a1;
end architecture eqn;
```

FIGURE 2.18   A circuit for a 2-to-1 multiplexer for 3-bit data sources (top), and a symbol for the multiplexer (bottom).

We can, of course, combine these two forms of expansion. If we need to select between N sources of data, each of which is encoded with m bits, we simply use m lots of N-to-1 multiplexers. The details are left as an exercise.

Before we leave the topic of multiplexers, it is interesting to note that all Boolean functions can be expressed in terms of multiplexers combined with negation. To illustrate, consider the function that we examined earlier, $f = (x + y) \cdot \bar{z}$, whose truth table is shown in Table 2.2. This function can be implemented using the circuit shown in Figure 2.19. Note the use of a literal 0 value for one input. This can be implemented by hard wiring the input to the 0V ground. We won't go into the general principles of how to implement Boolean functions using multiplexers here. We raise the topic since multiplexers can be very efficiently implemented in some fabrics. As an example, the basic circuit elements in FPGAs manufactured by Actel Corporation consist of two multiplexers and a small number of other associated components. However, the details of mapping arbitrary Boolean equations to multiplexers are generally handled by CAD tools.

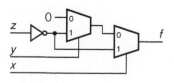

FIGURE 2.19   Implementing a Boolean function using multiplexers.

2.3.3 ACTIVE-LOW LOGIC

Thus far, we have focused on circuits in which a low logic level represents the falsehood of some condition and a high logic level represents truth of the condition. In Chapter 1, we identified this convention as *positive logic*, or *active-high logic*. In principle, the correspondence of low with falsehood and high with truth is largely arbitrary. We could just as well represent falsehood with a high logic level and truth with a low logic level, a convention that we referred to in Chapter 1 as *negative logic*, or *active-low logic*. Note that "positive" and "negative" in this context don't refer to the voltage polarity, but simply distinguish between the two conventions. We will use the terms "active high" and "active low" to avoid the confusion. We will also maintain the convention of associating 0 with a low logic level and 1 with a high logic level.

In a circuit that mixes both active-low and active-high logic, we could get confused about which convention is used for which signal. We should still label signals with the conditions they represent so that we can understand the intended function of the circuit. A commonly adopted approach is to label an active-low signal with the negation of the condition it represents. For example, an active-low signal representing the condition that a lamp is lit would be labeled $\overline{\text{lamp\_lit}}$, since the signal is 1 when the condition is false and 0 when the condition is true.

One reason for using active-low logic is that some kinds of digital circuits are able to sink more current when driving an output low than they can source when driving the output high. If such an output is used to activate some condition for which current flow is required, it would be better to use a low logic level rather than a high logic level.

EXAMPLE 2.19   Revise the night-light circuit from Figure 1.3 in Chapter 1 by connecting the lamp to the positive power supply instead of to ground.

SOLUTION   To make current flow in the lamp and light it, we need to drive the controlling signal low. Thus, we must use an active-low signal to implement the "lamp lit" condition. This is shown in Figure 2.20, in which the controlling signal is labeled $\overline{\text{lamp\_lit}}$. The gate performs the logical AND function of the lamp_enabled and dark signals, but its output must be negated to match the negation of the "lamp lit" condition. Hence, we use a NAND gate in place of the AND gate in the original circuit.

In general, this approach to dealing with active-low logic involves matching negation "bubbles" on components with active-low signals. When we do that, no negation of the logical condition represented by the signal is implied. Thus, we can interpret the circuit of Figure 2.20 as saying, "The lamp is lit when the lamp is enabled and it is dark." If we connect an active-low signal to a component without a bubble at

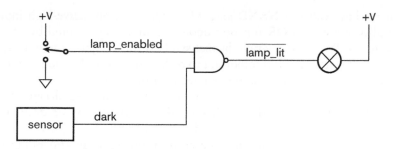

FIGURE 2.20   The night-light circuit using an active-low signal.

the connection point, we are implying negation of the logical condition represented by the symbol.

EXAMPLE 2.20   Returning to the original night-light circuit from Figure 1.3 in Chapter 1, think of the sensor as having an active-low output representing the condition "it is light." Redraw the circuit to take account of this change.

SOLUTION   As shown in Figure 2.21, we label the signal connected to the sensor $\overline{\text{light}}$, to show that it is active-low. We draw a bubble on the sensor output to indicate it is an active-low output. There is no negation implied by the connection at the sensor output, since we have a bubble output connected to an active-low signal. However, since there is no bubble on the AND gate input, logical negation is implied for its connection to $\overline{\text{light}}$. Thus, we can interpret the circuit as saying, "The lamp is lit when the lamp is enabled and it is not light."

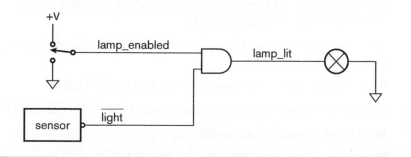

FIGURE 2.21   The night-light circuit with negation implied by connecting an active-low signal to an active-high input.

When we draw gate circuits for Boolean functions, it is important to use AND and OR gates as appropriate for the logical operations applied to conditions represented by signals. If any of those signals are active-low, and no implicit negation is intended, we should "draw a bubble" where the signal connects to a gate. We can make use of DeMorgan's laws to derive alternate views of gates. For example, Equation 2.18 tells us that the component

FIGURE 2.22   *Alternate logic symbols for a gate.*

that we have called a NAND gate when operating on active-high inputs can also perform an OR function upon conditions represented by active-low inputs. We can draw two distinct symbols for the gate component, as shown in Figure 2.22. It is important to realize, however, that both symbols represent the same circuit of interconnected transistors!

One of the problems we encounter when modeling designs with active-low signals in VHDL is that we don't have a way of drawing a negation bar over a signal name or drawing a bubble on a port. Instead, we usually adopt a textual naming convention, such as appending the suffix "_N" to a name, to indicate which signals and ports are active-low. For example, a VHDL model might give the active-low output of the sensor in Figure 2.21 the name light_N. A VHDL model for the sensor would assign '0' to the light_N signal when it is light and '1' when it is dark. The model for the AND gate assigns '1' to its output when both inputs are '1', and '0' to its output otherwise. Thus, the lamp_lit signal is assigned '1' when lamp_enabled is '1' ("the lamp is enabled") and light_N is '1' ("it is not light"). When we're dealing with active-low logic in VHDL models, we need to think carefully about which VHDL value represents truth or falsehood of each condition, and design accordingly.

KNOWLEDGE TEST QUIZ

1. For a decoder with inputs (a_2, a_1, a_0), write the Boolean equation for the output corresponding to the code word 100.

2. What would be the output of the encoder in Example 2.14 if both the Zone 2 and Zone 3 inputs were 1 at the same time? Would this output be correct?

3. What problem would arise if we did not include the valid output from the encoder in Example 2.14?

4. How does a priority encoder solve the problem of multiple inputs being 1 at the same time?

5. What decimal digit is represented by the BCD code 0101?

6. What is the 7-segment code corresponding to the BCD code 0011?

7. What is the purpose of a multiplexer?

8. How many select input bits are needed for a 6-to-1 multiplexer?

9. How can we construct a 2-to-1 multiplexer for 5-bit encoded data inputs?

10. What logic level would you expect on a signal labeled $\overline{\text{door\_closed}}$, connected to a door sensor, when the door is open?

11. If a VHDL signal named motor_on_N represents an active-low signal, what VHDL value would you assign it to turn the motor on?

2.4 VERIFICATION OF COMBINATIONAL CIRCUITS

In Section 1.5 we introduced a design methodology to guide us in the design and implementation of a digital system. The first task was to develop and enter a design description based on the application's requirements and constraints. In this chapter, we have seen examples of design descriptions, expressed as schematics and as VHDL models, for simple combinational circuits and components. Most systems are more involved and include sequential components as well as combinational subcircuits, so there is a limit to how much of the methodology we can demonstrate. Nonetheless, there are small-scale applications where combinational circuits are sufficient, so we will show how we can apply our design methodology to them.

The second step in our design methodology is functional verification, that is, ensuring that the design performs the operation required of it. Since, in a combinational circuit, the values of the outputs depend only on the current values of the inputs, we can simply verify that the circuit produces the required output for each combination of input values. For a design description expressed in VHDL, we can develop a *testbench* model that provides input values to the *design under verification* (DUV) and checks that the output values are correct. The DUV is also frequently called a *device under test* (DUT), but that usage may be confused with physical testing of manufactured devices. We will use the term DUV in this book to avoid the confusion. The testbench model is, itself, a VHDL model that we can execute using a simulator. However, it is not intended to describe hardware that will be built. Rather, its purpose is to apply a sequence of values, called *test cases*, to the input connections of the DUV, and to monitor the output connections to ensure that correct values are produced. The DUV is usually an instance of the VHDL entity and architecture that jointly describe the design. A simulator mimics the passage of time, executing the DUV and testbench models, and assigning values to signals at appropriate simulated times.

The difficult part of developing a testbench model is working out how to express the correctness conditions. If the requirements are expressed as Boolean equations, the design will probably implement those equations directly, so expressing the correctness conditions as Boolean equations gains nothing. A better approach is to determine some more abstract conditions that are required to hold, and to test that the design satisfies those conditions.

EXAMPLE 2.21 Develop a testbench model for the and_enable architecture of the traffic light control circuit of Example 2.11. Verify the conditions that, when the enable input is 1, the output is the same as the light input, and when the enable input is 0, all light outputs are inactive.

SOLUTION   The testbench model includes an instance of the design under verification, as well as code to apply test cases and to check for correct outputs. The organization of these components is shown in Figure 2.23.

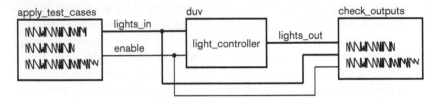

FIGURE 2.23   Organization of the testbench for the light controller.

Since the testbench is a VHDL model, it needs an entity declaration. However, since there are no external connections to the testbench, the entity has no ports. The entity declaration is

```
entity light_testbench is
end entity light_testbench;
```

The architecture is

```
library ieee; use ieee.std_logic_1164.all;

architecture verify of light_testbench is
  signal lights_out : std_logic_vector(1 to 3);
  signal lights_in  : std_logic_vector(1 to 3);
  signal enable      : std_logic;
begin

  duv : entity work.light_controller(and_enable)
    port map ( lights_out => lights_out,
               lights_in => lights_in, enable => enable );

  apply_test_cases : process is
  begin
    enable <= '0'; lights_in <= "000"; wait for 1 sec;
    enable <= '0'; lights_in <= "001"; wait for 1 sec;
    enable <= '0'; lights_in <= "010"; wait for 1 sec;
    enable <= '0'; lights_in <= "100"; wait for 1 sec;
    enable <= '1'; lights_in <= "001"; wait for 1 sec;
    enable <= '1'; lights_in <= "010"; wait for 1 sec;
    enable <= '1'; lights_in <= "100"; wait for 1 sec;
    enable <= '1'; lights_in <= "000"; wait for 1 sec;
    enable <= '1'; lights_in <= "111"; wait for 1 sec;
    wait;
  end process apply_test_cases;
```

(continued)

```
    check_outputs : process is
    begin
      wait on enable, lights_in;
      wait for 10 ms;
      assert (enable = '1' and lights_out = lights_in) or
             (enable = '0' and lights_out = "000");
    end process check_outputs;

  end architecture verify;
```

Note that we need to include the line that identifies the ieee library and the std_logic_1164 package before the architecture, since the line is not included before the entity declaration in this example. Within the architecture, the statement labeled duv is an instantiation of the light_controller entity and the corresponding and_enable architecture, taken from the library called work. In VHDL, work is the name of the library into which models are compiled. The input and output ports of the instance are connected to internal signals within the testbench architecture. Note that we have used *named port association* here, rather than *positional port association* as we did in Example 1.5. In named association, we write the name of the entity port on the left of the "=>" symbol and the signal to which it is connected on the right of the symbol. This allows us to write the connections in any order, rather than following the order of ports in the entity declaration. Given the advantages and clarity of named association, we will use it in models from now on.

Following the instantiation statement is a *process statement*, labeled apply_test_cases, that does as its label suggests. A process is a collection of VHDL statements that are executed one after another, much like statements in a programming language. When the last statement in a process has been executed, the process starts again from the first statement. In the first line, the apply_test_cases process makes an assignment to the enable signal, followed by an assignment to the lights_in signal. These two assignments constitute application of one test case to the inputs of the DUV. The process then executes a *wait statement* that suspends the process for 1 second of simulated time. (In VHDL, the word sec is the required abbreviation for seconds. Note also that a space is required between the number and the unit.) The process waits while other parts of the model, including the instance of the lights controller, continue executing. After the 1-second delay, the process continues with the second line, applying the next test case to the DUV inputs and then waiting a further second of simulated time. The process continues in this way until it reaches the last statement, which is just a wait statement with no time-out specified. This statement suspends the process "forever." Since the testing is complete at this point, we don't want the process to start again from the top.

The check_outputs process has the job of ensuring that the DUV outputs meet the requirements. In developing this process, we need to determine when to check the outputs. If we were to check them at the same time as changing the inputs, the DUV would not yet have responded to the input change, and the outputs would still reflect the previous inputs. In this example, we will wait for an interval of 10ms of simulated time after an input change before checking the outputs. The first statement in the check_outputs process waits for a change of value on either the enable signal or any of the bits of the lights_in signal. When that occurs, the process then moves on to the next statement, which waits for the 10ms interval. The statement after that is an assertion statement that checks whether the condition is true. If it is, execution simply proceeds. If the condition is false, the simulator executing the model issues an error message. In this model, the assertion condition embodies the conditions given in the requirements statement. Once the assertion has been checked, execution reaches the end of the process and starts again from the top of the process, waiting for the next change on the enable or lights_in input.

One thing to note about the test cases in this example is that not all possible input combinations are included. While it might be feasible to extend this testbench to be exhaustive, for larger designs, that would be intractable. Even if we wrote VHDL code to generate the input combinations automatically, rather than writing them out explicitly, a simulation would take too long to execute. That is because the number of test cases rises exponentially with the number of inputs. At issue here is the *functional coverage* of our testbench, that is, the proportion of the possible input combinations we have exercised. In the example, we have covered the usual operational cases and two unusual cases. In a larger model, we would have to be selective, and perhaps just cover a "typical" sample of normal cases plus a few unusual cases. We will return to the topic of coverage as part of our more detailed discussion of design methodology in Chapter 10.

Another thing to note about the test cases in the example is that the VHDL code is very repetitive. Each test case involves an assignment to the two inputs, followed by waiting for an interval. In larger models, there are more statements for each test case, and writing them repeatedly can be error prone. Fortunately, VHDL provides a feature that lets us abstract out the common parts of the test cases. We can write a *procedure* containing the common statements, and invoke the procedure once for each test case. We provide the particular values to use in each test case as *parameters* to the procedure.

EXAMPLE 2.22   Revise the testbench model of Example 2.21 to use a procedure for applying the test cases.

SOLUTION   The entity declaration is unchanged. The revised architecture is

```
library ieee; use ieee.std_logic_1164.all;

architecture verify1 of light_testbench is
  signal lights_out : std_logic_vector(1 to 3);
  signal lights_in  : std_logic_vector(1 to 3);
  signal enable      : std_logic;
begin

  duv : entity work.light_controller(and_enable)
    port map ( lights_out => lights_out,
               lights_in => lights_in, enable => enable );

  apply_test_cases : process is

    procedure apply_test
      ( enable_test    : in std_logic;
        lights_in_test : in std_logic_vector(1 to 3) ) is
    begin
      enable <= enable_test; lights_in <= lights_in_test;
      wait for 1 sec;
    end procedure apply_test;

  begin
    apply_test('0', "000");
    apply_test('0', "001");
    apply_test('0', "010");
    apply_test('0', "100");
    apply_test('1', "001");
    apply_test('1', "010");
    apply_test('1', "100");
    apply_test('1', "000");
    apply_test('1', "111");
    wait;
  end process apply_test_cases;

  check_outputs : process is
  begin
    wait on enable, lights_in;
    wait for 10 ms;
    assert (enable = '1' and lights_out = lights_in) or
           (enable = '0' and lights_out = "000");
  end process check_outputs;

end architecture verify1;
```

The difference between this testbench and the one in Example 2.21 is the
inclusion of the apply_test procedure declaration inside the apply_test_cases
process. The procedure declaration contains the statements needed to apply each

test case. The values to be applied are represented by the parameters enable_test and lights_in_test. Each of the parameter declarations looks similar to the declaration of an input port, specifying the name, direction (into the procedure in this case) and type for the parameter. We include the procedure declaration inside the process, between the process header and the begin keyword, since the procedure "belongs" to the process and performs actions as part of the process's activity.

In the statement part of the process, we invoke, or *call*, the procedure, once per test to be applied. Within parentheses in the procedure call, we supply the actual values to be used for the parameters for that call. The procedure then performs the statements in the procedure body, using those values in place of the parameter names. When the procedure statements finish, the procedure call is complete.

Having verified the functionality of the design, the next task in the design methodology is synthesis. To do that, we need to know what implementation fabric will be used, since synthesis involves refining the design to a structural implementation using primitive elements from the implementation fabric. We will discuss implementation fabrics, including those that can be used for combinational circuits, in more detail in Chapter 6. However, if the circuit is very simple, involving just a few gates, we may be able to use single gates packaged individually. This kind of circuit is sometimes needed as part of a larger system involving off-the-shelf ICs that must be connected together. If one of the ICs has outputs that differ slightly in function from the inputs of another, a small combinational circuit can deal with the differences.

EXAMPLE 2.23   A processor IC has three active-high outputs to control a memory that stores data: mem_en to enable operation of the memory, rd to control reading of data from the memory, and wr to control writing of data to the memory. A memory IC, however, has two active-high inputs: mem_rd to cause it to read data, and mem_wr, to cause it to write data. All other interconnections between the processor and memory are mutually compatible. Implement an interface circuit to compensate for the differences.

SOLUTION   The mem_rd input to the memory can be derived using an AND gate applied to the mem_en and rd outputs of the processor. Similarly, the mem_wr input can be derived using an AND gate applied to the mem_en and wr outputs. Thus, we just need two AND gates. These could be implemented using two 1G08 devices, each of which contains a single AND gate in a small 5-pin package that can be used on a printed circuit board. Given the simplicity of this circuit, we would synthesize it manually. That is, we would just instantiate AND-gate components in a structural model of the entire system.

---

1. What is the purpose of a testbench model?

2. Write VHDL statements that first apply a test case value of 0101 to a signal named s and then wait for 1ms.

3. What does a VHDL process do when execution reaches the last statement in the process?

4. Why should a process that checks outputs of a combinational circuit not check them at the same time that the inputs change?

5. When might it be appropriate to implement a combinational circuit using discrete logic gates in individual packages?

6. What is a PLD?

KNOWLEDGE
TEST QUIZ

2.5   CHAPTER SUMMARY

▶ A combinational circuit has outputs that depend only on its current inputs. Each output is a Boolean function of the inputs.

▶ Boolean functions can be defined by truth tables and by Boolean equations. Basic Boolean functions are AND, OR and negation. Other Boolean functions are NAND, NOR, XOR and XNOR. All of these have corresponding implementations as logic gates.

▶ A Boolean expression in sum-of-products form is the logical OR of product terms (p-terms), each of which is the logical AND of inputs, either directly or negated.

▶ Boolean expressions are equivalent if they have the same value for all combinations of input values. Optimization of combinational circuits involves choice among implementations of equivalent expressions for the function performed by the circuit.

▶ Buffers are gate components that perform the identity function. They are used to drive multiple loads from a single source.

▶ The don't care notation used for inputs in a truth table allows compaction of the truth table. The don't care notation used for outputs in a truth table expresses partial functions, and allows optimization of an implementation by choice of actual value for the function.

▶ The rules of Boolean algebra provide a formal basis for transforming circuits while maintaining equivalence. CAD tools perform optimization procedures based on the rules.

▶ VHDL models describe combinational circuits using assignment statements in architectures.

▶ Binary coding allows us to represent information with more than two values using multiple bits. An n-bit code can represent up to 2^n values. To represent information with N values, we need at least $\lceil \log_2 N \rceil$ bits.

▶ A one-hot code representing N values has N bits, with exactly one 1 bit in each code word.

▶ In VHDL, std_logic_vector signals can be used to represent binary coded information.

▶ Interference can cause bit flips in binary coded information, giving rise to invalid code words. A design can ignore them, fail safely, or take exception.

▸ Parity is an approach to error detection based on counting 1s in code words and augmenting the code with a parity bit. The value of the parity bit is set to ensure an odd number of 1s (odd parity) or an even number of 1s (even parity).

▸ A decoder derives a separate control signal for each code word of a binary coded input.

▸ An encoder derives a binary coded representation of whichever of a number of input bits is active. A priority encoder assigns relative priorities among its inputs, and encodes the active input with highest priority.

▸ Binary coded decimal (BCD) is a 4-bit binary code for decimal digits. A 7-segment decoder decodes a BCD input to control outputs for activating segments of a 7-segment display.

▸ A multiplexer chooses among two or more input sources to determine the value of its output. Multiplexers can be used in parallel for binary coded inputs.

▸ Active-low logic uses a high logic level to represent falsehood of a condition and a low logic level to represent truth of the condition. Bubbles on inputs and outputs of circuit symbols represent active-low connections.

▸ A VHDL testbench model is used to verify a design by applying test-case inputs and checking for correct outputs. Test cases are applied by processes containing assignments and wait statements. Outputs are checked by processes containing assertion statements.

▸ Simple combinational circuits can be implemented using discrete gates or in programmable logic devices (PLDs).

2.6 FURTHER READING

Discrete Mathematics, 5th Edition, K. R. Ross and C. R. B. Wright, Prentice Hall, 2003. Includes a rigorous presentation of Boolean algebra, and uses it as the basis for an introduction to digital logic.

Digital Design: Principles and Practices, 3rd Edition, John F. Wakerly, Prentice Hall, 2001. A textbook on basic digital logic design, including coverage of Karnaugh maps and other manual optimization methods.

The Student's Guide to VHDL, Peter J. Ashenden, Morgan Kaufmann Publishers, 1998. A supplementary reference showing how to model combinational circuits with VHDL.

Assertion-Based Design, Harry D. Foster, Adam C. Krolnik, David J. Lacey, Kluwer Academic Publishers, 2003. Presents a design methodology based on incorporating assertions into design to make verification more tractable.

Digital Logic Pocket Data Book, Texas Instruments, 2002. A listing of the manufacturer's digital logic components, including basic and complex gates. Available from www.ti.com.

EXERCISES

EXERCISE 2.1 Derive truth tables for the following Boolean expressions:

a) $a + b \cdot \bar{c}$

b) $x \oplus y \oplus z$

c) $(a + b) \cdot \overline{(c + d)}$

a	b	c	f
0	0	0	1
0	0	1	0
0	1	0	0
0	1	1	1
1	0	0	1
1	0	1	0
1	1	0	0
1	1	1	0

TABLE 2.10

EXERCISE 2.2 Draw schematic circuit diagrams for the combinational circuits described by each of the Boolean expressions in Exercise 2.1.

EXERCISE 2.3 Given the truth table in Table 2.10, write a Boolean expression for the function f, expressed as a sum of minterms.

EXERCISE 2.4 Draw a schematic circuit diagram for the combinational circuit described by the truth table in Table 2.10.

EXERCISE 2.5 Derive a truth table for the Boolean function implemented by the circuit in Figure 2.24.

FIGURE 2.24

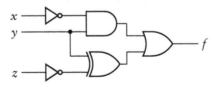

EXERCISE 2.6 Derive Boolean expressions for the circuit in Figure 2.24, both directly from the circuit and in the form of a sum of minterms from the truth table.

EXERCISE 2.7 Derive a truth table for the majority function M that is 1 when two or more of the inputs a, b and c are 1, and 0 otherwise.

EXERCISE 2.8 Show, using truth tables, that the two Boolean expressions in each of the following pairs are equivalent:

a) $x \cdot \overline{(y \cdot z)}$ and $x \cdot \bar{y} + x \cdot \bar{z}$

b) $\overline{x \oplus y}$ and $\overline{x} \oplus \overline{y}$

EXERCISE 2.9 Draw a schematic for a buffer tree to drive 12 inputs from a source, assuming that the source and each buffer can each drive at most three inputs.

EXERCISE 2.10 Reduce the size of the truth table in Table 2.11 by using the don't care notation for inputs.

EXERCISE 2.11 The truth table in Table 2.12 uses the don't care notion for the output. Add four columns to the truth table, one for each of the possible assignments of 0 or 1 as the actual output for the don't care combinations.

EXERCISE 2.12 Figure 2.9 shows circuits whose equivalence follows from the associative laws in Equations 2.3 and 2.4. Draw circuits that similarly follow from the distributive laws in Equations 2.5 and 2.6.

EXERCISE 2.13 Prove the identity laws (Equations 2.13 and 2.14) and the absorption laws (Equations 2.15 and 2.16) using just the axioms of Boolean algebra.

EXERCISE 2.14 Use the laws of Boolean algebra to transform the Boolean equation $\overline{(w + y)} \cdot (x + \overline{z})$ into sum-of-products form.

EXERCISE 2.15 Use the laws of Boolean algebra to prove that the Boolean expressions $\overline{a} \cdot b \cdot c + a \cdot \overline{b} \cdot c + a \cdot b \cdot \overline{c} + a \cdot b \cdot c$ and $a \cdot b + b \cdot c + a \cdot c$ are equivalent.

EXERCISE 2.16 For each of the following Boolean equations, write a VHDL model for a circuit that implements the equation.

a) $m = a \cdot b + b \cdot c + a \cdot c$

b) $s = \overline{(x + y)} \cdot (x + \overline{z})$

c) $y = (a \oplus b) \cdot (a + c)$

EXERCISE 2.17 Devise a minimal-length binary code to represent the state of a phone: on-hook, dial-tone, dialing, busy, connected, disconnected, ringing.

EXERCISE 2.18 Write a Boolean equation involving the bits of the code in Exercise 2.17 that determines when the phone is off-hook (that is, in a state other than on-hook or ringing).

EXERCISE 2.19 Devise a one-hot code for the state of a phone, described in Exercise 2.17.

EXERCISE 2.20 Develop a VHDL model for a circuit that has an input representing the state of the phone described in Exercise 2.17 and an output that is '1' when the phone is off-hook.

x	y	z	f
0	0	0	1
0	0	1	1
0	1	0	0
0	1	1	0
1	0	0	1
1	0	1	1
1	1	0	1
1	1	1	1

TABLE 2.11

a	b	c	f
0	0	0	0
0	0	1	1
0	1	0	0
0	1	1	–
1	0	0	–
1	0	1	1
1	1	0	1
1	1	1	0

TABLE 2.12

EXERCISE 2.21 Revise the Boolean equations of Example 2.12 so that no light is activated for invalid code words.

EXERCISE 2.22 Draw circuit diagrams of parity trees, similar to those in Figure 2.14, but generating and checking odd parity for an 8-bit code.

EXERCISE 2.23 Devise an example using an 8-bit code word to show that even parity and odd parity cannot be used to detect two separate bit flips in a code word.

EXERCISE 2.24 Write Boolean equations for a decoder for the code used in the burglar alarm of Example 2.14.

EXERCISE 2.25 Develop a VHDL model of a decoder for the code used in the burglar alarm of Example 2.14.

EXERCISE 2.26 Write Boolean equations for an ordinary (nonpriority) encoder for the code used in the ink-jet printer described in Example 2.9. For each pair of inputs, determine the code word output of the encoder if the two inputs are both 1.

EXERCISE 2.27 Develop a VHDL model of a priority encoder for the code used in the ink-jet printer described in Example 2.9.

EXERCISE 2.28 Write Boolean equations for a BCD decoder, that is, a decoder that has a BCD code word as input and that has outputs y_0 through y_9. Draw a circuit that uses AND and OR gates and inverters to implement the decoder.

EXERCISE 2.29 Develop a VHDL model of the BCD decoder described in Exercise 2.28.

EXERCISE 2.30 Write Boolean equations for a 2-to-1 multiplexer. Draw a circuit that uses AND and OR gates and inverters to implement the multiplexer.

EXERCISE 2.31 Use a 2-to-1 multiplexer to implement a circuit whose output is given by the Boolean expression $a \cdot (b + \overline{c})$ when $enable \cdot \overline{sel}$ is 1, and by the Boolean expression $x \oplus y$ otherwise.

EXERCISE 2.32 Develop a VHDL model of a circuit with the behavior described in Exercise 2.31.

EXERCISE 2.33 Draw a circuit diagram for a multiplexer that selects among four sources of data, each of which is encoded with three bits. The circuit should be implemented 4-to-1 multiplexers (see Figure 2.17).

EXERCISE 2.34 Develop a VHDL model of the multiplexer described in Exercise 2.33.

EXERCISE 2.35 Revise the vat buzzer circuit of Figure 1.5 so that the low-level sensor inputs and the buzzer output use active-low signals.

EXERCISE 2.36 Revise the VHDL models of the vat buzzer from Examples 1.5 and 1.6 so that the low level sensor input ports and the buzzer output port use active-low logic.

EXERCISE 2.37 Develop a VHDL testbench model for the vat buzzer from Example 1.5 and Example 1.6. Include test cases to ensure that the buzzer is activated when required, and not activated otherwise.

NUMERIC BASICS

3

One of the most common kinds of information processed by digital systems is numeric information. In this chapter, we will examine various binary codes for unsigned integers, signed integers, fixed-point fractions and floating-point real numbers. For each kind of code, we will describe how some arithmetic operations can be performed. We will also look at combinational circuits that implement arithmetic operations, and discuss trade-offs among different circuits that perform the same operation.

3.1 UNSIGNED INTEGERS

In many applications of digital electronics, we deal with signals that only take on nonnegative integer values. Some signals may be representations of real-world information, for example, the temperature set on a thermostat. Other signals may arise as a consequence of the way we organize the digital system, for example, as numeric indices for tables of information stored in the system's memory. In this section, we start with the most common representation for nonnegative integers, then describe arithmetic operations using that representation. We will finish the section by looking at an alternative representation that is used in some systems.

3.1.1 CODING UNSIGNED INTEGERS

We are all familiar with decimal positional representation of numbers. A decimal number such as 124_{10} denotes the sum of 1 hundred, 2 tens and 4 units. We use the subscript notation to specify that the number is to be interpreted as decimal, that is, base 10. The position of each digit in the number determines the power of 10 by which the digit is multiplied, starting with 10^0 for the right-most digit, 10^1 for the next digit to the left, and increasing by successive powers of ten for further digits from right to left. Thus, we write

$$124_{10} = 1 \times 10^2 + 2 \times 10^1 + 4 \times 10^0$$

In most applications that deal with nonnegative integers, the natural way to represent the numeric values is using *unsigned binary* numbers. Unsigned binary representation works in the same way as decimal representation, except that we only use the binary digits 0 and 1 and we multiply digits by powers of 2 instead of powers of 10. We can represent the same numeric value as 124_{10} in binary by determining the powers of two that sum to the number, namely,

$$124_{10} = 1 \times 2^6 + 1 \times 2^5 + 1 \times 2^4 + 1 \times 2^3 + 1 + 2^2 + 0 \times 2^1 + 0 \times 2^0$$

$$= 1111100_2$$

So, to represent this number in a digital system, we would need seven single-bit signals, each carrying one bit of the binary number. In general, we represent a number x using n bits $x_{n-1}, x_{n-2}, \ldots, x_0$, with

$$x = x_{n-1} 2^{n-1} + x_{n-2} 2^{n-2} + \cdots + x_0 2^0$$

EXAMPLE 3.1   What number is represented by the unsigned binary number 101101_2?

SOLUTION   Express the number as a sum of powers of two and calculate the result:

$$101101_2 = 1 \times 2^5 + 0 \times 2^4 + 1 \times 2^3 + 1 \times 2^2 + 0 \times 2^1 + 1 \times 2^0$$

$$= 1 \times 32 + 0 \times 16 + 1 \times 8 + 1 \times 4 + 0 \times 2 + 1 \times 1$$

$$= 45_{10}$$

Our discussion of binary codes in Section 2.2 applies equally to unsigned binary representation of numbers, since that is just one particular binary code. Thus, given an n-bit unsigned binary code, we can represent 2^n distinct numbers. The smallest number has all 0 bits, representing the number 0, and the largest number has all 1 bits, representing

$$1 \times 2^{n-1} + 1 \times 2^{n-2} + \cdots + 1 \times 2^1 + 1 \times 2^0 = 2^n - 1$$

Conversely, if we need to represent numbers between 0 and $N-1$, we need at least $\lceil \log_2 N \rceil$ bits for the unsigned binary representation. In computer systems, unsigned binary numbers are typically 8, 16 or 32 bits long, allowing representation of numbers up to 256, over 65,000, and over 4 billion, respectively. However, when we are designing a digital system with no other constraints applied to the number of bits, we would typically choose the smallest number of bits that can represent the range of numbers we expect to encode. There is no reason why this should not be a number of bits other than 8, 16 or 32, such as 5, 17 or 26.

EXAMPLE 3.2   Suppose we are designing a scientific instrument to measure the time interval between two random events very precisely, with a resolution of nanoseconds ($1\text{ns} = 10^{-9}$ seconds). Events may occur as much as a day apart. How many bits are needed to represent the interval as a number of nanoseconds?

SOLUTION   There are 10^9 nanoseconds per second, and $60 \times 60 \times 24 = 86{,}400$ seconds per day, so the largest number we need to allow for is 8.64×10^{13}. The number of bits needed is

$$\lceil \log_2(8.64 \times 10^{13}) \rceil = \left\lceil \frac{\log(8.64 \times 10^{13})}{\log 2} \right\rceil = \lceil 46.296\ldots \rceil = 47$$

So at least 47 bits are needed.

Unsigned Integers in VHDL

We saw in Section 2.1.3 that we can use signals of type std_logic_vector to model binary coded data. Since unsigned binary is just one form of binary code, we could use std_logic_vector signals for numeric data also. However, VHDL has a standard package of numeric operations that are useful for design and synthesis of arithmetic circuits, so it is best to use the types provided by that package. The package is called numeric_std, and it resides in the standard library of packages, ieee. The package declares a type named unsigned that represents unsigned integers as vectors of std_logic elements. We use the type in a similar manner to the type std_logic_vector, specifying ranges of index values for signals and ports, and using indexing to refer to individual bits. When we look at arithmetic operations on unsigned integers, we will see how they can be modeled in VHDL as operations on signals of type unsigned.

EXAMPLE 3.3   Develop a VHDL model of a 4-to-1 multiplexer that selects among four unsigned 6-bit integers.

SOLUTION   The entity declaration is

```
library ieee;

use ieee.std_logic_1164.all, ieee.numeric_std.all;

entity multiplexer_6bit_4_to_1 is
  port ( a0, a1, a2, a3 : in  unsigned(5 downto 0);
         sel            : in  std_logic_vector(1 downto 0);
         z              : out unsigned(5 downto 0) );
end entity multiplexer_6bit_4_to_1;
```

The first line identifies the ieee library and the two packages std_logic_1164 and numeric_std that we want to use in our model. The input ports a0 through a3 and the output port z are all 6-bit unsigned vectors, indexed from 5 down to 0. We choose this index range so that the index of each bit in a vector corresponds to the power of its binary weight. The input port sel, used to select among the inputs, is of type std_logic_vector, since we are not interpreting it as representing a number. The corresponding architecture is

```
architecture eqn of multiplexer_6bit_4_to_1 is
begin
  with sel select
    z <= a0 when "00",
         a1 when "01",
         a2 when "10",
         a3 when others;
end architecture eqn;
```

This is much the same as the multiplexer model that we saw in Section 2.3.2.

Octal and Hexadecimal Codes

We have seen that we need at least approximately $\log_2 N$ bits to represent the number N in unsigned binary form. The same number is represented in decimal with approximately $\log_{10} N$ digits. Now

$$\log_2 N = \log_{10} N / \log_{10} 2 = \log_{10} N / 0.301\ldots = \log_{10} N \times 3.32 \ldots$$

In other words, we need more than three times as many binary digits as decimal digits to represent a given number. While that is not necessarily a problem in terms of the digital system, it is cumbersome and error prone for us to write down and read the long strings of bits required for large numbers. For this reason, we often use *hexadecimal* (base 16) or, less commonly, *octal* (base 8) for those purposes. We will show how these representations work first, then discuss the advantages of using them.

Octal is just another form of positional number system, except that we use the digits 0 through 7 and multiply them by powers of 8 depending on their position. Thus, for example,

$$253_8 = 2 \times 8^2 + 5 \times 8^1 + 3 \times 8^0$$
$$= 2 \times 64 + 5 \times 8 + 3 \times 1$$
$$= 128 + 40 + 3 = 171_{10}$$

More important, for a given octal number, we can factor out powers of two in each digit and so very quickly determine the binary representation of the same number. For example,

$$253_8 = 2 \times 8^2 + 5 \times 8^1 + 3 \times 8^0$$

$$= (0 \times 2^2 + 1 \times 2^1 + 0 \times 2^0) \times 8^2 + (1 \times 2^2 + 0 \times 2^1 + 1 \times 2^0) \times 8^1$$
$$+ (0 \times 2^2 + 1 \times 2^1 + 1 \times 2^0) \times 8^0$$

$$= (0 \times 2^2 + 1 \times 2^1 + 0 \times 2^0) \times 2^6 + (1 \times 2^2 + 0 \times 2^1 + 1 \times 2^0) \times 2^3$$
$$+ (0 \times 2^2 + 1 \times 2^1 + 1 \times 2^0) \times 2^0$$

$$= (0 \times 2^8 + 1 \times 2^7 + 0 \times 2^6) + (1 \times 2^5 + 0 \times 2^4 + 1 \times 2^3)$$
$$+ (0 \times 2^2 + 1 \times 2^1 + 1 \times 2^0)$$

$$= 010101011_2$$

In general, given an octal number, we can replace each digit with the corresponding three binary digits to give the unsigned binary representation of the number. The three-bit patterns corresponding to the octal digits are

0: 000   1: 001   2: 010   3: 011   4: 100   5: 101   6: 110   7: 111

Note that we need to take care when using an octal number for an unsigned binary code if the code is not a multiple of three in length. We need to understand or specify explicitly how long the binary code is and drop unused bits from the left when converting from octal. For example, had we specified that the number 253_8 stood for an 8-bit binary number, we would have dropped the left-most bit to get 10101011_2. If any of the the bits we drop from the left are 1 rather than 0, the octal number is greater than the largest number that can be encoded in the given number of bits. Usually, this is considered an error.

We can also work in the reverse direction from an unsigned binary number. We divide the bits in to groups of three, starting from the right, and replace each group with the corresponding octal digit. For example, given the unsigned binary number 11001011, we can convert it to octal as follows:

$$11001011_2 \Rightarrow 11\ 001\ 011 \Rightarrow 313_8$$

Note that in this example, the number of bits is not a multiple of three, so we had to assume a 0 bit on the left. Again, we need to take care that the actual number of bits in the unsigned binary representation is understood or explicitly stated.

Hexadecimal is another form of positional number system, like octal, but based on powers of 16. The only minor problem we encounter is that we need digits with values from 0 through 15. We use the normal digits 0 through 9, but augment them with the letters A through F for the remaining digits. The correspondence is

$$A_{16} = 10_{10} \quad B_{16} = 11_{10} \quad C_{16} = 12_{10}$$
$$D_{16} = 13_{10} \quad E_{16} = 14_{10} \quad F_{16} = 15_{10}$$

Thus, for example,

$$3CE_{16} = 3 \times 16^2 + 12 \times 16^1 + 14 \times 16^0$$
$$= 3 \times 256 + 12 \times 16 + 14 \times 1$$
$$= 768 + 192 + 14 = 974_{10}$$

By similar arguments to those for octal numbers, we can arrive at a quick method for converting between hexadecimal and unsigned binary representations of a number. Whereas for octal, we formed groups of three bits (since $8 = 2^3$), for hexadecimal we form groups of 4 bits (since $16 = 2^4$). The 4-bit patterns corresponding to the hexadecimal digits are

0: 0000 1: 0001 2: 0010 3: 0011 4: 0100 5: 0101 6: 0110 7: 0111

8: 1000 9: 1001 A: 1010 B: 1011 C: 1100 D: 1101 E: 1110 F: 1111

Thus, for example, $3CE_{16} = 0011\ 1100\ 1110_2$. In the reverse direction:

$$11001011_2 \Rightarrow 1100\ 1011 \Rightarrow CB_{16}$$

As we mentioned earlier, nearly all computer systems use number representations that are 8, 16 or 32 bits long. Hence, the term *byte* for 8 bits of data has entered the common language. Since these are all multiples of 4 in length and not multiples of 3, hexadecimal is a more natural representation to convert to than octal. (Engineers sometimes use the term *nibble* to refer to 4 bits of data, punning on the fact that a nibble is a small bite.) With hexadecimal in these applications, we don't need to worry about assuming or dropping leading 0 bits. That's why programmers usually deal with hexadecimal and not octal. However, since we, as hardware designers, can select the number of bits that is best for our needs, we may find octal more useful in some cases, particularly if the number of bits is a multiple of 3.

3.1.2 OPERATIONS ON UNSIGNED INTEGERS

Since unsigned integers are binary coded, we can perform on them all of the operations on encoded data described in Section 2.3. A common application is to decode an n-bit unsigned binary number representing the location of information in a memory. The decoder has 2^n control outputs, which we can use to activate a particular memory location. We shall see this in more detail in Chapter 5. We can also use multiplexers in parallel, one per bit of an unsigned binary representation, to choose between multiple sources of numeric data. This was illustrated in Example 3.3. We should also expect to be able to perform arithmetic operations on numbers represented in unsigned binary. However, before we look at that, we will discuss some simpler operations.

Resizing Unsigned Integers

When we write numbers in decimal on paper, we usually don't write any leading insignificant zeros. We just use the least number of digits needed to represent the number. For example, we just write 123_{10}, and not 0123_{10} or 000123_{10}, although all represent the same number. We could do the same in binary, and just write 10110_2, and not 010110_2 or 00010110_2. However, in a digital circuit, each bit is implemented by a physical wire, and we choose the number of bits based on the largest value we expect to occur during operation of the circuit. Since wires do not come and go as values change, we normally do write leading insignificant zeros for unsigned binary numbers occurring in a digital circuit.

Recall that the largest value that can be represented with n bits is $2^n - 1$. Suppose we have some numeric data x represented with n bits:

$$x = x_{n-1}\,2^{n-1} + x_{n-2}\,2^{n-2} + \cdots + x_0 2^0$$

However, in order to perform some arithmetic operations, which may result in larger values than $2^n - 1$, we need to represent the same value in m bits, where $m > n$:

$$y = y_{m-1}\,2^{m-1} + \cdots + y_n\,2^n + y_{n-1}\,2^{n-1} + y_{n-2}\,2^{n-2} + \cdots + y_0 2^0$$

Since we want $y = x$, we can just set $y_i = x_i$, for $i = 0, 1, \ldots, n-1$, and $y_i = 0$, for $i = n, n+1, \ldots, m-1$. In other words, we just add leading insignificant 0 bits to the left of the n-bit representation to form the m-bit representation. In terms of circuit implementation, we simply add extra bit signals with their value hard-wired to 0, usually by connecting them to the circuit ground, as shown in Figure 3.1. This technique is called *zero extension*.

We can express zero extension in a VHDL model by concatenating a string of 0 bits to the left of a vector representing an unsigned integer. For example, given signals of type **unsigned** declared as

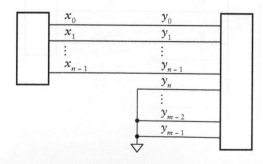

FIGURE 3.1   Implementation of zero extension in a circuit.

```
signal x : unsigned(3 downto 0);
signal y : unsigned(7 downto 0);
```

We can write the following assignment statement in an architecture to zero extend the value of x and assign it to y:

```
y <= "0000" & x;
```

The & operator that we have used here simply joins two vector values together to form a larger vector. For example, if x has the value "1010", the value assigned to y would be "00001010".

The converse operation to zero extension is *truncation*, in which we reduce the number of bits used to represent a numeric value from m to a smaller size, n. Recall again that the largest value representable in n bits is $2^n - 1$. Any m-bit value less than or equal to this value has 0 for all of the left-most $m - n$ bits. So to represent the value in n bits, we simply discard the left-most $m - n$ bits. The problem that might arise is that the value represented in m bits might be larger than $2^n - 1$, and so not be representable in n bits. Such a value has at least one of the left-most $m - n$ bits being 1. In most applications where we need to truncate, this situation does not arise, and we can discard the bits with impunity. We only reduce the number of bits when we know that the value must be within the range representable by the smaller number of bits. We might arrive at that conclusion by analyzing the arithmetic operations performed to derive the larger-sized value. In terms of circuit implementation, discarding bits does not mean physically removing anything from the circuit. Rather, we just leave the left-most bits unconnected, as illustrated in Figure 3.2.

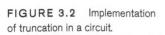

FIGURE 3.2 Implementation of truncation in a circuit.

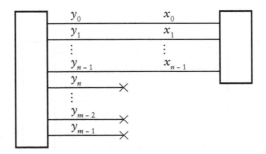

An alternative view of truncation of y from m bits to n bits is that it implements the operation $y \bmod 2^n$. We can demonstrate this as follows:

$$y \bmod 2^n$$

$$= (y_{m-1}2^{m-1} + \cdots + y_n2^n + y_{n-1}2^{n-1} + \cdots + y_02^0) \bmod 2^n$$

$$= ((y_{m-1}2^{m-n-1} + \cdots + y_n2^0)2^n + y_{n-1}2^{n-1} + \cdots + y_02^0) \bmod 2^n$$

$$= y_{n-1}2^{n-1} + \cdots + y_02^0$$

Thus, if we want to compute $y \bmod 2^n$, we just truncate y to n bits, regardless of the values of any of the discarded bits.

In a VHDL model, we express truncation of a value by picking out a *slice* of the signal representing the value. For example, given signals x and y declared as above, we can write the following assignment statement in an architecture to truncate the value of y and assign it to x:

```
x <= y(3 downto 0);
```

The range of values in parentheses specifies the index positions of the right-most elements that we want to use for the smaller representation. For example, if y has the value "00001110", the value assigned to x would be "1110".

An alternate way of expressing both zero extension and truncation of unsigned values is to use the resize operation defined in the numeric_std package. For example, the above assignments could be written as

```
y <= resize(x, 8);
```

and

```
x <= resize(y, 4);
```

Writing the operation in this way makes our intention clearer. However, the operation is only available for the types defined in the numeric_std package. Should we need to extend or truncate std_logic_vector values in order to implement some form of code conversion, we would have to use the concatenation operator or slicing.

Addition of Unsigned Integers

The addition operation on unsigned binary integers is analogous to the operation on decimal numbers. We start with the two least significant operand bits and add them to form the least significant sum bit and a carry into the next position. We then repeat until we reach the most significant position, forming the most significant sum bit and the carry out. The difference between doing this in binary and decimal is that, in binary, the sum of the two operand bits and the carry into a position is either 0, 1, 2 or at most 3. Since bits can only be 0 or 1, the case of the sum being 2 means the sum bit is 0 and the carry out is 1, and the case of the sum being 3 means the sum bit is 1 and the carry out is 1.

```
0 0 1 1 1 1 0 0 0 0
  1 0 1 0 1 1 1 1 0 0
  0 0 1 1 0 1 0 0 1 0
  ───────────────────
  1 1 1 0 0 0 1 1 1 0
```

FIGURE 3.3 Unsigned addition with carry out of 0.

```
  1 1 0 0 1
    0 1 0 0 1
    1 1 1 0 1
  ───────────
  1 0 0 1 1 0
```

FIGURE 3.4 Unsigned addition with carry out of 1.

EXAMPLE 3.4 Show the addition of the unsigned binary numbers 1010111100_2 and 0011010010_2.

SOLUTION The addition is shown in Figure 3.3. Here, we have included the carry-out bit from the most significant position. Since it is 0, the result can be represented in the same number of bits as the two operands.

EXAMPLE 3.5 Show the addition of the unsigned binary numbers 01001_2 and 11101_2.

SOLUTION The addition is shown in Figure 3.4. Again, we have included the carry out from the most significant position. However, this time it is 1, indicating that the result value cannot be represented in the same number of bits as the operands. If the design in which we are doing this addition requires the result to be five bits long, the carry out of 1 is an error condition. Alternatively, if the design allows us to use an extra bit for the result, we can use the carry-out bit as the extra most significant bit, as indicated in grey. This is the same as if we had zero extended the operands by one bit.

As these examples show, if we need to represent the result in the same number of bits as the operands (a not uncommon case), we can use the carry-out bit from the most significant position to indicate whether an *overflow* condition has occurred. When the bit is 1, the sum bits are incorrect.

Let's now look at how to design a digital circuit to perform addition upon unsigned binary numbers. Such a circuit is called, unsurprisingly, an *adder*. If we consider the method for addition described above, we see that for the least significant position, the sum (s_0) and carry-out (c_1) bits are Boolean functions of the two least significant operand bits (x_0, y_0). We can express the functions as Boolean equations:

$$s_0 = x_0 \oplus y_0 \qquad c_1 = x_0 \cdot y_0 \tag{3.1}$$

A circuit to implement these equations is called a *half adder*, and can be constructed with an XOR gate to produce the sum bit and an AND gate to produce the carry-out bit. The reason it's only half an adder will become clear in a moment.

For the remaining bits, at each position i, the sum (s_i) and carry-out (c_{i+1}) bits are Boolean functions of the operand (x_i, y_i) and carry-in (c_i) bits. The functions are as shown in the truth table in Table 3.1. They can also be expressed as Boolean equations, as follows:

$$s_i = (x_i \oplus y_i) \oplus c_i \tag{3.2}$$

$$c_{i+1} = x_i \cdot y_i + (x_i \oplus y_i) \cdot c_i \tag{3.3}$$

A circuit that implements these equations is called a *full adder*, since we can construct it from two half adders: one to add the two operand bits and one to add the result of that with the carry-in bit. A small amount of additional logic is needed to form the carry out. However, this form of full adder is largely of historical interest, since constraints that apply in most designs lead to different implementations.

One thing to note about the equations for a full adder is that, if the carry in, c_i, is 0, the equations simplify to those for a half adder. A consequence is that we can use a full adder for the least significant position instead of a half adder simply by setting the carry-in bit to 0. This allows us to treat all positions uniformly, and will also afford another advantage that we shall see when we get to signed integer addition and subtraction. Thus, a complete structure for an adder for unsigned integers consists of a full adder cell for each bit position, with carry outs chained to carry ins of adjacent positions, as shown in Figure 3.5. (For arithmetic circuits, we usually arrange components left-to-right in order of decreasing significance, to match the left-to-right order of bits of a number. The arrows on the carry connections in Figure 3.5 indicate that carry values flow from right to left, contrary to our usual convention of left-to-right flow.) The carry out of the most significant position can be used as the most significant sum bit if the sum is allowed to be longer than the operands. Otherwise, it can be used as an overflow condition signal.

This kind of adder structure is called a *ripple-carry adder*. We can see why it has this name by considering the flow of information through

x_i	y_i	c_i	s_i	c_{i+1}
0	0	0	0	0
0	0	1	1	0
0	1	0	1	0
0	1	1	0	1
1	0	0	1	0
1	0	1	0	1
1	1	0	0	1
1	1	1	1	1

TABLE 3.1 Truth table for sum and carry bits.

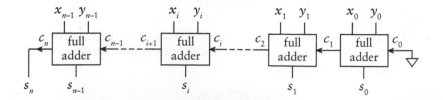

FIGURE 3.5 Structure of an adder for unsigned integers using full adder cells.

the structure. At each bit position, the values of the sum and carry outputs depend not only on the two operand bit inputs, but also on the carry from the adjacent less significant position. We can also see this by examining the Boolean equations for the full adder. They form a recurrence relation, so that, ultimately, each sum bit and the final carry-out bit depend on all of the less significant operand bits. When two operand values arrive at the adder inputs, each full adder determines a transient value for its sum and carry-out outputs. However, the full adders have some propagation delay, since they are just logic circuits. Thus, the carry out from the least significant position acts as an input to the next position after the propagation delay, possibly affecting the output of that position. Its carry out, after another propagation delay, may affect the output of the third position. In this way, carry values "ripple" from least significant to most significant position, possibly affecting sum-bit values along the way.

In the worst case, the delay from operand values arriving to the sum value settling is the product of each full adder's propagation delay and the number of bits in the unsigned binary representation. If the performance constraints of the application allow for an addition to be done slowly, a ripple-carry adder is a simple and effective adder structure. However, many applications require that arithmetic operations have high performance in order to meet timing constraints. In those cases, we can find alternate adder structures that have less delay, though at the expense of greater circuit area and power consumption.

We will now outline a couple of ways in which we can improve the adder performance over that of a ripple-carry adder. As the basis of our discussion, let's return to Equations 3.2 and 3.3 and to the truth table in Table 3.1. For a given position i, we can see the following properties.

▸ If x_i and y_i are both 0, then $c_{i+1} = 0$, regardless of the value of c_i. In this case, any carry in to the position is *killed*. We define a signal for this condition:

$$k_i = \overline{x_i} \cdot \overline{y_i} \tag{3.4}$$

▸ If one of x_i and y_i is 1 and the other is 0, then $c_{i+1} = c_i$. In this case, the carry in is *propagated* to the next position. A signal for this condition is

$$p_i = x_i \oplus y_i \tag{3.5}$$

▸ If x_i and y_i are both 1, then $c_{i+1} = 1$, regardless of the value of c_i. In this case, a carry out is *generated* for the next position. We define a signal for this condition:

$$g_i = x_i \cdot y_i \tag{3.6}$$

Substituting Equations 3.5 and 3.6 into Equations 3.2 and 3.3 gives

$$s_i = p_i \oplus c_i \tag{3.7}$$

$$c_{i+1} = g_i + p_i \cdot c_i \tag{3.8}$$

One way in which these reformulated equations help is by exposing a way of determining the carry values at each position more quickly than the ripple-carry method. Note that the k_i, p_i and g_i signals only depend on the operand bit values at their respective positions, so they can be determined quickly after the operand values arrive at the adder inputs. If a carry is killed or generated at a given position, we don't need to wait for the carry in from less significant positions; we can drive a 0 or 1 carry-out value immediately. On the other hand, if carry is to be propagated, we can switch the carry in to the carry out very quickly. These observations form the basis for the structure of a *fast-carry-chain adder*, sometimes also called a *Manchester adder*.

Figure 3.6 shows two alternate implementations of the full-adder cell used in such an adder. In the implementation on the left, the box at the top derives the propagate signal, which drives the select input of a multiplexer. If p_i is 0, then the carry is either generated (x_i and y_i are both 1) or killed (x_i and y_i are both 0). So either of the input bits can be selected to derive the carry out, without having to wait for the carry in. If p_i is 1, then the carry out is the same as the carry in. Like the ripple-carry adder, in the worst case, the carry has to propagate from the least significant to the most significant position. However, if the implementation fabric provides fast multiplexers (which many do), the propagation delay along this carry chain is much less than that of a chain of gate circuits based on Equation 3.3. As an example, several FPGA families manufactured by Xilinx include fast-carry chains using multiplexers, allowing fast-carry-chain adders to be implemented.

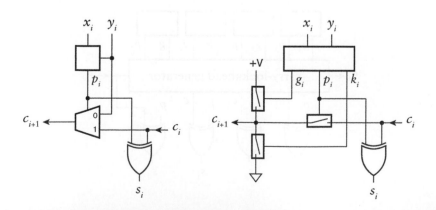

FIGURE 3.6 Fast-carry-chain full-adder cells.

The full-adder cell shown at the right of Figure 3.6 is very similar. The box at the top derives all of the generate, propagate and kill signals. These are used to drive the control inputs of electronic switches to derive the carry-out bit. If g_i is 1, the carry-out bit is switched to 1; if k_i is 1, the carry-out bit is switched to 0; and if p_i is 1, the carry-out bit is switched from the carry-in input. Again, in the worst case, a carry may have to propagate from the least significant to the most significant position. However, fabrics such as custom or standard-cell ASICs include switch components that have very small propagation delay, allowing fast-carry-chain adders to be implemented in this way.

Another way in which we can use the reformulated equations is to solve Equation 3.8 as a recurrence relation and determine all of the carry bits at once. Equation 3.8 gives us the equation for c_1 directly. We can substitute this back into Equation 3.8 to get the equation for c_2:

$$c_2 = g_1 + p_1 \cdot (g_0 + p_0 \cdot c_0) = g_1 + p_1 \cdot g_0 + p_1 \cdot p_0 \cdot c_0$$

We can repeat substitution and similarly get the equations for c_3 and c_4:

$$c_3 = g_2 + p_2 \cdot g_1 + p_2 \cdot p_1 \cdot g_0 + p_2 \cdot p_1 \cdot p_0 \cdot c_0$$

$$c_4 = g_3 + p_3 \cdot g_2 + p_3 \cdot p_2 \cdot g_1 + p_3 \cdot p_2 \cdot p_1 \cdot g_0 + p_3 \cdot p_2 \cdot p_1 \cdot p_0 \cdot c_0$$

Note that each of these expressions is a function of only c_0 and the operand input bits (since the generate and propagate signals are functions only of the operand bits). This gives us a way to determine the carry bit at each position without having to wait for carries to propagate up from less significant positions. We can then use the carry bit to derive the sum bits according to Equation 3.2. An adder based on this formulation is called a *carry-lookahead adder*. A 4-bit version of such an adder is illustrated in Figure 3.7. Each of the boxes at the top

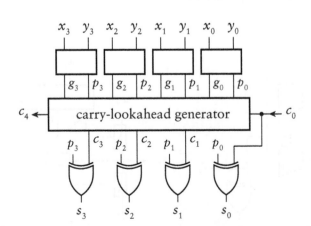

FIGURE 3.7   A 4-bit carry-lookahead adder.

derives the generate and propagate signals for the corresponding bit position. The *carry-lookahead generator* implements the equations shown above to derive the carry signals. These are combined with the propagate signals to derive the sum bits. The trade-off for getting the sum bits faster is the area and power consumed by the carry-lookahead generator circuitry.

We have shown a carry-lookahead generator for 4 bits, since that is about as large as we can practically make it. In principle, we could continue substituting in Equation 3.8 to get further carry bits. However, a more practical approach for wider adders is to use 4-bit carry-lookahead adders for segments of 4 bits, and to use a second level of carry-lookahead generators to derive the carry-in bits for each segment. There are also other forms of adders that build upon the reformulated expressions to compute carry bits in different ways. The choice among them is a question of making trade-offs among circuit area, power and performance, constrained by the resources available in implementation fabrics. A full discussion of these adder structures is beyond the scope of this book, but there are many references that go into detail.

In all of our discussion of adders so far, we have not yet described how to model them in VHDL. We could simply translate the Boolean expressions in the various forms we have discussed into VHDL. However, doing so would disguise our design intent of adding unsigned binary numbers. In particular, a CAD tool would just try to implement the model as combinational circuitry, and may not readily be able to recognize the opportunity to use any specialized circuit resources, such as fast-carry chains, available in an implementation fabric. A much better approach is to use the unsigned data type from the numeric_std package that we introduced in Section 3.1.1, since the package also provides an addition operation that we can apply to signals of the unsigned type. A synthesis CAD tool can then implement the addition operation using the most appropriate form of adder provided by the target fabric to meet design constraints. Alternatively, we could develop a structural model, selecting the most appropriate form of adder from a library of arithmetic components, and verify that the structural model produces the same results as a behavioral model using the addition operator.

EXAMPLE 3.6   Given the VHDL declaration of three signals:

```
signal a, b, s: unsigned(7 downto 0);
```

write a VHDL statement to assign the sum of a and b to s.

SOLUTION   The required statement is

```
s <= a + b;
```

This assumes that the enclosing entity or architecture is preceded by the line

```
library ieee; use ieee.numeric_std.all;
```

as described earlier. The + operator works on two unsigned values to produce an unsigned result of the same length. It does not produce a carry out, so if there is an overflow, it remains undetected.

EXAMPLE 3.7   Revise the statements to produce a carry-out bit, c.

SOLUTION   We can do this by zero extending a and b by one extra bit before doing the additions, in order to get a 9-bit result. The carry out is then the most significant bit of that result, and the 8-bit sum is the remaining bits. We need to declare a signal for the 9-bit intermediate result and for the carry bit:

```
signal tmp_result : unsigned(8 downto 0);
signal c          : std_logic;
```

The required statements are

```
tmp_result <= ('0' & a) + ('0' & b);
c          <= tmp_result(8);
s          <= tmp_result(7 downto 0);
```

The above example shows how we can use signals of type unsigned when we need to access the individual bits of the binary code. Often, we can raise the level of abstraction in our VHDL model by considering only the numeric aspects of data and not their binary encoding. VHDL allows us to do so using the type natural for numbers that only take on nonnegative values. We can declare a signal to be of type natural, constrained to a certain range, as follows:

```
signal n : natural range 0 to 63;
```

The range specification gives the least and greatest values allowed for the signal. A synthesis CAD tool can thus infer the exact number of bits needed to implement the signal in a digital circuit. For example, since 63 is $2^6 - 1$, a tool would implement the signal n with six bits.

EXAMPLE 3.8   Revise the declaration and statement in Example 3.6 to use natural signals instead of unsigned signals.

SOLUTION   The revised declaration is

```
signal a, b, s: natural range 0 to 255;
```

since 255 is $2^8 - 1$. The revised statement is

```
s <= a + b;
```

which looks exactly like the original. The difference is that the + operator produces a natural result, since its operands are natural values. There is no notion of carry out, since we're not dealing with binary representation. The values are abstract numbers. Should the result of addition overflow during simulation, the simulator will stop with an error. A synthesis tool will infer an 8-bit adder with no overflow checking. Note that the type natural and its + operator are predefined in VHDL, so we don't need to identify the library or package in which they're defined.

We may use a mixture of unsigned and natural signals in a model if we need to access the bits of the binary encoding of some values but want other values to be in abstract form. Where necessary, we can convert between the representations. For example

```
signal a1, a2 : natural                -- implies a
                 range 0 to 2**12 - 1;  -- 12-bit range
signal x, y   : unsigned( 7 downto 0);
signal z      : unsigned(15 downto 0);

a1 <= to_integer(x);
a2 <= a1 + to_integer(y);
z  <= to_unsigned(a2, 16);
```

(Note the use of the value 2**12, which is the way we write 2^{12} in VHDL. The ** operator performs exponentiation.) In the first assignment, the operation to_integer(x) converts the value of x from a binary-coded vector value to an abstract numeric value. The second statement shows how we can use such conversions within expressions. The conversion to_unsigned(a2, 16) works in the reverse direction, from an abstract integer to a binary-coded vector. The number 16 specifies how many bits to provide in the result vector. In this example, since the vector has to be the same size as the signal to which we assign it, it would be better to write

```
z <= to_unsigned(a2, z'length);
```

The notation z'length means "the length of the vector z." This way, if we change the declaration of z to give it a different length, we don't have to chase down all references to its length in the model and change them.

Subtraction of Unsigned Integers

We can work out how to perform subtraction of unsigned binary integers by following a process similar to that for addition. First, we devise the steps for binary subtraction, bit by bit, analogously to subtraction of decimal digits. Recall that, in decimal, if we subtract a larger digit from a smaller digit, we borrow from the next column. We do the same in binary, borrowing if we subtract 1 from 0.

$$
\begin{array}{ll}
b: & 0\ 1\ 0\ 1\ 1\ 0\ 0\ 0 \\
x: & 1\ 0\ 1\ 0\ 0\ 1\ 1\ 0 \\
y: & -\ 0\ 1\ 0\ 0\ 1\ 0\ 1\ 0 \\
\hline
d: & 0\ 1\ 0\ 1\ 1\ 1\ 0\ 0
\end{array}
$$

FIGURE 3.8   Unsigned subtraction.

EXAMPLE 3.9   Show the subtraction of the unsigned binary numbers 10100110_2 and 01001010_2.

SOLUTION   The subtraction is shown in Figure 3.8. Here, we have included the borrow-out bit from the most significant position. Since it is 0, the result can be represented in the same number of bits as the two operands.

Next, we look at how to design a *subtracter* circuit to perform subtraction upon unsigned binary numbers. For the least significant position, the difference (d_0) and borrow-out (b_1) bits are Boolean functions of the two least significant operand bits. The Boolean equations are

$$
d_0 = x_0 \oplus y_0 \quad b_1 = \overline{x_0} \cdot y_0
$$

For the remaining bits, at each position i, the difference (d_i) and borrow-out (b_{i+1}) bits are Boolean functions of the operand (x_i, y_i) and borrow-in (b_i) bits, with the truth table shown in Table 3.2. They can also be expressed as Boolean equations, as follows:

$$
d_i = (x_i \oplus y_i) \oplus b_i \tag{3.9}
$$

$$
b_{i+1} = \overline{x_i} \cdot y_i + \overline{(x_i \oplus y_i)} \cdot b_i \tag{3.10}
$$

As we did in the case of the adder, we can set the borrow in for the least significant position to 0 and just use Equations 3.9 and 3.10 uniformly for all positions. We could now go ahead and develop circuits for these equations.

x_i	y_i	b_i	d_i	b_{i+1}
0	0	0	0	0
0	0	1	1	1
0	1	0	1	1
0	1	1	0	1
1	0	0	1	0
1	0	1	0	0
1	1	0	0	0
1	1	1	1	1

TABLE 3.2   Truth table for difference and borrow bits.

However, many systems that need a subtracter also need an adder, and choose whether to add or subtract the operands. A little algebraic manipulation will expose a trick that allows us to use the same circuit to perform either addition or subtraction. Notice that the equation for the difference is the same as that for the sum in an adder, and that the equation for the borrow is similar to that for the carry. The trick lies in using the complemented form of the borrow bits. If we do that, we can rewrite the equations as

$$d_i = (x_i \oplus \overline{y_i}) \oplus \overline{b_i} \tag{3.11}$$

$$\overline{b_{i+1}} = x_i \cdot \overline{y_i} + (x_i \oplus \overline{y_i}) \cdot \overline{b_i} \tag{3.12}$$

Proof of this is left to Exercise 3.27. If we compare these equations with Equations 3.2 and 3.3, we see that they are identical in form, but with $\overline{y_i}$ replacing y_i and $\overline{b_i}$ replacing c_i. Consequently, we can use an adder circuit to perform subtraction simply by negating each bit of the second operand and using a negated form of borrow. For the least significant position, we set the negated borrow-in bit to 1. We can use the negated borrow out from the most significant position to indicate underflow: if it is 0, indicating a borrow, the true difference is negative, and so cannot be represented as an unsigned integer.

Now let's see how to modify an adder circuit to perform both addition and subtraction. Suppose we have a control signal that is 0 when we want the circuit to perform addition and 1 when we want it to perform subtraction. Since addition requires a 0 value for the least significant carry in and subtraction requires a 1 for the least significant negated borrow in, we can just use the control signal as the carry in/negated borrow in. We could also use the control signal to control an n-bit 2-to-1 multiplexer selecting between the second operand and its negation as the second input to the circuit. However, another part of the trick is to notice that $y_i \oplus 0 = y_i$ and $y_i \oplus 1 = \overline{y_i}$. So we can connect each bit of the second operand to an XOR gate with the control signal as the other gate input, and connect the gate outputs to the adder. The final circuit for an adder/subtracter is shown in Figure 3.9. The adder can be any of the circuits we described earlier: ripple-carry or optimized for the application's requirements and constraints.

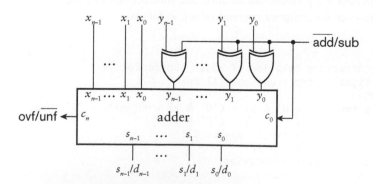

FIGURE 3.9 Adapting an adder to perform addition and subtraction.

As with VHDL models that perform addition, we normally write models that apply the subtraction operator to signals of type unsigned, rather than directly implementing the Boolean equations for a subtracter. That way, we can let the synthesis CAD tool decide on an appropriate subtracter circuit to use depending on constraints that apply. Moreover, if the system we are designing performs both addition and subtraction, the tool can decide whether to use separate circuits for the operations, or to share a single adder/subtracter between the operations. Naturally, it can only share the circuit if operations are to be done at different times. We shall see in later chapters how to control sequencing of operations. For now, we will just consider combinational circuits that assume the existence of a control signal for selecting between addition and subtraction operations.

EXAMPLE 3.10   Develop a VHDL behavioral model of an adder/subtracter for 12-bit unsigned binary numbers. The circuit has data inputs x and y, a data output s, a control input mode that is 0 for addition and 1 for subtraction, and an output ovf_unf that is 1 when an addition overflow or a subtraction underflow occurs.

SOLUTION   The entity declaration is

```
library ieee;
use ieee.std_logic_1164.all, ieee.numeric_std.all;

entity adder_subtracter is
  port ( x, y    : in  unsigned(11 downto 0);
         mode    : in  std_logic;
         s       : out unsigned(11 downto 0);
         ovf_unf : out std_logic );
end entity adder_subtracter;
```

The architecture performs the addition and subtraction using the + and − operators on the unsigned operand values, as follows:

```
architecture behavior of adder_subtracter is
  signal s_tmp : unsigned(12 downto 0);
begin
  s_tmp   <= ('0' & x) + ('0' & y) when mode = '0' else
             ('0' & x) - ('0' & y);
  s       <= s_tmp(11 downto 0);
  ovf_unf <= s_tmp(12);
end architecture behavior;
```

The first assignment in the architecture uses the mode input to choose between addition and subtraction of the operands. Since we want to use the carry-out or borrow-out bit for the ovf_unf output, we zero extend the operands by one bit and assign the result to an internal signal, s_tmp, that is 13 bits long. We then use the least significant 12 bits as the sum or difference result and the most significant bit as the ovf_unf value. In the case of addition, the most significant bit is the carry out: 1 for overflow, or 0 otherwise. In the case of subtraction, the most significant bit is the borrow out, not negated: 1 for underflow, or 0 otherwise. Thus, we can use this bit for the ovf_unf output.

EXAMPLE 3.11   Develop a verification testbench for the adder/subtracter that compares the result with the result of addition or subtraction performed on values of type natural.

SOLUTION   The entity, test_add_sub, has no ports, since it is a self-contained testbench:

```
entity test_add_sub is
end entity test_add_sub;
```

The architecture is

```
library ieee;
use ieee.std_logic_1164.all, ieee.numeric_std.all;

architecture compare of test_add_sub is

  signal x, y, s        : unsigned(11 downto 0);
  signal mode, ovf_unf : std_logic;

begin

  duv : entity work.adder_subtracter(behavior)
    port map ( x => x, y => y, s => s,
               mode => mode, ovf_unf => ovf_unf );

  apply_tests : process is

    procedure apply_test ( x_test, y_test : in natural;
                           mode_test       : in std_logic ) is
    begin
      x <= to_unsigned(x_test, x'length);
      y <= to_unsigned(y_test, y'length);
      mode <= '0'; wait for 10 ns;
    end procedure apply_test;
```

(continued)

```
begin
  apply_test(    0,    10, '0');
  apply_test(    0,    10, '1');
  apply_test(   10,     0, '0');
  apply_test(   10,     0, '1');
  apply_test(2**11, 2**11, '0');
  apply_test(2**11, 2**11, '1');
  -- ... further test cases
  wait;
end process apply_tests;

check_outputs : process is
  variable x_num, y_num, s_num : natural;
begin
  wait on x, y, mode;
  wait for 5 ns;
  x_num := to_integer(x); y_num := to_integer(y);
  s_num := to_integer(s);
  if mode = '0' then
    if x_num + y_num > 2**12 - 1 then
      assert ovf_unf = '1';
    else
      assert ovf_unf = '0' and s_num = x_num + y_num;
    end if;
  else
    if x_num - y_num < 0 then
      assert ovf_unf = '1';
    else
      assert ovf_unf = '0' and s_num = x_num - y_num;
    end if;
  end if;
end process check_outputs;

end architecture compare;
```

We precede the architecture with the library and use clauses for the std_
logic_1164 and numeric_std packages, since the architecture makes use of the
std_logic and unsigned types. The architecture declares signals to connect to the
inputs and outputs of the adder/subtracter instance, duv. The process apply_tests
contains a procedure to apply individual test cases. The process makes successive calls to the procedure to assign a sequence of input values to the inputs,
exercising both addition and subtraction with cases that produce normal results,
overflow and underflow.

The process check_outputs waits for changes of input values to the adder/
subtracter, then waits for the adder/subtracter to produce outputs. The process then converts the unsigned input values to numeric values of type natural
and assigns them to x_num, y_num and s_num. Note that these are *variables*,
declared within the process before the begin keyword. VHDL variables are

different from signals. They can only be declared within processes, and we use *variable assignment*, denoted by :=, to update them. The process then checks the value of the mode input. If it is '0', indicating addition, the process checks the numeric sum of the operands. Since it does this using the numeric variables, the result is not limited to the range representable in 12 bits. Hence, the process can compare the true sum with the largest value representable in 12 bits, namely, $2^{12} - 1$. If the sum is larger, the process asserts that the ovf_unf output should be '1'. Otherwise, the process asserts that the ovf_unf output should be '0' and that the sum result should equal the computed numeric sum. If mode is '1', indicating subtraction, the process performs similar checks, but compares the numeric difference between the operands with 0.

Note that we write the checks for condition and the choice between consequent actions in the process using VHDL *if statements*. Each if statement has the form

```
if condition then
  one or more statements
else
  one or more statements
end if;
```

In the check_outputs processes, there is an outer if statement (if mode='0' ...) that has nested if statements for each of the alternatives.

Incrementing and Decrementing Unsigned Integers

There are two further arithmetic operations that we may perform on unsigned binary integers and that are related to addition and subtraction. The *increment* operation involves adding the constant value 1, and the *decrement* operation involves subtracting the constant value 1. These operations arise quite frequently in digital systems, particularly as part of counters, which generate increasing or decreasing sequences of numbers.

A straightforward way to design an increment circuit would be to use an adder with one operand input hard wired to the unsigned binary representation of 1, namely, 0 ... 001. Alternatively, we could hard wire one input to the representation of 0 and the carry in to 1. However, since one input is a constant value, we can simplify the circuit considerably. To see how, let's return to the Boolean equations for an adder, Equations 3.2 and 3.3. If we substitute $y_i = 0$, we can simplify to the equations

$$s_i = x_i \oplus c_i \qquad c_{i+1} = x_i \cdot c_i$$

FIGURE 3.10 Structure of an incrementer for unsigned integers using half adder cells.

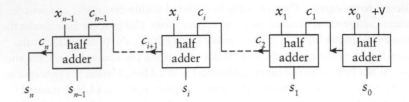

which are essentially those for a half adder (Equation 3.1 on page 98). In other words, an incrementer can be formed using a chain of half adders, as shown in Figure 3.10. The carry out of the most significant bit can be used for an overflow condition signal. A decrementer can be formed similarly by simplifying the equations for a subtracter with one input hard wired to the representation of 0 and the negated borrow in hard wired to 0.

Note that the incrementer of Figure 3.10 is a ripple-carry circuit, and so has similar delay characteristics to a ripple-carry adder. In the same way that we improved the performance of adders and subtracters, we could improve the performance of incrementers and decrementers, for example, using fast carry chains or carry-lookahead.

In VHDL models, we can express the increment or decrement operation by adding or subtracting the literal value 1 to an operand. For example, given signals declared as

```
signal x, s: unsigned(15 downto 0);
```

we could assign the incremented value of x to s with the statement

```
s <= x + 1;
```

and we could assign the decremented value with the statement

```
s <= x - 1;
```

Note that the value 1 is the numeric value, not the bit value '1'. The VHDL + and − operators are able to mix an unsigned operand and a natural operand in this way.

Comparison of Unsigned Integers

In some applications, it may be necessary to compare two unsigned binary integers for equality or inequality. Since there is exactly one code word for each numeric value, we can test for equality of two unsigned binary integers by testing whether the corresponding bits of each are the same. When we introduced the XNOR gate in Section 2.1.1, we mentioned that it is also called an equivalence gate, since its output is 1 only when its two inputs are the same. Thus, we can test for equality of two unsigned binary numbers using the circuit of Figure 3.11, called an *equality comparator*. In practice, an AND gate with many inputs is not workable, so we would modify this circuit to better suit the chosen implementation fabric. Better yet, we would express the comparison in a VHDL model and let the synthesis tool choose the most appropriate circuit from its library of cells.

Comparing two unsigned binary integers for inequality (greater than or less than comparison) is somewhat more complicated. To test whether a number x is greater than another number y, we can start by comparing the most significant bits, x_{n-1} and y_{n-1}. If $x_{n-1} > y_{n-1}$, we know immediately that $x > y$. Similarly, if $x_{n-1} < y_{n-1}$, we know immediately that $x < y$. In both cases, the final result is completely determined by comparing just the most significant bits. If $x_{n-1} = y_{n-1}$, the result depends on the remaining bits, and is true if and only if $x_{n-2\ldots0} > y_{n-2\ldots0}$. We can now apply the same argument recursively, examining the next pair of bits, and, if they are equal, continuing to less significant bits. Note that $x_i > y_i$ is only true for $x_i = 1$ and $y_i = 0$, that is, if $x_i \cdot \overline{y_i}$ is true. These considerations lead to the circuit of Figure 3.12, called a *magnitude comparator*. We can use the same circuit to test for less than inequality simply by exchanging the operands at the inputs.

In VHDL, we can express comparison operations on unsigned values using the =, >, and < operators. We can also use /= for "not-equal," <= for "less-than or equal," and >= for "greater-than or equal." However, one thing we need to be aware of is that these operators do not yield a '0' or '1' result of type std_logic. Rather, they yield a false or true result of the VHDL type boolean. This is convenient if the comparison occurs in the condition part of an if statement, since a boolean result is expected in

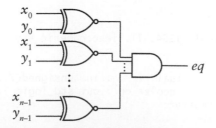

FIGURE 3.11 Circuit for an equality comparator.

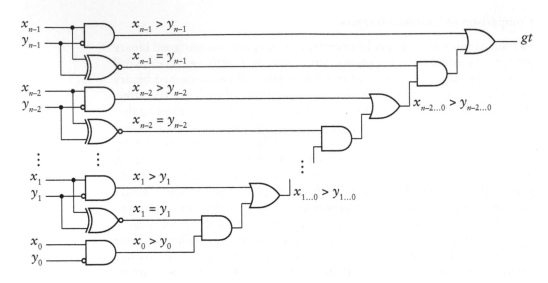

FIGURE 3.12 A magnitude comparator to test for greater than inequality.

that context. On the other hand, if we want to assign the result to a signal of type std_logic, with '0' representing falsehood of the condition and '1' representing truth of the condition, we need to convert the boolean result to type std_logic. The usual way to do this is with an assignment of the following form in an architecture

```
gt <= '1' when x > y else '0';
```

EXAMPLE 3.12   Develop a VHDL model for a thermostat that has two 8-bit unsigned binary inputs representing the target temperature and the actual temperature in degrees Fahrenheit (°F). Assume that both temperatures are above freezing (32°F). The detector has two outputs: one to turn a heater on when the actual temperature is more than 5°F below target, and one to turn a cooler on when the actual temperature is more than 5°F above target.

SOLUTION   The entity declaration is

```
library ieee;
use ieee.std_logic_1164.all, ieee.numeric_std.all;

entity thermostat is
  port ( target, actual       : in  unsigned(7 downto 0);
         heater_on, cooler_on : out std_logic );
end entity thermostat;
```

The architecture is

```
architecture rtl of thermostat is
begin
  heater_on <= '1' when actual < target - 5 else '0';
  cooler_on <= '1' when actual > target + 5 else '0';
end architecture rtl;
```

The assignments use the subtraction and addition operators to calculate the thresholds for turning the heater and cooler on. They use the $<$ and $>$ operators for performing the comparisons against the thresholds.

Scaling by a Constant Power of 2

Before we turn to multiplying unsigned integers in a general way, let's look at the specific case of scaling an unsigned integer by a given constant value that is a power of 2. The simplest case is multiplying by 2. Recall that the value x represented by the n bits $x_{n-1}, x_{n-2}, \ldots, x_0$ is

$$x = x_{n-1}2^{n-1} + x_{n-2}2^{n-2} + \cdots + x_0 2^0 \qquad (3.13)$$

If we multiply both sides by 2, we get

$$2x = x_{n-1}2^n + x_{n-2}2^{n-1} + \cdots + x_0 2^1 + (0)2^0$$

which is an $n+1$ bit number consisting of the bits of x, shifted left by one position, and a 0 bit appended as the least significant bit. If we are working with fixed-length integers, we can truncate the most significant bit to yield an n-bit number, provided the truncated bit is 0. This operation is called a *logical shift left* by one position. We can take this form of scaling further. To scale by a factor of 2^k, we repeat the scaling-by-2 process k times. That is, we shift the bits left by k positions and append k bits of 0 to the least significant end. If we need to truncate to an n-bit result, the k truncated bits must all be zero; otherwise an overflow has occurred.

Dividing by 2 works similarly. If we divide both sides of Equation 3.13 by 2 we get

$$x/2 = x_{n-1}2^{n-2} + x_{n-2}2^{n-3} + \cdots + x_1 2^0 + x_0 2^{-1}$$

Since 2^{-1} is the fraction ½, and we are dealing with integers only, we can discard the last term in this equation. The result is an $n-1$ bit number consisting of the bits of x, except for the least significant bit, shifted right by one position. If we are working with fixed-length integers, we can append a 0 to the most significant end to maintain the value. This operation is called a *logical shift right* by one position.

We can take this further also. To divide by 2^k, we shift the bits right by k positions, discarding the k least significant bits and appending k bits of 0 at the most significant end. If any of the discarded bits were nonzero, the true result of the division is truncated toward 0.

VHDL provides two operations for shifting the bits of an unsigned value. The shift_left operation performs a logical shift left, and the shift_right operation performs a logical shift right. For example, if the unsigned signal s has the value "00010011", representing the value 19_{10}, the VHDL expression

```
shift_left(s, 2)
```

would yield the value "01001100", representing the value 76_{10}. The expression

```
shift_right(s, 2)
```

would yield the value "00000100", representing the value 4_{10}.

Multiplication of Unsigned Integers

The final arithmetic operation on unsigned integers that we shall examine is multiplication. A straightforward approach for multiplying x by y is to expand the product out as follows:

$$xy = x(y_{n-1}2^{n-1} + y_{n-2}2^{n-2} + \cdots + y_0 2^0)$$
$$= y_{n-1}x2^{n-1} + y_{n-2}x2^{n-2} + \cdots + y_0 x 2^0$$

The largest value of the product is the product of the largest values of the operands. For n-bit operands, that is

$$(2^n - 1)(2^n - 1) = 2^{2n} - 2^n - 2^n + 1 = 2^{2n} - (2^{n+1} - 1)$$

which requires $2n$ bits to represent. If we provide this many bits for the product, there is no possibility of overflow.

Each of the terms in the expanded product equation is called a *partial product*, and consists of the product of a bit y_i, the number x and 2^i. Recall that $x2^i$ is just the bits of x shifted left by i positions. Also, y_i is either 0 or 1. If it is 0, the partial product is 0. If it is 1, the partial product is just the shifted version of x. Thus the partial product can be formed by AND-ing each bit of x with y_i and adding it, shifted i places to the left, into the final product. The addition of the partial products can be performed by a series of adders, as shown in Figure 3.13. This is a basic form of *combinational multiplier*, so called because it

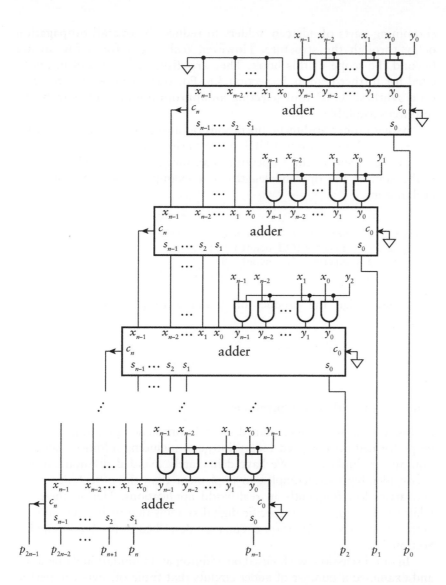

FIGURE 3.13 A combinational multiplier constructed from adders for partial products.

is a combinational circuit (albeit a large one). In Chapter 4, we will look at techniques that allow us to construct a *sequential multiplier*, in which we add partial products one at a time in successive clock cycles. A sequential multiplier trades off reduced area against time taken to yield the product.

In the multiplier circuit of Figure 3.13, we have not specified what kind of adder to use. We could use any of the adders we discussed earlier, with the choice depending on the performance requirements and area constraints that apply. We could also optimize the circuit by combining parts of adjacent adders to reduce the overall propagation delay through the structure. However, techniques for doing so are beyond the scope of this book. They are discussed in detail in books cited for further reading in Section 3.6. For our purposes, we will rely on a synthesis CAD tool selecting an appropriate multiplier from the resources available to it.

As with other arithmetic operations on unsigned binary integers, we represent multiplication in VHDL models using an operator on unsigned values. The result of the * operator is an unsigned vector whose length is the sum of the operand lengths. For example, given the following declarations:

```
signal x : unsigned( 7 downto 0);
signal y : unsigned(13 downto 0);
signal p : unsigned(21 downto 0);
```

we could assign the product of x and y to p with the following statement:

```
p <= x * y;
```

Summary of Arithmetic Operations

In this section, we have examined several arithmetic operations that can be performed on unsigned binary integers, including addition, subtraction and multiplication. We have deliberately avoided division, since it is considerably more complex to implement than the other operations, and arises less frequently in real-world applications. Hence, there are relatively few application-specific digital systems that include circuits for performing division. Division circuits are described in the books cited in Section 3.6.

In our discussion, we focused on addition as a foundational operation and examined a number of adder circuits that trade off between performance and circuit area. This is a recurring theme in digital design, and is well illustrated through consideration of adder circuits. We return to it throughout this book.

For each operation, we also discussed how to represent the operation in VHDL models that use the unsigned type provided by the numeric_std package. This approach allows us to abstract away from the details of the digital circuits that implement the arithmetic operations, relying on synthesis CAD tools to choose appropriate circuits from libraries of cells that can be implemented in the target fabric. As we shall see when we describe our implementation methodology in more detail, we separate the concerns of specifying the circuit behavior in VHDL and constraining the implementation. We provide speed and area constraints for use by the synthesis tool to determine an appropriate implementation. This approach helps us manage the complexity of designing systems to perform numerical computation.

3.1.3 GRAY CODES

The binary code that we have considered so far in this section is not the only code for unsigned integers, though it is the most natural code to use when we need to perform arithmetic operations. However, it has some disadvantages in other applications. Consider a scenario in which we are to design a system that uses a binary code to represent the angular position of a rotating shaft. A common way to measure the position is with a shaft encoder, illustrated in Figure 3.14. The disk attached to the shaft has a number of concentric bands, each of which has opaque parts and transparent parts. For each band, there is a light emitter and a detector. The detector output is 1 when the light shines through the transparent part of the band and 0 when the light is obscured by the opaque part of the band. The collection of four decoder outputs forms a binary code for the angular position of the shaft.

FIGURE 3.14 An optical shaft encoder.

The pattern of transparency and opacity in the bands on the disk is shown in Figure 3.15, and corresponds to a 4-bit *Gray code*, in which adjacent code words differ by only one bit. A complete rotation is divided into 16 segments, and between any two adjacent segments, exactly one band changes between transparent and opaque. This prevents any minor error in positioning of the detectors from causing incorrect position codes. Suppose, in contrast, that we used the unsigned binary code of Section 3.1.1 for the angular position. This would give a code word of 0011 for segment 3 and 0100 for segment 4. A minor error in position of the detector for the second band might cause it to sense the change from 0 to 1 before the detectors for the right two bands sense the changes from 1 to 0. This would give a code word of 0111, representing segment 7, for the angular position close to the boundary between segments 3 and 4. It is difficult to manufacture mechanical components with sufficient precision to avoid this kind of error. The Gray code, on the other hand, is much more tolerant of positioning error, and so is widely used in electromechanical components that measure position.

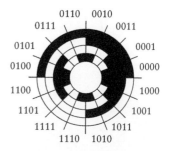

FIGURE 3.15 Gray code pattern on a shaft-encoder disk.

The 4-bit Gray code we have used in this example scenario is listed, along with the corresponding decimal and unsigned binary codes, in Table 3.3. Note how adjacent Gray code words differ in only one bit position, unlike the corresponding unsigned binary code words. This is not the only 4-bit Gray code; there are others that also have the property of single-bit difference between adjacent code words. The code we have used

DECIMAL	UNSIGNED BINARY	GRAY CODE
0	0000	0000
1	0001	0001
2	0010	0011
3	0011	0010
4	0100	0110
5	0101	0111
6	0110	0101
7	0111	0100
8	1000	1100
9	1001	1101
10	1010	1111
11	1011	1110
12	1100	1010
13	1101	1011
14	1110	1001
15	1111	1000

TABLE 3.3 4-bit Gray code, compared to unsigned binary code.

here is generated by the following rules, which allow us to generate an n-bit Gray code:

▶ A 1-bit Gray code has the two code words 0 and 1.

▶ The first 2^{n-1} code words of an n-bit Gray code consist of the code words of an $(n-1)$-bit Gray code, in order, each with a 0 bit appended as the left-most bit.

▶ The last 2^{n-1} code words of an n-bit Gray code consist of the code words of an $(n-1)$-bit Gray code, in reverse order, each with a 1 bit appended as the left-most bit.

EXAMPLE 3.13 Develop a VHDL model of a code converter to convert the 4-bit Gray code to a 4-bit unsigned binary integer.

SOLUTION For the Gray-code input to the converter, we use a port of type std_logic_vector, not unsigned, since the value is not represented as a binary-coded integer. The output port, on the other hand, is of type unsigned. The entity declaration is

```
library ieee;
use ieee.std_logic_1164.all, ieee.numeric_std.all;

entity gray_converter is
  port ( gray_value   : in  std_logic_vector(3 downto 0);
         numeric_value : out unsigned(3 downto 0) );
end entity gray_converter;
```

The architecture is

```
architecture table of gray_converter is
begin
  with gray_value select
    numeric_value <= "0000" when "0000", "0001" when "0001",
                     "0010" when "0011", "0011" when "0010",
                     "0100" when "0110", "0101" when "0111",
                     "0110" when "0101", "0111" when "0100",
                     "1000" when "1100", "1001" when "1101",
                     "1010" when "1111", "1011" when "1110",
                     "1100" when "1010", "1101" when "1011",
                     "1101" when "1001", "1111" when others;
end architecture table;
```

The assignment statement takes the form of a truth table. It uses the Gray-code value to select which unsigned numeric value to assign to the output.

KNOWLEDGE TEST QUIZ

1. How is a number x represented in binary as a sum of powers of 2?

2. What range of values can be represented as an n-bit unsigned binary number?

3. Write a VHDL declaration for a signal x of type unsigned to represent unsigned numbers in the range 0 to 8191.

4. Write the binary number 01011101 in octal and in hexadecimal.

5. Resize the unsigned binary number 10010011 to 12 bits and to 6 bits. In each case, does the result correctly represent the same value as the original number?

6. Add the two 8-bit unsigned binary numbers 01001010 and 01100000 to get an 8-bit result. Does the addition overflow?

7. What distinguishes a ripple-carry adder from a carry-lookahead adder?

8. Write VHDL assignments to add two signals s1 and s2 of type unsigned(15 downto 0) to get a result signal s3 of the same type as s1 and s2 and a carry-out signal c_out.

9. Perform the 8-bit unsigned binary subtraction $01001010 - 01100000$ to get an 8-bit result. Does the subtraction underflow?

10. Given a control signal $\overline{\text{add}}/\text{sub}$, how can we adapt an unsigned adder to perform both addition and subtraction?

11. Write a VHDL assignment that compares two unsigned signals a and b and assigns '1' to a signal smaller if $a < b$, or '0' otherwise.

12. How is an unsigned binary number multiplied by 16? How is it divided by 16?

13. How many bits are required for the product of two n-bit unsigned binary numbers?

14. Why are Gray codes often used in electromechanical position sensors?

3.2   SIGNED INTEGERS

While many applications deal only with nonnegative integers, there are others that deal with integers that range over both positive and negative values. In this section we will explore a binary code for signed integers and see how to implement operations on these encoded values.

3.2.1   CODING SIGNED INTEGERS

The predominant encoding used in digital systems for signed integers is called *2s complement*. It is a special case of *radix complement* representation in which the radix (the base used for positional representation) is 2. We will refer to the Further Reference books for details of general radix complement representations, and focus our attention here just on 2s complement.

A signed number is represented in 2s-complement form as a weighted sum of powers of two, in a similar way to unsigned binary representation. The difference is that, for an n-bit signed number, the weight of the leftmost bit is negative. An n-bit number x represents the value

$$x = -x_{n-1}2^{n-1} + x_{n-2}2^{n-2} + \cdots + x_0 2^0 \qquad (3.14)$$

This representation has a number of interesting and useful properties that we will now explore. First, the most negative number that can be represented has $x_{n-1} = 1$ and all other bits 0, giving the value -2^{n-1}. The most positive number has $x_{n-1} = 0$ and all other bits 1, giving the value $2^{n-1} - 1$. If x_{n-1} is 1, the number represented is negative, since the sum of all the positively weighted powers of 2 is less than 2^{n-1}. Thus, x_{n-1} serves as a sign bit: if it is 1, the number is negative, and if it is 0, the number is zero or positive. The range of numbers that can be represented is not symmetric about zero, since the negation of -2^{n-1} is one more than the most positive number that can be represented.

EXAMPLE 3.14   What values are represented by the 8-bit 2s-complement numbers 00110101 and 10110101?

SOLUTION   The first number is

$$1 \times 2^5 + 1 \times 2^4 + 1 \times 2^2 + 1 \times 2^0 = 32 + 16 + 4 + 1 = 53$$

The second number is

$$-1 \times 2^7 + 1 \times 2^5 + 1 \times 2^4 + 1 \times 2^2 + 1 \times 2^0 = -128 + 32 + 16 + 4 + 1 = -75$$

While 2s-complement representation for signed integers predominates, there are other forms that are useful in some applications. One form, *signed magnitude*, is analogous to our conventional decimal representation for signed integers, in which we write a sequence of decimal digits for the magnitude of a number, preceded by a + or − sign to indicate whether the number is positive or negative. In signed magnitude binary representation, we represent a signed number with a sequence of binary digits (bits), preceded by a binary code for the sign of the number. Usually, we would encode a − sign with 1 and a + sign with 0. While some early digital computers used signed magnitude representation, there are a number of disadvantages that make it uncommon in modern digital systems. For this reason, we will not describe in any further detail, and instead refer to the books listed in Section 3.6, Further Reading, for more information.

Representing Signed Integers in VHDL

We saw in Section 3.1.1 that the numeric_std package defines the type unsigned and a number of arithmetic operations for dealing with unsigned integers. The package also defines the type signed, along with arithmetic operations, for signed integers. The operators assume 2s-complement representation, with the sign bit being the left-most bit in a vector and the least significant bit being the right-most bit.

An important point to note is that, even though unsigned and signed are both vectors of std_logic values, they are distinct types. It is illegal in VHDL to assign a signed value to an unsigned signal, or vice versa. For example, given the following declarations:

```
signal s1 : unsigned(11 downto 0);
signal s2 :   signed(11 downto 0);
```

it is illegal to write the following assignment, even though the signals are the same length:

```
s1 <= s2; -- illegal, since signals are of different types
```

The rationale is that the two types represent different kinds of abstract values, and that a given vector of bit values represents different numeric values when interpreted as an unsigned or signed binary value. If we really want to make the assignment, for example, if we know that a signed value is nonnegative and can thus be treated as an unsigned value, we must explicitly convert the type of the value. We would rewrite the above assignment as

```
s1 <= unsigned(s2); -- s2 is known to be nonnegative
```

or, similarly, an assignment of s1 to s2 as

```
s2 <= signed(s1); -- s1 is known to be less than 2**11
```

Just as we have an abstract numeric type, natural, for nonnegative numbers, we also have the abstract numeric type integer for numbers that can be positive or negative. We can perform arithmetic operations on values of type integer, and we can convert between integer and signed values. We would use a mixture of signed and integer signals in a model if we need to access the bits of the 2s-complement encoding of some values but want other values to be in abstract form. For example

```
signal n1, n2 : integer                    -- implies an
                   range -2**7 to 2**7-1; -- 8-bit range
signal x, y   : signed( 7 downto 0);
signal z      : signed(11 downto 0);
signal z_sign : std_logic;

n1     <= to_integer(x);
n2     <= n1 + to_integer(y);
z      <= to_signed(n2, z'length);
z_sign <= z(z'left);
```

The operation to_integer, applied to a signed value, converts from a 2s-complement vector value to an abstract numeric value. The conversion to_signed works in the reverse direction, from an abstract integer to a 2s-complement vector. The notation z'left in the last assignment means "the left-most index of the vector z." By writing the index value this way, if we change the declaration of z to give it a different length, we don't have to chase down all references to its index range in the model and change them.

Octal and Hexadecimal Codes for Signed Integers

We saw in Section 3.1.1 that we could use octal or hexadecimal codes for unsigned integers. We can also use octal and hexadecimal for 2s-complement signed integers. However, when we do so, we don't usually think in terms of signed octal or signed hexadecimal numbers. Instead, we just use octal or hexadecimal as a shorthand notation for the vector of bits. We divide the vector into groups of three bits (for octal) or four bits (for hexadecimal) and substitute the corresponding octal or hexadecimal digit for each group.

EXAMPLE 3.15   The 12-bit 2s-complement representation of 844_{10} is 001101001100. Express the bit vector in hexadecimal.

SOLUTION   Dividing into groups of four bits, we get 0011 0100 1100. Substituting hexadecimal digits for the 4-bit groups gives $34C_{16}$.

EXAMPLE 3.16   The 10-bit 2s complement representation of -42 is 1111010110. Express the bit vector in octal.

SOLUTION   Dividing into groups of three bits, we get 1 111 010 110. Substituting octal digits for the 3-bit groups gives 1726_8. When reading this octal number, we need to understand that it represents 10 bits. The right-most three digits represent 9 bits, and the left-most digit represents just one bit, the sign bit. Since the sign bit is 1, the number is negative, even though the octal number does not include a $-$ sign.

3.2.2   OPERATIONS ON SIGNED INTEGERS

As with unsigned numbers and binary codes in general, we can perform operations on signed integers that don't rely on their numeric interpretation, such as selecting among several encoded numbers using multiplexers. In this section, we will describe operations that relate to the numeric interpretation, such as arithmetic operations. Most of these operations are implemented in a similar way to their counterparts for unsigned integers.

Resizing Signed Integers

The resizing operation on unsigned integers simply involved appending or truncating leading zeros to reach the desired length of representation while maintaining the same numeric value. With 2s-complement numbers, however, the left-most bit is the sign bit, so appending or truncating leading zeros will not work in general. Let's consider the two cases of nonnegative and negative numbers, respectively.

For nonnegative numbers, the sign bit is 0, and the remaining bits constitute the magnitude of the number. In this case, the 2s-complement representation is the same as the unsigned representation, and zero extending it maintains the same value. We can also truncate leading zeros, as we did for unsigned numbers, provided both that none of the truncated bits is 1 and that the left-most bit of the result is 0. Were the left-most bit of the result 1, that would imply a negative result, which would be incorrect. For example, the 8-bit 2s-complement representation of 41_{10} is 00101001. Truncating this to 6 bits would give 101001, which, interpreted as a 2s-complement number, is -23. The problem is that 41_{10} cannot be represented in 6-bit 2s-complement.

For negative numbers, the sign bit is 1. We can extend an n-bit negative number to m bits by appending leading 1 bits. To see that this conserves the negative numeric value, consider the value represented by a negative number x:

$$x = -2^{n-1} + x_{n-2}2^{n-2} + \cdots + x_0 2^0 \tag{3.15}$$

Extending this with leading 1 bits gives the 2s-complement number

$$-2^{m-1} + 2^{m-2} + \cdots + 2^{n-1} + x_{n-2}2^{n-2} + \cdots + x_0 2^0 \tag{3.16}$$

We can make use of the following identity:

$$2^k = 2^{k-1} + 2^{k-2} + \cdots + 2^0 + 1 \tag{3.17}$$

Expanding the first term in Equation 3.16 using this identity gives

$$-2^{m-2} - \cdots - 2^{n-1} - 2^{n-2} - \cdots - 2^0 - 1$$

$$+ 2^{m-2} + \cdots + 2^{n-1} + x_{n-2}2^{n-2} + \cdots + x_0 2^0$$

$$= -2^{n-2} - \cdots - 2^0 - 1 + x_{n-2}2^{n-2} + \cdots + x_0 2^0$$

$$= -(2^{n-2} + \cdots + 2^0 + 1) + x_{n-2}2^{n-2} + \cdots + x_0 2^0$$

$$= -2^{n-1} + x_{n-2}2^{n-2} + \cdots + x_0 2^0 = x$$

We can argue similarly to show that, for a negative number, we can truncate to a smaller length by truncating leading 1 bits, provided the left-most bit of the result is 1.

In summary, for a 2s-complement signed integer, extending to a greater length involves replicating the sign bit to the left. This is called *sign extension*, and preserves the numeric value, be it positive or negative. A circuit to implement sign extension of an n-bit signal x to an m-bit

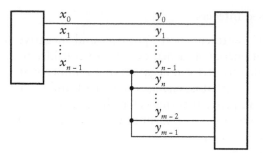

FIGURE 3.16 An implementation of sign extension in a circuit.

signal y is shown in Figure 3.16. We can truncate by discarding the left-most bits, provided all of the discarded bits and the resulting sign bit are the same as the original sign bit. The circuit implementation for truncation from m bits to n bits is the same as for truncation of an unsigned value, shown in Figure 3.2, and just involves leaving the left-most $m - n$ bits unconnected. The problem that might arise is that the value represented in m bits might be larger in magnitude than can be represented in n bits. Usually, this situation does not arise, since we only reduce the number of bits when we know that the value must be within the range representable by the smaller number of bits. We might arrive at that conclusion by analyzing the arithmetic operations performed to derive the larger-sized value.

We can express sign extension or truncation of a signed value in a VHDL model by using the resize operation. This is easier and clearer to understand than trying to use concatenation to replicate the sign bit. For example, given signals of type signed declared as

```
signal x : signed( 7 downto 0);
signal y : signed(15 downto 0);
```

we can write the following assignment statement in an architecture to sign extend the value of x and assign it to y:

```
y <= resize(x, y'length);
```

Similarly, we can write the following assignment to truncate the value of y and assign it to x:

```
x <= resize(y, x'length);
```

Negating Signed Integers

Since we can represent both positive and negative numbers using 2s-complement encoding, it makes sense to consider negating a number. The steps needed to perform negation of a number x are first to complement each bit of x (that is, change each 0 to 1 and each 1 to 0), and then to add 1. We can prove that this yields the 2s-complement representation of $-x$. We need to use the bit identity $\bar{x}_i = 1 - x_i$ together with the identity in Equation 3.17. The proof is

$$\bar{x} + 1 = -(1 - x_{n-1})2^{n-1} + (1 - x_{n-2})2^{n-2} + \cdots + (1 - x_0)2^0 + 1$$

$$= -2^{n-1} + x_{n-1}2^{n-1} + 2^{n-2} - x_{n-2}2^{n-2} + \cdots + 2^0 - x_0 2^0 + 1$$

$$= -(-x_{n-1}2^{n-1} + x_{n-2}2^{n-2} + \cdots + x_0 2^0)$$

$$\qquad -2^{n-1} + 2^{n-2} + \cdots + 2^0 + 1$$

$$= -x - 2^{n-1} + 2^{n-1} = -x$$

EXAMPLE 3.17   Determine the 8-bit 2s-complement representation of -43.

SOLUTION   The 8-bit 2s-complement representation of 43 is 00101011. Complementing this gives 11010100. Adding 1 gives 11010101, which is the required result.

Recall that the range of numbers representable in 2s-complement form is not symmetric about zero. Consider what happens if we try to complement and add 1 to the representation of -2^{n-1}, which is $100\ldots0$. Complementing gives $011\ldots1$. Adding 1 to this gives $100\ldots0$, which is the negative number we started with. So if we are to negate a 2s-complement number, we need either to sign extend it by one bit to allow for this case, or be sure that the value -2^{n-1} cannot occur as input.

In VHDL models, we express negation of a value of type signed or integer with the prefix $-$ operator. For example, to assign the negation of a signal x to a signal y, we would write

```
y <= -x;
```

Addition of Signed Integers

We can add two 2s-complement numbers x and y using much the same procedure that we used for unsigned binary numbers. The main difference lies in the way we deal with the sign bit, which has a negative weight of -2^{n-1}.

In order to understand how 2s-complement addition works, we can think of each number as the sum of the weighted sign part, which is either 0 or -2^{n-1}, and a positive offset, which is less than 2^{n-1}. That is,

$$x = -x_{n-1}2^{n-1} + x_{n-2}\ldots_0 \qquad y = -y_{n-1}2^{n-1} + y_{n-2}\ldots_0$$

and

$$x + y = -(x_{n-1} + y_{n-1})2^{n-1} + x_{n-2}\ldots_0 + y_{n-2}\ldots_0$$

We will do a case analysis of combinations of sign-bit values for the two n-bit operands.

First, consider the case of adding two nonnegative numbers. The sign bits are both 0, and can be added to give a result sign bit of 0 with no carry. The bits of the offsets are all positively weighted and can be added using the procedure for unsigned numbers, provided the carry out from position $n-2$ is 0, as in the first example in Figure 3.17. On the other hand, if the carry out from position $n-2$ is 1, as in the second example in Figure 3.17, the positive magnitude of the result would be larger than can be represented in n-bit 2s-complement form; that is, it would overflow.

Next, consider the case of adding two negative numbers, with both sign bits being 1. Adding the sign bits gives 0 with a carry out of 1 from the sign position. This corresponds to adding the weighted sign parts to give -2^n. So we need the sum of the positive offsets to yield a carry out of 1, with weight 2^{n-1}, to add to this to give -2^{n-1}. We can just add the carry out from the offsets to the sum of the sign bits to give a final sign bit of 1, as in the third example in Figure 3.17. On the other hand, if the sum of the positive offsets yields a carry out of 0, as in the fourth example in Figure 3.17, the result is more negative than can be represented in n-bit 2s-complement form; that is, it would overflow in the negative direction.

Finally, consider the case of adding one positive number (sign bit is 0) and one negative number (sign bit is 1). No overflow can occur in this case. Adding the two sign bits gives 1 with a carry out of 0. This corresponds to adding the weighted sign parts to give -2^{n-1}. If the sum of the positive offsets is less than 2^{n-1}, the carry out from position $n-2$ is 0, as in the fifth example in Figure 3.17, and the final result is negative. If the sum of the positive offsets is greater than or equal to 2^{n-1}, the carry out from position $n-2$ is 1, and the final result is nonnegative, as in the sixth example in Figure 3.17. We can add the carry out from position $n-2$ into the sign position to give a final sign bit of 0 and a carry out of 1 from the sign position.

So in all cases, we can perform 2s-complement addition using exactly the same process as unsigned addition, including adding the carry out from position $n-2$ into the sign position. Overflow is indicated when the carry into the sign position is different from the carry out of that position. We have circled these two bits to highlight them in each of the examples in Figure 3.17. It follows that we can use exactly the same circuit to add

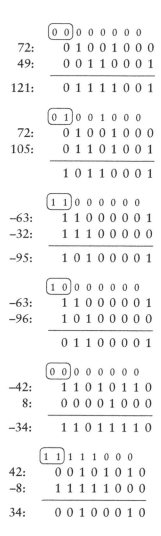

FIGURE 3.17 Examples of signed addition. In each case, the addition overflows if the left-most two carry bits differ.

unsigned numbers or 2s-complement numbers. We use the carry out from the most significant position to indicate overflow for unsigned addition, and the exclusive OR of the carry in and carry out of the most significant position to indicate overflow for signed addition.

In VHDL, we express addition of values of type signed or integer using the + operator, just as we did for unsigned and natural values. For signed values, if we want to allow for a result that would overflow if represented using the same number of bits as the operands, we can resize the operand values. For example, given the declarations

```
signal v1, v2 : signed(11 downto 0);
signal sum    : signed(12 downto 0);
```

we can add the two 12-bit values and get a 13-bit result using the assignment

```
sum <= resize(v1, sum'length) + resize(v2, sum'length);
```

Developing a VHDL model that represents the sum using the same number of bits as the operands and that derives the overflow condition is somewhat more involved. Referring back to our case analysis of the signs of the operands, we see that overflow only occurs if both operands are nonnegative and the carry in to the sign position is 1 (yielding an apparently negative result), or if both operands are negative and the carry in to the sign position is 0 (yielding an apparently nonnegative result). Given this observation and the declarations

```
signal x, y, z : signed(7 downto 0);
signal ovf     : std_logic;
```

we can write the following assignments to derive the required sum and overflow condition bit:

```
z <= x + y;
ovf <= ( not x(7) and not y(7) and     z(7) ) or
       (     x(7) and     y(7) and not z(7) );
```

Subtraction of Signed Integers

Now that we have seen how to perform addition and negation on 2s-complement numbers, subtraction follows from the identity

$$x - y = x + (-y) = x + \bar{y} + 1$$

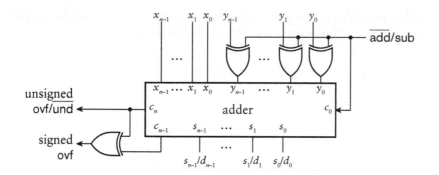

FIGURE 3.18 An adder/subtracter for both unsigned and 2s-complement numbers.

This suggests that we can use the same adder/subtracter, shown in Figure 3.9, that we described for unsigned numbers. The revised form that deals with both kinds of numbers, unsigned and 2s-complement, is shown in Figure 3.18. For signed numbers, when the $\overline{\text{add}}$/sub control input is 0, the y operand is passed through the XOR gates unchanged and the carry in to the adder is 0. When the $\overline{\text{add}}$/sub input is 1, the y operand is complemented by the XOR gates, and the carry in is 1. Thus the circuit subtracts by adding to x the complement of y and 1. Depending on whether the operands are interpreted as unsigned or signed operands, we use one or the other of the overflow condition outputs.

In VHDL, we express subtraction of values of type signed or integer using the − operator. For signed values, if we want to allow for a result that would overflow if represented as the same number of bits as the operands, we can resize the operand values, as we described for signed addition. Thus, given the declarations

```
signal v1, v2 : signed(11 downto 0);
signal diff   : signed(12 downto 0);
```

we can calculate the 13-bit difference between the two 12-bit values using the assignment

```
diff <= resize(v1, diff'length) - resize(v2, diff'length);
```

Again, a VHDL model that represents the difference using the same number of bits as the operands and that derives the overflow condition is somewhat more involved. Since $x - y$ is the same as $x + (-y)$, and the sign of $-y$ is the complement of the sign of y (except when y is zero), we can work out the overflow condition by examining sign bits in a way similar to that for addition. We just need to use the logical negation of the sign bit of y in the overflow expression. Thus, for the declarations

```
signal x, y, z : signed(7 downto 0);
signal ovf     : std_logic;
```

we can write the following assignments to derive the required difference and overflow condition bit:

```
z <= x - y;
ovf <= ( not x(7) and     y(7) and     z(7) ) or
       (     x(7) and not y(7) and not z(7) );
```

The case of y being zero is handled correctly by this expression, since in that case, the result z is the same as x, and so the sign of z is the same as the sign of x.

A further case to consider is subtraction of two unsigned numbers to give a signed result, rather than underflowing when the difference is negative. In order to determine the size to use for the result, we can consider the range of possible result values. Suppose we are subtracting n-bit unsigned values. The greatest result arises from subtraction of zero from the greatest unsigned value, giving $2^n - 1$. The least (most negative) result arises from subtraction of $2^n - 1$ from zero, giving $-2^n + 1$. This range is encompassed by a result with $n + 1$ bits. So the simplest way to express the subtraction is to zero extend the operands by one bit, treat them as signed, and then apply the signed subtraction operation. In VHDL, given 8-bit operands and a 9-bit result declared as

```
signal v1, v2, : unsigned(7 downto 0);
signal diff    :   signed(8 downto 0);
```

we could write the subtraction as

```
z <= signed( resize(v1, diff'length) ) -
     signed( resize(v2, diff'length) );
```

Other Arithmetic Operations on Signed Integers

As part of our examination of unsigned integers, we saw that we could use simplified forms of adder and subtracter to implement the increment and decrement operations. The same argument applies to incrementing and decrementing 2s-complement signed integers. However, we won't go into the details here. As with unsigned integers, we can use the $+$ operator in VHDL models to add 1 to a signed or integer value to increment, and use the $-$ operator to subtract 1 to decrement the value.

Comparison of signed integers is also done similarly to comparison of unsigned integers. The main difference arises from the negative weight for the sign bit. Hence, instead of using $x_{n-1} \cdot \overline{y_{n-1}}$ to compare the most significant bits in the comparator for $x > y$, we substitute $\overline{x_{n-1}} \cdot y_{n-1}$ to compare the sign bits. This follows, since a nonnegative number, with a sign bit of 0 is greater than a negative number with a sign bit of 1. We make the corresponding adjustment in a comparator for $x < y$. The VHDL comparison operators, $<$, $>$, $<=$, and $>=$, all work on signed and integer values in an analogous way to unsigned integers.

Scaling a signed integer by a constant power of 2 is slightly different for signed integers than for unsigned integers. Multiplying by 2^k involves shifting to the left by k positions and appending k bits of 0 to the least significant end. This is the same logical shift left operation that we say for unsigned numbers. However, if we need to represent the result in the same number of bits as the original unscaled number, we must truncate using the resizing rules for 2s-complement described earlier. Thus, the truncated bits must all be the same as the original sign bit, and the sign of the result must also have that same sign. Dividing by 2^k involves shifting the bits right by k positions, discarding the k least significant bits and appending k copies of the original sign bit at the most significant end. This operation is called an *arithmetic shift right*. It differs from a logical shift right in the replication of the sign bit instead of filling with 0 bits. Proof that these operations correctly implement scaling is left to Exercise 3.54.

In VHDL, we can apply the shift_left and shift_right operations to signed operands. In this case, while shift_left still performs a logical shift left, shift_right performs an arithmetic shift right. For example, if the signed signal s has the value "11110011", representing the value -13_{10}, the VHDL expression

```
shift_left(s, 2)
```

would yield the value "11001100", representing the value -52_{10}. The expression

```
shift_right(s, 2)
```

would yield the value "11111100", representing the value -4_{10}.

The final operation that we discussed in the context of unsigned integers was multiplication. Extending the multiplier design that we described there to deal with 2s-complement signed numbers gets quite complicated, since we need to deal with sign extension within partial products. In real designs, signed multipliers are based on transformations of this basic approach to reduce the amount of circuitry required and to improve performance. We will not go into detail here, but refer to the books listed in Section 3.6, Further Reading. In any case, using our design methodology, we can simply express multiplication in VHDL using the * operator on signed values and let synthesis CAD tools choose an appropriate multiplier circuit to use.

KNOWLEDGE TEST QUIZ

1. What is the difference in representation between unsigned binary and 2s-complement signed binary?

2. What is the range of values that can be represented using 12-bit 2s-complement signed binary form?

3. Write a VHDL declaration for a signal that represents a number in the range -512 to 511 in 2s-complement signed form.

4. Resize the 2s-complement numbers 01110001 and 11110011 to 12 bits and 6 bits. In each case, does the result correctly represent the same value as the original?

5. Negate the 2s-complement signed number 11110010.

6. How is a signed adder used to perform signed subtraction?

7. How is a 2s-complement signed number multiplied by 16? How is it divided by 16?

3.3   FIXED-POINT NUMBERS

While many applications deal with integer data, there is a growing list of applications that also deal with fractional numeric data. Many such applications involve digital signal processing, in which time-varying analog signals are sampled, converted to a digital representation and subject to numerical operations. For example, most modern audio devices deal with sampled audio signals and perform operations such as filtering, amplification and equalization. The audio samples are approximations to real numbers within a given range. The circuits representing and operating upon the samples need to deal with fractional values, that is, values that lie between integers. In this section, we will introduce the notion of fixed-point representation of nonintegral values.

3.3.1   CODING FIXED-POINT NUMBERS

Suppose we need to represent numeric values that lie in the range -12.0 to $+12.0$. Since there are an infinite number of real numbers in that range, we cannot represent all of them. Instead, we determine a precision, based on the requirements of our application, and approximate values with a multiple of that precision. For example, if our chosen precision is 0.01, we would round each value to the nearest multiple of 0.01. Thus an original value of 10.23683 would be approximated with a value of 10.24.

When we write decimal numbers in this way, we are extending the positional notation that we described for integers in Section 3.1. We use the decimal point to mark the boundary between digits whose weight is a nonnegative power of 10 and digits whose weight is a negative power of ten. For example, the number 10.24_{10} is

$$10.24_{10} = 1 \times 10^1 + 0 \times 10^0 + 2 \times 10^{-1} + 4 \times 10^{-2}$$

We can extend this idea to binary, in which the digits are weighted with powers of 2 and each binary digit (each bit) is 0 or 1. Thus, the binary number 101.01_2 is

$$101.01_2 = 1 \times 2^2 + 0 \times 2^1 + 1 \times 2^0 + 0 \times 2^{-1} + 1 \times 2^{-2}$$

Since we are dealing with nonintegral numbers, we use negative powers of 2 for the fractional part. We refer to the period dividing the binary number into its integral and fractional parts as the *binary point.*

When we come to implement nonintegral numbers in digital systems, the question arises of how to represent the binary point. The *fixed-point* representation relies on the position of the binary point being implicit. We just represent the bits, as we did for integral values, as a vector with one element per bit position. Thus, the number 101.01_2 could be represented by the bit vector 10101, with the assumption that the binary point lies two places from the right.

EXAMPLE 3.18   What number is represented by the fixed-point binary number 01100010, assuming the binary point is four places from the right?

SOLUTION   The number is

$$0110.0010_2$$

$$= 0 \times 2^3 + 1 \times 2^2 + 1 \times 2^1 + 0 \times 2^0 + 0 \times 2^{-1} + 0 \times 2^{-2} + 1 \times 2^{-3}$$
$$+ 0 \times 2^{-4}$$

$$= 0 + 4 + 2 + 0 + 0 + 0 + \frac{1}{8} + 0 = 6.125_{10}$$

In general, we write an n-bit unsigned fixed-point number with m bits before the assumed binary point and f bits after the assumed binary point, where $n = m + f$. The number x represented by the bits $x_{m-1}, \ldots, x_0, x_{-1}, \ldots, x_{-f}$ is

$$x = x_{m-1}2^{m-1} + \cdots + x_0 2^0 + x_{-1}2^{-1} + \cdots + x_{-f}2^{-f}$$

The smallest number representable using such a code is 0, with a code word of all 0 bits. The largest number representable has a code word of all 1 bits, and represents $2^m - 2^{-f}$. In between those bounds, numbers are represented as multiples of the precision, 2^{-f}.

Note that a code with no digits before the assumed binary point is permissible, and indeed, practical. This would correspond to a code with $m = 0$. In such a code, all of the bits represent the fractional part of the number, so the range is between 0 and $1 - 2^{-f}$. We can even go so far as to have the assumed binary point several positions to the left of the left-most bit, that is, for m to be negative. For example, a code with $m = -3$ and $f = 13$ would be a 10-bit code with values ranging from 0 to $2^{-3} - 2^{-13}$ in steps of 2^{-13}, or in decimal, from 0 to $0.12487\ldots$ in steps of $0.000122\ldots$

Similarly, we can have a fixed-point code with no digits to the right of the binary point, that is, with $f = 0$. Numbers represented in such a code are, in fact, unsigned integers. If we substitute $f = 0$ in the expressions for the upper bound and precision, we get an upper bound of $2^m - 1$ and a precision of 1, as we would expect for integers. Thus, integers are just a special case of fixed-point representation.

We can also use fixed-point representation for signed fractional numbers. We use the same approach as we did for integers, changing the weight of the most significant digit to be negative. This gives us a 2s-complement fixed-point signed representation. In this case, the number x represented with m bits before and f bits after the assumed binary point is

$$x = -x_{m-1} 2^{m-1} + \cdots + x_0 2^0 + x_{-1} 2^{-1} + \cdots + x_{-f} 2^{-f}$$

The range of numbers represented using this form is from -2^{m-1} to $2^{m-1} - 2^{-f}$, with a precision of 2^{-f}. Again, we can have a code with m being zero or negative. Since the left-most bit in a signed fixed-point representation is the sign bit, a code that represents values between -1 and just less than 1 has $m = 1$, with the single bit before the binary point being the sign bit.

EXAMPLE 3.19   What number is represented by the signed fixed-point binary number 111101, assuming the binary point is four places from the right?

SOLUTION   The number is

$$11.1101_2$$

$$= -1 \times 2^1 + 1 \times 2^0 + 1 \times 2^{-1} + 1 \times 2^{-2} + 0 \times 2^{-3} + 1 \times 2^{-4}$$

$$= -2 + 1 + \frac{1}{2} + \frac{1}{4} + 0 + \frac{1}{16} = -0.1875_{10}$$

Having described how we can represent fixed-point numbers with a given range and precision, the question arises of determining what range and precision to use in a given application. The answer is not simple, and depends on the application. In digital signal processing applications, where fixed-point numbers are used to represent samples of analog signals, the range of the representation affects the dynamic range (the ratio of maximum to minimum amplitude) of signals that can be processed, and the precision affects the signal-to-noise ratio (a measure of quality or fidelity) of the system. If the system is to perform arithmetic operations on the fixed-point values to implement some processing algorithm, the precision affects the numerical behavior of the algorithm. The finite precision of the representation means that analog signal values are only represented approximately, thus, there is an inherent error in the representation. Some numerical processing steps can magnify the effect of the error. Also, processing steps might yield intermediate values whose range differs from that of the samples, requiring a greater range, and thus more bits, for their representation. Mathematical analysis of the behavior and sensitivity of numerical computations is beyond the scope of this book. Nonetheless, it is a vital early design step in applications that implement numerical processing procedures. More information is provided in the reference books cited in Section 3.6, Further Reading.

Fixed-Point Representation in VHDL

We can represent fixed-point numbers in VHDL using the package fixed_pkg defined in library ieee. The package defines two types: ufixed for unsigned fixed-point numbers and sfixed for signed 2s-complement fixed-point numbers. Both types are vectors of std_logic elements, but are distinct types from each other and from the std_logic_vector type. For both types, we specify the left and right index bounds, indicating the power of two for the weights of the most significant and least significant bits, respectively. The binary point is assumed to be between indices 0 and -1, whether those indices actually occur in a given vector or not.

EXAMPLE 3.20   Write a VHDL entity declaration for a code converter that has an input representing an unsigned number in the range 0 to 48 with a precision of at least 0.01, and an output representing a signed number in the range -100 to 100 with a precision of at least 0.01.

SOLUTION   For the input, we need 6 bits before the binary point, since $\lceil \log_2 48 \rceil = 6$. We need a precision that is smaller than 0.01. Since $\log_2 0.01 \approx -6.64$, we need 7 bits after the binary point. For the output, $\lceil \log_2 100 \rceil = 7$, so we need 7 bits, plus one for the sign bit, giving 8 bits before the binary point. We just need to extend the 6 pre-binary-point input bits with two zero bits to get the 8 pre-binary-point output bits. Since we need the same output precision as the input, we use the same number of bits after the binary point, namely, 7. The entity declaration is

```
library ieee; use ieee.fixed_pkg.all;

entity fixed_converter is
  port ( input  : in  ufixed(5 downto -7);
         output : out sfixed(7 downto -7) );
end entity fixed_converter;
```

In our discussion of integers, we mentioned that VHDL provides the types natural and integer for abstract representation of numbers when we don't need access to the bits of their binary codes. Unfortunately, VHDL does not provide corresponding types for abstract representation of fixed-point numbers, though we could use the floating-point type real, described in Section 3.4.2. Abstract fixed-point types could, in principle, be included in the language, as has been done in the Ada programming language, for example. While we might hope that abstract fixed-point types might be included in a future version of VHDL as applications become more common, for now, we will just make use of the ufixed and sfixed types.

At the time of writing, the fixed-point package is still under development and review within the IEEE committees responsible for the VHDL standard. It may take some time before CAD tool vendors uniformly provide the package for use with their tools. We would expect that simulation tools will make the package available earlier, since the package is written in VHDL and can thus be interpreted directly by a simulator. Vendors of other tools, such as synthesis tools, may initially treat the package in the same way as the user's code and implement the fixed-point operations using operations from the numeric_std package. Ultimately, they may provide build-in implementations of the fixed-point operations. We should consult each tool vendor's documentation to see whether and to what extent they support use of the package. In this book, we will proceed on the assumption that the package is available, in the expectation of widespread adoption by the digital design community.

3.3.2 OPERATIONS ON FIXED-POINT NUMBERS

We now turn to implementation of arithmetic operations on fixed-point numbers. We have already covered most of what we need in our discussion of arithmetic operations on integers, since fixed-point numbers can be viewed as scaled integers. For example, if x and y are fixed-point numbers with the binary point f positions from the right, then $x \times 2^f$ and $y \times 2^f$

are integers represented by the same bit vectors as x and y, respectively. Furthermore,

$$x + y = (x \times 2^f + y \times 2^f)/2^f$$

We know how to add the two integers, and dividing by 2^f simply consists of moving the binary point f places to the left, giving us the result in the same fixed-point format as x and y. Thus, we can use the same kinds of adder circuits for fixed-point numbers as for integers. Similar arguments hold for subtraction, incrementing, decrementing, scaling by constant powers of 2, and resizing.

One issue we need to be aware of is that a design might represent different signals as fixed-point numbers of different lengths or with the binary point in different positions. When we perform operations such as addition or subtraction, we need to ensure that we add or subtract the bits with corresponding binary weights, wherever they occur in a vector. We may need to resize one operand to align it with the other. If we need to add or truncate on the left-hand end of a fixed-point number, the same considerations apply for resizing integers. Thus, in the case of unsigned fixed-point numbers, we add 0 bits to the left to extend the number, and we truncate 0 bits to reduce its size. In the case of 2s-complement signed numbers, we replicate the sign bit to extend the number, and we truncate bits to reduce the number, provided the truncated bits and the resulting sign bit are all the same as the original sign bit. If we need to add or truncate on the right-hand end of a number, things are simpler, since the right-most bits all have positive weight. For both unsigned and 2s-complement representations, we add 0 bits to extend and truncate bits to reduce the size.

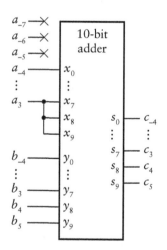

FIGURE 3.19 Alignment of operands for fixed-point addition.

EXAMPLE 3.21   Show how to use an adder for two signed fixed-point signals: a, with 4 pre-binary-point and 7 post-binary-point bits, and b, with 6 pre-binary-point and 4 post-binary-point bits. The result c should have 6 pre-binary-point and 4 post-binary-point bits.

SOLUTION   The operand a needs to be sign extended by two bits on the left-hand end and can be truncated by three bits on the right-hand end. A 10-bit adder is needed, connected as shown in Figure 3.19.

The VHDL $+$ and $-$ operators applied to ufixed and sfixed operands take care of alignment and resizing. They yield results in which the number of pre-binary-point bits is one more than the larger of the numbers of pre-binary-point bits of the operands, and the number of post-binary-point bits is the larger of the numbers of post-binary-point bits of the operands.

EXAMPLE 3.22   Write VHDL declarations and an assignment to perform the addition described in Example 3.21.

SOLUTION   The declarations for the signals a, b and c are

```
signal a     : sfixed(3 downto -7);
signal b, c : sfixed(5 downto -4);
```

We could try the following assignment as a first attempt:

```
c <= a + b;
```

Since a has 4 pre-binary-point bits and b has 6, the result of the + operation has 7 pre-binary-point bits, indexed from 6 down to 0. We need to resize the result to match the number of pre-binary-point bits in c. Furthermore, a has 7 post-binary-point bits and b has only 4, so the result would have 7 post-binary-point bits. Since we only want 4, we should truncate a before applying the + operator. We can revise the assignment using the resize operation and slicing, as shown in the following assignment:

```
c <= resize( a(3 downto -4) + b, c );
```

The fixed-point resize operation determines the result size from the left and right bounds of the second operand, c in this case.

Another related issue to be aware of is the position of the binary point in the result of a multiplication. We can appeal to the way in which we do multiplication of decimals for an analogy. Suppose, for example, that we wish to multiply 23.76 by 3.128. We first multiply the digits without regard to the decimal points to get 7432128. We then add the number of post-decimal digits in the operands, namely, 2 and 3, to get the number of post-decimal digits in the result, namely, 5. Thus the product is 74.32128.

By analogy, multiplying two fixed-point binary numbers with m_1 and m_2 pre-binary-point bits and f_1 and f_2 post-binary-point bits, respectively, gives us a product with $m_1 + m_2$ pre-binary-point bits and $f_1 + f_2$ post-binary-point bits. For example, multiplying 1.101_2 by 10.1_2 gives 100.0001_2. The * operator declared in the fixed_pkg package for ufixed and sfixed values yields a result with this number of pre- and post-binary-point bits.

KNOWLEDGE
TEST QUIZ

1. How is a nonnegative number x represented as a sum of powers of 2 in fixed-point form?

2. What range of values can be represented as signed fixed-point numbers with m pre-binary-point bits and f post-binary-point bits?

3. Write a VHDL declaration for a signal x of type ufixed to represent numbers in the range 0.0 to 359.9 with a precision of 0.1.

4. Write a VHDL assignment to subtract the value of a signal s2 from the value of a signal s1, where both are of type sfixed(7 downto –7), to get a result signal s3 of the same type. No overflow detection is required.

5. How many bits are required for the product of two *fixed-point* numbers with 5 pre-binary-point bits and 9 post-binary-point bits?

3.4 FLOATING-POINT NUMBERS

The final number representation that we will discuss in this chapter is floating-point, which is another representation for approximating real numbers. They allow for representation of a greater range of numbers than a fixed-point representation with the same number of bits. However, implementation of arithmetic operations is considerably more complex. Indeed, most circuits for floating-point arithmetic are not combinational, since they would otherwise be too complex and reduce overall system performance. Since we have deferred detailed discussion of sequential circuit design to a later chapter, we will not go into circuits for floating-point arithmetic here. For completeness of our survey of numeric representations in this chapter, we will just introduce floating-point format and describe how we can model floating-point design in VHDL.

3.4.1 CODING FLOATING-POINT NUMBERS

Floating-point representation in digital systems is based on the same ideas as scientific notation for decimal numbers. We can write numbers that are very small or very large as the product of a fixed-point decimal fraction and a power of 10. This saves us from writing long strings of leading or trailing zeros and makes the number much easier to read and understand. Examples of numbers expressed in scientific notation are $6.02214199 \times 10^{23}$ (Avogadro's number) and $1.60217653 \times 10^{-19}$ (the charge, in Coulombs, of an electron). We call the fractional part before the $\times$ sign the *mantissa* and the power to which 10 is raised the *exponent*.

Floating-point representations adopt these ideas, but use binary instead of decimal. The mantissa is expressed as a fixed-point binary number, the base of the exponent is 2, and the exponent is a signed binary number. Within these general guidelines, there are many alternative floating-point representations, and, historically, several have been implemented in computer designs. However, modern general-purpose computers have almost universally adopted a floating-point representation standardized as IEEE Standard 754, the so called *IEEE floating-point format*. In this section,

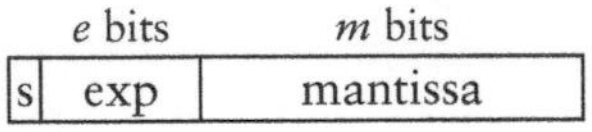

e bits	m bits
s	exp	mantissa

FIGURE 3.20   Floating-point format.

we will describe this format and formats that differ from it only in the number of bits used for the mantissa and exponent.

A floating-point number is represented as a vector of bits arranged as shown in Figure 3.20. The mantissa is represented using a sign bit, s, located in the left-most bit of the vector, and the unsigned magnitude, located in the right-most m bits of the vector. The exponent is represented using e bits between the sign bit and the mantissa magnitude. The IEEE floating-point standard defines two standard floating-point sizes: 32-bit single precision, with $m = 23$ bits and $e = 8$ bits; and 64-bit double precision, with $m = 52$ bits and $e = 11$ bits. These are implemented by most computers. However, if we are designing custom digital circuits for specific applications, we need not be constrained to these sizes. We can choose smaller or larger sizes in order to meet the requirements and constraints of the application. After we've explored some more of the details of the way in which numbers are represented, we will see how the sizes of the exponent and mantissa affect the range and precision of numbers represented.

A floating-point number is usually *normalized*, meaning that the magnitude of the mantissa is greater than or equal to 1.0_{10} (that is, 1.0_2) and less than 2.0_{10} (that is, less than or equal to, $1.111\ldots1_2$), with the exponent being adjusted to give the required value for the number. The mantissa magnitude could be represented as a fixed-point fraction with the binary point located just to the right of the most significant bit. However, as a consequence of normalizing, the most significant bit is always 1. So we can gain an extra bit of precision by not explicitly representing the most significant bit, but assuming that it is 1. This implicit bit in the floating-point format is called the *hidden bit*. Note that the mantissa is not represented using 2s-complement encoding, even though it is a signed value. The sign/magnitude representation turns out to have several advantages, including simplification of circuits for some arithmetic operations. We won't go into details here.

Similarly, though the exponent is a signed number, it also is not represented in 2s-complement form. Rather, it is represented in *excess* form. That is, for a given actual exponent value E, we represent it with the e-bit unsigned binary code for $E + 2^{e-1} - 1$. The value $2^{e-1} - 1$ is called the *bias*, and is chosen so that a symmetric range of positive and negative actual exponent values can be represented. For example, if 5 bits are used for the exponent, the bias would be $2^4 - 1 = 15$, that is, 01111_2. An actual exponent value of 3 would be represented using the 5-bit unsigned binary code for $3 + 15 = 18$, that is 10010_2. The reason for using excess coding is that all exponent codes are unsigned. Given the position of the exponent within a floating-point code word, and the fact that numbers with smaller exponents are smaller than numbers with larger exponents (due to normalization), floating-point numbers can be compared using the same hardware as for comparing integers. This is a useful trick for saving cost and execution time in floating-point arithmetic hardware.

Let's now consider the range and precision of values that can be represented using floating-point format. As with fixed-point numbers, the range and precision are important factors that influence the numerical behavior of computations. The range of values is determined by the length of the exponent, since the most positive exponent determines the largest value and the most negative exponent determines the smallest value. The IEEE floating-point format reserves two exponent encodings for special purposes: the largest encoding, $2^e - 1$, with all 1 bits; and the smallest encoding, with all 0 bits. We will return to these shortly. Setting them aside, the smallest exponent has an encoding of 1, representing an actual exponent value of $-2^{e-1} + 2$. Putting this together with the smallest mantissa magnitude of 1.0 gives us the smallest representable value of $\pm 1.0 \times 2^{-2^{e-1}+2}$. The largest exponent has an encoding of $2^e - 2$, representing an actual exponent value of $2^{e-1} - 1$. Putting this together with the largest mantissa magnitude of just under 2.0 gives us the largest representable value of just under $\pm 2.0 \times 2^{2^{e-1}-1}$, that is, $\pm 2^{2^{e-1}}$. For IEEE single-precision format, this corresponds to a range of approximately $\pm 1.2 \times 10^{-38}$ to $\pm 3.4 \times 10^{38}$, and for IEEE double-precision format, a range of approximately $\pm 2.2 \times 10^{-308}$ to $\pm 1.8 \times 10^{308}$. A custom floating-point representation with a 5-bit exponent, on the other hand, would give us a range of approximately $\pm 6.1 \times 10^{-5}$ to $\pm 6.6 \times 10^4$.

When considering the precision of floating-point numbers, we usually talk about relative precision, since absolute precision varies with the exponent. The relative precision is determined by the number of bits in the mantissa magnitude. All of the bits are significant, since there are no leading zeros in the mantissa (taking into account the hidden bit). So the relative precision remains the same across the full range of values, and is approximately 2^{-m}. Another way of thinking about precision is to specify the number of significant decimal digits, which is approximately $m \times \log_{10} 2$, that is $m \times 0.3$ digits. For example, IEEE single-precision format gives a precision of approximately 7 decimal digits, and IEEE double-precision format gives approximately 16 decimal digits. A custom format with 16 bits of mantissa magnitude would give a precision of approximately 5 decimal digits.

We can return now to the special exponent encodings that we mentioned above. First, the smallest exponent encoding, all zeros, is used for *denormal* numbers, in which the hidden bit is 0. The actual exponent is still represented using excess form, and so has a value of $-2^{e-1} + 1$. Thus, denormal numbers are all smaller in magnitude than the smallest normalized number, though they have fewer significant bits. They allow for *gradual underflow* in a computation, where the results diminish toward 0.0 once the limit of precision has been reached. This feature of the representation improves the numerical behavior of some algorithms. If all the mantissa bits in a denormal number are 0, we get $\pm 0.0 \times 2^{-2^{e-1}+1}$. Thus, there are two alternate representations for 0.0, one with a sign bit of 0 and the other with a sign bit of 1. The IEEE standard specifies that a zero result in most cases be represented by the nonnegative version, but that in any case, the two versions should be deemed equal.

The other special exponent encoding, all 1s, has two uses. If the mantissa magnitude bits are all 0 (not counting the hidden bit), the number represents an infinite value. The value of the sign bit determines whether it is a positive or negative infinity. Operations that overflow generally yield an infinite result, which is maintained in subsequent computations. This avoids having to check for overflow until completion of a multistep computation, thus improving performance. If the exponent encoding is all 1s and the mantissa magnitude is other than all 0s, the value is said to represent *not a number* (NaN). NaN results arise from computations such as division of 0 by 0, and can also be maintained through a multistep computation.

In addition to the representation for floating-point numbers, the IEEE standard also specifies how arithmetic operations are to be performed, provides options for specifying how operations are to be rounded, and specifies the conditions under which exceptions may occur. (A system may abort a computation or take recovery action when an exception occurs.) The details are beyond the scope of this book, but can be found in the Further Reading references.

For a given number of bits of representation, floating-point representation can give a larger range of values than fixed-point, albeit at the expense of precision. The choice between floating-point and fixed-point in a given application will depend largely on the range of values that must be represented, both for the input and output signals, as well as for intermediate results during computation. There is also a trade-off with the complexity of circuits needed to perform the computations. Fixed-point circuits are generally simpler, but if significantly more bits are needed to get the required range, the circuits may consume more area. In many cases, the choice will only be made after thorough exploration of the numerical behavior of the computations to be performed and comparison of implementation complexities of alternate representations. This exploration will usually be performed by a system architect early in the development process. The result of the exploration will be a design specification that includes details of number representations to be used within the system. In a circuit that is customized for a particular application, a floating-point representation can use exponent and mantissa sizes other than those defined by the IEEE standard, thus reducing cost and potentially improving performance.

3.4.2 FLOATING-POINT REPRESENTATION IN VHDL

VHDL provides the float_pkg package in library ieee for modeling floating point numbers. The package defines the type float for floating-point numbers whose exponent and mantissa sizes we wish to specify. The type is a vector of std_logic elements, but is distinct from other types with std_logic elements. To represent a floating-point number with *e* exponent bits and *m* mantissa magnitude bits, we declare a signal of type float with *e* as the left index bound and $-m$ as the right index bound. The sign bit is then the element at index *e*, the exponent is the slice from $e-1$ down to 0, and the mantissa magnitude (without the hidden bit) is the slice from -1 down to $-m$. For example, a floating-point signal with 5 exponent bits and 10 mantissa-magnitude bits would be declared as

```
signal fp_num : float(5 downto -10);
```

The sign bit would then be fp_num(5), the exponent fp_num4(4 downto 0), and the mantissa magnitude fp_num(–1 downto –10).

The package also defines the types float32 and float64 for IEEE single-precision and double-precision format and float128 for a 128-bit representation. For these types, the exponent and mantissa sizes are set by the type definitions in the package. For float32, the index range is 8 down to -23, for float64 it is 11 down to -52, and for float128 it is 15 down to -112. As an example, a signal of type float32 would be declared as

```
signal fp_num32 : float32;
```

EXAMPLE 3.23   Write a VHDL model for a code converter that has an input representing a signed number in the range -100 to 100 with a precision of at least 0.01, and an output representing a floating-point number with a range of $\pm 10^{-4}$ to $\pm 10^4$ and a precision of at least 5 decimal digits.

SOLUTION   The input can use the same fixed-point format as the output described in Example 3.20, namely, 8 bits before the binary point and 7 bits after the binary point. For the floating-point output, we need to represent a range of $\pm 2^{-14}$ to $\pm 2^{14}$, so we need an exponent length of 5. This gives an actual range encompassing the required range. We can achieve the required precision with 16 bits of mantissa magnitude, as described above. Thus, the index range for the output is 5 down to -16. The entity declaration is

```
library ieee; use ieee.fixed_pkg.all, ieee.float_pkg.all;

entity float_converter is
  port ( input  : in sfixed(7 downto -7);
         output : out float(5 downto -16) );
end entity float_converter;
```

The architecture can make use of the to_float conversion operation defined in float_pkg:

```
architecture eqn of float_converter is
begin
  output <= to_float(input, output);
end architecture eqn;
```

The to_float operation uses the first operand value, input, as the value to convert. The numerical value of the second operand, output, is not used. The operand is just provided so that to_float can determine the exponent and mantissa sizes of the conversion result.

As we have previously mentioned, VHDL provides the types natural and integer for abstract representation of integers when we don't need access to the bits of their binary codes. VHDL also provides the type real for abstract representation of floating-point values. The VHDL standard requires that numbers of type real be implemented by CAD tools using at least IEEE double-precision format. While we can use type real as an abstraction for floating-point and fixed-point numbers, we don't have the fine control over the range and precision afforded by types ufixed, sfixed and float. Nonetheless, using type real can be valuable for exploration of numerical algorithms in the early design stages, especially since simulators will perform computations on real values much faster than on ufixed, sfixed or float values.

The definition of type real is built into VHDL, so we don't need to reference any library or package to use it. We can perform arithmetic operations on values of type real, and we can convert between real and float values. We would use a mixture of float and real signals in a model if we need to access the bits of the floating-point encoding of some values but want other values to be in abstract form. Some examples are

```
signal r1, r2 : real;
signal x, y   : float(5 downto -16);
signal z      : float(8 downto -16);
signal z_sign : std_logic;

r1     <= to_real(x);
r2     <= r1 / to_real(y);
z      <= to_float(r2, z);
z_sign <= z(z'left);
```

The operation to_real used here converts from a floating-point vector value to an abstract numeric value. The conversion to_float works in the reverse direction, from an abstract numeric value to a floating-point vector. The target of the assignment, z, is provided as the second operand to the conversion so that its size can be used to determine the index bounds for the conversion result.

At the time of writing, the floating-point package, like the fixed-point package, is still under development and review within the IEEE committees responsible for the VHDL standard, so the same caveats as mentioned in Section 3.3.1 apply. We should consult each tool vendor's documentation to see whether and to what extent they support use of the floating-point package. In this book, we will proceed on the assumption that the package is available, in the expectation of widespread adoption by the digital design community.

1. Express the number 4.5_{10} in floating-point format with 5 bits of exponent and 12 bits of mantissa magnitude.

2. What values are represented by the following bit vectors, interpreted in floating-point format with 4 bits of exponent and 11 bits of mantissa magnitude: 0000000000000000, 0111100000000000 and 0100010000000000?

3. Write a VHDL declaration for a signal x, representing a floating-point value in the range -100 to 100 with a precision of at least 4 decimal digits.

4. What range and precision are provided by values of the VHDL type float32?

KNOWLEDGE TEST QUIZ

3.5 CHAPTER SUMMARY

▶ A nonnegative integer x less than or equal to $2^n - 1$ is represented in n-bit unsigned binary form as

$$x = x_{n-1}2^{n-1} + x_{n-2}2^{n-2} + \cdots + x_0 2^0$$

▶ A signed integer x between -2^{n-1} and $2^{n-1} - 1$ inclusive is represented in n-bit 2s-complement form as

$$x = -x_{n-1}2^{n-1} + x_{n-2}2^{n-2} + \cdots + x_0 2^0$$

▶ Octal (base 8) and hexadecimal (base 16) are shorthand codes for binary codes.

▶ Unsigned integers are modeled in VHDL using the type unsigned from the numeric_std package, or using the built-in type natural. Signed integers are modeled using the type signed from numeric_std, or using the built-in type integer. Arithmetic operators can be used for these types.

▶ An unsigned number is zero-extended by adding 0s to the left, and is truncated by discarding leading 0s. A 2s-complement signed number is sign-extended by replicating the sign bit to the left, and is truncated by discarding leading copies of the sign bit.

▶ Addition of binary-coded integers is performed by an adder circuit. The simplest form of adder is a ripple-carry adder. Fast carry chain, carry-lookahead and other adder structures improve performance at the cost of circuit area and power.

▶ A 2s-complement signed integer is negated by complementing and adding 1.

▶ Subtraction of binary-coded integers can be implemented using an adder by complementing the second operand and setting the carry in to 1.

▶ A magnitude comparator compares two binary-coded integers for equality or inequality (greater than or less than comparison).

▶ Binary-coded integers are multiplied by a power of two by a logical shift left. Unsigned integers are divided by a power of 2 by a logical shift right. 2s-complement signed integers are divided by a power of 2 by an arithmetic shift right.

▶ A combinational multiplier forms partial products by multiplying one operand by each bit of the other operand, then adds the partial products to form the product.

▶ Gray codes change only in one bit position between adjacent code words. They are commonly used in electromechanical position sensors.

▶ A fractional number can be represented in fixed-point binary form by assuming a fixed position for the binary point. Arithmetic circuits for integers can be used, since fixed-point numbers can be interpreted as scaled integers.

▶ Fixed-point numbers are modeled in VHDL using the types ufixed and sfixed (for unsigned and signed numbers, respectively) from the fixed_pkg package. Arithmetic operators can be used for these types.

▶ A fractional number can be represented in floating-point binary form with a signed mantissa and an exponent. IEEE format specifies sign/magnitude representation for the mantissa and excess representation for the exponent. Special representations are provided for denormal numbers, infinities and not-a-number values.

▶ Floating-point numbers are modeled in VHDL using the type float from the float_pkg package, or using the built-in type real. Arithmetic operators can be used for these types.

▶ Modeling a design using types and operations from the numeric_std, fixed_pkg and float_pkg packages allows a synthesis tool to choose arithmetic components optimized for the target fabric, subject to performance requirements and constraints.

3.6 FURTHER READING

Digital Arithmetic, Miloš D. Ercegovac and Tomás Lang, Morgan Kaufmann Publishers, 2004. A comprehensive reference on numerical representations and algorithms and circuit structures for arithmetic operations.

Understanding Digital Signal Processing, Richard G. Lyons, Prentice Hall, 2001. An introduction to the theory of digital signal processing (DSP), including a discussion of the effects of finite fixed-point representation.

IEEE Standard for Binary Floating-Point Arithmetic, IEEE Std 754-1985. This standard defined the representation for single-precision (32-bit) and double-precision (64-bit) and extended-precision floating-point numbers. It also specifies how arithmetic operations on such numbers are to be performed.

The Designer's Guide to VHDL, 2nd Edition, Peter J. Ashenden, Morgan Kaufmann Publishers, 2002. Includes a description of the types and operations provided by the IEEE standard numeric packages.

EXERCISES

EXERCISE 3.1 Express the following decimal numbers in 8-bit unsigned binary form: 5, 83 and 240.

EXERCISE 3.2 What decimal numbers are represented by the following 8-bit unsigned binary numbers: 00100101 and 11000000?

EXERCISE 3.3 What range of numbers can be represented in unsigned binary form in 6 bits, in 14 bits and in 30 bits?

EXERCISE 3.4 How many bits are required to represent numbers in each of the following ranges in unsigned binary form: 0 to 31, 0 to 100, 0 to 1000 and 0 to 8191?

EXERCISE 3.5 How many bits would be required to represent:

a)   An angle in degrees between 0° and 360°.

b)   The milage on a car odometer, assuming six decimal digits.

c)   The delay in a radar echo in ns, assuming a maximum delay of 1ms.

EXERCISE 3.6 Express the following unsigned binary numbers in octal: 001110010, 00000000 and 1111011111.

EXERCISE 3.7 Express the following unsigned binary numbers in hexadecimal: 10000101, 01111101, 1111001001 and 000011111.

EXERCISE 3.8 What numbers, expressed in unsigned binary form, are represented by the following octal numbers: 7024 and 0001?

EXERCISE 3.9 What numbers, expressed in 8-bit unsigned binary form, are represented by the following octal numbers: 055 and 307?

EXERCISE 3.10 Is 2901 a valid octal number? If so, what unsigned binary number does it represent? If not, why not?

EXERCISE 3.11 What numbers, expressed in unsigned binary form, are represented by the following hexadecimal numbers: 7F39BA, C108 and 7024?

EXERCISE 3.12 What numbers, expressed in 10-bit unsigned binary form, are represented by the following hexadecimal numbers: 06C and 307?

EXERCISE 3.13 Is 2GA1 a valid hexadecimal number? If so, what unsigned binary number does it represent? If not, why not?

EXERCISE 3.14 Resize the following unsigned binary numbers to 8 bits: 01101, 111000, 0001011001, 0011110000 and 000110001001. In which cases does the result not represent the same numeric value as the original number?

EXERCISE 3.15   Perform the following unsigned binary additions. In each case, can the result be represented in the same number of bits as the operands, or is an extra bit required?

a)   01011001 + 01011110

b)   11110001 + 01110100

c)   10000010 + 11000001

EXERCISE 3.16   Perform the following unsigned binary additions to produce 8-bit results. In each case does the addition overflow or not?

a)   00111000 + 10010000

b)   11110000 + 00010010

c)   11111100 + 10000111

EXERCISE 3.17   Draw circuits composed of gates that implement a half adder and a full adder.

EXERCISE 3.18   Identify two unsigned binary numbers for which a ripple-carry adder exhibits its worst-case delay. Show how the propagation of signals through the components of the adder cause the delay to be the worst-case delay.

EXERCISE 3.19   For the addition of the two 14-bit unsigned binary numbers 01110001010101 and 11100011000110, with $c_0 = 1$, determine the values of k_i, p_i and g_i for each bit position i, and thus determine the values of s_i and c_{i+1}. What is the longest chain of propagated carries in this addition?

EXERCISE 3.20   Draw a circuit composed of gates that implements the 4-bit-wide carry-lookahead generator shown in Figure 3.7.

EXERCISE 3.21   We have shown that addition of two n-bit unsigned binary numbers requires $n + 1$ bits for the result to be represented without overflow. Show that addition of three n-bit unsigned binary numbers also requires no more than $n + 1$ bits.

EXERCISE 3.22   Write a VHDL model of a circuit that adds three 12-bit unsigned binary numbers to produce a 13-bit result with no overflow detection.

EXERCISE 3.23   Develop a VHDL testbench model for the adder described in Exercise 3.22.

EXERCISE 3.24   Write a VHDL model of a circuit that adds three 12-bit unsigned binary numbers to produce a 12-bit result with overflow detection.

EXERCISE 3.25   Develop a VHDL testbench model for the adder described in Exercise 3.24.

EXERCISE 3.26   Perform the following unsigned binary subtractions to produce 8-bit results. In each case does the subtraction underflow or not?

a) $10111000 - 01010000$

b) $01110000 - 00110010$

c) $01111100 - 10000111$

EXERCISE 3.27   Prove Equations 3.11 and 3.12.

EXERCISE 3.28   Determine the Boolean equations for an unsigned decrementer by simplifying Equations 3.11 and 3.12 for an unsigned subtracter.

EXERCISE 3.29   Revise the equality comparator of Figure 3.11, for 16-bit x and y inputs, by replacing the n-input AND gate with two-input NAND and NOR gates. Hint: DeMorgan's law tells us that $\overline{a + b} = \overline{a} \cdot \overline{b}$.

EXERCISE 3.30   The magnitude comparator of Figure 3.12 suffers from similar worst-case delay behavior to that of a ripple-carry adder. The final result may require the result of comparison of the least significant bits to ripple up the chain of AND and OR gates. Devise a 4-bit lookahead magnitude comparator that avoids the ripple behavior.

EXERCISE 3.31   Perform a logical shift left by 4 positions on each of the following unsigned numbers to form a 12-bit result: 000111000110 and 000010110100. In each case, does an overflow occur?

EXERCISE 3.32   Perform a logical shift right by 4 positions on each of the following unsigned numbers to form a 12-bit result: 100101010000 and 000101001000. In each case, does the result exactly represent division by 16?

EXERCISE 3.33   Perform the following unsigned binary multiplication to form a 12-bit result: 101001×010101.

EXERCISE 3.34   Suppose we use a 4-bit unsigned binary representation to encode the rotational position of the shaft in Figure 3.14 instead of a Gray code. Identify all cases where more than one bit changes between adjacent code words.

EXERCISE 3.35   Devise a 5-bit Gray code using the scheme described in Section 3.1.3.

EXERCISE 3.36   Develop a VHDL model of a converter that converts from a 4-bit unsigned binary code input to a 4-bit Gray coded output.

EXERCISE 3.37   Express the following decimal numbers in 8-bit 2s-complement signed form: 5, 83 and −120.

EXERCISE 3.38   What decimal numbers are represented by the following 8-bit 2s-complement signed numbers: 00100101 and 11000000?

EXERCISE 3.39   What range of numbers can be represented in 2s-complement signed form in 6 bits, in 14 bits and in 30 bits?

EXERCISE 3.40   How many bits are required to represent numbers in each of the following ranges in 2s-complement signed form: −32 to 31, −100 to 100, −1000 to 1000 and −8192 to 8191?

EXERCISE 3.41   How many bits would be required to represent:

a)   A temperature in °C between absolute zero (−273°C) and 5000°C.

b)   An altitude in meters, between −5000 (below sea level) and 20,000 (above sea level).

EXERCISE 3.42   Resize the following 2s-complement signed numbers to 8 bits: 01101, 111000, 0001011001, 0011110000 and 111110001001. In which cases does the result not represent the same numeric value as the original number?

EXERCISE 3.43   Negate the following 2s-complement signed numbers: 00111010, 11101111 and 00000000.

EXERCISE 3.44   Perform the following 2s-complement signed additions to produce 8-bit results. In each case, can the result be represented correctly in the same number of bits as the operands?

a)   01011001 + 01011110

b)   11110001 + 01110100

c)   11111100 + 11110010

d)   10000010 + 11000001

EXERCISE 3.45   Write a VHDL model of a circuit that adds three 12-bit 2s-complement signed numbers to produce a 12-bit result with overflow detection.

EXERCISE 3.46   Develop a VHDL testbench model for the adder described in Exercise 3.45.

EXERCISE 3.47   Perform the following 2s-complement signed subtractions to produce 8-bit results. In each case does the subtraction overflow?

a) $10111000 - 01010000$

b) $01110000 - 00110010$

c) $01111100 - 10000111$

d) $11110001 - 10001010$

EXERCISE 3.48   Show how a 2s-complement adder/subtractor can be used to compute the absolute value of number. Hint: $y = 0 + y$ and $-y = 0 - y$.

EXERCISE 3.49   Draw a circuit composed of gates, similar to that of Figure 3.12, that implements a 2s-complement signed magnitude comparator.

EXERCISE 3.50   Perform a logical shift left by 4 positions on each of the following 2s-complement signed numbers to form a 12-bit result: 000111000110, 111111100101 and 000000110100. In each case, does an overflow occur?

EXERCISE 3.51   Perform an arithmetic shift right by 4 positions on each of the following 2s-complement signed numbers to form a 12-bit result: 100101010000 and 000101001000. In each case, does the result exactly represent division by 16?

EXERCISE 3.52   Write a VHDL model of a circuit that calculates the average of four 16-bit 2s-complement signed numbers, without checking for overflow. Hint: use a shift operation to perform the division by 4.

EXERCISE 3.53   Develop a VHDL testbench model for the averager described in Exercise 3.52.

EXERCISE 3.54   Prove that the shift-left and arithmetic shift-right operations described in Section 3.2.2 correctly implement scaling by a power of 2.

EXERCISE 3.55   What numbers are represented by the following unsigned fixed-point binary numbers, assuming the binary point is three places from the right: 1001001 and 0011110?

EXERCISE 3.56   What is the range and precision of each of the following unsigned fixed-point representations, with m pre-binary-point and f post-binary-point bits:

a)   12 bits, with $m = 5$ and $f = 7$

b)   10 bits, with $m = -2$ and $f = 12$

c)   8 bits, with $m = 12$ and $f = -4$

EXERCISE 3.57   How many pre-binary-point and post-binary-point bits would be required to represent numbers in the range 0.0 to 12.0 with a precision of 0.003?

EXERCISE 3.58   What numbers are represented by the following signed 2s-complement fixed-point numbers, assuming the binary point is four places from the right: 00101100 and 11111101?

EXERCISE 3.59   What is the range and precision of each of the following signed 2s-complement fixed-point representations, with m pre-binary-point and f post-binary-point bits:

a)   14 bits, with $m = 6$ and $f = 8$

b)   8 bits, with $m = -4$ and $f = 12$

EXERCISE 3.60   How many pre-binary-point and post-binary-point bits would be required to represent numbers in the range -5.0 to $+5.0$ with a precision of 0.02?

EXERCISE 3.61   Write a VHDL entity declaration for a component that calculates the square of a signed fixed-point number with 4 pre-binary-point and 6 post-binary-point bits. The result is unsigned, with 8 pre-binary-point and 6 post-binary-point bits.

EXERCISE 3.62   Show how a 16-bit signed integer adder/subtracter can be used to add two signed operands, a, with 3 pre-binary-point and 9 post-binary-point bits, and b, with 7 pre-binary-point and 5 post-binary-point bits, to produce a result with 7 pre-binary-point and 9 post-binary-point bits and no carry or overflow outputs.

EXERCISE 3.63   Given a floating-point representation with 7 exponent bits and 16 mantissa bits:

a)   How would the following numbers be represented: $+5.625$ and -0.3125?

b)   What values are represented by the following floating point numbers, shown in hexadecimal shorthand: 44F000 and BC4000?

c)   What is the range and precision of the floating-point representation?

EXERCISE 3.64   Determine the smallest floating-point representation that includes values whose absolute value is in the range 10^{-6} to 10^{+6} and that has a precision of 6 decimal digits.

SEQUENTIAL BASICS

4

Sequential circuits are the mainstay of digital systems. In this chapter, we start by examining several sequential circuit elements that are widely used in digital systems for storing information and for counting events. We then see how a system can be built from two main sections: a datapath and a control section. We complete the chapter with a discussion of a clocked synchronous timing methodology based on the abstraction of discrete time. This methodology is central to design of complex digital systems.

4.1 STORAGE ELEMENTS

In Chapter 1, we briefly introduced the idea of sequential circuits. We described a sequential circuit as one whose outputs depend not only on the current values of inputs, but also on the previous values of inputs. Such circuits have some form of memory, or storage, of the history of input values. We mentioned that sequential circuits are commonly regulated by a periodic clock signal that divides the passage of time into discrete clock cycles. We also showed one of the simplest elements for storing values, a D flip-flop, that can store one bit of information. In this section, we will look at further uses of the D flip-flop and other storage elements.

4.1.1 FLIP-FLOPS AND REGISTERS

As a reminder, the symbol for a D flip-flop is shown in Figure 4.1, and a timing diagram is shown in Figure 4.2. The flip-flop is edge-triggered, meaning that on each rising edge of the clk input, the current value of the D input is stored within the flip-flop and reflected on the Q output. We illustrated use of D flip-flops in sequential circuits in Example 1.2, where we stored the previous two values of an input signal on successive clock edges so that we could detect a given sequence of input values.

While it is possible to implement a flip-flop as a combination of gates, it is not very instructive to do so. Moreover, flip-flops are provided as

FIGURE 4.1   A D flip-flop.

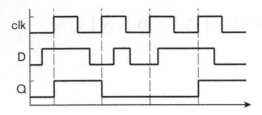

FIGURE 4.2   Timing diagram for a D flip-flop.

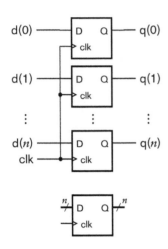

FIGURE 4.3   A register composed of D flip-flops (top), and the symbol for the register (bottom).

primitive elements in most implementation fabrics, so we would only need to implement one using gates in very exceptional circumstances. Advanced books on IC design typically include more detailed treatment of flip-flop implementation (see Section 4.6, Further Reading).

In most digital circuits, flip-flops are not used individually, but in groups to store binary-coded values. A group of flip-flops used in this way is called a *register*. Each flip-flop in the register stores one bit of the code word of the stored value, as shown in Figure 4.3. The circuit at the top of the figure shows that each bit of an input and an output signal is connected to the input and output, respectively, of one of the flip-flops, and that the clock signal is connected in common to the clock input of all of the flip-flops. When there is a rising edge on the clock input, each flip-flop in the register updates its stored bit from the signal connected to its data input and drives the new value on its data output. The symbol for the register is shown at the bottom of Figure 4.3. The difference, compared to the symbol for a single flip-flop, is in the thick lines used for the data input and output, denoting multiple bits. We can think of this as a more abstract component that has similar behavior to a D flip-flop, except that it stores a complete code word rather than a single bit.

We can model simple D flip-flops and registers in VHDL using a process of the form

```
reg: process (clk) is
begin
  if rising_edge(clk) then
      q <= d;
  end if;
end process reg;
```

This is the first of a small number of process templates that we will introduce for modeling sequential circuits. It is important that we adhere to the template structures, since synthesis tools can generally only synthesize sequential circuits that use the templates. A complete description of the templates and the way synthesis tools process them is included in Appendix C.

We would place a process statement representing a flip-flop or register in the statement part of an architecture. The reference to the clock signal clk in parentheses after the process keyword is called the process's *sensitivity list*. A change of value of any signal mentioned in the sensitivity list triggers the process to perform the statements between the begin and end keywords. In this case, the statements check whether the value change is a rising edge of the clk signal, and if so, assign the current value of the data input d to the data output q. Since this assignment only happens on rising edges of clk, and the value of q remains unchanged between rising edges, the process models the behavior we described for an edge-triggered D flip-flop or a register. The distinction between the two arises from the types of d and q. If the signals are of type std_logic, the process models a D flip-flop, storing just a single bit of data. If the signals are of a vector type, such as std_logic_vector, unsigned, sfixed, and so on, the process models a register.

One use for a register constructed from simple D flip-flops is as a *pipeline register* in a sequential design. We will discuss this in further detail in Chapter 9, focusing on the use of pipelining as a technique for improving performance of a digital system. For now, consider the circuit outlined at the top of Figure 4.4. Successive values of data arriving at the input are processed by a number of combinational subcircuits, for example, by arithmetic subcircuits built from components described in Chapter 3. The total propagation delay of the circuit is the sum of the propagation delays of the individual subcircuits. This total delay must be less than the interval between arriving data values, otherwise data values may be lost. If the total delay is too long, we can divide the circuit into segments by inserting a register after each subcircuit, as shown at the bottom of Figure 4.4. This arrangement is called a *pipeline*, as it allows data and intermediate results to flow through over several clock cycles. A new input value arrives at the beginning of each clock cycle. During a clock cycle, each subcircuit uses the value from the preceding register (or from the input, in the case of the first subcircuit) to perform its combinational function and to yield an intermediate result. On the next

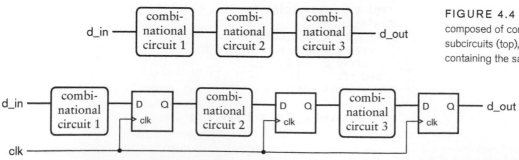

FIGURE 4.4 A circuit composed of combinational subcircuits (top), and a pipeline containing the same subcircuits.

rising clock edge, the intermediate results are stored in the registers at the outputs of the subcircuits. Each intermediate result is then used by the next subcircuit during the next clock cycle. Computation is thus performed in assembly-line fashion. A new final result reaches the output on each clock edge, having taken several clock cycles to be computed.

EXAMPLE 4.1   Develop a VHDL model for a pipelined circuit that computes the average of corresponding values in three streams of input values, a, b and c. The pipeline consists of three stages: the first stage sums values of a and b and saves the value of c; the second stage adds on the saved value of c; and the third stage divides by three. The inputs and output are all signed fixed-point numbers indexed from 5 down to −8.

SOLUTION   The entity declaration is

```
library ieee;
use ieee.std_logic_1164.all, ieee.fixed_pkg.all;

entity average_pipeline is
  port ( clk    : in  std_logic;
         a, b, c : in  sfixed(5 downto -8);
         avg     : out sfixed(5 downto -8) );
end entity average_pipeline;
```

The architecture body is

```
architecture rtl of average_pipeline is

  signal a_plus_b, sum, sum_div_3 : sfixed(5 downto -8);
  signal saved_a_plus_b,
         saved_c, saved_sum       : sfixed(5 downto -8);
begin

  a_plus_b <= a + b;

  reg1 : process (clk) is
  begin
    if rising_edge(clk) then
      saved_a_plus_b <= a_plus_b;
      saved_c  <= c;
    end if;
  end process reg1;

  sum <= saved_a_plus_b 1 saved c;
```

(continued)

```
reg2 : process (clk) is
begin
  if rising_edge(clk) then
    saved_sum <= sum;
  end if;
end process reg2;

sum_div_3 <= saved_sum *
             to_sfixed(1.0/3.0,
                       sum_div_3'left, sum_div_3'right);

reg3 : process (clk) is
begin
  if rising_edge(clk) then
    avg <= sum_div_3;
  end if;
end process reg3;

end architecture rtl;
```

The signals declared before the begin keyword are used for the intermediate results of the arithmetic operations and for the values saved in registers. The simple assignment statements model the arithmetic operations (two additions and a multiplication). We express the division by three as a multiplication by one-third, as multipliers are generally simpler circuits than dividers. Moreover, some implementation fabrics have built-in multipliers that can be used. The to_sfixed operation converts the value 1.0/3.0 to an sfixed vector whose index bounds are the same as those of sum_div_3. The processes reg1, reg2 and reg3 model the pipeline registers storing the intermediate results. Note that reg1 actually stores two values together: the sum of a and b, and the input value c. If c were not saved in this way, the wrong value from the input stream c would be added by the second adder, rather than the value corresponding to the saved sum of a and b. Also note that reg3 assigns directly to the output avg, as the value saved by the third register is the value required at the output.

The D flip-flop that we have considered so far is somewhat limited in its use, since it stores a new value on every rising edge of the clock input. Many systems only require a flip-flop to store a value when some controlling condition arises. For that, we can use an enhanced form of D flip-flop with a *clock-enable* input (sometimes call a *load-enable* input), illustrated in Figure 4.5. This flip-flop only updates the stored value when the CE input is 1 at the time of a rising clock edge. If the CE input is 0 on a rising clock edge, the flip-flop maintains the stored value unchanged. This behavior is shown in the timing diagram in Figure 4.6. As we mentioned in Section 1.3.6, the value on the data input must be stable for the setup time before and the hold time after the clock edge. A similar constraint

FIGURE 4.5 A D flip-flop with clock-enable input.

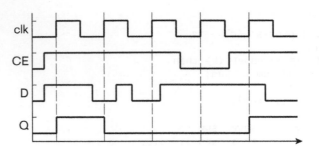

FIGURE 4.6   Timing diagram for a D flip-flop with clock enable.

applies to the clock-enable input. We say that the clock-enable input is a *synchronous control input,* meaning that it must be stable around a clock edge, and its effect is only acted upon when a clock edge occurs.

As with the simple D flip-flop, we can use multiple flip-flops with clock enable in parallel to form a register with clock enable. This form of register is probably the most common used in sequential digital systems, as it allows for storage of an intermediate result computed during one clock cycle to be used as an input to a subsequent computation any number of clock cycles later. We will see in Section 4.3 how we can develop control conditions that govern when data is stored in registers.

We can model flip-flops and registers with clock enable inputs by extending the process template used to model simple D flip-flops and registers. The revised template is

```
reg: process (clk) is
begin
  if rising_edge(clk) then
    if ce = '1' then
        q <= d;
    end if;
  end if;
end process reg;
```

The difference between this and the previous template is the addition of the nested if statement. When a rising edge occurs on the clk input, the output signal is only updated if the ce input is '1'; otherwise, the stored value is unchanged. As before, the types of d and q determine whether the process models a single-bit flip-flop or a multibit register.

A further extension to the simple flip-flop involves adding an input to reset the stored value to 0. This is useful for ensuring that the flip-flop is initialized to a known state when power is first applied to a sequential circuit or when the circuit must be restarted from an initial state. Some circuits include a push button to allow the user to reset the circuit, for example, when it has encountered an error condition from which it cannot recover. Figure 4.7 shows a symbol for a flip-flop with

FIGURE 4.7   A D flip-flop with clock-enable and reset inputs.

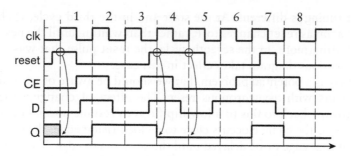

FIGURE 4.8 Timing diagram for a flip-flop with clock-enable and synchronous reset inputs.

both a clock-enable input and a reset input. The reset input overrides the clock-enable and data inputs. That is, when reset is 1, the stored value and the output Q are both changed to 0, regardless of the values on the CE and D inputs.

An important question to consider is the timing of changes on the reset input and when the reset operation occurs. There are two alternative behaviors, and a flip-flop with reset exhibits one or the other. The first reset behavior is called *synchronous reset*, and treats the reset input as a synchronous control input. This behavior is illustrated in Figure 4.8, in which the reset input causes the flip-flop to be reset on the first, fourth and fifth rising clock edges. Notice that, during the seventh clock cycle, reset changes to 1, but then changes back to 0 before a clock edge occurs. Since reset is 0 at the time of the next clock edge, the flip-flop is not reset. Notice also that we have shown the initial value of the Q output as neither 0 nor 1, but some unknown value, denoted by the grey shading. The fact that reset is 1 at the first clock edge forces the output to the known 0 value. Finally, we have ensured that the value of reset, like other data and control inputs is stable around each clock edge.

The second reset behavior for flip-flops is called *asynchronous reset*. In this case, the reset input is treated as an *asynchronous control input*, that is, when it changes to 1, it has an immediate effect regardless of the value of the clock or occurrence of clock edges. Moreover, the effect continues for as long as the reset input is 1. This behavior is illustrated in Figure 4.9. The timing of the inputs is the same as in Figure 4.8, but the

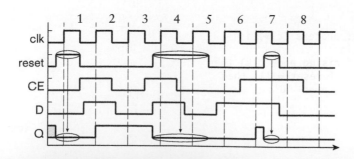

FIGURE 4.9 Timing diagram for a flip-flop with clock-enable and asynchronous reset inputs.

output timing is different. At the start and in the third cycle, Q changes to 0 as soon as reset changes to 1, rather than waiting until the next clock edge. Furthermore, in the seventh cycle, the reset pulse that was ignored in the previous diagram takes effect in this case.

There is a potential problem that we should be aware of when designing circuits with asynchronous reset. The effect of changing the reset input from 1 back to 0 is to allow flip-flops to resume normal operation. However, if the change occurs close to a clock rising edge, the effect may occur at that edge or be delayed until the subsequent edge. This can cause problems in a system with numerous flip-flops, all of which are connected to the same clock and reset signals. Differences in the wiring delays can cause the change of reset from 1 to 0 to occur at slightly different times relative to clock edges for different flip-flops. Consequently, some flip-flops may be released from reset and resume storing values at one clock edge, whereas others might not resume until the subsequent clock edge, resulting in incorrect circuit operation. The solution to this problem is to ensure that the release of the reset signal from 1 to 0 always occurs synchronously with the clock; that is, to ensure that the change occurs sufficiently before a clock edge that the reset signal is stable around the edge for all flip-flops in the system.

The choice between synchronous and asynchronous reset may be influenced by the implementation fabric used for a design. Some fabrics only provide flip-flops with one or the other form of reset. Others, such as many FPGAs, allow us to program each flip-flop to use one or the other form of reset. Alternatively, the choice between the two forms of reset may be made by a system architect based on requirements for the design or the timing practices adopted for the design project. In that case, the chosen form of reset would be incorporated as a design specification for the subcircuits of the larger system. Generally, we should simplify the timing of a design by adopting one form of reset, either synchronous or asynchronous, uniformly throughout the design.

Just as we can use simpler flip-flops in parallel to form registers, so we can use flip-flops with reset in parallel. The result is a register that can be reset to a code word of all 0s. We can model flip-flops and registers with reset in VHDL by extending our previous process templates. The template for a flip-flop with synchronous reset and clock enable is

```
reg: process (clk) is
begin
  if rising_edge(clk) then
    if reset = '1' then
      q <= '0';
```

(continued)

```
      elsif ce = '1' then
          q <= d;
      end if;
   end if;
end process reg;
```

On a rising clock edge, the process first checks whether the reset input is active, since this input has priority over all of the other logic in the flip-flop. If the reset input is active, the output is reset to '0'. If we are modeling a multibit register, we would change the assignment to something like

```
q <= "000000";
```

to clear all output bits. The length of the string will, of course, depend on the number of elements in the vector output signal. The remainder of the process template, after the test for reset, is the same as before. Only if reset is inactive does the process check the clock-enable input.

　　If we need to model a flip-flop or register with asynchronous reset, we need to take account of the fact that the reset input has an effect regardless of the value of the clock input. The process template for this kind of flip-flop is

```
reg: process (clk, reset) is
begin
  if reset = '1' then
      q <= '0';
  elsif rising_edge(clk) then
    if ce = '1' then
        q <= d;
    end if;
  end if;
end process reg;
```

We have included the reset input in the sensitivity list of the process, since the process may need to update the outputs on a change of value of the reset input, not just on a change of value of the clock input. The revised process checks the value of the reset input first, before it looks at the clock input. If the reset input is '1', the process clears the output immediately. Only if the reset input is '0' does the process proceed to check for a rising clock edge. As before, we can change the assignment to the output to reflect the difference between a single-bit flip-flop and a multibit register.

EXAMPLE 4.2   Develop a VHDL model for an accumulator that calculates the sum of a sequence of fixed-point numbers. Each input number is signed with 4 pre-binary-point and 12 post-binary-point bits. The accumulated sum has 8 pre-binary-point and 12 post-binary-point bits. A new number arrives at the input during a clock cycle when the data_en control input is 1. The accumulated sum is cleared to 0 when the reset control input is 1. Both control inputs are synchronous.

SOLUTION   The entity declaration requires a clock input, two control inputs, a data input and a data output, as follows:

```
library ieee;
use ieee.std_logic_1164.all, ieee.fixed_pkg.all;

entity accumulator is
  port ( clk, reset: in  std_logic;
         data_en    : in  std_logic;
         data_in    : in  sfixed(3 downto -12);
         data_out   : out sfixed(7 downto -12) );
end entity accumulator;
```

The architecture is

```
architecture rtl of accumulator is
  signal sum, new_sum : sfixed(7 downto -12);
begin
  new_sum <= resize(sum + data_in, sum);
  reg: process (clk) is
  begin
    if rising_edge(clk) then
      if reset = '1' then
          sum <= (others => '0');
      elsif data_en = '1' then
          sum <= new_sum;
      end if;
    end if;
  end process reg;
  data_out <= sum;
end architecture rtl;
```

The first assignment in the architecture models the addition of the accumulated sum and the data input. The result of the addition is resized to match the size of the sum. The data input is to match the size of the sum. The process models the register used to accumulate the sum. It is based on the template for a register with synchronous reset and clock enable. When reset is 1, the process clears

the register output, represented by the signal sum. The notation (others => '0')
denotes a vector with all 0 elements. This is a convenient alternative to writing
a fixed-length string of 0 bits. If we need to change the length of the sum vector as the design evolves, we don't need to modify the assignment. If reset is '0',
the process checks whether a new data value has arrived and been added to the
sum. In that case, the register output is updated with the new sum; otherwise, it
is unchanged. The final assignment mirrors the sum value on the data_out port.
We need a separate internal signal for the sum, rather than using the output port
directly, since the architecture needs to read the sum to add further input values.
Recall that VHDL does not permit us to read output ports within the architecture.

We have now covered the main aspects of flip-flops and registers.
There are other extensions, but they are just variations on the themes we
have seen. One such variation is the addition of a control input to preset
a flip-flop to 1. This is much like a reset control input, and may be either
synchronous or asynchronous. Another variation is for the reset control
input to use active-low logic, that is, for a 0 on the reset input to clear the
stored data and output. Likewise, a preset control input might use active-
low logic. A further variation is to use active-low logic for the clock input.
This involves triggering a change of stored value on a falling edge of the
clock signal rather than on a rising edge.

EXAMPLE 4.3   The symbol in Figure 4.10 shows a negative-edge-triggered
flip-flop with clock enable, negative-logic asynchronous preset and clear, and
both active-high and active-low outputs. It is illegal for both preset and clear to
be active together. Develop a VHDL model for this flip-flop.

SOLUTION   The entity and architecture are

FIGURE 4.10 A negative-edge-triggered flip-flop.

```
library ieee; use ieee.std_logic_1164.all;

entity flip_flop_n is
  port ( clk_n, CE,
         pre_n, clr_n, D : in std_logic;
         Q, Q_n          : out std_logic );
end entity flip_flop_n;

architecture behavior of flip_flop_n is
  signal Q_tmp : std_logic;
begin
  ff : process ( clk_n, pre_n, clr_n ) is
  begin
    assert not ( pre_n = '0' and clr_n = '0')
```

(continued)

```
        report "Illegal inputs: pre_n and clr_n both '0'";
      if pre_n = '0' then
         Q_tmp <= '1';
      elsif clr_n = '0' then
         Q_tmp <= '0';
      elsif falling_edge(clk_n) then
        if CE = '1' then
           Q_tmp <= D;
        end if;
      end if;
    end process ff;
   Q <= Q_tmp; Q_n <= not Q_tmp;
end architecture behavior;
```

We adopt the convention of appending "_n" to a name to indicate active-low logic. The process ff models the flip-flop behavior. Since the pre_n and clr_n inputs are asynchronous control inputs, we include them, along with the clock input, in the sensitivity list of the process. Within the process, we include an assertion statement to verify that the illegal condition described in the specification does not arise during use of the flip-flop in a circuit. The remainder of the process is based on the template for a flip-flop with asynchronous control. In this case, we have two asynchronous control inputs, so we test them, one after the other, before checking for a clock edge. Since the clock uses active-low logic, we check for a falling edge using the falling_edge(clk_n) notation. Thus, the stored value is only updated when the clock changes from 1 to 0, and then only when the clock-enable input is 1.

4.1.2   SHIFT REGISTERS

A register, as we have seen, stores data and makes it available at the output unchanged. A *shift register*, on the other hand, can perform a shift operation on the stored data. We described shift operations in Chapter 3, and showed how a shift operation has the effect of scaling a numeric value by a power of 2. As we will see in Chapter 8, shift operations are also used to implement serial transfer of data, that is, transfer one bit at a time over a single wire, instead of using separate wires for each of the bits of data. For now, we will just focus on use of shift registers to combine arithmetic scaling with storage functions.

Figure 4.11 shows a symbol for a shift register, and Figure 4.12 shows how it can be implemented with D flip-flops and multiplexers. The shift register is updated on a rising clock edge when CE is 1. In that case, when the load_en signal is 1, the multiplexers select new data on the D(n−1) through D(0) inputs for updating the register. Alternatively, when CE is 1 and load_en is 0, the multiplexers select the existing data, shifted right by

FIGURE 4.11   A symbol for a shift register.

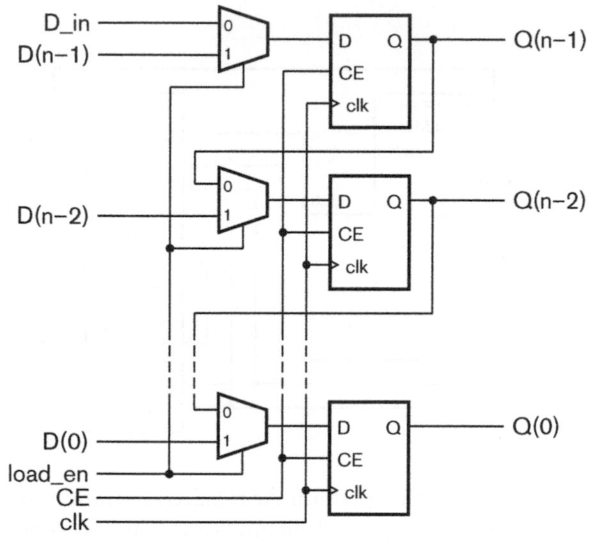

FIGURE 4.12 A shift register implemented with D flip-flops and multiplexers.

one place. The least significant bit is discarded, and the most significant bit is updated with the value of the D_in signal. If we tie D_in to 0, the shift register performs a logical shift right operation on the stored data. Alternatively, if we connect the most significant output bit back to D_in, the shift register performs an arithmetic shift right operation. We will see in Chapter 8 how we connect the D_in input and the Q(0) output for serial transfer of data.

EXAMPLE 4.4   In Chapter 3, we showed how to perform multiplication of unsigned integers by addition of partial products. Construct a multiplier for two 16-bit operands containing just one adder that adds successive partial products over successive clock cycles. The final product is 32 bits.

SOLUTION   In order to perform the operation over multiple cycles, we need a number of registers to hold intermediate results, as shown in Figure 4.13. The x operand is stored in an ordinary register whose output connects to an array of 16 AND gates that form a partial product. The y operand is stored in a shift register whose least significant bit, Q(0), controls the AND gates. The y operand is shifted on successive cycles, thus giving the 16 successive partial products. The sum of the partial products are accumulated in a 17-bit ordinary register and a 15-bit shift register. Since the shift register is never required to load data other than through the D_in connection, the data and load_en inputs are absent. On each clock cycle, the least significant bit of the ordinary register is shifted into the shift register, and the remaining bits of the ordinary register are added with the next partial product. By shifting the accumulated sum in this way, partial products are added at successively more significant positions of the result.

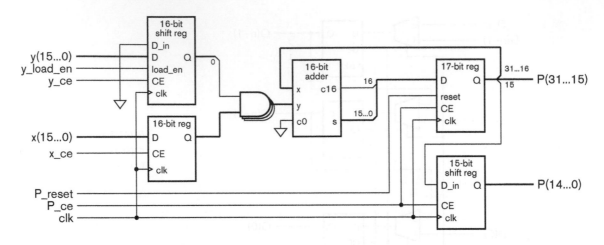

FIGURE 4.13 Registers, shift registers and other components used to form a sequential multiplier.

Making the sequential multiplier perform the required operations over successive clock cycles requires a separate control circuit. We will discuss control sequencing in detail in Section 4.3, and leave detailed design of the multiplier control to Exercise 4.20.

4.1.3 LATCHES

FIGURE 4.14 Symbol for a latch.

As we have seen, a flip-flop is a basic sequential circuit element that stores one bit. Most digital circuits use edge-triggered flip-flops that store a new data value when the clock signal changes from 0 to 1. No further values are stored while the clock remains at 1, nor when the clock returns to 0. Some systems, however, use sequential elements called *latches*, with slightly different timing for storage of values. Figure 4.14 shows a symbol for a latch, and Figure 4.15 shows the timing behavior.

The latch has two inputs, a data input, D, and a latch-enable input, LE. It also has a data output, Q. When the latch-enable input is 1, the value at the data input is stored in the latch and transmitted through to the output. As the timing diagram shows, provided the data input remains unchanged for the entire time that the latch-enable input is 1, the behavior is the same as that of a flip-flop. However, if the data input changes while the latch-enable input is 1, the changed value is transmitted to the output. When the latch-enable input eventually changes to 0, the value stored

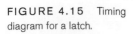

FIGURE 4.15 Timing diagram for a latch.

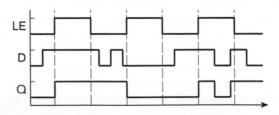

in the latch just before the change is maintained in the latch and at the output. The fact that data is transmitted through to the output while the latch-enable input is 1 leads us also to use the name *transparent latch* for this component. While the latch-enable input is 1, what we see on the output is the value present on the input, so the latch appears to be transparent.

We can model a latch in VHDL using a process of the form

```
latch: process (LE, D) is
begin
  if LE = '1' then
    q <= d;
  end if;
end process latch;
```

This process includes both the latch-enable input and the data input in the sensitivity list. Thus, the process responds to changes on either input. However, it only updates the output Q when LE is '1'. If the D input changes while the LE input is '1', the change on D is reflected on the output, modeling the transparent state of the latch. On the other hand, if D changes while LE is '0', the output is not assigned and maintains its previous value.

Just as we can implement multibit registers with flip-flops connected in parallel, so we can implement multibit latches with single-bit latches connected in parallel. The result is a latch in which multiple data bits flow through when the latch-enable input is 1 and are stored when the latch-enable input is 0.

While latch circuits are relatively simple to implement in many fabrics, the fact that data can flow through them transparently can make it harder to design complex systems with correct timing behavior. The usual solution is to use two-phase nonoverlapping clock signals. Since this approach is not widely used now, the details are beyond the scope of this book. (See the books in Section 4.6, Further Reading.) However, we do need to consider how latching behavior can arise inadvertently from VHDL models, since it is a common design error.

First, let's return to our definition of a combinational logic circuit. We said that such a circuit is one whose outputs are defined purely as a function of the current input values, and that have no dependence on previous input values. The way in which a circuit's output can depend on previous input values is for the circuit to have a feedback path, that is, a cycle of connections from the output of a gate through other gates and back to the input of the gate. Perhaps the simplest such circuit is an inverter whose output is connected to its input, as shown at the top of Figure 4.16. Since the output of the inverter is the logical negation of its input, the output

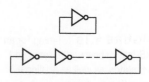

FIGURE 4.16 Inverters connected in feedback loops.

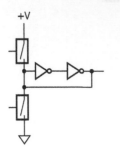

FIGURE 4.17 Using switches to force a node of an inverter ring to 0 or 1.

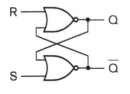

FIGURE 4.18 Cross-coupled RS-latch.

will oscillate between 0 and 1 with a frequency that is dependent on the propagation delay through the inverter. (Alternatively, the inverter may exhibit analog circuit behavior and reach an intermediate voltage level that is neither a valid logic low nor a valid logic high.) If we extend the feedback loop with more inverters to give an odd number of inverters in total (as shown at the bottom of Figure 4.16), we reduce the overall frequency of oscillation. This form of oscillator is called a *ring oscillator*. If we extend the ring to have an even number of inverters, the circuit will reach a stable state in which alternate inverters have a 0 at their output and the others have a 1. There are two possible stable states for such a ring of inverters. We could force the ring into one or other of the states by forcing a given node to 0 or 1, for example, by using switches as shown in Figure 4.17. (This is an idealization. In a real circuit, the switches would have some series resistance, thus avoiding damage to the output of the second inverter.) When both switches are open, the circuit retains the state into which is was forced. Hence, its output depends on the previous input value. This is a basic form of one-bit storage, called a *reset-set latch*, or *RS-latch* for short.

A more common implementation of an RS-latch uses cross-coupled gates, as shown in Figure 4.18. The timing behavior of the RS-latch is shown in Figure 4.19. Normally, the reset input R and the set input S are both 0. Assume initially that Q is 0 and $\overline{Q}$ is 1. This is a stable state, called the *reset state*. If the R input changes to 1 in this state, neither output changes and the latch stays in the reset state. However, if the S input changes to 1, $\overline{Q}$ changes to 0. This value is fed back to the other gate, which causes Q to change to 1. This is also a stable state, called the *set state*. When S returns to 0, the latch stays in the set state. Further changes of S to 1 while the latch is in the set state make no difference. However, if R goes to 1, the feedback causes the latch to change back to the reset state. Thus, which state the latch is in at any time depends on which of the S or R inputs was 1 most recently. Note that if both R and S are 1 at the same time, both Q and $\overline{Q}$ are 0. This is usually considered an illegal operating condition for an RS-latch.

FIGURE 4.19 Timing for an RS-latch, showing the reset and set states, as well as an illegal operating condition.

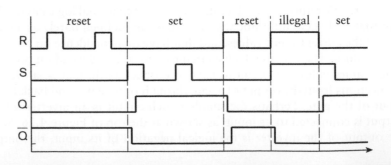

Now that we have seen ways in which feedback can cause latching behavior, let's see how feedback can arise in VHDL models. In Chapter 2, we showed how a combinational circuit is modeled using an assignment statement in an architecture. Normally, we include the inputs to the circuit in the expression on the right-hand side of the assignment symbol and the output of the circuit on the left-hand side. However, if we have an assignment with a given signal appearing both on the left-hand side and on the right-hand side, we imply a feedback loop from the output to the input. Most synthesis CAD tools will not synthesize such circuits without complaint, since the timing is not readily predictable and correct operation is not guaranteed. For example, if we write the following in a model:

```
a <= a + b;
```

we imply an adder with the output feeding back directly into an input. In this sense, assignments modeling combination hardware in VHDL are different from assignments to variables, both in VHDL and in programming languages. Depending on the propagation delay through the synthesized and implemented circuit, we may add the value of b to itself once, twice, or more times within a given time interval. Moreover, if the delays are different for different bits, the result may not correspond to addition of the value of b at all. Most synthesis tools would either issue a warning or reject an assignment in the above form as erroneous.

A feedback loop can also be implied by a number of assignments in combination, where there is a cycle of dependencies between them. For example, consider the following assignments:

```
x <= y + 1;
y <= x + z;
```

Due to the first assignment, the value of x depends on the value of y. Due to the second assignment, the value of x depends on y, and thus indirectly on x itself. A synthesis tool should also issue a warning or flag this as erroneous.

The fact that synthesis tools object to feedback loops in combinational circuits can make it hard to model circuits in which we deliberately include such loops. For example, a VHDL model of the cross-coupled RS-latch of Figure 4.18 might be written as

```
Q   <= R nor Q_n;
Q_n <= S nor Q;
```

These assignments imply a cyclic dependency between Q and Q_n, which is exactly what we want in the synthesized circuit. An alternative way of modeling this behavior is to use a process and an assignment, as follows:

```
RS_latch : process (R, S) is
begin
  if R  = '1' then
     Q <= '0';
  elsif S = '1' then
     Q <= '1';
  end if;
end process RS_latch;

Q_n <= not Q;
```

The assignment simply negates the value of Q, which is generated by the process. In the process, we have included the R and S inputs in the sensitivity list. Thus, the process will be reactivated whenever either input changes. If R is 1, the process updates the Q output to represent the reset state, and if S is 1, the process updates the output to represent the set state. Note that, if neither input is 1, the process makes no assignment to Q. In that case, the outputs remain unchanged; that is, it stores the previously updated state. In general, if there is any execution path through a process where we do not update an output, then the process represents latching behavior for that output, since the output maintains its previous value. If this is intended, as in the process modeling the RS-latch, we don't have a problem. However, it is a common VHDL modeling error to inadvertently omit an assignment to an output in an execution path, for example, in one alternative of a complex if statement. The unintended latching behavior for that output can be most perplexing until the error is located and corrected.

EXAMPLE 4.5   The following process statement is intended to model multiplexer circuitry that selects between a number of inputs to assign to outputs z1 and z2. Identify the error in the process and describe the behavior that results.

```
mux_block : process (sel, a1, b1, a2, b2)
begin
  if sel = '0' then
     z1 <= a1; z2 <= b1;
  else
     z1 <= a2; z3 <= b2;
  end if;
end process mux_block;
```

SOLUTION   The assignment to z3 in the "else" part of the if statement should assign to z2. As a consequence, z2 is not updated on that execution path and z3 is not updated on the execution path in which sel is '0'. Thus, the process implies transparent latches for z2 and z3. The latch for z2 is transparent when sel is '0' and stores a value when sel is '1'. The latch for z3 is transparent when sel is '1' and stores a value when sel is '0'. This unintended behavior can be corrected simply by changing the target of the assignment from z3 to z2, as it should be.

KNOWLEDGE
TEST QUIZ

1. Write a VHDL process for a simple rising-edge-triggered register.

2. What do we call an arrangement of combinational subcircuits and registers that operate in assembly-line-like fashion?

3. What effect does a clock-enable input have on a register?

4. What is the distinction between an asynchronous reset and a synchronous reset?

5. What additional function does a shift register provide compared to an ordinary register?

6. What is meant by the term "transparent" with respect to a latch?

7. What problem is caused by omitting an assignment to an output in a VHDL process that models combinational logic?

4.2   COUNTERS

A counter is a sequential component that increments or decrements a stored value. Counters occur in many digital circuit applications. For example, if an application requires a given operation to be performed on a number of items of data or to be repeated a number of times, a counter can be used to keep track of how many items have been processed or how many times the operation has been performed. Counters are also used as timers, by counting the number of intervals of a fixed duration that have passed.

A simple form of counter is composed of an edge-triggered register and an incrementer, as shown in Figure 4.20. The value stored in the register is interpreted as an unsigned binary integer. The incrementer can be implemented using the circuit we described for an unsigned incrementer in Section 3.1.2 on page 111. The counter increments the stored value on every clock edge. When the stored count value reaches its maximum value ($2^n - 1$, for an n-bit counter), the incrementer yields a result of all zeros, with the carry out being ignored. This result value is stored on the next clock edge. Thus, the counter acts like the odometer in a car, rolling over to zeros after reaching its maximum value. Mathematically speaking, the

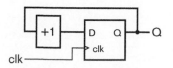

FIGURE 4.20   A simple counter composed of a register and an incrementer.

counter increments modulo 2^n. The counter goes through all 2^n unsigned binary integer values in order every 2^n clock cycles. One use for such a counter is in conjunction with a decoder to produce periodic control signals.

EXAMPLE 4.6   Design a circuit that counts 16 clock cycles and produces a control signal, ctrl, that is 1 during every eighth and twelfth cycle.

SOLUTION   We need a 4-bit counter, since $16 = 2^4$. The counter counts from 0 to 15 and then wraps back to 0. During the eighth cycle, the counter value is 7 (0111_2), and during the twelfth cycle, the counter value is 11 (1011_2). We can generate the control signal by decoding the two required counter values and forming the logical OR of the decoded signals. The required circuit is shown in Figure 4.21.

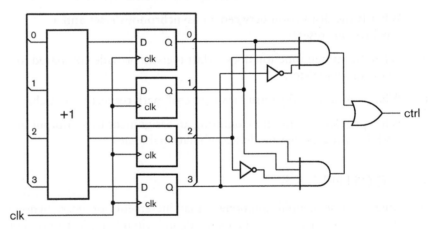

FIGURE 4.21   A counter with decoded outputs.

EXAMPLE 4.7   Develop a VHDL model of the circuit from Example 4.6.

SOLUTION   The entity and architecture are

```
library ieee;
use ieee.std_logic_1164.all, ieee.numeric_std.all;

entity decoded_counter is
  port ( clk  : in std_logic;
         ctrl : out std_logic );
end entity decoded_counter;

architecture rtl of decoded_counter is
  signal count_value : unsigned(3 downto 0);
```

(continued)

```
begin
  counter : process (clk) is
  begin
    if rising_edge(clk) then
      count_value <= count_value + 1;
    end if;
  end process counter;

  ctrl <= '1' when count_value = "0111" or
                   count_value = "1011" else '0';
end architecture rtl;
```

The architecture contains a process that represents the counter. It is similar in form to a process for an edge-triggered register. The difference is that the value assigned to the count_value output on a rising clock edge is the incremented count value. The assignment to count_value represents the update of the value stored in the register, and the addition of 1 represents the incrementer. The final assignment statement in the architecture represents the decoder.

The counter that we have described so far is free running, increment-ing the count value on every clock cycle. We can modify the counter to make it useful in applications that require more control over the count value. Two simple modifications involve adding a clock-enable and a reset input to the storage register within a counter. The clock enable input allows us to control when the counter increments its value, so this input is often called a *count-enable* input. The reset input allows us to clear the count value back to zero. A counter modified in this way is shown in Figure 4.22. This form of counter is very useful for counting occurrences of events. We would connect a signal indicating event occurrence to the count-enable input of the counter. If we need to count events over several intervals, we can reset the counter at the start of each interval.

Another modification is a *terminal-count* output. This is simply a decoded output that is 1 when the counter reaches is maximum, or ter-minal, value. For the counters we have described above, the maximum value of $2^n - 1$ is represented by a count value with all 1 bits. We can use an n-input AND gate to generate the terminal count output, as shown in Figure 4.23. For a free-running counter, the terminal-count output is 1 for a single clock cycle every 2^n clock cycles; that is, it is a periodic signal whose frequency is the input clock frequency divided by 2^n.

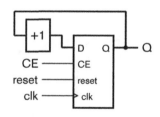

FIGURE 4.22   A counter with clock-enable and reset inputs.

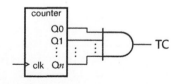

FIGURE 4.23   A counter with terminal-count output.

EXAMPLE 4.8   A digital alarm clock needs to generate a periodic signal at a frequency of approximately 500 Hz to drive the speaker for the alarm tone. Use a counter to divide the system's master clock signal, with a frequency of 1 MHz, to derive the alarm tone.

SOLUTION   We need to divide the master clock signal by approximately 2000. We can use a divisor of $2^{11} = 2048$, which gives us an alarm tone frequency of 488 Hz, which is close enough to 500 Hz. Thus, we could use the terminal-count output of an 11-bit counter for the tone signal. However, the duty cycle (the ratio of time for which the signal is 1 to the time for which it is 0) would only be 1/2048, which would have very low AC energy. We can rectify this by dividing the master clock by 2^{10} with a 10-bit counter, and using the terminal-count output as the count-enable input to a divide-by-2 counter. A circuit is shown in Figure 4.24, and a timing diagram in Figure 4.25. The output of the divide-by-2 counter alternates between 0 and 1 for every pulse on its clock-enable input. The output thus has a 50% duty cycle, which will drive a speaker much more efficiently.

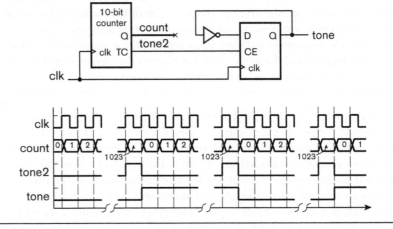

FIGURE 4.24   An alarm clock frequency divider.

FIGURE 4.25   Timing diagram for an alarm clock frequency divider.

Not all free-running counter applications need to divide by a power of 2. If we need to divide by some other value, k, we need the counter to wrap back to 0 after reaching a terminal count of $k - 1$. Mathematically speaking, the counter increments modulo k. We can construct such a counter by decoding the unsigned binary code word for $k - 1$ and using that as the terminal count output. We can feed the terminal count signal back to a synchronous reset input to the storage register within the counter.

EXAMPLE 4.9   Design a circuit for a modulo 10 counter, otherwise known as a *decade counter*.

SOLUTION   The maximum count value is 9, so we need 4 bits for the counter. The unsigned binary code word for 9 is 1001_2. We can decode this value and use it to reset to counter to 0 on the next clock cycle. The circuit is shown in Figure 4.26.

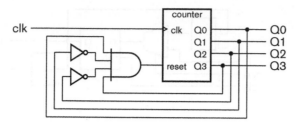

FIGURE 4.26   A decade counter.

EXAMPLE 4.10   Develop a VHDL model for the decade counter of Example 4.9.

SOLUTION   The entity and architecture are

```vhdl
library ieee;
use ieee.std_logic_1164.all, ieee.numeric_std.all;

entity decade_counter is
  port ( clk : in  std_logic;
         q   : out unsigned(3 downto 0) );
end entity decade_counter;

architecture rtl of decade_counter is
  signal count_value : unsigned(3 downto 0);
begin
  count : process (clk) is
  begin
    if rising_edge(clk) then
      if count_value = 9 then
        count_value <= "0000";
      else
        count_value <= count_value + 1;
      end if;
    end if;
  end process count;
  q <= count_value;
end architecture rtl;
```

We model the output port using type unsigned, since it is represents a binary-coded integer value. We need to declare an internal signal for the count value, since the process needs to read it in order to increment it. (Recall that VHDL does not allow us to read an output port, only assign to it.) On a rising clock edge, the process checks whether the counter has reached the terminal count value. If so, the count value wraps back to 0; otherwise, the process adds 1 to yield the new count value.

FIGURE 4.27 A down
counter with synchronous load.

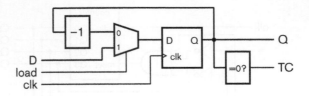

Another form of counter that is useful in timing applications is a
down counter with load. This counter is loaded with an input value, and
then decrements the count value. The terminal count output is activated
when the count value reaches zero. A circuit for the counter is shown in
Figure 4.27. It consists of a register whose input comes either from the
input value to be loaded or from the decremented count value. In this
case, the loading of input data is synchronous, since it occurs on a rising
clock edge.

If the clock input to the counter is a periodic signal with period t and
the counter is loaded with a value k, the terminal count is reached after
an interval of $k \times t$. Thus, this form of counter can be used as an *interval
timer*, where the terminal-count output signal is used to trigger an activity
after expiration of a given time interval.

EXAMPLE 4.11   Develop a VHDL model for an interval timer that has
clock, load and data input ports and a terminal-count output port. The timer
must be able to count intervals of up to 1000 clock cycles.

SOLUTION   The data input and counter need to be 10 bits wide, since
that is the minimum number of bits needed to represent 1000. The entity and
architecture are

```
library ieee;
use ieee.std_logic_1164.all, ieee.numeric_std.all;

entity interval_timer is
  port ( clk, load : in  std_logic;
         data       : in  unsigned(9 downto 0);
         tc         : out std_logic );
end entity interval_timer;

architecture rtl of interval_timer is
  signal count_value : unsigned(9 downto 0);
begin
  count : process (clk) is
```

(*continued*)

```
    begin
      if rising_edge(clk) then
        if load = '1' then
          count_value <= data;
        else
          count_value <= count_value - 1;
        end if;
      end if;
    end process count;
    tc <= '1' when count_value = 0 else '0';
end architecture rtl;
```

On a rising clock edge, the process uses the load input to determine whether to update the count value with the data input or the decremented count value. The decrement operation is performed using an unsigned subtraction without borrow out. So after reaching zero, the count value wraps back to the largest 10-bit value, namely, 1023. The final assignment in the architecture drives the terminal count to 1 when the count value reaches zero.

EXAMPLE 4.12   Modify the interval timer so that, when it reaches zero, it reloads the previously loaded value rather than wrapping around to the largest count value.

SOLUTION   We need to use a separate register to store the data value to load into the counter. When the load input is activated, a new data value is loaded into the storage register as well as into the counter. When the terminal count is reached, the counter should be loaded from the storage register. The inputs and outputs of the revised interval timer are the same, so we don't need to change the entity declaration. The revised architecture is

```
architecture repetitive of interval_timer is
  signal load_value, count_value : unsigned(9 downto 0);
begin
  count : process (clk) is
  begin
    if rising_edge(clk) then
      if load = '1' then
        load_value <= data;
        count_value <= data;
      elsif count_value = 0 then
        count_value <= load_value;
      else
        count_value <= count_value - 1;
      end if;
```

(continued)

```
        end if;
      end process count;
      tc <= '1' when count_value = 0 else '0';
    end architecture repetitive;
```

In this architecture, we have added a separate signal, load_value, to represent the storage register. The process is revised so that, when load is 1 on a rising clock edge, both the load_value signal and the count_value signal are updated from the data input. Also, when the count value is 0 on a rising clock edge (provided load is not 1), the count value is updated from the load_value signal. Otherwise, the count value is decremented as before.

The last kind of counter that we will describe in this section is a *ripple counter* (distinct from ripple carry used in an incrementer of a counter), shown in Figure 4.28. It is somewhat different in structure from the synchronous counters we have previously examined. Like those counters, it has a collection of flip-flops for storing the count value. However, unlike them, the clock signal is not connected in common to all of the flip-flop clock inputs. Rather, the clock input just triggers the flip-flop for the least significant bit, causing it to toggle between 0 and 1 on each rising clock edge. When the Q output changes to 0, the $\overline{Q}$ output changes to 1, triggering the next flip-flop to toggle between 0 and 1. This flip-flop behaves similarly, causing the third flip-flop to toggle when it (the second flip-flop) changes from 1 to 0. In general, we can think of the flip-flops for bits 0 to $i - 1$ as forming an i-bit counter. The most significant bit of this counter changes from 1 to 0 when it overflows. When that happens, the next flip-flop, for bit i, toggles between 0 and 1. This behavior is shown in the timing diagram of Figure 4.29.

An important timing issue arises from the fact that the flip-flops in a ripple counter are not all clocked together. Each flip-flop has a propagation delay between a rising edge occurring on its clock input and the outputs changing value. These propagation delays are shown in Figure 4.29.

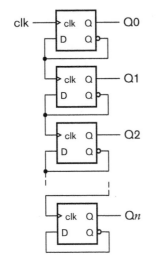

FIGURE 4.28 Structure of a ripple counter.

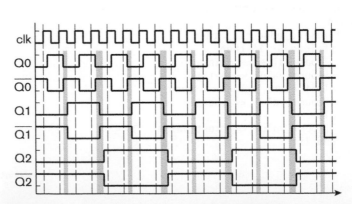

FIGURE 4.29 Timing diagram for a ripple counter.

Since each flip-flop is clocked from the output of the previous flip-flop, the propagation delays accumulate. The outputs of the counter don't all change at once on a change of the counter's clock input. Instead, the output changes "ripple" along the counter as they propagate through the flip-flops; hence, the name of this kind of counter. The shaded areas in the timing diagram show intervals where the count value is not correct, due to changes not having propagated completely through the counter. Whether this lack of synchronization among output changes is a problem or not depends on the particular application under consideration. Some factors to consider include:

▸ The length of the counter. For longer counters, there are more flip-flops through which changes have to propagate, making the maximum accumulated delay larger. For short counters, the delay may be acceptable.

▸ The period of the input clock relative to the propagation delays of the counter. For a short clock period, the accumulated delay may exceed the clock period. In that case, there will be clock cycles during which the counter outputs don't reach the correct value before the end of the cycle. For systems with long clock periods, the count value will settle early in the clock cycle.

▸ The tolerance for transient incorrect count values. If the count value may be sampled before it has settled, incorrect operation may result. However, if the count value is not sampled until it is guaranteed settled, operation is correct.

The main advantages of a ripple counter are that it uses much less circuitry in its implementation (since an incrementer is not required) and that it consumes less power. Hence, it is useful in those applications that are sensitive to area, cost and power and that have less stringent timing constraints. As an example, a digital alarm clock might use ripple counters to count the time, since changes occur infrequently relative to the propagation delay (seconds compared to nanoseconds).

1. Show in a diagram how an incrementer and a register can be connected to form a simple counter.

2. What is the maximum count value for an n-bit counter? What value does it then advance to?

3. How is a modulo k counter constructed?

4. What is a decade counter?

5. What is an interval timer?

6. Why might a long ripple counter be unsuitable for an application with a fast clock?

KNOWLEDGE
TEST QUIZ

4.3 SEQUENTIAL DATAPATHS AND CONTROL

We have now arrived at a key point in our discussion of digital logic design. We have seen how information can be binary encoded, how encoded information can be operated upon using combinational circuits, and how encoded information can be stored using registers. We have also seen that registers are needed both to avoid feedback loops in combinational circuits and to deal with data that arrives at the inputs sequentially. We have discussed counters as examples of combining registers and combinational circuits to perform sequential operations, that is, operations that proceed over a number of discrete intervals of time. We are now in a position to take a more general view of sequential operations. This general view will form the basis of our subsequent discussions of digital systems and embedded systems.

In many digital systems, the operations to be performed on input data are expressed as a combination of simpler operations, such as arithmetic operations and selection between alternative data values. Our general view of a digital system divides the circuit that implements the operations into a *datapath* and a *control section*. The datapath contains the combinational circuits that implement the basic operations and the registers that store intermediate results. The control section generates *control signals* that govern the operation of the datapath elements: selecting operations to be performed and enabling registers. In particular, the control section ensures that control signals are activated in the right order and at the right times to cause the datapath to perform the required operations on the data flowing through it. Hence, we say that the control section performs *control sequencing*. In many cases, the control section makes use of *status signals* generated by the datapath. The status signals indicate whether certain conditions of interest are true, for example, whether data has certain values, or whether input data is available. The values of the status signals can influence the control sequence.

One of the most challenging tasks in digital design is designing a datapath and corresponding control section to meet the given requirements and constraints. There are usually many alternative datapaths that could meet the functional requirements. Choosing among them usually involves trading off between area and performance.

EXAMPLE 4.13   Develop a datapath to perform a complex multiplication of two complex numbers. The operands and product are all in Cartesian form. The real and imaginary parts of the operands are represented as signed fixed-point numbers with 4 pre-binary-point and 12 post-binary-point bits. The real and imaginary parts of the product are similarly represented, but with 8 pre-binary-point and 24 post-binary-point bits. The complex multiplier is subject to constraints that strongly limit the circuit area.

SOLUTION   Given two complex numbers $a = a_r + ja_i$ and $b = b_r + bj_i$, the complex product is

$$p = ab = p_r + jp_i = (a_r b_r - a_i b_i) + j(a_r b_i + a_i b_r) \qquad (4.1)$$

This computation requires four fixed-point multiplications, one subtraction and one addition. If we were to implement the complex multiplier as a combinational circuit, separate components would be needed for each of these operations, consuming a large amount of circuit area. Since area is a strong constraint, we can reduce the area by using one multiplier to perform the four multiplications in sequence, and one adder/subtracter to form the real and imaginary parts of the product. We will need registers to store the intermediate results. The full computation will take place over several clock cycles.

A datapath to perform the sequential complex multiplication is shown in Figure 4.30. Since the multiplier is shared, multiplexers at the multiplier inputs are needed to select the operands. The result of a given multiplication is stored in one or other of the partial-product registers. To form the real part of the complex product, two partial products are subtracted by the adder/subtracter. To form the imaginary part, two partial products are added. In each case, the part of the complex product is stored in an output register.

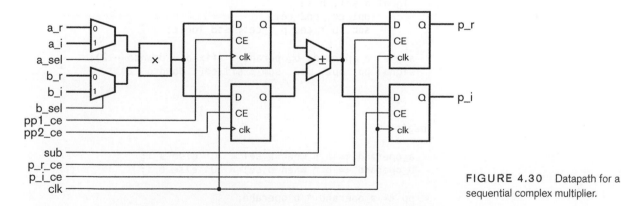

FIGURE 4.30 Datapath for a sequential complex multiplier.

In the diagram, the signals upon which data flows are drawn with thicker lines, since they carry multibit binary-coded values. The remaining signals, drawn with lighter weight lines, are the clock and the control signals. They include select signals for the multiplexers, clock-enable signals for the registers, and a signal to choose the operation to be performed by the adder/subtracter. The values of the control signals are driven by a separate control section, not shown on the diagram.

EXAMPLE 4.14   Develop a VHDL model of the complex multiplier datapath.

SOLUTION   We will start with the entity declaration. It includes ports for the data inputs and outputs, as well as clock and reset inputs and an input to indicate the arrival of new data. We will return to the last of these inputs later. The entity declaration is

```vhdl
library ieee;
use ieee.std_logic_1164.all, ieee.fixed_pkg.all;

entity multiplier is
  port ( clk, reset          : in std_logic;
         input_rdy           : in std_logic;
         a_r, a_i, b_r, b_i  : in sfixed(3 downto -12);
         p_r, p_i            : out sfixed(7 downto -24) );
end entity multiplier;
```

The architecture is

```vhdl
architecture rtl of multiplier is

  signal a_sel, b_sel,
         pp1_ce, pp2_ce,
         sub, p_r_ce, p_i_ce : std_logic;

  signal a_operand, b_operand : sfixed(3 downto -12);
  signal pp, pp1, pp2, sum : sfixed(7 downto -24);

  ...

begin

  a_operand <= a_r when a_sel = '0' else a_i;
  b_operand <= b_r when b_sel = '0' else b_i;

  pp <= a_operand * b_operand;

  pp1_reg : process (clk) is
  begin
    if rising_edge(clk) then
      if pp1_ce = '1' then
        pp1 <= pp;
    end if;
```

(continued)

```
      end if;
  end process pp1_reg;

  pp2_reg : process (clk) is
  begin
    if rising_edge(clk) then
      if pp2_ce = '1' then
        pp2 <= pp;
      end if;
    end if;
  end process pp2_reg;

  sum <= pp1 + pp2 when sub = '0' else pp1 - pp2;

  p_r_reg : process (clk) is
  begin
    if rising_edge(clk) then
      if p_r_ce = '1' then
        p_r <= sum;
      end if;
    end if;
  end process p_r_reg;

  p_i_reg : process (clk) is
  begin
    if rising_edge(clk) then
      if p_i_ce = '1' then
        p_i <= sum;
      end if;
    end if;
  end process p_i_reg;

  ...

end architecture rtl;
```

The signals declared before the begin keyword represent the control signals and the internal data connections. There are further signal declarations for the control section that we will return to later. In the statement part of the architecture, the assignments to a_operand and b_operand represent the multiplexers, and the assignment to pp represents the multiplier. The processes pp1_reg and pp2_reg represent the partial-product registers. The assignment to sum represents the adder/subtracter, and the processes p_r_reg and p_i_reg represent the output registers. We will return to further statements that represent the control section later.

EXAMPLE 4.15   Design a control sequence for the control signals of the sequential complex multiplier.

SOLUTION   We first need to determine a sequence of operations to be performed by the datapath to implement the required function expressed in Equation 4.1. There are many possible sequences, but we must ensure that there is no conflict for resources; that is, we must ensure that we don't try to use an element of the datapath for more than one operation at a time. One possible sequence, initiated by input_rdy being 1, is:

1. Multiply a_r and b_r, and store the result in partial product register 1.

2. Multiply a_i and b_i, and store the result in partial product register 2.

3. Subtract the partial product register values and store the result in the product real part register.

4. Multiply a_r and b_i, and store the result in partial product register 1.

5. Multiply a_i and b_r, and store the result in partial product register 2.

6. Add the partial product register values and store the result in the product imaginary part register.

This sequence would take six clock cycles to complete. In each cycle, only one of the arithmetic components is used, so there is no conflict for resources. However, we can reduce the number of cycles required, without creating conflict, by using the multiplier and the adder/subtracter concurrently. Specifically, we can merge steps 3 and 4 into one step, in which we subtract partial products to form the real part of the product and we multiply a_r and b_i to form a further partial product.

Given this 5-step sequence, the control signals that need to be activated in each step are shown in Table 4.1. The combination of control signal values in each step cause the datapath components to perform the required operations for that step. Note that in some steps, the multiplexers and adder/subtracter are not used. We don't care what values are driven for the control signals governing those components in those steps.

TABLE 4.1   Control sequence for the complex multiplier.

step	a_sel	b_sel	pp1_ce	pp2_ce	sub	p_r_ce	p_i_ce
1	0	0	1	0	–	0	0
2	1	1	0	1	–	0	0
3	0	1	1	0	1	1	0
4	1	0	0	1	–	0	0
5	–	–	0	0	0	0	1

4.3.1 FINITE-STATE MACHINES

Example 4.15 describes a control sequence for a sequential datapath, but we have yet to show how to design a circuit for the control section that generates the control sequence. We will introduce an abstraction called a *finite-state machine* for this purpose. There is a substantial body of mathematical theory underlying finite-state machines. Some of the useful results from this theory are implemented in CAD tools that transform finite-state machines to optimize sequential circuits. However, we will take a pragmatic approach, focusing on the design of control sections to sequence the operation of datapaths.

In general terms, a finite-state machine is defined by a set of *inputs*, a set of *outputs*, a set of *states*, a *transition function* that governs transitions between states, and an *output function*. The states are just abstract values that mark steps in a sequence of operations. The machine is called "finite-state" because the set of states is finite in size. The finite-state machine has a *current state* in a given clock cycle. The transition function determines the *next state* for the next clock cycle based on the current state and, possibly, the values of inputs in the given clock cycle. The output function determines the values of the outputs in a given clock cycle based on the current state and, possibly, the values of inputs in the given clock cycle.

Figure 4.31 shows a schematic representation of a finite-state machine. The register stores the current state in binary coded form. One of the states in the state set is designated the *initial state*. When the system is reset, the register is reset to the binary code for the initial state; thus, the finite-state machine assumes the initial state as its current state. During each clock cycle, the value of the next state is computed by the next state logic, which is a combinational circuit that implements the transition function. Also, the outputs are driven with the value computed by the output logic, which is a combinational circuit that implements the output function. The outputs are the control signals that govern operation of a datapath. On the rising clock edge marking the beginning of the next clock cycle, the current state is updated with the computed next-state value. The next state may be the same as the previous state, or it may be a different state.

Finite-state machines are often divided into two classes. In a *Mealy* finite-state machine, the output function depends on both the current state

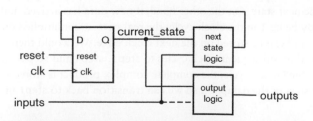

FIGURE 4.31 Circuit structure for a finite-state machine.

and the values of the inputs. In such a machine, the connection drawn with a dashed line in Figure 4.31 is present. If the input values change during a clock cycle, the output values may change as a consequence. In a *Moore* finite-state machine, on the other hand, the output function depends only on the current state, and not on the input values. The dashed connection in Figure 4.31 is absent in a Moore machine. If the input values change during a clock cycle, the outputs remain unchanged. In theory, for any Mealy machine, there is an equivalent Moore machine, and *vice versa*. However, in practice, one or the other kind of machine will be most appropriate. A Mealy machine may be able to implement a given control sequence with fewer states, but it may be harder to meet timing constraints, due to delays in arrival of inputs used to compute the next state. As we present examples of finite-state machines, we will identify whether they are Mealy or Moore machines.

In many finite-state machines, there is an idle state that indicates that the system is waiting to start a sequence of operations. When an input indicates that the sequence should start, the finite-state machine follows a sequence of states on successive clock cycles, with the output values controlling the operations in a datapath. Eventually, when the sequence of operations is complete, the finite-state machine returns to the idle state.

EXAMPLE 4.16   Design a finite-state machine to implement the control sequence for the complex multiplier described in Example 4.15. The control sequence is initiated by input_rdy being 1 during the clock cycle in which new data arrives at the datapath inputs.

SOLUTION   Our finite-state machine needs five states, one for each of the steps of the control sequence. Let's call them step1 through step5. We also need to deal with the case of waiting for input data to arrive. We could consider a separate idle state for that case. When, in the idle state, input_rdy is 1, we would then transition to state1 to start the multiplication; otherwise, we would stay in the idle state. The problem with this is that it wastes a clock cycle, since we would not perform the first multiplication until after the cycle in which data arrived.

The alternative is to use step1 as the idle state. If it turns out that new data has not arrived in a given clock cycle while in this state, we simply repeat step1 as the next state. On the other hand, if new data has arrived, indicated by input_rdy being 1 in the clock cycle, the real parts are multiplied during that clock cycle and can be stored on the next clock edge. We would then transition to step2, and on subsequent clock cycles to step3, step4 and step5. At the end of the step5 clock cycle, the complete complex product is stored in the output registers of the datapath, so we can transition back to step1 in the next clock cycle.

In summary, our finite-state machine has the signal input_rdy as its single input, and the control signals listed in Example 4.15 as outputs. The state set is {step1, step2, step3, step4, step5}, with step1 being the initial state. The transition function is defined in Table 4.2. The output function is defined in Table 4.1. Since the output function depends only on the current state and not on the input value, this finite-state machine is a Moore machine.

current_ state	input_ rdy	next_ state
step1	0	step1
step1	1	step2
step2	–	step3
step3	–	step4
step4	–	step5
step5	–	step5

TABLE 4.2 The transition function for the complex multiplier finite-state machine.

An important issue to consider when designing a finite-state machine is how to encode the state values. We glossed over that in Example 4.16 by treating the states as abstract values. As we discussed in Chapter 2, if we have N states, we need at least $\lceil \log_2 N \rceil$ bits in our code. However, we may choose to have more if that simplifies circuitry that uses encoded states. In particular, while a longer than minimal code length requires more flip-flops in the state register and more wires for the state signals, it may make the next-state and output logic circuits simpler and smaller. In general choosing an optimal state encoding is a complex mathematical problem. However, synthesis CAD tools incorporate methods for choosing a state encoding, so we may be able to let a tool make the choice for us. One aspect of state encoding is the choice of a code word to represent the initial state. In many cases, a good choice is a code word with all 0 bits, since that allows us to use a simple register with reset for the state register. If some other code word is chosen for the initial state, that code word must be loaded into the register on system reset.

Modeling Finite-State Machines in VHDL

Since a finite-state machine is composed of a register, next-state logic and output logic, a straightforward way to model a finite-state machine is to use the VHDL features that we already know for modeling registers and combinational logic. The only aspect we have not addressed is how to represent the state set, particularly when we want to take an abstract view and leave state encoding to the synthesis tool. Fortunately, VHDL provides a way of specifying a set of abstract values without implying any encoding. We can define an *enumeration type* that just defines a set of values. For example, we can define an enumeration type for the states in Example 4.16 as follows:

```
type multiplier_state is
  (step1, step2, step3, step4, step5);
```

This is a user-defined type, as opposed to the standard types such as integer and std_logic. The type definition specifies that multiplier_state is a set of five abstract values, denoted by the names step1 through step5.

We can declare a signal to represent the current state of a state machine using this type as follows:

```
signal current_state : multiplier_state;
```

This specifies that current_state is a signal that takes on values from the set denoted by multiplier_state. So, for example, we could make the following assignment in a process:

```
current_state <= step4;
```

to assign the value step4 to the signal.

EXAMPLE 4.17   Develop a VHDL model of the finite-state machine in Example 4.16.

SOLUTION   We will augment the architecture declaration of Example 4.14 with the VHDL representation of the control section. The additional declarations of a type for the set of states and signals for the current and next state are

```
type multiplier_state is (step1, step2, step3, step4, step5);
signal current_state, next_state : multiplier_state;
```

The additional statements added to the architecture are

```
state_reg : process (clk, reset) is
begin
  if reset = '1' then
    current_state <= step1;
  elsif rising_edge(clk) then
    current_state <= next_state;
  end if;
end process state_reg;

next_state_logic : process (current_state, input_rdy) is
begin
  case current_state is
    when step1 =>
      if input_rdy = '0' then
```

(continued)

```
              next_state <= step1;
          else
              next_state <= step2;
          end if;
      when step2 =>
          next_state <= step3;
      when step3 =>
          next_state <= step4;
      when step4 =>
          next_state <= step5;
      when step5 =>
          next_state <= step1;
    end case;
  end process next_state_logic;

  output_logic : process (current_state) is
  begin
    a_sel <= '0'; b_sel <= '0';  pp1_ce <= '0'; pp2_ce <= '0';
    sub <= '0';    p_r_ce <= '0'; p_i_ce <= '0';
    case current_state is
      when step1 =>
        pp1_ce <= '1';
      when step2 =>
        a_sel <= '1'; b_sel <= '1'; pp2_ce <= '1';
      when step3 =>
        b_sel <= '1'; pp1_ce <= '1'; sub <= '1'; p_r_ce <= '1';
      when step4 =>
        a_sel <= '1'; pp2_ce <= '1';
      when step5 =>
        p_i_ce <= '1';
    end case;
  end process output_logic;
```

The process state_reg models the state storage for the finite-state machine. It is based on the template for a register with asynchronous reset. When the reset input is active, the process resets the current state to the initial state, step1. Otherwise, on a rising clock edge, the process updates the current state with the computed next state.

The next state is computed by the process next_state_logic, which models the transition function of Table 4.2. Since the next state depends on the value of current_state and input_rdy, these signals are included in the sensitivity list of the process. The statement inside the process is a *case statement*. It uses the value of the current_state signal to choose among alternatives for updating next_state. The alternative for step1 uses a nested if statement to determine whether to proceed to step2 or stay in step1, depending on the value of input_rdy. All other alternatives simply advance the state unconditionally.

The output values are computed by the process output_logic, which models the output function of Table 4.1. Since this is a Moore machine, the outputs depend only on current_state, so that is the only signal included in the sensitivity list. This process also includes a case statement that chooses alternatives for assigning values to the outputs depending on the value of current_state. Rather than including an assignment for every output in each alternative of the case statement, we precede the case statement with a default assignment of '0' for each output, and only include overriding assignments of '1' in those alternatives where they are required. This style for modeling the output function usually makes the process more succinct, and helps to avoid inadvertent introduction of latches due to omission of an output assignment in an alternative.

State Transition Diagrams

A *state transition diagram* is an abstract diagrammatic representation of a finite-state machine. It uses a circle, or "bubble," to represent each state. Directed arcs between state bubbles represent transitions from one state to another. An arc may be labeled with a combination of input values that allow the transition to occur. To illustrate, Figure 4.32 shows a state transition diagram for a finite-state machine with states s1, s2 and s3. Each arc is labeled with the values of two inputs, a_1 and a_2, that are required for the transition. Thus, when the finite-state machine is in state s1 and the inputs are both 1, the state of the machine in the next clock cycles is s3. If the machine is in state s1 and both inputs are 0, the machine stays in state s1. From state s1, if the inputs are 0 and 1, or 1 and 0, the machine transitions to state s2. Note that we have omitted a label on the arc from s2 to s3. This is a common convention to indicate an unconditional transition; that is, when the machine is in state s2, the next state is s3 regardless of the input values. Another important point is that all possible combinations of input values are accounted for in each state, and that no combination is repeated on more than one arc from a given state.

A bubble diagram may also be labeled with the values of outputs. Since Moore-machine outputs depend only on the current state, we attach the labels for such outputs to the state bubbles. This is shown on the augmented bubble diagram in Figure 4.33. For each state, we list the values of two Moore-style outputs, x_1 and x_2, in that order.

Mealy-machine outputs, on the other hand, depend on both the current state and the current input values. Usually, the input conditions are the same as those that determine the next state, so we usually attach Mealy-output labels to the arcs. This does not imply that the outputs change at the time of the transition, only that the output values are driven when the current state is the source state of the arc and the input values are those of the arc label. If the inputs change while in the source state, the outputs change to those listed on some other arc labeled with

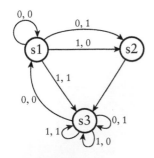

FIGURE 4.32   A state transition diagram.

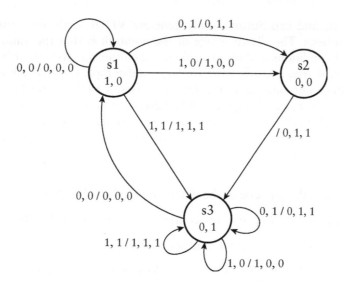

FIGURE 4.33 A state transition diagram augmented with Moore- and Mealy-style output values.

the new input values. Mealy-style outputs are also shown on the arcs in Figure 4.33. In each case, the output values are listed after the "/" in the order y_1, y_2 and y_3.

EXAMPLE 4.18 Draw a state transition diagram for the finite-state machine of Example 4.16. Include the output values in the order of their occurrence in Table 4.1.

SOLUTION The diagram is shown in Figure 4.34. There is a transition from step1 to step2 that occurs when input_rdy is 1, and a transition from step1 back to itself when input_rdy is 0. All other transitions are unconditional. Since it is a Moore machine, the output values are all drawn in the state bubbles.

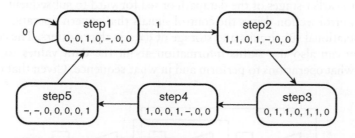

FIGURE 4.34 State transition diagram for the complex multiplier.

In many applications, a state transition diagram is a useful notation, since it graphically conveys the control organization of a sequential design. Many CAD tools provide graphical editors for entering state transition diagrams, and can automatically generate VHDL code for simulation and synthesis. The disadvantage of the notation is that the annotations of input conditions and output values can clutter the diagram, obscuring the control organization. Also, for large and complex state machines, the diagram can become unwieldy. In those cases, a VHDL model in textual form may be more intelligible. Ultimately, since state transition diagrams and VHDL models of state machines encapsulate the same information, it is a question of personal preference or project guidelines that determine the method to use.

KNOWLEDGE TEST QUIZ

1. What is the purpose of the datapath in a digital system?

2. What is the purpose of the control section in a digital system?

3. What are control signals and status signals?

4. What is the distinction between a Moore and a Mealy finite-state machine?

5. Write a VHDL type declaration for the set of states s0, s1, s2 and s3.

6. In a state transition diagram, where are labels written for Mealy-style outputs and for Moore-style outputs?

4.4 CLOCKED SYNCHRONOUS TIMING METHODOLOGY

We now have a general view of a digital system, shown in Figure 4.35. It comprises a datapath that stores and transforms binary-coded information and a control section that sequences operations within the datapath. The datapath, in turn, includes combinational subcircuits that perform operations on the data and registers that store the data. Stored data can be fed back to earlier stages of the datapath or fed forward to subsequent stages. The control section drives the control signals that govern operation of the combinational subcircuits and storage of data in the registers. The control section can also use status information about the data values to determine what operations to perform and in what sequence. Given that data is

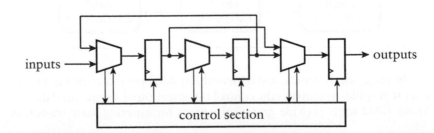

FIGURE 4.35 A general view of a digital system.

transferred between registers through combinational subcircuits, this view of a system is often called a *register-transfer level* (RTL) view. The word "level" refers to the level of abstraction. Register-transfer level is more abstract than a gate-level view, but less abstract than an algorithmic view.

In Chapter 1, we identified division of time into discrete intervals as a key abstraction for managing the complexity of timing in digital systems. We also described some of the specific timing characteristics of flip-flops (and hence registers) over which the discrete-timing approach abstracts. Now that we have seen some more complex digital systems, we can begin to see the value of the discrete-timing abstraction. It is based on driving all of the registers shown in Figure 4.35 with a common periodic clock signal. We say that the registers are all clocked *synchronously* on each rising clock edge. The combinational subcircuits perform their operations in the interval between one clock edge and the next, called a *clock cycle*. This *clocked synchronous timing methodology* helps us ensure that operations are completed by combinational subcircuits by the time their results are needed, and simplifies composition of large systems from smaller subsystems.

Since registers are composed of flip-flops connected in parallel, we can derive the timing characteristics of registers from those of flip-flops. We will make the simplifying assumption that all of the flip-flops in a given register have the same timing characteristics, or that any differences are negligible. We can thus identify the setup time (t_{su}), hold time (t_h) and clock-to-output delay (t_{co}) of a register as being the same as those characteristics of the constituent flip-flops. All of the bits of data to be stored in a register must be stable at the input for at least the setup time before a clock edge and for at least the hold time after the clock edge. We can only guarantee that all bits of the stored data will be available at the output after the clock-to-output delay following the clock edge.

These considerations lead us to the register-to-register timing for a path in the system shown in Figure 4.36. Q1 is the output of one register that feeds into a combinational subcircuit. D2 is the output of the subcircuit, feeding into the next register. The timing parameters are illustrated in Figure 4.37. After a clock rising edge, Q1 changes to the new stored

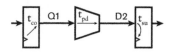

FIGURE 4.36   A register-to-register path.

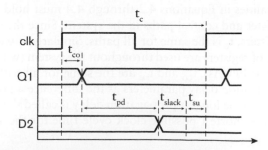

FIGURE 4.37   Register-to-register timing.

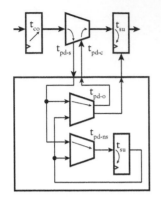

FIGURE 4.38
Control path in a digital system.

value and stabilizes by the end of the interval t_{co}. The new value then propagates through the combinational subcircuit, stabilizing at the output D2 by the end of the interval t_{pd}, the propagation delay of the subcircuit. The value on D2 must be stable at least t_{su} before the next clock edge, so there is a slack period, t_{slack}, where nothing changes. The diagram shows that the sum of these intervals must be equal to the clock cycle time, t_c. Alternatively, we can express this as an inequality:

$$t_{co} + t_{pd} + t_{su} < t_c \qquad (4.2)$$

Another important path in the digital system is the control path shown in Figure 4.38. At the top of the figure is a register-to-register section of the datapath, and at the bottom is the finite-state machine in the control section. The status signals driven by the combinational subcircuit are inputs to the output logic and next-state logic in the control section. The control signals driven by the output logic govern the operation of the combinational subcircuit and the target register. (In general, status signals from one combinational subcircuit would influence operation of some other combinational subcircuit, but the same timing considerations apply.) Our timing analysis for these control paths is similar to that for the register-to-register datapath. We simply aggregate the combinational propagation delays through the combinational subcircuit and output logic to derive the inequality:

$$t_{co} + t_{pd\text{-}s} + t_{pd\text{-}o} + t_{pd\text{-}c} + t_{su} < t_c \qquad (4.3)$$

Here, t_{pd-s} is the propagation delay through the combinational subcircuit to drive the status signals, t_{pd-o} is the propagation delay through the output logic to drive the control signals, and t_{pd-c} is the propagation delay through the combinational subcircuit for a change in the control signal to affect the output data. For a Moore-style control signal that does not depend on a status input, we can ignore the parameter t_{pd-s} in this inequality. In a similar way, we can derive the following inequality for the path that generates the next-state value:

$$t_{co} + t_{pd\text{-}s} + t_{pd\text{-}ns} + t_{su} < t_c \qquad (4.4)$$

where t_{pd-ns} is the propagation delay through the next-state logic.

The inequalities in Equations 4.2 through 4.4 must hold for all of the register-to-register and control paths in the system. Since the clock is common to all registers, t_c is the same for all paths. Similarly, if we assume that the same kinds of registers are used throughout the system (which is the case in fabrics such as FPGAs), t_{co} and t_{su} are the same for all paths. That only leaves the propagation delay parameters as the difference among paths.

The path with the longest propagation delay is called the *critical path*. It determines the shortest possible clock cycle time for the system. Since all operations are performed in times determined by the clock, the critical path determines the overall system performance. Hence, if we need to address performance issues, we need to identify which combinational sub-circuit is on the critical path and attempt to reduce its delay. In most systems, the critical path will be a register-to-register path in the datapath of the system. For example, if there is such a path that performs an arithmetic operation or that includes a counter, the carry chain may be the critical path. Alternatively, if a system uses a Mealy finite-state machine and a control path corresponding to Equation 4.3 is on the critical path, it may be possible to use an equivalent Moore machine to avoid the status-signal delay in the control path. Of course, once the delay on the critical path is reduced below that of another path, that other path becomes the critical path. Hence, attention may need to be paid to several paths in a system to address performance issues.

Depending on the requirements and constraints for the system, we can interpret Equations 4.2 through 4.4 in two ways. One interpretation involves treating the propagation delays as independent parameters and determining the resulting minimum clock period. The system can then be operated with any clock period greater than the minimum. This interpretation is appropriate for systems where high performance is not a requirement.

The other interpretation involves treating the clock cycle time as the independent parameter and determining the propagation delays from it. We might be given a target clock cycle time by a system architect or our marketing department and be asked to design the system to meet that target. In that case, the inequalities place constraints on the propagation delays through the combinational data and control paths. If we meet the constraints with plenty of slack, we might try to optimize the design to reduce cost, for example, by using subcircuits with less area. If we don't meet the constraints, we need to focus attention on the critical path or paths to reduce their delay. It may be that we have designed the system with too much computation to be performed in one or more combinational subcircuits to allow sufficient reduction of the critical path propagation delay. In that case, we could divide the computation into a number of smaller steps that can be done sequentially or in parallel. The combinational subcircuits for the simpler steps should have smaller propagation delay than the original. Thus, even if more steps are required overall to perform the system's operation, the fact that the clock cycle time is reduced may allow us to meet our performance target.

EXAMPLE 4.19   Suppose we have designed a system that includes a multiplication operation on 16-bit unsigned binary-coded integers. The system is required to operate at 50 MHz (a clock cycle time of 20 ns). We have included a combinational multiplier to perform the multiplication, but its propagation delay is 35 ns. All other data and control paths have plenty of slack with the 20 ns clock cycle time. The result of the multiplication is not needed until 20 cycles after the operands are available. Describe how use of the sequential multiplier of Example 4.4 could help us meet our timing requirement.

SOLUTION   The sequential multiplier performs the multiplication operation in 17 steps with one adder. In the first step, we store the operands and reset the output register to zero. Then on each of the 16 subsequent steps, we add the partial products. Each step involves only an AND operation and an addition. Thus, the combinational subcircuit between the operand registers and the product output registers will have significantly smaller propagation delay than the 35 ns delay of the full combinational multiplier. This reduction should allow the clock period to be reduced to meet the timing constraint.

Further timing considerations arise from the way the clock signal is connected to all of the registers in a circuit. Suppose, in a register-to-register path, the clock signal to the target register is connected via a long wire with significant delay, as shown in Figure 4.39. A rising clock edge arrives at the source register earlier than at the target register. This phenomenon is called *clock skew*. If the propagation delay through the combinational subcircuit is small (for example, if the subcircuit is just a direct connection to the target register with negligible delay), the value from the previous cycle may not remain stable for the hold time after the clock edge, as shown in Figure 4.40. In most implementation fabrics, the hold time is very small, or may even be negative, thus reducing the likelihood of this problem. (A negative hold time simply means that the data may start changing before the clock edge.) However, if we don't take care to minimize clock skew in a design, the circuit may operate unreliably. Given the importance of minimizing skew across the clock connection network, together with the need for buffers to drive the large number of flip-flop clock inputs as described in Section 2.1.1, we usually leave implementation of the clock signal to CAD tools. As part of the physical design, a tool will insert clock buffers into the circuit and route the connections so as to minimize skew. In FPGA fabrics, dedicated buffer and wiring resources for clock distribution are built into the chip.

The timing parameters and constraints that we have considered so far apply to the datapath and control section within an integrated circuit chip. When we use that chip as a component of a larger system, we also need to take account of the effect of the input and output pins that connect the chip to other components via wires on a printed circuit board. The inputs have internal buffers that protect the chip from excessive voltage swings and static discharge, and the outputs have buffers to drive the relatively large capacitances and inductances that occur outside the chip. These buffers, together with the associated wiring connecting the integrated

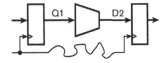

FIGURE 4.39   A register-to-register path with clock skew.

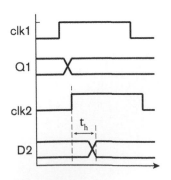

FIGURE 4.40   A timing problem arising from clock skew.

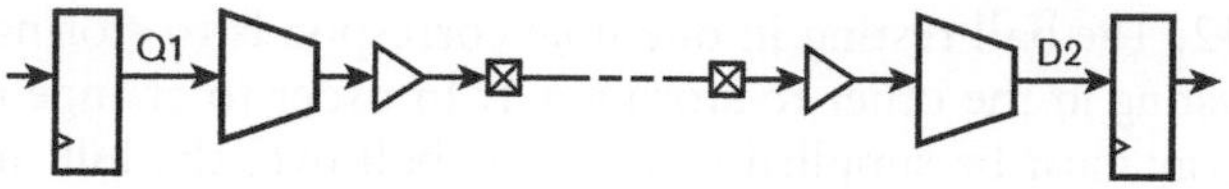

FIGURE 4.41
A register-to-register path
between chips.

circuit chip to the package pins, introduce propagation delays. So when we analyze the timing behavior of the complete system, we need to include the pin and wiring delays. We can apply the same path-based analysis that we used for internal paths. Figure 4.41 shows a register-to-register path between a source register on one chip and a target register on another. The path includes output combinational logic, the output buffer and pin, the printed-circuit-board wiring, the input pin and buffer, and input combinational logic. The sum of the propagation delays plus the register clock-to-output and setup times must be less than the system's clock cycle time. For high-speed systems, this can be a difficult constraint to meet. In such systems, we usually avoid having any combinational input or output logic. An input that connects directly to an input register is often called a *registered input*, and an output that is driven directly from an output register is called a *registered output*. High-speed design methodologies often require registered inputs, registered outputs or a combination of both. Using both allows a whole clock cycle for inter-chip transmission.

4.4.1 ASYNCHRONOUS INPUTS

Our clocked synchronous timing methodology requires us to ensure that inputs to registers are stable during an interval around each clock edge. For those signals that are generated within the circuit, we can ensure that we meet this constraint. However, most circuits must deal with some inputs that are generated externally, either by transducers whose outputs represent real-world quantities or events, or by separate systems that do not share a common clock. We call such signals *asynchronous inputs*. We have no control over the times at which they change value; hence, we cannot guarantee that they meet our timing constraints for register inputs.

Before we describe how to deal with asynchronous inputs, let's examine the behavior of a register, or more specifically, a flip-flop, when its input can change at any time. A flip-flop circuit internally uses a combination of charge storage and positive feedback to store a 0 or a 1 value. Figure 4.17 on page 172 gives a general idea of how this might work in a latch. A D flip-flop circuit elaborates on this structure to make storage edge-triggered. In order to change from storing a 0 to storing a 1, or *vice versa*, some energy input is required. A common analogy is to consider a ball resting in one of two holes, with a hill in between, as shown in

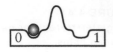

FIGURE 4.42   An analogy for the behavior of a flip-flop.

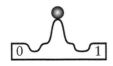

FIGURE 4.43   An analogy for the behavior of a flip-flop.

Figure 4.42. The ball resting in one hole corresponds to storing a 0, and the ball resting in the other to storing a 1. In order to change the stored value, energy must be supplied to push the ball over the hill. In the case of a D flip-flop, a pulse of energy is sampled from the D input when the clock rises. If the input is 0, the ball is pushed toward the 0 hole, and if the input is 1, the ball is pushed toward the 1 hole.

Now if the input changes close to the time the clock rises, insufficient energy may be sampled. For example, if the ball is in the 0 hole and the input changes to 1, there may be insufficient energy to push the ball to the 1 hole. The ball may get close to the top of the hill then fall back again. This corresponds to the flip-flop output starting to change from 0 to 1, but then reverting to 0. A particularly significant case arises if there is just sufficient energy to push the ball to the top of the hill, as shown in Figure 4.43, but not to push it straight over. In that case, the ball teeters on the top for some time before falling one way or the other. The time for which it teeters and the direction in which it falls are unpredictable. This condition is called *metastability*. The behavior of a real flip-flop in a *metastable state* depends on the details of the internal electrical and physical design of the flip-flop. Some flip-flops may delay a change between 0 and 1, some may oscillate, and others may have an invalid logic level at the output for some time. The problem is not so much the indeterminate behavior of the flip-flop output while the metastable state persists, but the fact that the delay until the output is stable is not bounded. As a consequence, we can't guarantee that the timing constraints for the circuits connected to the flip-flop output will be met.

Mathematical models of flip-flop behavior can be developed to help us understand how asynchronous inputs affect circuit operation. The details of these models are beyond the scope of this book, so we just summarize the conclusions here. Suppose an asynchronous input changes with a frequency of f_1 and the clock frequency of the system is f_2. We sample the output value of the flip-flop to which the asynchronous input is connected after a period t. Occasionally, the sampled value will be incorrect due to metastability in the flip-flop, and that will cause some form of failure. The mathematical model gives us the mean time between failures (MTBF):

$$\mathrm{MTBF} = \frac{e^{k_2 t}}{k_1 f_1 f_2} \qquad (4.5)$$

The constants k_1 and k_2 are measured for a particular flip-flop. Since the MTBF is inversely proportional to the frequencies, higher frequencies lead to shorter MTBF, that is, to more frequent failure. More significant, however, is that the MTBF is nonlinearly related to the time before sampling. The value of k_2 is typically large and positive, so a small increase in the time before sampling yields a significant increase in the MTBF.

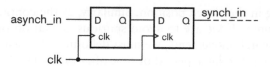

FIGURE 4.44   A synchronizer for an asynchronous input.

The usual approach to dealing with asynchronous inputs is to connect them to a *synchronizer*, and to use the output of the synchronizer in the rest of the system. A simple synchronizer is shown in Figure 4.44. The first flip-flop samples the value of the asynchronous input at each clock edge. Usually, the value is passed on to the flip-flop's output within the clock-to-output delay of the flip-flop and sampled on the next clock edge by the second flip-flop. The output of the second flip-flop is used in the rest of the system. On those occasions where the asynchronous input changes close to a clock edge, the first flip-flop may enter the metastable state. However, its output is not sampled for an entire clock cycle, giving the flip-flop time to resolve the metastability. In terms of Equation 4.5, the sampling interval t is one clock cycle period, t_c.

It is only in fairly recent times that component manufacturers have developed a complete understanding of metastability and its effects on system reliability. Earlier than 15 years or so ago, published data on the metastability characteristics of flip-flops was hard to find. Since then, manufacturers have improved both their device behavior and their published data. For most applications using modern implementation fabrics, the simple synchronizer shown in Figure 4.44 is sufficient to give a MTBF that is much longer than the lifetime of the system. However, for those applications in which reliability is a key requirement and that have many asynchronous inputs, we should study the published data for implementation fabric we use and follow the manufacturer's advice on synchronizing inputs.

Switch Inputs and Debouncing

We mentioned that externally generated signals are often asynchronous inputs to a system. A common example is connection of switches that form a user interface to the system. This includes push-button, slider, toggle and rotary switches. A user can change a switch position at random times, so we cannot assume synchronization with a clock signal. Similarly, a microswitch used to sense mechanical input may change asynchronously. There is a further problem that we must also deal with. Switches are electromechanical devices containing electrical contacts that open and close a circuit in response to mechanical movement. As the contacts close, they *bounce*, causing the circuit to open and close one or more times before finally setting in the closed position. Similarly, as the contacts open, they may also bounce. If we are to avoid spurious activation of the system's response to switch movements, we must debounce the switch input. This involves waiting for some period of time after an initial change in circuit closure is detected before treating the switch input as valid. For most switches, the time taken to settle is of the order of a few millisecond, so a debounce delay of up to 10ms is common practice. Delaying too long causes the user to notice the lag in response to switch activation. A response time of less than 50ms is generally imperceptible.

There are probably as many solutions to switch debouncing as there are design engineers. One simple approach is shown in Figure 4.45. It

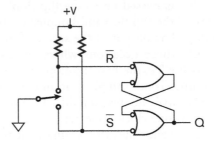

FIGURE 4.45 A switch debouncer using an RS-latch.

uses an RS-latch with negative-logic inputs and a double-throw switch. When the switch is in the position shown, it holds the reset input of the latch active, producing a 0 at the Q output. When the switch is toggled, we assume that one contact is opened before the other contact is closed. (This is sometimes called "break before make.") Bouncing on the contact to be opened simply leaves the latch in the reset state. When the first bounce occurs on the contact to be closed, the set input is activated, causing the Q output to change to 1. Subsequent bounces leave the latch in the set state. The behavior is similar when the switch is toggled in the other direction.

While this approach is very effective, it has two drawbacks. First, it requires two inputs to the digital system for what is really just one input. Second, it requires a double-throw switch, whereas many low-cost applications require a single-throw switch consisting of two contacts that are shorted together by a push button. Simple circuits for debouncing single-throw switches generally rely on analog circuit design techniques and require components external to the main digital chip. We will not discuss them here, but refer to Section 4.6, Further Reading. Instead, we will outline a fully digital approach to debouncing that can be designed into the main digital circuit of a system.

A simple way of connecting a single-throw or momentary-contact switch to a digital circuit input is shown in Figure 4.46. When the switch is open, the input is pulled to 1, and when the switch is closed, the input

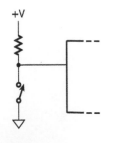

FIGURE 4.46 Simple switch input connection.

is pulled to 0. A change of switch position causes the input to toggle between 0 and 1 until the bouncing stops and the input settles at its final value. Rather than using the input value directly within the system, we sample it at intervals longer than the bounce time. When we get two successive samples that have the same value, we use that value as the stable state of the switch input.

EXAMPLE 4.20   Develop a VHDL model of a debouncer for a pushbutton switch that uses a debounce interval of 10ms. Assume the system clock frequency is 50MHz.

SOLUTION   The entity declaration and architecture are

```
library ieee; use ieee.std_logic_1164.all;
entity debouncer is
  port ( clk, reset    : in  std_logic;
         pb            : in  std_logic;
         pb_debounced  : out std_logic );
end entity debouncer;

architecture rtl of debouncer is
  signal count500000 : integer range 0 to 499999;
  signal clk_100Hz : std_logic;
  signal pb_sampled    : std_logic;
begin
  div_100Hz : process (clk, reset) is
  begin
    if reset = '1' then
      count500000 <= 499999;
    elsif rising_edge(clk) then
      if clk_100Hz = '1' then
        count500000 <= 499999;
      else
        count500000 <= count500000 - 1;
      end if;
    end if;
  end process div_100Hz;

  clk_100Hz <= '1' when count500000 = 0 else '0';

  debounce_pb : process (clk) is
  begin
    if rising_edge(clk) then
      if clk_100Hz = '1' then
        if pb = pb_sampled then
          pb_debounced <= pb;
```

(continued)

```
        end if;
      pb_sampled <= pb;
    end if;
   end if;
  end process debounce_pb;
 end architecture rtl;
```

The process div_100 Hz represents a down counter that divides the clock by 500,000. The assignment following the process decodes the terminal count to derive a sampling clock that pulses to '1' every 10ms. When the sampling clock is '1', the debounce_pd process compares the current push-button input value (pb) with a previously sampled value (pb_sampled). If they are the same, the process updates the debounced output with the current value. If they are not the same, the output is unchanged. Also, when the sampling clock is '1', the process updates the sampled value with the current value.

It is important to note that, even though the debouncer of Example 4.20 uses much more circuitry than the simple debouncer of Figure 4.45, it will probably be cheaper to implement. It uses a simple single-throw switch and only a single resistor external to the integrated circuit, and only requires one input pin. The saving in packaging resources and printed circuit board assembly costs would be more significant in a large-volume application than the expense of additional circuit resources used within the integrated circuit. We might also consider implementing the debounce operation in software run on an embedded processor, if the application requires a processor to be included anyway. If the processor has sufficient time in its task schedule to perform debouncing, that might be a more efficient use of resources. The lesson to learn is that, when we make these trade-off decisions, we must consider all of the costs and resources for the entire system, not just for one aspect in isolation.

4.4.2 VERIFICATION OF SEQUENTIAL CIRCUITS

Now that we have described the design of clocked sequential circuits and the timing constraints that apply, we can return to the verification steps outlined in the design methodology in Section 1.5. We need to consider functional verification (that the sequential circuit performs its function correctly) and timing verification (that the circuit meets timing constraints). We outlined in Section 1.5 how tools perform static timing analysis to verify timing constraints. Here, we will discuss functional verification using VHDL models, expanding on the ideas introduced in Section 2.4 relating to verification of combinational circuits.

When verifying a combinational circuit, we saw that we need to wait for some time after applying a test case to the circuit's inputs before checking the circuit's outputs, to allow for the propagation delay of the circuit. Similarly, when verifying a sequential circuit, we need to take account of the fact that operations take one or more clock cycles to complete. We need to ensure that the process that checks the output is synchronized with the stimulus process, and knows how many clock cycles after application of a test case to wait before checking the output. If all operations complete in the same number of cycles, and only one operation takes place at a time, this is relatively straightforward. On the other hand, if operations take varying numbers of cycles to complete, the checker needs to check both that the operation completes at the correct time and that the correct result is produced. If multiple operations can take place concurrently, for example, if the datapath is a pipeline, the checker needs to ensure that all operations that start also complete, and that no spurious results are produced.

Developing testbench models for complex sequential circuits can itself become a complex endeavor. We will discuss some of the general techniques that can be used in Chapter 10. For now, we will illustrate a simulation-based approach for verifying circuits that we introduced in previous examples.

EXAMPLE 4.21   Develop a testbench model for the sequential multiplier of Example 4.14. Verify that the result computed by the multiplier is the same (within the limits of the precision of the operands) as that produced using multiplication with type complex defined in the IEEE standard package math_complex.

SOLUTION   The testbench has no external connections, and so the entity declaration is

```
entity multiplier_testbench is
end entity multiplier_testbench;
```

The architecture is

```
library ieee;
use ieee.std_logic_1164.all, ieee.fixed_pkg.all,
  ieee.math_complex.all;

architecture verify of multiplier_testbench is
  constant t_c : time : = 50 ns;
```

(continued)

```vhdl
  signal clk, reset : std_logic;
  signal input_rdy  : std_logic;
  signal a_r, a_i, b_r, b_i   : sfixed(3 downto -12);
  signal p_r, p_i   : sfixed(7 downto -24);

  signal a, b : complex;

begin

  duv : entity work.multiplier(rtl)
    port map ( clk => clk, reset => reset,
               input_rdy => input_rdy,
               a_r => a_r, a_i => a_i,
               b_r => b_r, b_i => b_i,
               p_r => p_r, p_i => p_i );

  clk_gen : process is
  begin
    wait for t_c / 2; clk <= '1';
    wait for t_c - t_c / 2; clk <= '0';
  end process clk_gen;

  reset <= '1', '0' after 2 * t_c;

  apply_test_cases : process is

    procedure apply_test ( a_test, b_test : in complex ) is
    begin
      a <= a_test; b <= b_test; input_rdy <= '1';
      wait until falling_edge(clk); input_rdy <= '0';
      for i in 1 to 5 loop
        wait until falling_edge(clk);
      end loop;
    end procedure apply_test;

  begin
    wait until falling_edge(clk) and reset = '0';
    apply_test(cmplx(0.0, 0.0), cmplx(1.0, 2.0));
    apply_test(cmplx(1.0, 1.0), cmplx(1.0, 1.0));
    -- further test cases ...
    wait;
  end process apply_test_cases;

  a_r <= to_sfixed(a.re, a_r'left, a_r'right);
  a_i <= to_sfixed(a.im, a_i'left, a_i'right);
  b_r <= to_sfixed(b.re, b_r'left, b_r'right);
  b_i <= to_sfixed(b.im, b_i'left, b_i'right);

  check_outputs : process is
    variable p : complex;
```

(continued)

```
  begin
    wait until rising_edge(clk) and input_rdy = '1';
    p := a * b;
    for i in 1 to 5 loop
      wait until falling_edge(clk);
    end loop;
    assert abs (to_real(p_r) - p.re) < 2.0**(-12) and
           abs (to_real(p_i) - p.im) < 2.0**(-12);
  end process check_outputs;

end architecture verify;
```

The architecture is preceded by a line that identifies the **ieee** library and the **std_logic_ 1164, fixed_pkg** and **math_complex** packages that are used in the testbench. Within the architecture, we have instantiated the **multiplier** entity with its rtl architecture as the device under verification. The instance is connected to testbench signals declared in the architecture.

Since the multiplier is clocked, we need to generate a clock signal to drive it. This is done by the **clk_gen** process. It uses a constant, called **t_c**, for the clock cycle time. The constant is declared before the **begin** keyword of the architecture. Using a constant like this allows us to change the clock cycle time without having to chase down every number that varies as a consequence of the change. The **clk_gen** process waits for half a clock cycle time, sets the clock to '1', waits a further half a clock cycle time, then sets the clock to '0'. (The expression for the duration of the second half clock cycle time is structured so as to compensate for any rounding that may occur in the expression for the first half cycle duration.) After that, the process repeats from the beginning. We also need to generate a reset pulse for the device under verification. This is done by the assignment following the clock generator process. The assignment sets **reset** to '1' immediately, then back to '0' after a delay of two clock cycles.

The **apply_test_cases** process stimulates the device under verification with input data. The process contains a procedure to abstract out the common operations in applying each test case. Rather than generating fixed-point values directly, the process generates test-case operands of type **complex** on the signals a and b. The operation **cmplx** creates a complex number given the real and imaginary parts. For example, **cmplx**(1.0, 2.0) creates the number $1.0 + j2.0$. We can access the real and imaginary parts of a as real numbers using the notations a.re and a.im. The assignments following the **apply_test_cases** process use this notation to assign test-case values to the input inputs of the device under verification.

Within the **apply_test_cases** process, we must ensure that we generate input stimulus values that meet the timing requirements of the device under verification. The operand values and the **input_rdy** signal must be set up before a clock edge. The operand values must be held for four cycles while the operation proceeds. To satisfy these requirements, we wait until the first falling clock edge after reset has returned to '0'. The statement wait until *condition* causes the process to suspend activity until the next time a change in any of the signals mentioned in the condition causes the condition to be true. Thus, the statement will wait, even if the condition is already true. The call to the apply_test procedure then assigns the first test-case operands to the inputs and sets input_rdy to '1'. Next, the procedure waits for the subsequent falling clock edge before resetting input_rdy back to '0'. It then waits a further five cycles, giving the device under verification time to produce its output. After that, subsequent calls to the procedure repeat these steps with the further test-case operands.

The check_output process verifies that the multiplier produces the correct results. It must synchronize with the input stimulus to ensure that it checks the results at the right time. It waits on the same condition as the multiplier's controller finite-state machine, namely, input_rdy being '1' on a rising clock edge. When that occurs, the process reads the stimulus operand values from the signals a and b, multiplies them using the operator defined in the math_complex package, and saves the product in the local variable p. The process then waits until the fifth subsequent falling clock edge, by which time the device under verification has stored its result in its output registers. The results are available on the p_r and p_i signals. The process converts them to real form and compares them with the real and imaginary parts saved in p. Since the types real and sfixed are discrete approximations to mathematical real numbers, an exact equality test is unlikely to succeed. Instead, we check whether the absolute value of the difference is within the required precision, in this case, the precision of the input-operand representation.

4.4.3 ASYNCHRONOUS TIMING METHODOLOGIES

We will close this section on timing methodology with a brief discussion of some alternative approaches. While the clocked synchronous approach yields significant simplifications, there are some applications where it breaks down. Two key assumptions are that the clock signal is distributed globally (that is, across the entire system) with minimal skew, and that the propagation delay between registers is less than a clock cycle. In large high-speed systems, these assumptions are very difficult to maintain. For example, in a large integrated circuit operating with a clock frequency of several GHz, the time taken for a change of signal value to propagate along a wire that stretches across the chip may be a large proportion of a clock cycle, or even more than a clock cycle.

One emerging solution is to reconsider the assumption of a single global clock signal for the entire chip or system. Instead, the system is divided into several regions, each with its own local clock. Where signals connect from one region to another, they are treated as asynchronous inputs. The timing for the system is said to be *globally asynchronous, locally synchronous* (GALS). The benefit of this approach is that it makes the constraints on clock distribution and timing within each region simpler to manage. The downside is that inter-region connections must be synchronized, thus adding delay to communication. The challenge for the system architect is to find a partitioning for the system that minimizes the amount of communication between regions, or that avoids sensitivity to delay in inter-region communication.

The difficulty in distributing high-speed clock signals and managing timing is even greater in the context of a complete circuit board consisting of several integrated circuits, or a large system consisting of several circuit boards. It is simply not practical to distribute a high-speed clock across a large system. Instead, a slower clock is often used externally to high-speed chips, and operations between chips are synchronized to that external clock. The internal clocks operate at a frequency that is a multiple of the external clock, allowing for synchronization of clock edges. The separate boards in a high-speed system typically are not synchronized, but have independent clocks. Data transmitted from one board to another is treated as an asynchronous input by the receiving board.

Another aspect of timing in clocked synchronous systems is that all register-to-register operations take one clock cycle, whether the combinational subcircuit is on the critical path or not. In principle, the slack time in a clock cycle is wasted; all operations are held back to the time taken by the slowest. It is possible to design *asynchronous* circuits in which completion of one operation triggers dependent operations. Such circuits are also called *delay insensitive*, since they operate as fast as the components and the data allow. However, appropriate design techniques are far less mature than those for clocked circuits, and there is negligible CAD tool support for asynchronous methodologies. Hence, products using asynchronous circuits are very uncommon.

A separate issue with the clocked approach is that clocked circuits consume significant amounts of power. Even if a flip-flop does not change its stored value, changing the clock input between 0 and 1 involves switching transistors on and off, thus consuming extra power. In applications with very low power budgets, such as battery powered mobile devices, this waste of power is unacceptable. One approach to dealing with it is to avoid clocking parts of a system that are inactive. *Clock gating*, as it is called, is becoming a more common design technique as the number of low-power applications increases. Asynchronous circuits are an alternative, since logic levels only change when data values change. If there is no new data to operate upon, the circuit becomes quiescent. A few low-power products using asynchronous circuits have been successfully fielded. Low-power applications may be a more significant motivation for asynchronous design than the potential performance gains.

KNOWLEDGE
TEST QUIZ

1. What is meant by the term *register transfer level*?

2. Write the timing condition that must apply on a register-to-register path.

3. What is the critical path in a system?

4. How does the critical path delay affect the clock cycle time of the system?

5. If a given clock cycle time is required, but the critical path delay is too long to achieve it, where should optimization effort be focused?

6. What is meant by the term *clock skew*?

7. Why are registered inputs and outputs used in high-speed systems?

8. What problem can be caused in input registers by asynchronous inputs?

9. Why must inputs from electromechanical switches be debounced?

10. What is the main difference between a testbench for a combinational circuit and a testbench for a sequential circuit?

11. What is meant by the term *globally asynchronous, locally synchronous* (GALS)?

4.5 CHAPTER SUMMARY

▶ Registers are storage components composed of flip-flops. Simple registers can be augmented with clock-enable, reset and preset control inputs.

▶ Synchronous control inputs are acted upon on a clock edge. Asynchronous control inputs are acted upon immediately.

▶ Latching behavior is produced by feedback paths in digital circuits. A transparent latch passes data through while the enable input is 1 and stores data when the enable input is 0.

▶ A simple free-running counter consists of an incrementer and a register. Substituting a decrementer for the incrementer causes the counter to count down instead of up. Adding a clock-enable input to the register allows control over when the counter increments. Adding a reset input to the register allows the count value to be cleared to 0.

▶ An n-bit counter counts modulo 2^n; that is, it counts to $2^n - 1$ then wraps to 0. A modulo k up counter decodes the value $k - 1$ and uses it to reset the counter. A modulo k down counter decrements down to 0 and then reloads the value $k - 1$.

▶ A ripple counter uses the output of one flip-flop to trigger the next flip-flop. It uses less circuitry and consumes less power than a synchronous counter, and can be used in applications where timing constraints allow and power constraints are significant.

▶ A digital system, in general, consists of a datapath and a control section. The datapath contains combinational subcircuits for operating on data and registers for storing data. The control section sequences operations in the datapath by activating control signals at various times. The control section uses status signals to influence the control sequence.

▶ A finite-state machine (FSM) has a set of inputs, a set of outputs, a set of states, a transition function and an output function. For a given clock cycle, the FSM has a current state. The transition function determines the next state given the current state and the input values. The output function determines the output values given just the current state (Moore machine), or given the current state and the input values (Mealy machine).

▶ The state encoding of an FSM can influence the complexity of the next-state and output logic. Synthesis CAD tools are usually able to optimize the state encoding.

- ▶ A state transition diagram represents a finite state machine with bubbles for states, arcs for transitions, and labels for input conditions and output values. Labels for Moore-style outputs are written in the bubbles, and labels for Mealy-style outputs are written on arcs.

- ▶ At the register-transfer level of abstraction, operation of a system is described in terms of transfer of data between registers through combinational circuits that operate on the data.

- ▶ The clocked synchronous timing methodology involves a common clock for all registers, and operation on data by combinational circuits between clock edges.

- ▶ For each path from register output to register input, the sum of the clock-to-output delay, combinational propagation delay and setup time must be less than the clock cycle time. The path with the least slack time is the critical path.

- ▶ The critical path delay places a lower bound on the clock cycle time. Alternatively, a required clock cycle time places an upper bound on the critical path delay.

- ▶ Clock skew is the difference in arrival time of a clock edge at different flip-flops in a system. Clock skew must be minimized to ensure that clocked synchronous circuits operate correctly. CAD tools typically implement clock distribution to minimize skew.

- ▶ Registered inputs and outputs reduce combinational delays in interchip register-to-register paths, and thus help in meeting timing constraints.

- ▶ Asynchronous inputs are those that are not guaranteed to be stable around clock edges. They can cause metastability in input registers. Synchronizers are required to avoid system failure due to metastability.

- ▶ Testbenches for clocked sequential circuits must ensure that stimulus inputs are applied so as to meet timing constraints, and must wait until outputs are valid before checking them.

- ▶ A globally asynchronous, locally synchronous (GALS) system has regions with local clocks, and treats inter-region connections as asynchronous inputs.

4.6 FURTHER READING

Digital Design: Principles and Practices, 3rd Edition, John F. Wakerly, Prentice Hall, 2001. Describes flip-flops and latches in detail, presents detailed low-level design procedures for finite-state machines, provides an analysis procedure for feedback circuits, and discusses metastability and synchronizers in detail.

CMOS VLSI Design: A Circuits and Systems Perspective, 3rd Edition, Neil H. E. Weste and David Harris, Addison-Wesley, 2005. Among many other aspects of CMOS circuit design, this book discusses detailed design of flip-flops and latches and addresses both single-phase and two-phase clocking schemes.

The Student's Guide to VHDL, Peter J. Ashenden, Morgan Kaufmann Publishers, 1998. A supplementary reference showing how to model digital systems with VHDL.

Asynchronous Circuit Design, Chris J. Myers, Wiley-Interscience, 2001. An in-depth treatment of theory and practice.

A Guide to Debouncing, Jack G. Ganssle, The Ganssle Group, 2004, www.ganssle.com/debouncing.pdf. Presents empirical data on switch bounce behavior, and describes hardware and software approaches to debouncing.

Comprehensive Functional Verification: The Complete Industry Cycle, Bruce Wile, John C. Goss and Wolfgang Roesner, Morgan Kaufmann Publishers, 2005. Describes strategies and techniques for stimulus generation and result checking in simulation-based verification.

EXERCISES

EXERCISE 4.1 Draw a schematic for a 6-bit register, constructed from D flip-flops, that updates the stored value on every clock cycle.

EXERCISE 4.2 Write a VHDL model for a 12-bit register that stores an unsigned integer value.

EXERCISE 4.3 Develop a VHDL model of a pipelined circuit that computes the maximum of corresponding values in three streams of input values, a, b and c. The pipeline should have two stages: the first stage determines the larger of a and b and saves the value of c; the second stage finds the larger of c and the maximum of a and b. The inputs and outputs are all 14-bit signed 2s-complement integers.

EXERCISE 4.4 Revise the schematic of Exercise 4.1 to include a clock enable and a reset input to the register, using flip-flops with clock-enable and reset inputs.

EXERCISE 4.5 Write a VHDL model for a register with clock-enable and synchronous reset that stores a 16-bit 2s-complement signed integer value.

EXERCISE 4.6   Draw a datapath for a pipelined complex multiplier. Unlike the sequential multiplier in Example 4.13 that takes five cycles to do each multiplication, the pipelined multiplier should take just two cycles for each pair of complex operands: one cycle for the four multiplications and one cycle for the subtraction and addition. The multiplier should accept new operand inputs on each clock cycle and produce a product on each clock cycle.

EXERCISE 4.7   Develop a VHDL model for a peak detector that finds the maximum value in a sequence of 10-bit unsigned integers. A new number arrives at the input during a clock cycle when the data_en input is 1. If the new number is greater than the previously stored maximum value, the maximum value is updated with the new number; otherwise, it is unchanged. The stored maximum value is cleared to zero when the reset control input is 1. Both data_en and reset are synchronous control inputs.

EXERCISE 4.8   Write a VHDL model of a flip-flop with a negative-logic synchronous clock-enable input, positive-logic asynchronous preset and reset inputs, and both positive- and negative-logic data outputs.

EXERCISE 4.9   Suppose we replaced the edge-triggered registers in the pipeline of Figure 4.4 with transparent latches, with the latch-enable inputs all connected to the clock signal. Describe how the circuit would operate, and whether it would still function as a pipeline.

EXERCISE 4.10   Draw a circuit for a free-running counter that counts 32 clock cycles and produces a control signal that is 1 during every 4th, 20th and 24th cycle.

EXERCISE 4.11   Develop a VHDL model of the counter of Exercise 4.10.

EXERCISE 4.12   Draw a circuit that uses counters to divide a master clock of 20.48MHz to generate a signal with 50% duty cycle and a frequency of exactly 5kHz.

EXERCISE 4.13   Design a circuit for a modulo 12 counter, similar to the decade counter of Figure 4.26.

EXERCISE 4.14   Develop a VHDL model of the modulo 12 counter of Exercise 4.13.

EXERCISE 4.15   Develop a VHDL model of a 12-bit up counter with synchronous count-enable, reset and load-enable inputs, and a terminal-count output.

EXERCISE 4.16   The schematic in Figure 4.47 shows a ripple counter connected to a decoder. Augment the timing diagram of Figure 4.29 to show the values on the decoder outputs, including any spurious pulses that occur when the counter increments.

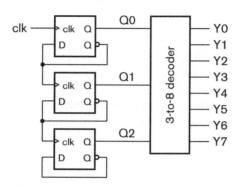

FIGURE 4.47

EXERCISE 4.17   Revise the complex multiplier datapath of Example 4.13 to include two fixed-point multiplier components instead of just one. How can the control sequence described in Example 4.15 be revised as a consequence to reduce the time taken to perform a complex multiplication?

EXERCISE 4.18   Develop a finite-state machine to implement the revised control sequence from Exercise 4.17. Show the transition and output functions both in tabular form and using a state transition diagram.

EXERCISE 4.19   Develop a VHDL model of the complex multiplier as revised in Exercises 4.17 and 4.18.

EXERCISE 4.20   Identify the control steps required for sequential multiplication using the datapath described in Example 4.4, and develop a finite-state machine for the control section. Assume that the x and y operand values are valid on a cycle when a control signal, start, is 1. Generate a control signal, done, that is 1 when the multiplication is complete. Use a 4-bit counter to count the successive accumulation steps.

EXERCISE 4.21   Develop a VHDL model of the sequential multiplier in Exercise 4.20, including both the datapath and the control section.

EXERCISE 4.22   An arbiter is a circuit that allows at most one subsystem at a time to use a shared resource. A four-way arbiter is shown in Figure 4.48. Each subsystem sets its request signal to 1 when it wants to use the resource. When the arbiter sets the grant signal to 1, the subsystem uses the resource. The subsystem sets its request back to 0 when it has finished, and waits for grant to be 0 before

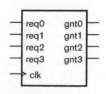

FIGURE 4.48

starting a subsequent request. While a subsystem is granted use of the resource, other requests must wait, rather than pre-empting the active subsystem.

a) Develop a FSM for a priority arbiter, in which subsystem 0 has highest priority and subsystem 3 has least priority. A pending request from a higher-priority subsystem takes precedence over a pending request from a lower-priority system.

b) Develop a FSM for a round-robin arbiter. Subsystems are granted requests in order, starting with 0, then 1, 2, 3 and back to 0. A subsystem is skipped if it has no pending request.

EXERCISE 4.23 Suppose a clocked synchronous system uses registers with setup time of 150ps and clock-to-output delay of 400ps. Three register-to-register paths in the datapath have propagation delays of 600ps, 900ps and 1.3ns, respectively.

a) What is the maximum clock frequency at which the datapath can be operated?

b) If the path with a delay of 1.3ns is optimized to reduce its delay to 800ps, what is the maximum clock frequency for the optimized datapath?

EXERCISE 4.24 Suppose a clocked synchronous system, in which registers have setup time of 100ps and clock-to-output delay of 200ps, has a timing constraint that the clock frequency be 800MHz. Propagation delays through combinational elements in the datapath and control section are shown in Figure 4.49. The control section uses a Mealy FSM.

a) Identify the critical path in the system.

b) Is the timing constraint on the clock frequency met?

c) If the FSM were changed to be a Moore FSM, would the critical path change, and would the constraint be met?

FIGURE 4.49

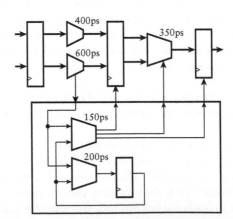

EXERCISE 4.25 For a system that operates at a high clock frequency and samples an asynchronous input that changes with high frequency, the simple synchronizer of Figure 4.44 may exhibit an unacceptable MTBF. Equation 4.5 indicates that doubling the sampling delay yields a disproportionate improvement in MTBF. Design an enhanced synchronizer that samples the input on alternate rising clock edges.

EXERCISE 4.26 The debouncer circuit of Figure 4.45 uses a "break before make" switch. What would happen if a "make before break" switch were used? Such a switch may close the new contact before opening the other contact.

EXERCISE 4.27 Develop a testbench model to verify operation of the debouncer described in Example 4.20.

EXERCISE 4.28 Develop a testbench model to verify the sequential multiplier of Example 4.4 with the control section as described in Exercise 4.20.

MEMORIES

5

Many digital systems use memories for storing information. Memory in general-purpose computers takes several forms, including semiconductor memory chips, magnetic disks (hard disks), and optical disks (CDs and DVDs). In this chapter, we describe the various types of semiconductor memories, since other forms of memory are much less frequently used in application-specific digital systems. We start by introducing the general concepts that are common to all kinds of semiconductor memory, and then focus on the particular features of each type. We complete the chapter with a discussion of techniques for dealing with errors in the stored data.

5.1 GENERAL CONCEPTS

In Chapter 4 we introduced registers as components for storing binary-coded information. We generally use separate registers when the number of items of information to store is small, or when we need to use many of the items concurrently. When there are numerous items that we can use one after another, we use *memory* components instead to store the information. In this section, we will discuss some of the general concepts that apply to all kinds of memory components. Then, in the next section, we will identify some of the specific kinds of memory that are used in different design scenarios.

A memory is conceptually an array of storage registers, or *locations*, each of which has a distinct *address*, which is a number identifying the location. Addresses for a memory typically start at 0 and increase by one for each location, up to one less than the number of locations. For most memory components, the number of locations is a power of 2. Thus, a memory with 2^n locations would have addresses ranging from 0 to $2^n - 1$, requiring an n-bit address. If each location stores m bits of encoded information, the total number of bits in the memory component is $2^n \times m$.

EXAMPLE 5.1   If a memory has 32,768 locations, each of 32 bits, what is the total capacity of the memory, and how many address bits does it require?

SOLUTION   The capacity is 1,048,576 bits, that is 2^{20} bits. Since $32,768 = 2^{15}$, the memory requires 15 address bits.

When referring to memory sizes, we usually use the following multiplier prefixes denoting powers of 2:

▶ Kilo (K): $2^{10} = 1,024$

▶ Mega (M): $2^{20} = 1,024 \times 2^{10} = 1,048,576$

▶ Giga (G): $2^{30} = 1,024 \times 2^{20} = 1,073,741,824$

Thus, the memory referred to in Example 5.1 has a capacity of 1M bit. Note that the multiplier values are close to, but slightly greater than, the decimal multiplier values with the same names. Note also that we use an uppercase "K" for the binary multiplier 2^{10}, compared with the lowercase "k" for the decimal multiplier 10^3. The context of referring to a memory size is usually assumed to indicate use of the binary multipliers rather than the decimal multipliers.

Given a memory of a certain capacity, we can organize it in different ways, varying the number of locations and the number of bits per location. For example, a 1M bit memory might be organized as a $32K \times 32$-bit memory, as shown in Example 5.1, or as a $16K \times 64$-bit memory, $64K \times 16$-bit memory, and so on. In practice, the number of locations and the size of each location are determined by the application requirements, dictating the memory capacity required.

The two basic operations performed by a memory are writing binary data to a location and reading the content of a location. For both operations, we need to provide the address of the location to be written or read on a set of input signals to the memory component. For a write operation, we provide the data to write as a further set of input signals, and for a read operation, the memory component provides the data as a set of output signals. We control the write operation using control signals generated by a control section of the digital system that contains the memory component. We will describe the particular control signals used by different kinds of memories in a later section. For now, we will just assume a simple form of memory component with simple control signals. The input and output signals are shown on a symbol for a memory component in Figure 5.1. The signal a is the address, encoded as an unsigned binary number. The signals d_in and d_out carry the data to be written and the data read, respectively. The encoding for these signals depends on the application. The control signals are en (enable) and wr (write). When en is 0, the memory simply maintains all of the stored

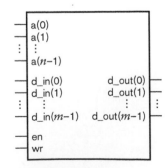

FIGURE 5.1   Symbol for a basic memory component.

data. When en is 1 and wr is 1, the memory writes data present on the d_in inputs at the location whose address is present on the a inputs. When en is 1 and wr is 0, the memory reads the content of the location whose address is present on the a inputs and drives the data value on the d_out outputs.

EXAMPLE 5.2   Design an audio echo effects unit that operates by delaying samples of an audio signal represented as a stream of 16-bit 2s-complement binary-coded values. The sample rate is 50kHz. Arrival of a new input sample is indicated by a control input, audio_in_en, being 1 for the clock cycle in which the sample arrives. The unit should indicate availability of an output sample using an output control signal, audio_out_en, in the same way. The delay time is determined by an 8-bit unsigned input representing the number of milliseconds of delay. The system clock frequency is 1MHz.

SOLUTION   We can delay the arriving audio sample values by storing them in a memory until they are required at the output. The maximum delay expressed by the 8-bit unsigned input is 255ms. Since samples arrive at a rate of 50kHz (that is, 50 per millisecond), we need to store up to $255 \times 50 = 12,750$ samples. A 16K × 16-bit memory, with 14-bit addresses (since $16K = 2^{14}$), will suffice. A diagram of the datapath including the memory and other components to compute addresses is shown in Figure 5.2. The figure shows the widths of each of the multibit signals.

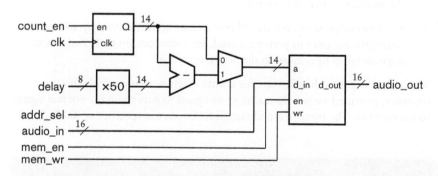

FIGURE 5.2   Datapath for an audio echo effects unit.

We need to use a 14-bit counter to keep track of where samples are stored in the memory. As each input sample arrives, we store it at the next available memory location, whose address is given by the counter. We next read from the memory the value written d milliseconds in the past (where d is the value of the delay input) and provide it at the output, then increment the counter to refer to the next location in memory. This behavior is illustrated in the timing diagram of Figure 5.3. The value written d milliseconds previously is stored $50 \times d$ locations prior to the current location given by the address counter. Thus, we can compute its address by multiplying d by 50 and subtracting the result from the value of the address counter. The counter will increment to the maximum address value

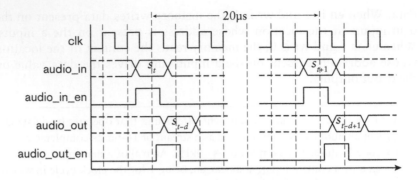

FIGURE 5.3 Timing diagram for the audio echo effects unit.

then wrap around to 0, effectively incrementing modulo 16K. Thus, once the memory is filled, old locations will be overwritten with newly arriving samples. However, they will have been written more than the maximum delay in the past, so they will no longer be needed. When we perform the subtraction, we can ignore the borrow output of the subtracter. The subtracter will yield the difference modulo 16K, and so give the correct address of the required delayed sample.

The control sequence for the unit involves two steps:

1. When a sample arrives (indicated by audio_in_en being 1), set the multiplexer to use the counter value as the memory address and enable the memory to perform a write.

2. Set the multiplexer to use the subtracter output as the memory address, enable the memory to perform a read, set audio_out_en to 1, and enable the counter to increment on the next clock edge.

We can use step 1 as the idle state for a state machine that controls this sequence, provided we use the audio_in_en signal to gate the write control signal to the memory. The transition and output functions are specified in Table 5.1.

TABLE 5.1 Transition and output functions for the echo unit control section.

STATE	audio_in_en	NEXT STATE	addr_sel	mem_en	mem_wr	count_en	audio_out_en
step 1	0	step 1	0	0	0	0	0
step 1	1	step 2	0	1	1	0	0
step 2	–	step 1	1	1	0	1	1

The mem_en and mem_wr signals are Mealy-style outputs, since they depend on both the state and the audio_in_en input, whereas the remaining control signals are all Moore-style outputs.

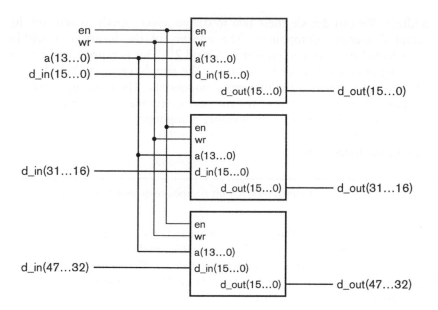

FIGURE 5.4 Connection of memory components in parallel to form a wider memory.

Manufacturers provide semiconductor memory components in a range of capacities, varying from a few Kbits through several Mbits and, at time of writing, up to 2G bits for separately package memory components. Typically, for a given capacity, a manufacturer provides components organized with differing widths (1, 4, 8 or 16 bits per location). If an application for which we are designing a system needs a memory of some other width, we need to use a number of memory components in parallel. For example, if we need a $16K \times 48$-bit memory for an application, we could construct it using three $16K \times 16$-bit memory components. We would connect the address and control signals together, as shown in Figure 5.4, and use the data input and output signals of each component for a slice of the overall data input and output signals.

Connecting multiple memory components together to construct a memory with more locations is somewhat more involved. We need to partition the total number of locations among the memory components. For each read and write operation we need to arrange for the component containing the required location to perform the operation, and for other components to remain passive. In many applications, the total number of locations is a power of 2, say 2^n, and each memory component has a smaller number of locations, 2^k. The number of memory components is $2^n/2^k$. The simplest approach to partitioning is to place the first 2^k locations in the first component, the second 2^k in the second component, and so on. If we number the individual memory components 0, 1, 2, and so on up to $(2^n/2^k)-1$, the component containing a location with address A is $\lfloor A/2^k \rfloor$. This is represented by the most significant $n-k$ bits of the address. We can decode these bits to derive select signals to activate the required memory component. The address of the location A within the selected memory component is $A \bmod 2^k$. This is represented by the least significant k bits of the address. We simply connect these bits of the address to each of the memory components. The data input signals are also connected to each of the memory components. The data output signals need to be driven by the memory component that is selected, so we use a multiplexer to choose the appropriate data value based on the most significant address bits.

EXAMPLE 5.3   Design a 64K × 8-bit composite memory using four 16K × 8-bit components.

SOLUTION   The complete composite memory is shown in Figure 5.5. Address bits 15 and 14 are decoded to select which of the four memory components is enabled for read and write operations. Those bits also control the multiplexer to select the output data from the enabled component during a read operation.

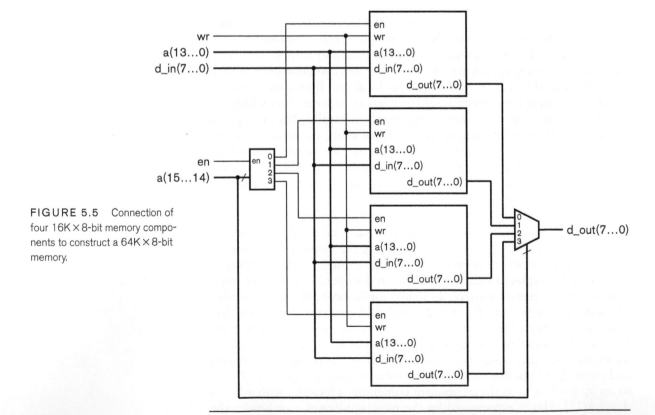

FIGURE 5.5   Connection of four 16K × 8-bit memory components to construct a 64K × 8-bit memory.

+V

output

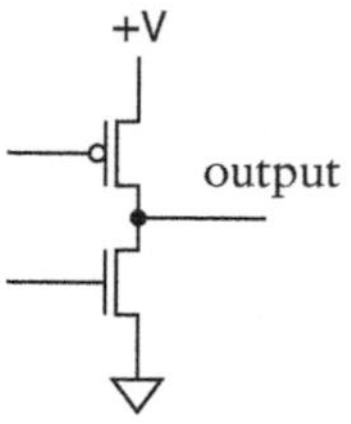

FIGURE 5.6   Output stage circuit.

Many manufacturers simplify the connection of memory components to form larger memories by using a special kind of output driver, called a *tristate* driver, for each of the data outputs. Tristate drivers are also used for buses that allow multiple data sources to provide data in a system. We will discuss tristate and other bus structures in more detail in Chapter 8 as part of our discussion of embedded computer systems. For now, we will focus on their use in memory components.

Unlike ordinary component outputs, which always drive either a low or high logic level, the output of a tristate driver can be turned off by placing it in a *high-impedance*, or *hi-Z*, state. ("*Z*" is commonly used as the symbol for impedance in a circuit.) Thus, a tristate driver has three output states: logic low, logic high and high impedance; hence the name. The output circuit of a CMOS digital component involves two transistor switches as shown in Figure 5.6. To drive the output with a low logic level, the component turns the bottom transistor on and the top transistor off, and to drive a high logic level, the component turns the top transistor on and the bottom transistor off. A tristate driver has the same output stage, but can turn both transistors off, effectively isolating the component from the output.

If we use memory components with tristate data outputs to construct a larger memory, we can omit the output multiplexer shown in Figure 5.5. Instead, we simply connect the data outputs of the memory components together. When a read operation is performed, only the selected memory component enables its data outputs; all of the disabled components leave their outputs in the high-impedance state.

Many memory components that have tristate data outputs also combine the data inputs and outputs into a single set of *bidirectional* connections, illustrated in Figure 5.7. This allows a composite memory to be constructed as shown in Figure 5.8. For memory components implemented as separate integrated circuits for use on printed circuit boards, the use of bidirectional connections results in significant cost savings, since there are fewer package

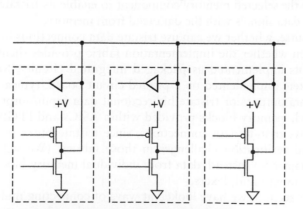

FIGURE 5.7   Bidirectional tristate data connections.

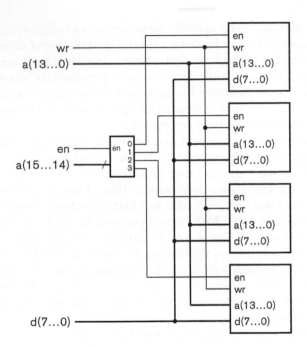

FIGURE 5.8   A composite memory constructed using components with common data inputs and outputs.

pins and interconnecting wires. As we shall see when we study embedded processors in more detail, this type of memory works well as part of an embedded computer system, since memory write and read operations are performed independently. When we perform a write operation, we drive the data signals with the data to be written. The selected memory component treats the data connections as inputs and accepts the data to be written. It keeps its tristate drivers disabled so as not to interfere with the logic levels in the data signals. When we perform a read operation, we ensure that all other drivers connected to the data signals are in the high-impedance state and allow the selected memory component to enable its tristate drivers. It drives the data signals with the data read from memory.

Of course, whether we can use tristate data connections in a memory depends on whether the implementation fabric provides them. Memory components implemented as packaged integrated circuits, for use in a larger system implemented on a printed circuit board, typically do have tristate data outputs or tristate bidirectional data input/outputs. On the other hand, memory blocks provided within ASICs and FPGAs typically do not have tristate data connections, since tristate buses present some design and verification challenges in those fabrics. (We will return to this in Chapter 8.) Instead, data from individual memory blocks must be combined using multiplexers.

In this section, we have looked at ways of connecting multiple memory components together to form a memory with wider or more storage locations than provided by a single chip. In each of these schemes, the memory performs just one operation at a time. In high performance systems, we can connect multiple memory components together in ways that permit multiple operations to proceed concurrently, thus increasing the total number of operations completed per second. These schemes usually involve organizing the memory into a number of banks, each of which can perform an operation in parallel with other banks. Successive addresses are assigned to different banks, since, in many systems, locations are often accessed in order. As an example, a system with four banks would assign locations 0, 4, 8, ... to bank 0; locations 1, 5, 9, ... to bank 1; 2, 6, 10, ... to bank 2; and 3, 7, 11, ... to bank 3. When a read operation is required for location 4, bank 0 would read that location. Moreover, the other banks would start a read, *prefetching* locations 5, 6 and 7. By the time a read operation is required for these locations (assuming access in order), the data would already be available from the memory. We will not describe these advanced memory organizations in any further detail in this book. Books on computer organization, particularly those concentrating on high-performance computers, are a good source of further information. (See Section 5.5, Further Reading.)

1. What is the capacity in bits of a memory with 4096 locations, each of 24 bits? How many address bits are required?

2. What is the effect of a write operation? What is the effect of a read operation?

3. How would we connect four $256M \times 4$-bit memory components to make a $256M \times 16$-bit memory?

4. How would we connect four $256M \times 8$-bit memory components to make a $1G \times 8$-bit memory?

5. Which memory component in Question 4 would contain the location with address $5FC0000_{16}$?

6. What are the three states of a tristate driver?

7. How do memory components with tristate data outputs simplify construction of large memories?

KNOWLEDGE
TEST QUIZ

5.2 MEMORY TYPES

In this section, we will introduce the various types of memory provided by manufacturers, either as individual integrated circuits or as resources within ASIC or FPGA fabrics. We will discuss the distinguishing properties of each kind of memory, including their timing characteristics and costs, and describe how to model some of them in VHDL. We will distinguish between memory that can be both read and written, called *random access memory* (RAM), and memory that can only be read, called *read-only memory* (ROM). We use the term RAM instead of read/write memory largely for historical reasons. Memories in very early computers enforced sequential access, that is, access to locations in increasing order of address, due to the physical medium on which the data was stored. The invention of memories in which locations could be read and written with equal facility in any order (that is, randomly) was a significant milestone, and so the term RAM has stuck.

5.2.1   ASYNCHRONOUS STATIC RAM

One of the simplest forms of memory is asynchronous static RAM. It is *asynchronous* because it does not rely on a clock for its timing. The term *static* means that the stored data persists indefinitely so long as power is applied to the memory component. Compare this with dynamic RAM, which we will describe later and which loses stored data if it is not periodically rewritten. Static RAM is *volatile*, meaning that it requires power to maintain the stored data, and loses data if power is removed. Since engineers are fond of abbreviations, the term static RAM is usually further shortened to SRAM.

Asynchronous SRAM internally uses 1-bit storage cells that are similar to the D-latch circuit that we described in Chapter 4. Within the memory component, the address is decoded to select a particular group of cells that comprise one location. For a write operation, the selected latch cells are enabled and the input data is stored. For a read operation, the address activates a multiplexer that routes the outputs of the selected latch cells to the data outputs of the memory component.

The external interface of an asynchronous SRAM is very close to our general description of a memory component in Section 5.1. For largely historical reasons, most manufacturers use active-low logic for the control signals. Further, since asynchronous SRAMs are usually only available as packaged integrated circuits, and not as blocks in ASIC libraries or FPGAs, they usually have bidirectional tristate data input/output pins. Figure 5.9 shows a symbol for a typical asynchronous SRAM. The address input and the data input/output are as we described in Section 5.1. The chip-enable input ($\overline{CE}$) is used to enable or disable the memory chip. We usually drive this input from a select control signal, for example, from an address decoder in a composite memory. The write-enable input ($\overline{WE}$) controls whether the memory, if enabled, performs a write or read operation. The output-enable input ($\overline{OE}$) controls the tristate data drivers during a read operation. When $\overline{OE}$ is low during a read, the drivers are enabled and can drive the read data onto the data pins. When $\overline{OE}$ is high, the drivers are in the high-impedance state.

FIGURE 5.9   Symbol for an asynchronous SRAM.

Given that the storage cells in an asynchronous SRAM are basically latches, it is not surprising that the timing is similar to that of a D-latch. The sequencing of signals to perform a write operation is shown at the left of Figure 5.10. The control section that sequences the datapath containing the memory must ensure that the address is stable before commencing the write operation and is held stable during the entire operation. Otherwise, locations other than the one to be updated may be affected. The control section selects the particular memory chip by driving $\overline{CE}$ low, activates the write operation by driving $\overline{WE}$ low, and ensures that the chip's tristate drivers are disabled by driving $\overline{OE}$ high. It also sets control signals to the datapath to provide data on the data signals. The data is stored transparently in the latch cells for the addressed location. The final data to be stored must be stable on the data signals a setup time before the rising edge of the $\overline{WE}$ signal or the $\overline{CE}$ signal, whichever occurs first. The data and the address must also remain stable for a hold time after the $\overline{WE}$ or $\overline{CE}$ signal goes high.

The typical sequencing of signals for a read operation is similar, and is shown at the right of Figure 5.10. The difference is that the $\overline{WE}$ signal is held high, and the $\overline{OE}$ signal is driven low to enable the memory chip's tristate drivers. While this sequence is typical for a read operation done in isolation, we can also perform back-to-back read operations simply by changing the address value. The read operation is essentially a combinational operation, involving decoding the address and multiplexing the selected latch-cell's value onto the data outputs. Changing the address simply causes a different cell's value to appear on the outputs after a propagation delay.

Manufacturers of asynchronous SRAM chips publish the timing parameters for write and read operations in data sheets. The parameters typically include setup and hold times for address and data values, and delays for turning tristate drivers on and off. One of the figures of merit of a memory chip is its *access time*, which is the delay from the start of a read

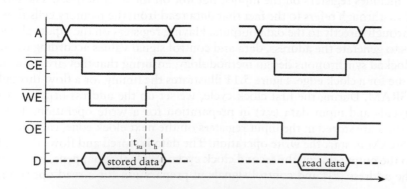

FIGURE 5.10 Timing for write and read operations in an asynchronous SRAM.

operation to having valid data at the outputs. Other performance-related parameters are the *write cycle time* and the *read cycle time*, which are the times taken to complete write and read operations, respectively. Manufacturers offer chips in different speed grades, with faster chips usually costing more. This allows us, as designers, to make cost/performance trade-offs in our designs.

While asynchronous SRAMs are conceptually simple and have simple timing behavior, the fact that they are asynchronous can make them difficult to use in clocked synchronous systems. The need to set up and hold address and data values before and after activation of the control signals and to keep the values stable during the entire cycle means that we must either perform operations over multiple clock cycles, or use delay elements to ensure correct timing within a clock cycle. The former approach reduces performance, and the latter approach violates assumptions inherent in the clocked synchronous methodology, and so complicates timing design and analysis. For these reasons, asynchronous SRAMs are usually used only in systems with low performance requirements, where their low cost is a benefit.

5.2.2   SYNCHRONOUS STATIC RAM

Given the difficulties associated with asynchronous SRAMs, many memory component vendors and implementation fabrics provide *synchronous SRAMs*, otherwise known as SSRAMs. The internal storage cells of SSRAMs are the same as those of asynchronous SRAMs. However, the interface includes clocked registers for storing the address, input data and control signal values, and in some cases, output data. In this section, we will describe two forms of SSRAMs in general terms. The details of control signals and timing will vary between SSRAMs provided by different component vendors and implementation fabrics. As always, we need to read and understand the data sheets before using a component in a design.

The simplest kind of SSRAM is often called a *flow-through* SSRAM. It includes registers on the inputs, but not on the data outputs. The term flow-through refers to the fact that data read from the memory cells flows through directly to the data outputs. Having registers on the inputs allows us to generate the address, data and control signal values according to our clocked synchronous design methodology, ensuring that they are stable in time for a clock edge. Figure 5.11 illustrates the timing for a flow-through SSRAM. During the first clock cycle, we set up the address (a_1), control signals and input data (xx) in preparation for a write operation. These values are stored in the input registers on the next clock edge, causing the SSRAM to start the write operation. The data is stored and flows through to the output during the second clock cycle. While that happens, we set up the address (a_2) and control signals in preparation for a read operation.

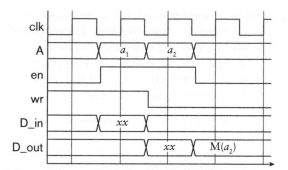

FIGURE 5.11   Timing for a flow-through SSRAM.

Again, these values are stored on the next clock edge, and during the third cycle the SSRAM performs the read operation. The data, denoted by $M(a_2)$, flows through from the memory to the output. Now, in the third cycle, we set the enable signal to 0. This prevents the input registers from being updated on the next clock edge, so the previously read data is maintained at the output.

EXAMPLE 5.4   Design a circuit that computes the function $y = c_i \times x^2$, where x is a binary-coded input value and c_i is a coefficient stored in a flow-through SSRAM. x, c_i and y are all signed fixed-point values with 8 pre-binary-point and 12 post-binary-point bits. The index i is also an input to the circuit, encoded as a 12-bit unsigned integer. Values for x and i arrive at the input during the cycle when a control input, start, is 1. The circuit should minimize area by using a single multiplier to multiply c_i by x and then by x again.

SOLUTION   A datapath for the circuit is shown in Figure 5.12. The 4K × 20-bit flow-through SSRAM stores the coefficients. A computation starts with the index value, i, being stored in the SSRAM address register, and the data

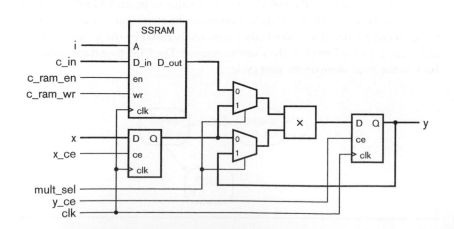

FIGURE 5.12   Datapath for a circuit to multiply the square of an input by an indexed coefficient.

input, x, being stored in the register shown below the SSRAM. On the second clock cycle, the SSRAM performs a read operation. The coefficient read from the SSRAM and the stored x value are multiplied, and the result is stored in the output register. On the third cycle, the multiplexer select inputs are changed so that the value in the output register is further multiplied by the stored x value, with the result again being stored in the output register.

For the control section, we need to develop a finite state machine that sequences the control signals. It is helpful to draw a timing diagram showing progress of the computation in the datapath and when each of the control signals needs to be activated. The timing diagram is shown in Figure 5.13, and includes state names for each clock cycle. An FSM transition diagram for the control section is

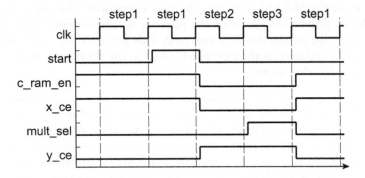

FIGURE 5.13 Timing diagram for the computation circuit.

shown in Figure 5.14. The FSM is a Moore machine, with the outputs shown in each state in the order c_ram_en, x_ce, mult_sel and y_ce. In the step1 state, we maintain c_ram_en and x_ce at 1 in order to capture input values. When start changes to 1, we change c_ram_en and x_ce to 0 and transition to the step2 state to start computation. The y_ce control signal is set to 1 to allow the product of the coefficient read from the SSRAM and the x value to be stored in the y output register. In the next cycle, the FSM transitions to the step3 state, changing the mult_sel control signal to multiply the intermediate result by the x value again and storing the final result in the y output register. The FSM then transitions back to the step1 state on the next cycle.

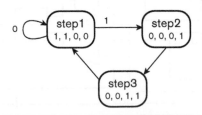

FIGURE 5.14 Transition diagram for the circuit control section.

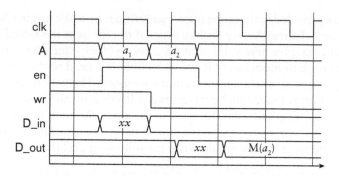

FIGURE 5.15   Timing for a
pipelined SSRAM.

Another form of SSRAM is called a *pipelined* SSRAM. It includes a register on the data output, as well as registers on the inputs. A pipelined SSRAM is useful in higher-speed systems where the access time of the memory is a significant proportion of the clock cycle time. If there is no time in which to perform combinational operations on the read data before the next clock edge, it needs to be stored in an output register and used in the subsequent clock cycle. A pipelined SSRAM provides that output register. The timing for a pipelined SSRAM is illustrated in Figure 5.15. Timing for the inputs is the same as that for a flow-through SSRAM. The difference is that the data output does not reflect the result of a read or write operation until one clock cycle later, albeit immediately after the clock edge marking the beginning of that cycle.

EXAMPLE 5.5   Suppose we discover that, in the datapath of Example 5.4, the combination of the SSRAM access time plus the delays through the multiplexer and multiplier is too long. This causes the clock frequency to be too slow to meet our performance constraint. We change the memory from a flow-through to a pipelined SSRAM. How is the circuit design affected?

SOLUTION   As a consequence of the SSRAM change, the coefficient value is available at the SSRAM output one cycle later. To accommodate this, we could insert a cycle into the control sequence to wait for the value to be available. Rather than wasting this time, we can use it to multiply the value of x by itself, and perform the multiplication by the coefficient in the third cycle. This change requires us to swap the input to the top multiplexer in Figure 5.12, so that it selects the stored x value when mult_sel is 0 in state step2 and the SSRAM output when mult_sel is 1 in step3. The FSM control sequence is otherwise unchanged.

VHDL Models of Synchronous Static Memories

In this section, we will describe how to model SSRAMs in such a way that synthesis CAD tools can infer a RAM and use the appropriate memory resources provided in the target implementation fabric. We saw in Chapter 4 that to model a register, we declare a signal to represent the stored register value and assign a new value to it on a rising clock edge. We can extend this approach to model an SSRAM in VHDL. We need to declare a signal that represents all of the locations in the memory. The way to do this is to declare an *array type*, which represents a collection of values, each with an index that corresponds to its location in the array. We then declare a signal of the array type to represent the stored data. For example, to model a 4K × 16-bit memory, we would write the following declarations:

```
type RAM_4Kx16 is array (0 to 4095)
                    of std_logic_vector(15 downto 0);
signal data_RAM : RAM_4Kx16;
```

The type declaration specifies a new type named RAM_4Kx16 that is an array with elements index from 0 to 4095. Each element is a 16-bit std_logic_vector value. The signal data_RAM is then declared to be of this type.

Once we have declared the signal representing the storage, we write a process that performs the write and read operations. The process is similar in form to that for a register. For example, a process to model a flow-through SSRAM based on the signal declaration above is

```
data_RAM_flow_through : process (clk) is
begin
  if rising_edge(clk) then
    if en = '1' then
      if wr = '1' then
        data_RAM(to_integer(a)) <= d_in;  d_out <= d_in;
      else
        d_out <= data_RAM(to_integer(a));
      end if;
    end if;
  end if;
end process data_RAM_flow_through;
```

On a rising clock edge, the process checks the enable input, and only performs an operation if it is '1'. If the write control input is '1', the process updates the element of the data_RAM signal indexed by the address using the data input. We need to convert the address from an unsigned vector value to an integer value to index the signal. The process also assigns the data input to the data output, representing the flow-through that occurs during a write operation. If the write control input is '0', the process performs a read operation by assigning the value of the indexed data_RAM element to the data output.

EXAMPLE 5.6   Develop a VHDL model of the circuit using flow-through SSRAMs, as described in Example 5.4.

SOLUTION   The entity declaration includes the address, data and control ports, as follows:

```
library ieee;
use ieee.std_logic_1164.all,
    ieee.numeric_std.all, ieee.fixed_pkg.all;

entity scaled_square is
  port ( clk, reset : in  std_logic;
         start      : in  std_logic;
         i          : in  unsigned(11 downto 0);
         c_in, x    : in  sfixed(7 downto -12);
         y          : out sfixed(7 downto -12) );
end entity scaled_square;
```

The architecture is

```
architecture rtl of scaled_square is

  signal c_ram_en, c_ram_wr, x_ce, mult_sel, y_ce : std_logic;
  signal c_out, x_out : sfixed(7 downto -12);
  signal y_out        : sfixed(7 downto -12);

  type c_array is array (0 to 4095) of sfixed(7 downto -12);
  signal c_RAM : c_array;

  type state is (step1, step2, step3);
  signal current_state, next_state : state;
begin

  c_ram_wr <= '0';

  c_RAM_flow_through : process (clk) is
  begin
    if rising_edge(clk) then
      if c_ram_en = '1' then
        if c_ram_wr = '1' then
          c_RAM(to_integer(i)) <= c_in;
          c_out <= c_in;
        else
          c_out <= c_RAM(to_integer(i));
        end if;
      end if;
    end if;
  end process c_RAM_flow_through;
```

(continued)

```
y_reg : process (clk) is
  variable operand1, operand2 : sfixed(7 downto -12);
begin
  if rising_edge(clk) then
    if y_ce = '1' then
      if mult_sel  = '0' then
        operand1 : = c_out;
        operand2 : = x_out;
      else
        operand1 : = x_out;
        operand2 : = y_out;
      end if;
      y_out <= resize(operand1 * operand2, y_out);
    end if;
  end if;
end process y_reg;

y <= y_out;

state_reg : process ...

next_state_logic : process ...

output_logic : process ...

end architecture rtl;
```

The architecture declares signals for the internal datapath connections and
control signals. It declares an array type and a corresponding signal to represent
the coefficient memory (c_array and c_RAM). It also declares a type for the state
of the control section finite-state machine, and signals for the current and next
state.

After the begin keyword, we include processes and assignments for the
datapath and control section. We omit the details of the finite-state machine.
They are based on the template we described in Chapter 4, and are available on
the companion website. The process c_RAM_flow_through represents the
coefficient SSRAM. It uses the i input as its address. The y_reg process
represents both the combinational circuits of the datapath and the output
register. If the y_ce signal is '1', the register is updated with the value computed
by the combinational circuits. We use intermediate local variables to divide the
computation into two parts, corresponding to the multiplexers and the multiplier, respectively.

Modeling a pipelined SSRAM in VHDL is somewhat more involved,
as we must represent the internal connection from the memory storage

to the output register and ensure that the pipeline timing is correctly represented. One approach, extending our previous process for a 16-bit-wide memory, is

```
data_RAM_pipelined : process (clk) is
  variable pipelined_en    : std_logic;
  variable pipelined_d_out : std_logic_vector(15 downto 0);
begin
  if rising_edge(clk) then
    if pipelined_en = '1' then
      d_out <= pipelined_d_out;
    end if;
    pipelined_en := en;
    if en = '1' then
      if wr = '1' then
        data_RAM(to_integer(a)) <= d_in;
        pipelined_d_out := d_in;
      else
        pipelined_d_out := data_RAM(to_integer(a));
      end if;
    end if;
  end if;
end process data_RAM_pipelined;
```

In this process, the variable pipelined_en saves the value of the enable input on a clock edge so that it can be used on the next clock edge to control the output register. Similarly, the variable pipelined_d_out saves the value read or written through the memory on one clock edge for assignment to the output on the next clock edge if the output register is enabled. Since there are many minor variations on the general concept of a pipelined SSRAM, it is difficult to present a general template, especially one that can be recognized by synthesis tools. A common alternative approach is to use a CAD tool that generates a memory circuit and a VHDL model of that circuit. We can then instantiate the generated model as a component in a larger system.

5.2.3 MULTIPORT MEMORIES

Each of the memories that we have looked at, both in Section 5.1 and previously in this section, is a *single-port* memory, with just one port for writing and reading data. It has only one address input, even though the data connections may be separated into input and output connections. Thus, a single-port memory can perform only one access (a write or a read operation) at a time. In contrast, a *multiport* memory has multiple address inputs, with corresponding data inputs and outputs. It can perform as

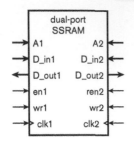

FIGURE 5.16   A dual-port memory.

many operations concurrently as there are address inputs. The most common form of multiport memory is a *dual-port* memory, illustrated in Figure 5.16, which can perform two operations concurrently. (Note that in this context, we are using the term "port" to refer to a combination of address, data and control connections used to access a memory, as distinct from a VHDL port.)

A multiport memory typically consumes more circuit area than a single-port memory with the same number of bits of storage, since it has separate address decoders and data multiplexers for each access port. Only the internal storage cells of the memory are shared between the multiple ports, though additional wiring is needed to connect the cells to the access ports. However, the cost of the extra circuit area is warranted in some applications, such as high performance graphics processing and high-speed network connections. Suppose we have one subsystem producing data to store in the memory, and another subsystem accessing the data to process it in some way. If we use a single-port memory, we would need to multiplex the addresses and input data from the subsystems into the memory, and we would have to arrange the control sections of the subsystems so that they take turns to access the memory. There are two potential problems here. First, if the combined rate at which the subsystems need to move data in and out of the memory exceeds the rate at which a single access port can operate, the memory becomes a bottleneck. Second, even if the average rates don't exceed the capacity of a single access port, if the two subsystems need to access the memory at the same time, one must wait, possibly causing it to lose data. Having separate access ports for the subsystems obviates both of these problems.

The only remaining difficulty is the case of both subsystems accessing the same memory location at the same time. If both accesses are reads, they can proceed. If one or both is a write, the effect depends on the characteristics of the particular dual-port memory. In an asynchronous dual-port memory, a write operation performed concurrently with a read of the same location will result in the written data being reflected on the read port after some delay. Two write operations performed concurrently to the same location result in an unpredictable value being stored. In the case of a synchronous dual-port memory, the effect of concurrent write operations depends on when the operations are performed internally by the memory. We should consult the data sheet for the memory component to understand the effect.

Some multiport memories, particularly those manufactured as packaged components, provide additional circuits that compare the addresses on the access ports and indicate when contention arises. They may also provide circuits to arbitrate between conflicting accesses, ensuring that one proceeds only after the other has completed. If we are using multiport memory components or circuit blocks that do not provide such features and our application may result in conflicting accesses, we need to include some form of arbitration as a separate part of the control section in our design. An alternative is to ensure that the subsystems accessing the memory through separate ports always access separate locations, for example, by ensuring that they always operate on different blocks of data stored in different parts of the memory. We will discuss block processing of data in more detail in Chapter 9.

EXAMPLE 5.7   Develop a VHDL model of a dual-port, 4K × 16-bit flow-through SSRAM. One port allows data to be written and read, while the other port only allows data to be read.

SOLUTION   The entity declaration is

```vhdl
library ieee;
use ieee.std_logic_1164.all, ieee.numeric_std.all;

entity dual_port_SSRAM is
  port ( clk      : in  std_logic;
         en1, wr1 : in  std_logic;
         a1       : in  unsigned(11 downto 0);
         d_in1    : in  std_logic_vector(15 downto 0);
         d_out1   : out std_logic_vector(15 downto 0);
         en2      : in  std_logic;
         a2       : in  unsigned(11 downto 0);
         d_out2   : out std_logic_vector(15 downto 0) );
end entity dual_port_SSRAM;
```

The clk input is common to both memory ports. The inputs and outputs with names ending in "1" are the connections for the read/write memory port, and the inputs and outputs with names ending in "2" are the connection for the read-only memory port. The architecture is

```vhdl
architecture synth of dual_port_SSRAM is
  type RAM_4Kx16 is array (0 to 4095)
                         of std_logic_vector(15 downto 0);
  signal data_RAM : RAM_4Kx16;
begin

  read_write_port : process (clk) is
  begin
    if rising_edge(clk) then
      if en1 = '1' then
        if wr1 = '1' then
          data_RAM(to_integer(a1)) <= d_in1; d_out1 <= d_in1;
        else
          d_out1 <= data_RAM(to_integer(a1));
        end if;
```

(continued)

```
      end if;
    end if;
  end process read_write_port;

  read_only_port : process (clk) is
  begin
    if rising_edge(clk) then
      if en2 = '1' then
        d_out2 <= data_RAM(to_integer(a2));
      end if;
    end if;
  end process read_only_port;

end architecture synth;
```

This is much like our earlier model of a flow-through SSRAM, except that there are two processes, one for each memory port. The declaration of the type and signal for the memory storage is the same, with the signal being shared between the two processes. The process read_write_port is identical in form to the process we introduced earlier. The process read_only_port is a simplified version, since it does not need to deal with updating the storage signal.

In this model, we make no special provision for the possibility of concurrent write and read accesses to the same address. During simulation of the model, one or other process would be activated first. If read_write_port is activated first, it updates the memory location, and the read operation yields the updated value. On the other hand, if read_only_port is activated first, it reads the old value before the location is updated. When the model is synthesized, the synthesis tool chooses a dual-port memory component from its library. The effect of a concurrent write and read would depend on the behavior of the chosen component.

One specialized form of dual-port memory is a *first-in first-out* memory, or FIFO. It is used to queue data arriving from a source to be processed in order of arrival by another subsystem. The data that is first in to the FIFO is the first that comes out; hence, the name. The most common way of building a FIFO is to use a dual-port memory as a *circular buffer* for the data storage, with one port accepting data from the source and the other port reading data to provide to the processing subsystem. Each port has an address counter to keep track of where data is written or read. Data written to the FIFO is stored in successive free locations. When the write-address counter reaches the last location, it wraps to location 0. As data is read, the read-address counter is advanced to the next available location, also wrapping to 0 when the last location is reached. If the write address wraps around and catches up with the read address, the FIFO is full and can accept no more data. If the read address catches up with the write address, the FIFO is empty and can provide no more data. This scheme is similar to that used for the audio echo effects unit in Example 5.2, except that the distance between the write and read addresses is not fixed. Thus, a FIFO can store a variable amount of data, depending on the rates of writing and reading data. The size of memory needed in a FIFO depends on the maximum amount by which reading of data lags writing. Determining the maximum size may be difficult to do. We may need to evaluate worst-case scenarios for our application using mathematical or statistical models of data rates or using simulation.

EXAMPLE 5.8   Design a FIFO to store up to 256 data items of 16 bits each, using a 256×16-bit dual-port SSRAM for the data storage. The FIFO should provide status outputs, as shown in the symbol in Figure 5.17, to indicate when the FIFO is empty and full. Assume that the FIFO will not be read when it is empty, nor be written to when it is full, and that the write and read ports share a common clock.

SOLUTION   The datapath for the FIFO, shown in Figure 5.18, uses 8-bit counters for the write and read addresses. The write address refers to the next free location in the memory, provided the FIFO is not full. The read address refers to the next location to be read, provided the FIFO is not empty. Both counters are cleared to 0 when the reset signal is active.

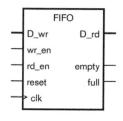

FIGURE 5.17   Symbol for a FIFO with empty and full status outputs.

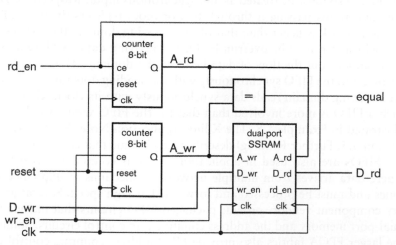

FIGURE 5.18   Datapath for a FIFO using a dual-port memory.

The FIFO being empty is indicated by the two address counters having the same value. The FIFO is full when the write counter wraps around and catches up with the read counter, in which case the counters have same value again. So equality of the counters is not sufficient to distinguish between the cases of the FIFO being empty or full. We could keep track of the number of items in the FIFO, for example, by using a separate up/down counter to count the number of items rather than trying to compare the addresses. However, a simpler way is to keep track of whether the FIFO is filling or emptying. A write operation

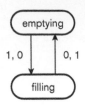

FIGURE 5.19 Transition diagram for the FIFO FSM.

without a concurrent read means the FIFO is filling. If the write address becomes equal to the read address as a consequence of the FIFO filling, the FIFO is full. A read operation without a concurrent write means the FIFO is emptying. If the write address becomes equal to the read address as a consequence of the FIFO emptying, the FIFO is empty. If a write and a read operation occur concurrently, the amount of data in the FIFO remains unchanged, so the filling or emptying state remains unchanged. We can describe this behavior using an FSM, as shown in Figure 5.19, in which the transitions are labeled with the values of the wr_en and rd_en control signals, respectively. The FSM starts in the emptying state. The empty status output is 1 if the current state is emptying and the equal signal is 1, and the full status output is 1 if the current state is filling and the equal signal is 1. Note that this control sequence relies on the assumption of a common clock between the two FIFO ports, since the FSM must have a single clock to operate.

One important use for FIFOs is to pass data between subsystems operating with different clock frequencies, that is, between different *clock domains*. As we discussed in Section 4.4.1, when data arrives asynchronously, we need to resynchronize it with the clock. If the clocks of two clock domains are not in phase, data arriving at one clock domain from the other could change at any time with respect to the receiving domain's clock, and so must be treated as an asynchronous input. Resynchronizing the data means passing it through two or more registers. If the sending domain's clock is faster than that of the receiving domain, the data being resynchronized may be overrun by further arriving data. A FIFO allows us to smooth out the flow of data between the domains. Data arriving is written into the FIFO synchronously with the sending domain's clock, and the receiving domain reads data synchronously with its clock. Control of such a FIFO is more involved than that for the FIFO with a single clock illustrated in Example 5.8. The Xilinx Application Note, XAPP 051 (see Section 5.5, Further Reading) describes a technique that can be used.

FIFOs are also used in applications such as computer networking, where data arrives from multiple network connections at unpredictable times and must be processed and forwarded at high speed. Several memory component vendors provide packaged FIFO circuits that include the dual-port memory and the address counting and control circuits. Some of the larger FPGA fabrics also provide FIFO address counting control circuits that can be used with built-in memory blocks. If we need a FIFO in a system implemented in other fabrics, we can either design one, as we did in Example 5.8, or use a FIFO block from a library or a generator tool.

5.2.4 DYNAMIC RAM

Dynamic RAM (DRAM) is another form of volatile memory that uses a different form of storage cell for storing data. We mentioned in Section 5.2.1 that static RAM uses storage cells that are similar to D-latches. In contrast, a storage cell for a dynamic RAM uses a single capacitor and a single transistor, illustrated in Figure 5.20. The DRAM cells are thus much smaller than SRAM cells, so we can fit many more of them on a chip, making the cost per bit of storage lower. However, the access times of DRAMs are longer than those of SRAMs, and the complexity of access and control is greater. Thus, there is a trade-off of cost, performance and complexity against memory capacity. DRAMs are most commonly used as the main memory in computer systems, since they satisfy the need for high capacity with relatively low cost. However, they can also be used in other digital systems. The choice between SRAM and DRAM depends on the requirements and constraints of each application.

A DRAM represents a stored 1 or 0 bit in a cell by the presence or absence of charge on the capacitor. When the transistor is turned off, the capacitor is isolated from the bit line, thus storing the charge on the capacitor. To write to the cell, the DRAM control circuit pulls the bit line high or low and turns on the transistor, thus charging or discharging the capacitor. To read from the cell, the DRAM control circuit precharges the bit line to an intermediate level, then turns on the transistor. As the charges on the capacitor and the bit line equalize, the voltage on the bit line either increases slightly or decreases slightly, depending on whether the storage capacitor was charged or discharged. A sensor detects and amplifies the change, thus determining whether the cell stored a 1 or a 0. Unfortunately, this process destroys the stored value in the cell, so the control circuit must then restore the value by pulling the bit line high or low, as appropriate, before turning off the transistor. The time taken to complete the restoration is added to the access time, making the overall read cycle significantly longer than than that for an SRAM.

Another property of a DRAM cell is that, while the transistor is turned off, charge leaks from the capacitor. This is the meaning of the term "dynamic" applied to DRAMs. To compensate, the control circuit must read and restore the value in each cell in the DRAM before the charge decays too much. This process is called *refreshing* the DRAM. DRAM manufacturers typically specify a period of 64ms between refreshes for each cell. The cells in a DRAM are typically organized into several rectangular arrays, called banks, and the DRAM control circuit is organized to refresh one row of each bank at a time. Since the DRAM cannot perform a normal write or read operation while it is refreshing a row, the refresh operations must be interleaved between writes and reads. Depending on the application, it may be possible to refresh all rows in a burst once every 64ms. Alternatively, we may have to refresh one row at a time between writes and reads, making sure that all rows are refreshed within 64ms. The important thing is to

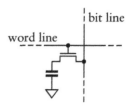

FIGURE 5.20   A DRAM storage cell.

avoid scheduling a refresh when a write or read is required and cannot be deferred.

Historically, timing of DRAM control signals used to be asynchronous, and management of refreshing was performed by control circuits external to the DRAM chips. More recently, manufacturers changed to synchronous DRAMs (SDRAMs) that use registers on inputs to sample address, data and control signals on clock edges. This is analogous to the difference between asynchronous and synchronous SRAMs, and makes it easier to incorporate DRAMs into systems that use a clocked synchronous timing methodology. Manufacturers have also incorporated refresh control circuits into the DRAM chips, also making use of DRAMs easier. Since applications with very high data transfer rate requirements may be limited by the relatively slow access times of DRAMs, manufacturers have more recently incorporated further features to improve performance. These include the ability to access a burst of data from successive locations without having to provide the address for each, other than the first, and the ability to transfer on both rising and falling clock edges (double-data rate, or DDR, and its successors, DDR2 and DDR3). These features are mainly motivated by the need to provide high-speed bursts of data in computer systems, but they can also be of benefit in noncomputer digital systems.

Because of the relative complexity of controlling DRAMs, we will not go into detail of the control signals required and their sequencing. For most implementation fabrics, we can incorporate a DRAM control block from a library, allowing us to connect external DRAMs to the sequential circuits in our chip. An example is the SDRAM controller, described in Xilinx Application Note XAPP134, that allows an FPGA-based system to connect to and control an external SDRAM memory (see Section 5.5, Further Reading).

5.2.5   READ-ONLY MEMORIES

The memories that we have looked at so far can both read the stored data and update it arbitrarily. In contrast, a *read-only memory*, or ROM, can only read the stored data. This is useful in cases where the data is constant, so there is no need to update it. It does, of course, beg the question of how the constant data is placed in the ROM in the first place. The answer is that the data is either incorporated into the circuit during its manufacture, or is programmed into the ROM subsequently. We will describe a number of kinds of ROM that take one or other of these approaches.

Combinational ROMs

A simple ROM is a combinational circuit that maps from an input address to a constant data value. We could specify the ROM contents in tabular form, with a row for each address and an entry showing the data value for that address. Such a table is essentially a truth table, so we could, in principle, implement the mapping using the combinational circuit design techniques we described in Chapter 2. However, ROM circuit structures are generally much denser than arbitrary gate-based circuits, since each ROM cell needs at most one transistor. Indeed, for a complex combinational function with multiple outputs, it may be better to use a ROM to implement the function than a gate-based circuit. For example, a ROM might be a good candidate for the next-state logic or the output logic of a complex finite-state machine.

EXAMPLE 5.9   Design a 7-segment decoder with blanking input, as described in Example 2.16 on page 68, using a ROM.

SOLUTION   The decoder has five input bits: four for the BCD code and one for the blanking control. It has seven output bits: one for each segment. Thus, we need a 32×7-bit ROM, as shown in Figure 5.21. The contents of the ROM are given in Table 5.2.

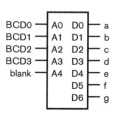

FIGURE 5.21   A 32×7-bit ROM used as a 7-segment decoder.

ADDRESS	CONTENT	ADDRESS	CONTENT
0	0111111	6	1111101
1	0000110	7	0000111
2	1011011	8	1111111
3	1001111	9	1101111
4	1100110	10–15	1000000
5	1101101	16–31	0000000

TABLE 5.2   ROM contents for the 7-segment decoder.

EXAMPLE 5.10   Develop a VHDL model of the 7-segment decoder of Example 5.9.

SOLUTION   The entity is the same as in Example 2.16. The architecture is

```
library ieee; use ieee.numeric_std.all;

architecture ROM_based of seven_seg_decoder is
  type ROM_array is array (0 to 31)
                    of std_logic_vector(7 downto 1);
  constant ROM_content : ROM_array
    := ( 0 => "0111111", 1 => "0000110",
         2 => "1011011", 3 => "1001111",
         4 => "1100110", 5 => "1101101",
```

(continued)

```
            6 => "1111101", 7 => "0000111",
            8 => "1111111", 9 => "1101111",
            10 to 15 => "1000000",
            16 to 31 => "0000000" );
   begin
     seg <= ROM_content(to_integer(unsigned(blank & bcd)));
   end architecture ROM_based;
```

We declare a type, ROM_array, in the same way as we do for a RAM. However, instead of including a signal declaration for the memory storage, we include a constant declaration, ROM_content. We specify the name and type for the constant, as well as the value after the ":=" symbol. In this case, since the value is an array, we list the values of the elements in parentheses. For elements 0 through 9, we specify the index and the corresponding element value. For elements 10 through 15 and 16 through 31, we specify the range of elements that have like values, along with the element value. In the statement part of the architecture, we have an assignment that constructs the ROM address by concatenating the blank input and the encoded bcd input. Since this results in a std_logic_vector value, we convert it to an unsigned value and then to an integer value to index the ROM_content constant.

In FPGA fabrics that provide SSRAM blocks, we can use an SSRAM block as a ROM. We simply declare a constant for the data instead of a signal, as in Example 5.10, and modify the process template for the memory to omit the part that updates the memory content. For example,

```
type ROM_512x20 is array (0 to 511)
                   of std_logic_vector(19 downto 0);
constant data_ROM : ROM_512x20
  := (X"00000", X"0126F", ...);
...

FPGA_ROM : process (clk) is
begin
  if rising_edge(clk) then
    if en = '1' then
      d_out <= data_ROM(to_integer(a));
    end if;
  end if;
end process FPGA_ROM;
```

The content of the memory is loaded into the FPGA as part of its programming when the system is turned on. Thereafter, since the data is not updated, it is constant.

Programmable ROMs

ROMs in which the contents are manufactured into the memory are suitable for applications where the number of manufactured parts is high and where we are sure that the contents will not need to change over the lifetime of the product. In other applications, we would prefer to be able to revise the ROM contents from time to time, or to use a form of ROM with lower costs for low-volume production. A *programmable ROM* (PROM) meets these requirements. It is manufactured as a separately packaged chip with no content stored in its memory cells. The memory contents are programmed into the cells after manufacture, either using a special programming device before the chip is assembled into a system, or using special programming circuits when the chip is in the final system.

There are a number of forms of PROMs. Early PROMs used fusible links to program the memory cells. Once a link was fused, it could not be replaced, so programming could only be done once. These devices are now largely obsolete. They were replaced by PROMs that could be erased, either with ultraviolet light (so called EPROMs), or electrically using a higher-than-normal power-supply voltage (so-called electrically erasable PROMs, or EEPROMs).

Flash Memories

Most new designs use flash memory, which is a form of electrically erasable programmable ROM. It is organized so that blocks of storage can be erased at once, followed by programming of individual memory locations. A flash memory typically allows only a limited number of erasure and programming operations, typically hundreds of thousands, before the device "wears out." Thus, flash memories are not a suitable replacement for RAMs.

There are two kinds of flash memories, NOR and NAND flash, referring to the organization of the transistors that make up the memory cells. Both kinds are organized as blocks (commonly of 16, 64, 128, or 256 Kbytes) that must be erased in whole before being written. In a NOR flash memory, locations can then be written (once per erasure) and read (an arbitrary number of times) in random order. The IC has similar address, data and control signals to an SRAM and can read data with a comparable access time, making it suitable for use as a program memory for an embedded processor, for storing configuration parameters to be used to control system operation, and for storing configuration information for FPGAs.

In a NAND flash memory, on the other hand, locations are written and read one page at a time, a page being typically 2 Kbytes. Read access to a given location would require reading the page containing the location, followed by selection of the required data, taking several microseconds. If all of the locations in a page are required, however, sequential reading is much faster, comparable in time to SRAM. Erasing a block and writing a page of data are significantly slower than SRAM access times.

For example, the data sheet for the Micron Technology MT29F16G08FAA 16G bit IC specifies a random read time of 25µs, a sequential read time of 25ns, a block erase time of 1.5ms, and a page write time of 220µs. Given their different access behavior, NAND flash memories have a different interface than SRAMs, making control circuits more involved. The advantage of NAND flash memory is that the density of storage cells is greater than that of NOR flash. Thus, NAND flash chips are better suited to applications in which large amounts of data must be stored cheaply. One of the largest applications of NAND flash memories is in memory cards for consumer devices such as digital cameras. They are also used in USB memory sticks for general purpose computers.

KNOWLEDGE TEST QUIZ

1. What is the difference between RAM and ROM?

2. What is meant by the terms volatile and nonvolatile?

3. What is the difference between static and dynamic RAM?

4. What is meant by the access time of a RAM?

5. Why are asynchronous SRAMs difficult to use in high-speed clocked synchronous designs?

6. What is the difference between flow-through and pipelined SSRAMs?

7. What VHDL type is required for a signal to represent memory storage?

8. What benefit does a multiport memory have over a single-port memory with multiplexed address and data connections?

9. How can we work out what will happen if we perform concurrent writes to a given location in a synchronous dual-port memory?

10. What does FIFO stand for?

11. How does a FIFO facilitate communication of data between clock domains?

5.3 ERROR DETECTION AND CORRECTION

In most of our discussions, we have assumed that digital circuits store and process information correctly, though in Section 2.2.2 we did introduce the idea of bit errors and some approaches to dealing with them. Bit errors can occur in memories from a number of causes. Some errors are *transient*, also called *soft errors*, and involve a bit flip in a memory cell without a permanent effect on the cell's capacity to store data. In DRAMs, soft errors are typically caused by high-energy neutrons generated by collision of cosmic rays with atoms in the earth's atmosphere. The neutrons collide with silicon atoms in the DRAM chip, leaving a stream of charge that can disrupt the storage or reading of charge in a DRAM cell. The frequency of soft-error occurrence, the *soft-error rate*, depends on the way in which DRAMs are manufactured and the location in which they operate. Hence, soft-error rates are highly variable between systems. Soft errors can also occur in DRAMs and other memories from electrical interference, the effects of poor physical circuit design and other causes.

Errors that persist in a memory circuit are called *hard errors*. They can result from manufacturing defects or from electrical "wear" after prolonged use. A memory cell or chip affected by a hard error is no longer able to store data. A read operation would always yield a 0 or a 1 value, regardless of the bit value that was previously written.

Given that memories are more susceptible to bit errors than logic circuits using flip-flops and registers for storage, due to the storage density and the longevity of data in memories, it is more common to include some form of error detection in memory circuits than in logic circuits. A common approach is to use parity, described in Section 2.2.2. Recall that parity involves counting the number of 1 bits in a code word and setting a parity bit to 1 or 0 to ensure that the total number of 1 bits is even (if we choose even parity) or odd (if we choose odd parity). In the case of memories, use of parity involves adding an extra bit cell to each memory location. When we write to a location, we compute the parity bit and store it in the extra cell. When we read a location, we check that the data, together with the parity bit, have the correct parity. If so, we assume the data is uncorrupted. Otherwise, we take appropriate action to deal with the error in the stored data.

The problem with using parity to check for errors, as we discussed in Section 2.2.2, is that it only allows us to detect a single bit flip in a stored code word. It does not allow us to identify which bit flipped, nor does it allow us to detect an even number of bit flips. If we could identify the particular bit that flipped, we could correct the error by flipping the bit back to its original value, and then continue operating as normal. We could also write the corrected data back to the memory on the assumption that the bit flip was a soft error. In order to be able to identify which bit flipped, we need to consider the invalid code words that result from flipping each bit of each valid code word. Provided all of those invalid code words are distinct, we can use the value of the invalid code word to identify the flipped bit.

One scheme for doing this is to use a form of *error correcting code* (ECC) known as a *Hamming code*. We will start with a single-error correcting Hamming code, that is, a code that allows us to correct a single bit flip within a code word. If our code word has N bits, we need $\log_2 N + 1$ additional check bits for the ECC. For example, if we have 8 data bits, we need 4 check bits, giving a total of 12 bits. The check bits are computed from the values of the data bits during a write operation, and the entire ECC word is written to the memory location.

FIGURE 5.22   Distribution
of data and check bits within an
ECC word.

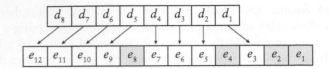

To illustrate how the check bits are computed, we will number the data bits of an 8-bit code word d_1 through d_8 and the ECC bits e_1 through e_{12}. (Normally, we've numbered bits starting from 0, but for this explanation, it's more convenient to number all index positions from 1.) The ECC bits whose indices are powers of 2 are used as check bits, and the remaining ECC bits are the data bits, in order, as shown in Figure 5.22. If we write the indices of the ECC bits in binary, the check bit with a 1 in position i of its index is the exclusive-OR (that is, the parity) of the data ECC bits that have a 1 in position i of their indices. For example, check bit e_2 (at index 0010_2) is the exclusive-OR of data bits e_3, e_6, e_7, e_{10} and e_{11} (at indices 0011_2, 0110_2, 0111_2, 1010_2 and 1011_2). Since each data ECC bit has at least two 1 bits in its binary index (otherwise it would be a check bit), each data bit is included in the computation of at least two check bits.

When the memory location is read, again, the entire ECC word is read. We recompute the values of the check bits from the data ECC bits and compare them, using a bit-wise exclusive OR, with the check bits read from memory. If the comparison result is 0000, the recomputed check bits match the read check bits, so all is well. However, if one of the stored ECC bits (either a data bit or a check bit) is flipped from the original, the comparison result, called the *syndrome*, will be other than 0000. It turns out to be the binary index of the ECC bit that has flipped. Thus, we can use the syndrome value to correct the error by flipping the indexed bit back.

EXAMPLE 5.11   Compute the 12-bit ECC word corresponding to the 8-bit data word 01100001.

SOLUTION   The check bits are:

$$e_1 = e_3 \oplus e_5 \oplus e_7 \oplus e_9 \oplus e_{11} = d_1 \oplus d_2 \oplus d_4 \oplus d_5 \oplus d_7 = 1 \oplus 0 \oplus 0 \oplus 0 \oplus 1 = 0$$

$$e_2 = e_3 \oplus e_6 \oplus e_7 \oplus e_{10} \oplus e_{11} = d_1 \oplus d_3 \oplus d_4 \oplus d_6 \oplus d_7 = 1 \oplus 0 \oplus 0 \oplus 1 \oplus 1 = 1$$

$$e_4 = e_5 \oplus e_6 \oplus e_7 \oplus e_{12} = d_2 \oplus d_3 \oplus d_4 \oplus d_8 = 0 \oplus 0 \oplus 0 \oplus 0 = 0$$

$$e_8 = e_9 \oplus e_{10} \oplus e_{11} \oplus e_{12} = d_5 \oplus d_6 \oplus d_7 \oplus d_8 = 0 \oplus 1 \oplus 1 \oplus 0 = 0$$

Thus the ECC word is 011000000110.

EXAMPLE 5.12   Determine whether there is an error in the ECC word 110111000110, and if so, correct it.

SOLUTION   The check bits computed from the data bits of the ECC word are

$e_1 = e_3 \oplus e_5 \oplus e_7 \oplus e_9 \oplus e_{11} = 1 \oplus 0 \oplus 1 \oplus 1 \oplus 1 = 0$

$e_2 = e_3 \oplus e_6 \oplus e_7 \oplus e_{10} \oplus e_{11} = 1 \oplus 0 \oplus 1 \oplus 0 \oplus 1 = 1$

$e_4 = e_5 \oplus e_6 \oplus e_7 \oplus e_{12} = 0 \oplus 0 \oplus 1 \oplus 1 = 0$

$e_8 = e_9 \oplus e_{10} \oplus e_{11} \oplus e_{12} = 1 \oplus 1 \oplus 0 \oplus 1 = 1$

The syndrome is $1010 \oplus 1010 = 0000$. Thus, there is no error in the read ECC.

EXAMPLE 5.13   Determine whether there is an error in the ECC word 000111000100, and if so, correct it.

SOLUTION   The check bits computed from the data bits of the ECC word are

$e_1 = e_3 \oplus e_5 \oplus e_7 \oplus e_9 \oplus e_{11} = 1 \oplus 0 \oplus 1 \oplus 1 \oplus 0 = 1$

$e_2 = e_3 \oplus e_6 \oplus e_7 \oplus e_{10} \oplus e_{11} = 1 \oplus 0 \oplus 1 \oplus 0 \oplus 0 = 0$

$e_4 = e_5 \oplus e_6 \oplus e_7 \oplus e_{12} = 0 \oplus 0 \oplus 1 \oplus 0 = 1$

$e_8 = e_9 \oplus e_{10} \oplus e_{11} \oplus e_{12} = 0 \oplus 0 \oplus 0 \oplus 1 = 1$

The syndrome is $1101 \oplus 1000 = 0101$. Thus, there is an error in bit e_5 of the read ECC. That bit should be flipped back from 0 to 1, giving the corrected ECC word 000111010100.

Note that we have assumed that only one bit of the stored ECC word could be in error. If two or more bits flip, the checking process may incorrectly identify a single bit as having flipped, or it may yield an invalid syndrome. The problem arises from the fact that we have insufficient invalid code words to distinguish between single-bit errors and double-bit errors. A simple remedy is to add further check bits. If we add a check bit that is the exclusive-OR of all of the data bits, the resulting error-checking code allows us to correct any single-bit error and to detect (but not correct) any double-bit error. If we assume that errors are independent, the probability of a double-bit error is very low, so this scheme suffices in many applications. If extreme reliability and resilience to errors is required, we can further extend the error-checking code to enable correcting of multiple-bit errors. The details of how we might do this are beyond the scope of this book, but are described in Section 5.5, Further Reading.

A final consideration in our discussion of error checking and correcting for memories is the storage overhead required. In our illustration of ECCs for 8-bit code words, we saw that correcting single-bit errors requires 4 check bits (a 50% overhead) and detecting double-bit errors requires 5 check bits (a 63% overhead). This is clearly a significant storage overhead, especially when compared to the single parity bit required just to detect single-bit errors (a 13% overhead). However, we noted that single-bit correction using Hamming codes needs $\log_2 N + 1$ check bits for N bits of data. Double-bit error detection needs $\log_2 N + 2$ check bits. If we provide checking and correction over longer data words, the relative storage overhead is less, as shown in Table 5.3. For larger data words, provision of this form of error detection and correction is increasingly attractive.

TABLE 5.3 Number of check bits and relative storage overhead for single-bit correction and additional double-bit detection of errors.

N	SINGLE-BIT CORRECTION		DOUBLE-BIT DETECTION	
	CHECK BITS	OVERHEAD	CHECK BITS	OVERHEAD
8	4	50%	5	63%
16	5	31%	6	38%
32	6	19%	7	22%
64	7	11%	8	13%
128	8	6.3%	9	7.0%
256	9	3.5%	10	3.9%

There are other, more elaborate, error correction and detection codes that we can use as alternatives to Hamming codes. However, they also add check bits to the data, and so require extra storage capacity and extra circuitry to detect and correct errors. They differ in the storage overhead and the complexity of the additional circuitry, as well as in the number of simultaneous errors they can deal with. This range of techniques allows us to make design trade-offs, depending on the reliability requirements and other constraints of our application. Since Hamming codes are one of the simplest ECCs, they are most often used in applications requiring moderately high reliability, such as network server computers. More complex ECCs are used in specialized high-reliability applications, such as aerospace computers and communications systems.

KNOWLEDGE TEST QUIZ

1. What is the distinction between a soft error and a hard error?

2. What is a common cause of soft errors in DRAMs?

3. What corrective action can we take when a parity error is detected?

4. Using a Hamming code, how many check bits are required for single-error correction and double-error detection for 4-bit data words?

5.4 CHAPTER SUMMARY

▸ A memory contains an array of storage locations, each with a unique address. A $2^n \times m$-bit memory has n-bit addresses that run from 0 to $2^n - 1$.

▸ A write operation stores a data value at a given location. A read operation yields the data value stored at a given location. Control signals govern write and read operations.

▸ We can connect multiple memory components in parallel to store wider data values. We can connect multiple memory components in banks, with a decoder to select among the banks, to provide more locations.

▸ Memories with tristate drivers on the data outputs simplify bank connection. At most one component drives data outputs at a time; the rest place their outputs in the high-impedance (hi-Z) state.

▸ Volatile memory only retains data for as long as power is applied. Nonvolatile memory retains data without power. The term RAM refers to volatile memory that can be written and read with equal facility in any order. ROM refers to memory that can only be read once it is manufactured or programmed.

▸ Data in static RAM (SRAM) persists for as long as power is supplied, whereas data in dynamic RAM (DRAM) must be periodically refreshed. Asynchronous SRAM does not rely on a clock for its timing. Synchronous SRAM (SSRAM) uses a clock to sample control, address and data signals, thus simplifying their incorporation into clocked synchronous systems. SSRAMs include flow-through and pipelined variants.

▸ The access time is the delay from starting a read operation to having valid data. The cycle time is the total time taken for a read or write operation.

▸ Multiport memories allow concurrent operations by different parts of a digital system. A first-in first-out (FIFO) is a dual-port memory used as a queue for data. An important use of FIFOs is to pass data between different clock domains.

▸ A ROM is a combinational circuit that maps from an address to a data value. It can be used to implement an arbitrary Boolean function.

▸ Programmable ROMs (PROMs) are programmed with data after manufacture. Flash memories can be erased and reprogrammed during system operation, and are useful for storing configuration information.

▶ Atmospheric neutrons and other effects can cause bit errors in data stored in a memory. The error may be transient (a soft error) or permanent (a hard error).

▶ Check bits can be stored along with data to detect and correct errors. A single parity bit can detect a single-bit error but not a double-bit error. Error correcting codes, such as Hamming codes, can correct single-bit errors and detect double-bit errors.

5.5 FURTHER READING

Advanced Semiconductor Memories: Architectures, Designs, and Applications, Ashok K. Sharma, Wiley-IEEE Press, 2002. Describes a range of memory devices, including SRAMS, DRAMS and nonvolatile memories.

Computer Organization and Design: The Hardware/Software Interface, David A. Patterson and John L. Hennessy, Morgan Kaufmann Publishers, 2005. This book contains a chapter on memory system design for computers, describing how alternative organizations can improve memory system performance.

Memory Systems: Cache, DRAM, Disk—A Holistic Approach to Design, Bruce Jacob, Spencer Ng, and David Wang, Morgan Kaufmann Publishers, 2007. Includes an extensive description of DRAM technology and its place in computer memory systems. Also describes error-correcting codes, including Hamming codes and more elaborate schemes, and the causes and frequency of occurrence of memory errors.

Synchronous and Asynchronous FIFO Designs, Peter Alfke, Xilinx Application Note XAPP051, 1996, http://direct.xilinx.com/bvdocs/appnotes/xapp051.pdf. Describes a FIFO control scheme for an FPGA in which the write and read clocks are different.

Synthesizable High-Performance SDRAM Controllers, Xilinx Application Note XAPP134, 2005, http://www.xilinx.com/bvdocs/appnotes/xapp134.pdf. This application note gives an overview of SDRAM operation and describes a controller subsystem that can be implemented as part of an FPGA-based design.

A Nonvolatile Memory Overview, Jitu J. Makwana and Dieter K. Schroder, 2004, http://aplawrence.com/Makwana/nonvolmem.html. Describes the circuit structures and operation of nonvolatile memory devices.

EXERCISE 5.1 A system requires storage for 1 second of video from a camera. The video data consists of 25 frames per second, with each frame containing 640×480 pixels of 24 bits. How much memory is required?

EXERCISE 5.2 Suppose 8-bit-wide memory ICs that are used to construct a memory have a read cycle time of 6ns. The application requires data to be processed at a rate of 400Mbyte/sec. How wide should the memory be to satisfy the performance requirement?

EXERCISE 5.3 Figure 5.23 shows a symbol for a $512K \times 8$-bit memory component. Draw a schematic diagram showing connection of these components to form a $512K \times 32$-bit memory.

EXERCISE 5.4 Draw a schematic diagram showing connection of the components of Figure 5.23 to form a $1M \times 8$-bit memory.

EXERCISE 5.5 Draw a schematic diagram showing connection of the components of Figure 5.23 to form a $2M \times 16$-bit memory.

EXERCISE 5.6 Figure 5.24 shows a symbol for a $512K \times 8$-bit memory component with tristate bidirectional data input/output connections. Draw a schematic diagram showing connection of these components to form a $2M \times 16$-bit memory.

EXERCISE 5.7 Suppose a datapath connected to an asynchronous SRAM has control signals addr_sel to select the memory address and d_out_en to enable the memory data tristate drivers. The control section generates these control signals as well as the control signals for the SRAM, as described in Section 5.2.1 and shown in Figure 5.10. Develop a control sequence for a write operation that ensures that the setup and hold constraints are met. How many clock cycles does a write operation take using your control sequence?

EXERCISE 5.8 Develop a VHDL model of the circuit described in Example 5.4 using a flow-through SSRAM.

EXERCISE 5.9 Develop a VHDL model of the revised circuit described in Example 5.5 using a pipelined SSRAM.

EXERCISE 5.10 Revise the circuit described in Example 5.4 to compute $y = c_i^2 \times x$ using a flow-through SSRAM.

EXERCISE 5.11 Develop a VHDL model of the revised circuit described in Exercise 5.10.

EXERCISE 5.12 Revise the circuit described in Examples 5.4 and 5.5 to compute $y = c_i^2 \times x$ using a pipelined SSRAM.

EXERCISES

FIGURE 5.23

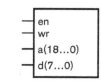

FIGURE 5.24

EXERCISE 5.13 Develop a VHDL model of the revised circuit described in Exercise 5.12.

EXERCISE 5.14 Design a circuit that computes the product of two 128-element vectors, a and b; that is, a vector p such that $p_i = a_i \times b_i$. The elements of a and b are stored in separate flow-through SSRAMs, and the result is to be written into a third flow-through SSRAM. Assume that computation is started by a control signal, go, being 1 during a clock cycle. An output control signal, done, is to be set to 1 during the cycle when the computation is complete.

EXERCISE 5.15 Develop a VHDL model of the circuit described in Exercise 5.14.

EXERCISE 5.16 Revise the circuit of Exercise 5.14 to use pipelined SSRAMs instead of flow-through SSRAMs.

EXERCISE 5.17 Develop a VHDL model of the revised circuit described in Exercise 5.16.

EXERCISE 5.18 Revise the model of Example 5.7 so that, in the case of concurrent write and read operations to the same address, the read operation always yields

a)   the original value before the write

b)   the value written

EXERCISE 5.19 Revise the model of Example 5.7 to allow writes through both ports. Use a simulator to observe the effect of concurrent writes to the same address.

EXERCISE 5.20 Design an arbiter for a dual-port flow-through SSRAM. For each port, the arbiter provides a busy signal that indicates the operation on the port cannot proceed. Concurrent read and write operations proceed without either busy being activated, provided the addresses are different or both are reads. For a read and a write occurring at the same address concurrently, the write proceeds, and the reading port has its busy signal activated. For two writes occurring at the same address concurrently, port 1 proceeds and port 2 has its busy signal activated.

EXERCISE 5.21 Develop a VHDL model of the FIFO described in Example 5.8.

EXERCISE 5.22 Suppose a system includes a data source that provides a stream of 16-bit data values and a processing unit that operates on the stream, as shown in Figure 5.25. The source provides successive values at irregular intervals, sometimes faster than they can be processed, and sometimes slower. It has a ready output that is 1 during a clock cycle when a data item is available. The process-

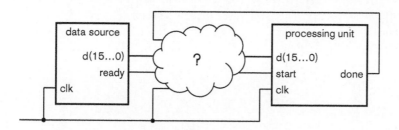

FIGURE 5.25

ing unit has a start control input to initiate processing and a done output that is set to 1 for a cycle when the data item is processed. Show how the source and processing unit can be connected using the FIFO of Example 5.8, including any control sequences required. Assume that if the FIFO is full when a new data item is provided by the source, the data item is dropped from the stream.

EXERCISE 5.23 The data sheet for a Micron Technolog MT48LC128M4A2 512M bit SDRAM describes the device as consisting of four banks, each containing 8,192 rows by 4,096 columns of 4 bits. A refresh operation refreshes a given row of all four banks at once. Locations must be refreshed every 64ms. What is average interval between refresh operations? If the cycle time for data accesses is 7.5ns, what proportion of accesses would conflict with refresh operations?

EXERCISE 5.24 Develop a VHDL model of a 4-bit Gray code to unsigned binary converter implemented using a combinational ROM. (See Section 3.1.3 on page 119 for details of the Gray code.)

EXERCISE 5.25 Using the Hamming code described in Section 5.3, compute the 12-bit ECC word corresponding to the 8-bit data words 10010110 and 01101001.

EXERCISE 5.26 Using the Hamming code described in Section 5.3, determine whether there is an error in each of the following ECC words, and if so, determine the corrected ECC word and the original data value.

a)   100100011010

b)   000110111000

c)   111011011101

EXERCISE 5.27 Draw a diagram, similar to Figure 5.22, showing the distribution of data and check bits within an ECC word for 16-bit data. Write the Boolean equations for the check bits.

IMPLEMENTATION FABRICS

6

The hardware of a digital system is implemented using integrated circuits connected together on printed circuit boards. In this chapter, we describe the range of integrated circuits that are used for digital systems. We also discuss some of the important characteristics of integrated circuits and printed circuit boards that give rise to constraints on our designs.

6.1 INTEGRATED CIRCUITS

The history of digital logic circuits predates the invention of integrated circuits. Early digital systems were constructed using discrete switching components, such as relays, vacuum tubes, and transistors. However, the ability to manufacture a complete circuit on the surface of a silicon wafer brought about a tremendous cost reduction. Invention of the integrated circuit is credited to Jack Kilby at Texas Instruments in 1958. The techniques were refined by several developers, and the market for ICs grew rapidly during the 1960s. As digital ICs became commodity parts, adoption of digital logic circuits became widespread.

It is instructive to review the history of development of digital IC technology for two reasons. First, we sometimes need to deal with *legacy systems*, that is, systems designed some time ago but that are still in operation and needing maintenance. Where obsolete parts are unavailable, we need to design replacement circuits to keep the system operating. Hence, we need to understand the operation of legacy components and the constraints under which they operate. Second, we need to realize that circuit technology is continually evolving. It's not sufficient for us to learn how to design using current components, since they will be obsolete at some stage in the future. Instead, we need to understand technology evolution and trends, so that we can "future proof" our designs. Understanding history is important for projecting into the future.

In this section, we will review the history of digital logic components and survey the components that are available to us now. We will also

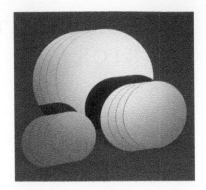

FIGURE 6.1   An ingot of crystalline silicon (left), and sawn wafers (right).

consider the trends affecting near-term evolution of these components and implementation fabrics.

6.1.1 INTEGRATED CIRCUIT MANUFACTURE

Implementation fabrics for digital systems are based on integrated circuits (ICs), which are manufactured on the surface of a wafer of pure crystalline silicon using a sequence of photographic and chemical process steps. A number of identical rectangular ICs are manufactured together, and then broken apart for individual packaging. Hence, we often use the name *silicon chip* to refer to a piece of a silicon on which an IC is manufactured.

In preparation for chip manufacture, a cylindrical ingot of silicon is formed (Figure 6.1, left) and then sawn into wafers less than a millimeter thick and finely polished (Figure 6.1, right). Early wafers were 50mm in diameter, but, since then, improvements in manufacturing processes have yielded successively larger wafer sizes. Now, chips can be manufactured on 300mm diameter wafers. This allows more chips to be manufactured at once, and reduces the waste at the edges.

The process of manufacturing a circuit on the wafer surface involves a number of steps that change the properties of certain areas of the surface silicon, or add a surface layer of some material in certain areas. There are several kinds of processing steps that can be applied to selected areas of the wafer, including

▶ Ion implantation: exposing the surface to a plasma of impurity ions that diffuse into the silicon, thus altering its electrical properties in controlled ways.

▶ Etching: chemically eroding an underlying film of material that has been deposited onto the surface. Films include insulating materials, such as silicon dioxide; semiconducting materials, such as polycrystalline silicon (also known as polysilicon); and conducting materials, such as aluminum and copper.

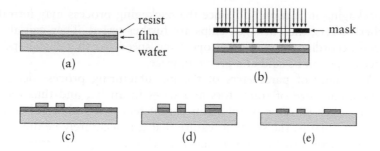

FIGURE 6.2 Steps in photolithographic etching: (a) the wafer and film coated with resist; (b) exposure through a photomask; (c) the resist developed; (d) etching of the underlying film; (e) the remaining resist stripped.

The key to selecting which areas are affected is *photolithography*, which means using a photographic process to draw on the surface (see Figure 6.2, showing selective etching of a film). The surface is coated with a thin layer of *photoresist*, a chemical whose resistance to chemical reaction is changed by exposure to light. The surface is then exposed to light through a mask that has opaque and transparent areas in the pattern of features to be drawn. The resist is then developed, dissolving either the exposed areas or the unexposed areas, depending on the kind of resist. The areas that are now uncoated can be processed, and then the remaining resist is stripped off.

Manufacturing circuits requires several different masks to form the circuit layers, as illustrated in Figure 6.3. MOS transistors are formed in the bottom layers with channel areas containing diffused impurity ions and gates formed from polysilicon lines. Wiring is formed in higher layers using etched metal conductors. A complete wafer contains between a few hundred to a few thousand circuits, depending on the individual circuit size, as shown in Figure 6.4.

Once the circuits on a wafer have been manufactured, they must be tested to determine which ones work and which fail due to defects. Small defects can be introduced into an individual IC by stray particles obscuring light during photolithography, by impurities occurring in chemical process steps, or by particles impinging on wafers during handling in the manufacturing process. IC foundries are meticulous in their cleanliness, using chemicals of high purity and operating in special clean rooms. Nonetheless, stray particles and impurities cannot be completely avoided. A defect can prevent an IC on a wafer from working. The *yield* is the proportion of manufactured ICs that work. Since a whole wafer-lot of ICs is manufactured together, the cost of the discarded defective ICs must be amortized over those that work. Larger ICs have an increased chance of being defective, so it is important for designers to constrain IC area to reduce cost.

After testing the ICs on a wafer, the wafer is broken into individual chips, which are then packaged. We will describe the different kinds

FIGURE 6.3 Graphical representation of the layers of an IC.

FIGURE 6.4 A complete wafer containing multiple ICs.

of packaging in Section 6.3. Since the packaging process may introduce further defects, the packaged chips are further tested. They can also be graded according to maximum operating speed, so that higher performance chips can be sold at a premium price.

A number of parameters of the manufacturing process determine the minimum size of transistors and wires in an IC, and thus the size of a complete IC. One of the main parameters is the photolithography resolution, that is, the smallest feature that can accurately be drawn and processed. Much of the improvement in IC technology is attributable to advances in photolithography, including use of higher-resolution masks and shorter wavelengths of light. Reducing the feature size has a number of benefits. It results in smaller chips for a given function, thus reducing cost. It allows more circuitry to be placed on a chip of a given size, thus increasing functionality. It also reduces circuit delays, allowing higher operating speed. Feature size, along with several other parameters, have been improving exponentially as manufacturing technology has matured. Currently, use of extreme ultraviolet light allows us to manufacture circuits with 90nm feature sizes. Further improvements will allow for 65nm and smaller sizes. These trends are expected to continue for some time yet. The publication *Exponential Trends in the Integrated Circuit Industry* (see Section 6.6) summarizes the trends.

6.1.2 SSI AND MSI LOGIC FAMILIES

While many early ICs were developed for specific applications, in 1961 Texas Instruments introduced a family of logic components that designers could use as building blocks for larger circuits. Three years later, they introduced the 5400 and 7400 families of TTL (transistor-transistor logic) ICs that became the basis of logic design for many years. The 5400 family components were manufactured for high-reliability military applications, requiring operation over large temperature ranges, whereas the 7400 family components were for commercial and industrial applications. Other IC manufacturers also provided compatible components, thus making the 7400 family a *de facto* standard.

The components in the 7400 family are numbered according to the logic functions they provide. For example, a 7400 component provides four NAND gates, a 7427 provides three NOR gates, and a 7474 provides two D flip-flops. Since these components integrate a relatively small number of circuit elements, they are referred to as *small-scale integrated* (SSI) components. As manufacturing techniques improved, larger circuits could be integrated, leading to what we now call *medium-scale integrated* (MSI) components. Examples include the 7490 4-bit counter, and the 7494 4-bit shift register. The boundary between SSI and MSI is somewhat arbitrary. For example, it's not clear whether a 7442 BCD decoder is SSI or MSI.

In addition to extending the range of functions available within the 7400 family, manufacturers developed alternative versions of the components with different internal circuitry and correspondingly different electrical characteristics. One variation reduced the power consumed by components, at the cost of reduced switching speed. Components in this family are identified by inclusion of the letter "L" in the part number, for example, 74L00 and 74L74. Another variation used Schottky diodes within the internal circuits to reduce switching delays, albeit at the expense of increased power consumption. These components include the letter "S" in the part number, for example, 74S00 and 74S74. One of the most popular variations, the 74LS00 family, combined the lower-power circuits with Schottky diodes to yield a good compromise between power and speed. Later variations included the 74F00 "fast" family components and the 74ALS00 "advanced low-power Schottky" family.

One of the problems with TTL circuits is that they use bipolar transistors, which have relatively high power consumption even when not switching. In previous chapters, we have described an alternative circuit structure, CMOS, which uses field-effect transistors. It was originally developed around the same time as TTL. One of the earliest CMOS logic families was the 4000 family, which provided SSI and MSI functions, but with much lower power consumption. They could also operate over a much larger power supply range (3V−15V), compared to TTL's nominal 5V, but were much slower and had logic levels that were incompatible with TTL components. Hence, they did not gain widespread use.

Later, in the 1980s, some manufacturers introduced a new family of CMOS logic components, the 74HC00 family, that were compatible with TTL components. They provided the same functionality, but with lower power consumption and comparable speed. Subsequent variations, such as the 74AHC00 family, offered improved speed and electrical characteristics.

One important characteristic of CMOS circuits is that the power consumption and speed are dependent on the power-supply voltage. By reducing the voltage, as well as by making the internal circuit features smaller, speed is increased and power consumption is reduced. These considerations led the electronics industry to agree on a lower standard power supply voltage of 3.3V. Manufacturers subsequently developed component families to operate at the lower voltage with reduced logic thresholds (74LVC00 family) or with TTL-compatible thresholds (74LVT00 family). They also developed advanced variations, such as the 74ALVC00 and 74ALVT00 family.

As a result of these evolutionary steps, we now have numerous logic families from different manufacturers, with an alphabet soup of letters between the "74" prefix and the number that denotes the logic function. Each family has different trade-offs in power consumption, speed and logic-level thresholds. The data books published by the manufacturers document the characteristics of each family. As designers, we need to understand the power, speed and compatibility constraints of an application and choose components from a family that meets the constraints.

Another aspect of evolution of these families is a change in the logic functions provided. Early components (generally those with smaller numbers) provided gates and simple combinational and sequential functions from which more complex systems could be built. However, during the 1970s, IC technology developed to the level of *large-scale integration* (LSI), at which it became feasible to provide a small computer, a *microprocessor*, on a single IC. Embedded systems using microprocessors became more cost-effective in many applications than systems constructed from SSI and MSI components. The 7400 family components were then commonly used as *glue logic*, that is, simple logic circuits for interconnecting LSI components. As a consequence, new functions added to the later logic families were more oriented toward glue and interconnection functions, such as multibit tristate drivers and registers. These more recent components in the CMOS logic families are the only ones we would consider for new designs, with older components and TTL families being used only for maintenance of legacy systems. Programmable logic devices (described in Section 6.2) and ASICs have almost completely supplanted other families.

EXAMPLE 6.1   Use the following components to design a 4-digit decimal counter with a 7-segment LED display: two 74LS390 dual decade counters, four 74LS47 BCD to 7-segment decoders, four 7-segment displays, plus any additional gates required.

SOLUTION   The 74LS390 component contains two counters, each as shown in Figure 6.5. Internally, the counter consists of a single-bit counter clocked on the falling edge of CP0, and a 3-bit divide-by-five counter clocked on the falling edge of CP1. A decade (divide-by-ten) counter can be formed by using the single-bit counter for the least significant bit and connecting the Q0 output externally to the CP1 input. When Q0 changes from 1 to 0, it causes the more significant bits to count up. The MR input to the counter is a master reset input. When 1, it forces the counter outputs to 0000.

We can cascade the 74LS390 decade counters together, using the outputs of each decade to generate a clock for the next decade. The outputs of a given decade changing from 1001 (the binary code for 9) to 0000 should cause the next decade to count up. The only time this occurs is when Q3 and Q0 of the given decade both change to 0, so we can use an AND gate to generate the clock for the next decade, as shown in Figure 6.6.

FIGURE 6.5   A symbol for each of the decade counters in a 74LS390 component.

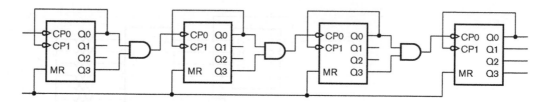

The 74LS47 component is shown in Figure 6.7. The inputs A through D are the BCD value, with A being the least significant bit and D the most significant bit. The segment outputs a through g are active-low, allowing them to drive the cathodes of a 7-segment display. When the lamp test input, LT, is low, it turns all segments on. We can tie it to a high logic level in this application, since we don't need to use it. The ripple-blank input (RBI) and ripple-blank output (RBO) are use to turn off any leading zero digits in the displayed value. When the RBI input to a decoder is low and the BCD value is 0000, all of the segments are turned off and RBO is driven low. We tie the RBI input of the most significant digit low, and chain the RBO of all digits to the RBI of the next digit (except the least significant digit, which we always want to display something).

Our complete circuit for the counter with display is shown in Figure 6.8. The two 74LS390 components are connected as shown in Figure 6.6 using three AND gates. These gates can be implemented using three of the four AND gates in a 74LS08 component. Each counter output drives a 74LS47 decoder, which in turn drives a 7-segment LED display. The resistors are required to limit the current flowing in each segment LED. The value of the resistor depends on the required display brightness. Information on current versus brightness can be found from manufacturer's data sheets. If we assume 2mA is sufficient, that the decoder output has a low-level voltage of 0.4V at 2mA, and that the voltage drop across the segment LED is 1.6V, we need the resistor to drop $5.0 - 1.6 - 0.4 = 3.0$V at 2mA. Thus, a 1.5kΩ resistor will suffice.

FIGURE 6.6 Four 74LS390 decade counters cascaded to make a 4-digit counter.

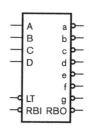

FIGURE 6.7 A symbol for the 74LS47 BCD to 7-segment decoder.

6.1.3 APPLICATION-SPECIFIC INTEGRATED CIRCUITS (ASICS)

The development of IC technology beyond the LSI level led to *very large scale integrated* (VLSI) circuits. At that stage, it became clear that the industry would soon run out of superlatives to prefix to "LSI." Thus, the term VLSI came to refer more to the way in which ICs were designed than the number of transistors they carried. The term now usually means the detailed circuit design of integrated circuits, as opposed to system-level design. The widespread availability of CAD tools for VLSI design and the growth of the IC manufacturing service industry has now made it practical to develop ICs for a wide range of applications. We use the term *application-specific integrated circuit*, or *ASIC*, to refer to an IC

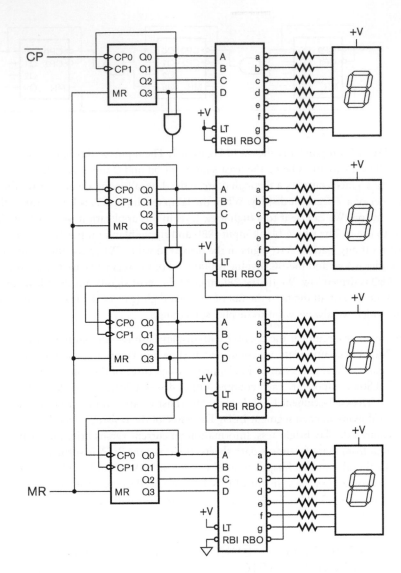

FIGURE 6.8 The complete circuit for the 4-digit counter with display.

manufactured for a particular application. This is not to say that an ASIC is necessarily manufactured just for one customer or project. Rather, it is designed to meet a particular set of requirements, and so contains circuits customized for those requirements. It may be designed for a particular end product provided by one manufacturer, for example, a portable music player, a toy, an automobile, a piece of military equipment, or an industrial machine. Alternatively, it may be designed for use in a range of products provided by manufacturers in a particular market segment. These kinds of ASIC are sometimes called *application-specific standard products*, or *ASSPs*, since they are treated as a standard part within the market segment, but are not of use outside that segment. Examples include ICs for cell phones, which are used by a number of competing cell-phone manufacturers, but which are not of use in, say, automobile control circuits.

One of the main reasons we would develop an ASIC for a product is that, being customized for that application, it has lower cost per IC than a programmable component such as an FPGA (see Section 6.2). However, in order to achieve that level of customization, we need to invest much more design and verification effort. We must amortize the *non-recurring engineering* (NRE) cost over all of the product units sold. Hence, it only makes sense to use an ASIC if our product sales volume is sufficiently large. The amortized NRE cost per unit should be less than the cost difference between an ASIC and a programmable part. This, of course, assumes that it is feasible to use a programmable component. If the application requires a level of performance that cannot be achieved with an FPGA, then an ASIC or an ASSP is our only real option, and the higher NRE is a necessary part of the product cost.

There are two main design and manufacturing techniques for ASICs, differing in the degree of customization for the application. We will describe them briefly in this section, deferring in-depth discussion to advanced references on VLSI design. First, *fully custom* integrated circuits involve detailed design of all of the transistors and connections in an ASIC. This allows the most effective use of the hardware resources on an IC and yields higher performance, but has high NRE cost and requires advanced VLSI design expertise within the design team. As a consequence, fully custom ASICs are usually only designed for high-volume products, such as CPUs and ICs for consumer appliances. Second, *standard cell* ASICs involve selection of basic cells, such as gates and flip-flops, from a library to form the circuit. The cells have been previously designed by an IC manufacturer or an ASIC vendor, and are used by the synthesis tool during the design process to implement the design. The value of this approach is that the NRE cost for each ASIC design is significantly reduced, since the cost of designing the cell library is amortized over a number of ASIC designs. The compromise is that the ASIC may not be as dense or have the performance of a fully custom ASIC.

1. What is *photolithography* in IC manufacture?

2. How do IC area and defect density on a wafer affect IC cost?

3. What do the "L" and "S" in the part name 74LS47 stand for?

4. What is meant by the term *glue logic*?

5. What do the terms *ASIC* and *ASSP* stand for?

6. Would it make sense to design an ASIC for a customized building security system to be installed in a new office building? Why (or why not)?

KNOWLEDGE
TEST QUIZ

7. Similarly, would it make sense to design an ASIC for an engine control system in a car? Why (or why not)?

6.2 PROGRAMMABLE LOGIC DEVICES

The components in SSI and MSI families and ASICs all have fixed functions, determined by the logic circuit for each component. *Programmable logic devices* (PLDs), on the other hand, can be programmed after manufacture to have different functions. In this section, we will look at the evolution of PLDs, leading to FPGAs that are in widespread use today.

6.2.1 PROGRAMMABLE ARRAY LOGIC

One of the first successful families of PLDs was introduced in the late 1970s by Monolithic Memories, Inc., and called *programmable array logic* (PAL) components. These components were an evolution of earlier PLDs, but were simpler to use in many applications. A simple representative component in the family is the PAL16L8, whose circuit is shown in Figure 6.9. The component has 10 pins that are inputs, 2 pins that are outputs, and 6 pins that are both inputs and outputs. This gives a total of 16 inputs and 8 outputs (hence the name "16L8"). The symbol at each input in Figure 6.9 represents a gate that is a combination of a buffer and an inverter. Thus, the vertical signals carry all of the input signals and their negations. The area in the dashed box is the *programmable AND array* of the PAL. Each horizontal signal in the array represents a p-term of the inputs, suggested by the AND-gate symbol at the end of the line. (Recall that a p-term, or product term, is the logical AND of a number of signals; see Section 2.1.1.) In the unprogrammed state, there is a wire called a *fusible link*, or *fuse*, at each intersection of a vertical and horizontal signal wire, connecting those signal wires. The PAL component can be programmed by blowing some of the fuses to break their connections, and leaving other fuses intact. This is done by a special programming instrument before the component is inserted into the final system.

On the diagram of Figure 6.9, we draw an X at the intersection of a vertical and a horizontal signal to represents an intact fuse. An intersection without an X means that the intersecting signals are not connected. So, for example, the horizontal signal numbered 0 has connections to the vertical signals numbered 24 and 31, which are the signals $\overline{I8}$ and $\overline{I10}$. Some of the p-terms are connected to the enable control signals for the inverting tristate output drivers. Others are connected to the 7-input OR gates. So, for each output, we can form the AND-OR-INVERT function of inputs, with up to 7 p-terms involved. In the circuit shown in Figure 6.9, output O1 implements the function $I1 \cdot I2 + I3 \cdot \overline{I10}$, with the output enabled by the condition $I8 \cdot \overline{I10}$.

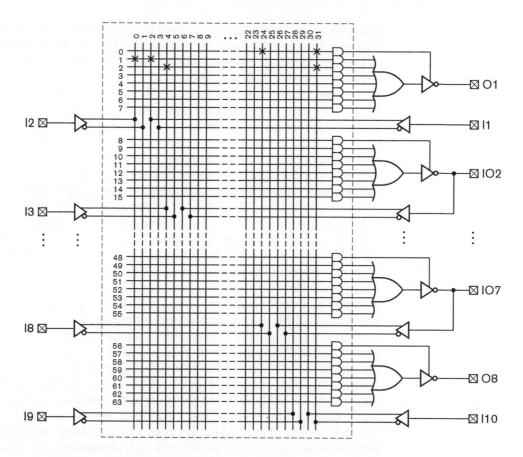

FIGURE 6.9   The internal circuit of a PAL16L8 component.

A PAL component such as the PAL16L8 can be programmed to implement a variety of combinational functions. Other PAL components also include registers, allowing us to implement simple sequential circuits. As an example, the output circuit of a PAL16R8 component is shown in Figure 6.10. The feedback from the register output to the programmable AND array is useful for implementation of FSMs. Even if a circuit is very simple, requiring only a handful of gates and flip-flops, there is often

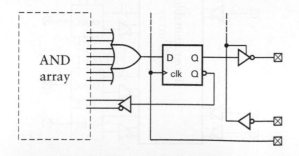

FIGURE 6.10   Registered output circuit of a PAL16R8.

a benefit in cost and reliability gained from combining several separate packages into one PAL package. We typically describe the functions to be implemented in a PAL component in terms of Boolean equations expressed in an HDL. We then use a synthesis tool to transform the equations into a gate-level circuit, which we can verify using the same testbench that we used to verify the Boolean-equation model. Finally, we use physical-design CAD tools to transform the gate-level model into a *fuse map*, that is, a file used by the programming instrument to determine which fuses to blow. If our functions are too complex to be expressed using the resources available in a given PAL component, we would either need to use a larger component, or divide the functions across several components.

As manufacturers developed PLD technology, they found it more convenient to provide fewer generic components in a family, rather than a larger number of variants of a given organizational theme. Most of the variation in earlier families arose in the resources provided for the outputs, for example, whether outputs were inverting or not, whether registers were provided, and whether outputs could be fed back as inputs. In contrast, generic array logic (GAL) components provide *output logic macrocells* (OLMCs) that replace the combinations of OR gates, registers and tristate drivers in PAL output circuits. Each OLMC includes circuit elements together with programmable multiplexers, allowing the output functionality to be determined as part of programming the component. As an example, Figure 6.11 shows the internal circuit organization of the GAL22V10 component, now manufactured by Lattice Semiconductor Corporation, and Figure 6.12 shows the OLMC circuit for each section. The OLMC has p-term inputs from an AND array with the same organization as that of a PAL component. The number of p-terms ranges from 8 for some sections to 16 for others. The output of the OR gate connects to a D flip-flop that has clock, asynchronous reset

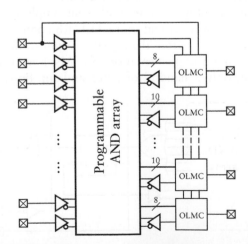

FIGURE 6.11   Internal circuit organization of a GAL220V10 component.

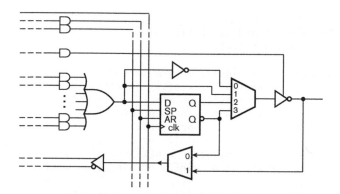

FIGURE 6.12 Output logic macrocell of a GAL22V10 component.

and synchronous preset signals in common with other OLMCs in the component. The select inputs of the multiplexers are set by programming the component. The four-input multiplexer allows selection of registered or combinational output, either inverting or noninverting. The two-input multiplexer allows either registered or combinational feedback, or, if the output driver is in the high-impedance state, direct input from the component pin. By appropriate programming, a GAL component can emulate any of the PAL components, including the PAL16L8 and PAL16R8 components we have shown here.

In modern designs, we would typically use PLDs such as GAL components for simple combinational glue logic, and for relatively simple sequential circuits. As with PAL components, we would describe the required functionality in terms of Boolean equations and use a CAD tool to determine the programming for the component. The circuitry of the original GAL families was based on similar technology to EPROMs, allowing them to be programmed, erased using ultraviolet light, and subsequently reprogrammed. Current components can be erased electrically and programmed *in situ* in the final circuit.

EXAMPLE 6.2   Design a priority encoder that has 16 inputs, I(0 to 15); a four-bit encoded output, Z(3 downto 0); and a valid output that is 1 when any input is 1. Input I(0) has the highest priority, and I(15) the lowest priority. The design is to be implemented in a GAL22V10 component.

SOLUTION   The Boolean equations for the encoder, expressed in VHDL, are

```
win(0) <= I(0);
win(1) <= I(1) and not I(0);
win(2) <= I(2) and not I(1) and not I(0);

...
```

(continued)

```
win(15) <= I(15) and not I(14) and not I(13) and ...
           and not I(0);

Z(3) <= win(15) or win(14) or win(13) or win(12)
           or win(11) or win(10) or win(9) or win(8);
Z(2) <= win(15) or win(14) or win(13) or win(12)
           or win(7) or win(6) or win(5) or win(4);
Z(1) <= win(15) or win(14) or win(11) or win(10)
           or win(7) or win(6) or win(3) or win(2);
Z(0) <= win(15) or win(13) or win(11) or win(9)
           or win(7) or win(5) or win(3) or win(1);

valid <= I(15) or I(14) or I(13) or ... or I(0);
```

Each of the win elements can be implemented as a p-term in a row of the GAL AND array. Each Z output is thus the OR of 8 p-terms. Since each OLMC in a GAL22V10 component has at least 8 p-term inputs, these equations will fit in any of the sections.

The valid output is the OR of 16 inputs, so it could fit in either of the two sections that have 16 p-term inputs to the OLMC. However, we can rewrite the equation for valid using the DeMorgan law (see Section 2.1.2) as

```
valid  <= not (not I(15) and not I(14) and not I(13) and ...
               and not I(0));
```

By programming the OLMC for the valid output to negate the OR result, we can use just one p-term. This allows us to place the valid output in any section of the GAL component rather than only in those sections that have 16 OR-gate inputs. The flexibility afforded by this transformation reduces the constraints on choice of output pins for the component, and may thus simplify connection of the component in a larger circuit.

6.2.2 COMPLEX PLDS

A further evolution of PLDs, tracking advances in integrated circuit technology, led to the development of so-called *complex programmable logic devices* (CPLDs). We can think of a CPLD as incorporating multiple PAL structures, all interconnected by a programmable network of wires, as shown in Figure 6.13. (This gives a general idea of CPLD organization. The actual organization varies between components provided by different manufacturers.) Each of the PAL structures consists of an AND array and a number of embedded macrocells (M/Cs in the figure). The macrocells contain OR gates, mutiplexers and flip-flops, allowing choice among

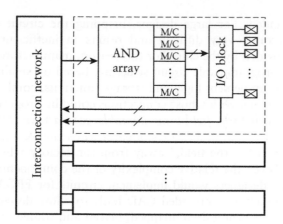

FIGURE 6.13   The internal organization of a CPLD.

combinational or registered connections to other elements within the component, with or without logical negation, choice of initialization for flip-flops, and so on. They are essentially expanded forms of the simple macrocell shown in Figure 6.12, but without the direct connections to external pins. Instead, the external pins are connected to an I/O block, which allows selection among macrocell outputs to drive each pin. The network interconnecting the PAL structures allows each PAL to use feedback connections from other PALs as well as inputs from external pins.

As well as providing more circuit resources than simple PLDs, modern CPLDs are typically programmed differently. Rather than using EPROM-like technology, they use SRAM cells to store configuration bits that control connections in the AND-OR arrays and the select inputs of multiplexers. Configuration data is stored in nonvolatile flash RAM within the CPLD chip, and is transferred into the SRAM when power is applied. Separate pins are provided on the chip for writing to the flash RAM, even while the chip is connected in the final system. Thus, designs using CPLDs can be upgraded by reprogramming the configuration information.

Manufacturers provide a range of CPLDs, varying in the number of internal PAL structures and input/output pins. A large CPLD may contain the equivalent of tens of thousands of gates and hundreds of flip-flops, allowing for implementation of quite complex circuits. Whereas it might be feasible to manually determine the programming for a simple PLD, it would be quite intractable to do so for a CPLD. Hence, we would use CAD tools to synthesize a design from an HDL model and to map the design to the resources provided by a CPLD.

6.2.3 FIELD-PROGRAMMABLE GATE ARRAYS

As we saw in the last section, manufacturers were able to provide larger programmable implementation fabrics by replicating the basic PAL structure on a chip. However, there is a limit to how far this structure can be expanded. For large designs, mapping the circuit onto CPLD resources becomes very difficult and results in inefficient use of the resources provided by the chip. For this reason, manufacturers turned to an alternate programmable circuit structure, based on smaller programmable cells to implement logic and storage functions, combined with an interconnection network whose connections could be programmed. They named such structures *field-programmable gate arrays* (FPGAs), since they could be thought of as arrays of gates whose interconnection could be programmed "in the field," away from the factory where the chips were made. Given the relative complexity of the components, it was not expected that designers would implement circuits for FPGAs manually. Instead, manufacturers provided CAD tools to allow designs expressed in an HDL to be synthesized, mapped, placed and routed automatically, though with designer intervention if necessary. Since their introduction, FPGAs have grown in capacity and performance, and are now one of the main implementation fabrics for designs, particularly where product volumes do not warrant custom integrated circuits.

Most FPGAs available today are organized along the lines shown in Figure 6.14. They include an array of logic blocks that can be programmed to implement simple combinational or sequential logic functions; input/ output (I/O) blocks that can be programmed to be registered or nonregistered, as well as implementing various specifications for voltage levels, loading and timing; embedded RAM blocks; and a programmable interconnection network. The more recent FPGAs also include special circuits for clock generation and distribution. The specific organization, as well as the names used for the blocks, varies between manufacturers and FPGA families.

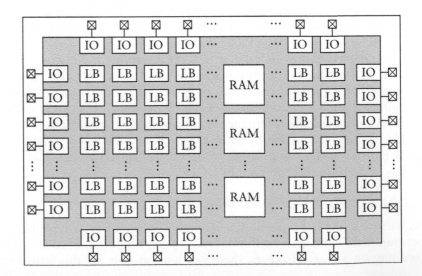

FIGURE 6.14 The internal organization of an FPGA consisting of logic blocks (LB), input/output blocks (IO), embedded RAM blocks (RAM) and programmable interconnections (shown in gray).

In many FPGA components, the basic elements within logic blocks are small 1-bit-wide asynchronous RAMs called *lookup tables* (LUTs). The LUT address inputs are connected to the inputs of the logic block. The content of an LUT determines the values of a Boolean function of the inputs, in much the same way as we discussed in Section 5.2.5. By programming the LUT content differently, we can implement any Boolean function of the inputs. The logic blocks also contain one or more flip-flops and various multiplexers and other logic for selecting data sources and for connecting data to adjacent logic blocks.

As an illustration, Figure 6.15 shows the circuit for a slice within a logic block of a Xilinx Spartan-II FPGA. The logic block contains two such slices, together with a small amount of additional logic. Each slice consists of two 4-input LUTs, each of which can be programmed to implement any function of the four inputs. The carry and control logic consists of circuitry to combine the LUT outputs, an XOR gate and an AND gate for implementing adders and multipliers, as well as multiplexers that can be used to implement a fast carry chain (see Section 3.1.2). Additional components, not shown in the figure, allow programming for various signals to be negated. A number of the connections within the control and carry logic are governed by the programming of the FPGA. The logic block contains SRAM cells for these programming bits.

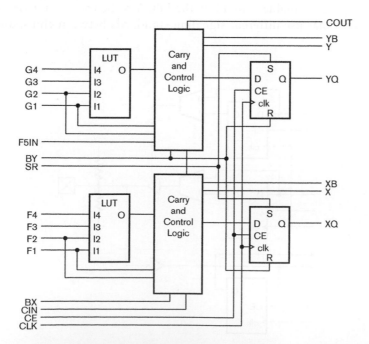

FIGURE 6.15 The circuit of a slice of a Xilinx Spartan-II FPGA logic block.

In contrast to LUT-based logic blocks, which can implement relatively complex functions, some FPGAs have more fine-grained logic blocks. For example, the logic block of Actel ProASIC3 FPGAs contains just enough gates, multiplexers and switches to implement combinational functions of three inputs, or a flip-flop with set or reset. Since each logic block is smaller and simpler, CAD software that maps a design into the FPGA resources may find it easier to perform its task without leaving parts of logic blocks unused. However, a given design will require more logic blocks, and consequently denser interconnection between them. This may make the place and route software's task more difficult.

The I/O block of an FPGA is typically organized as shown in Figure 6.16, but with some variation between components from different vendors. The select inputs of the multiplexers are programmed to control whether the output is registered or combinational. The top flip-flop and multiplexer control the high-impedance state of the tristate driver that drives the pin as an output, and the middle flip-flop and multiplexer drive the output value. The output driver is programmable, allowing selection of logic levels (regular 5V TTL, low voltage TTL, or others) and control of the slew rate, that is, rate of voltage change at the output. (We will discuss why slew rate control is important in Section 6.4.) The input buffer is likewise programmable, allowing selection of threshold voltage and other characteristics. The pull-up and pull-down resistors are programmable, allowing them to be connected and their resistance to be selected. The reason for making all of these characteristics programmable is to allow the FPGA to be used in a wide range of systems that use different signaling standards between chips, and to

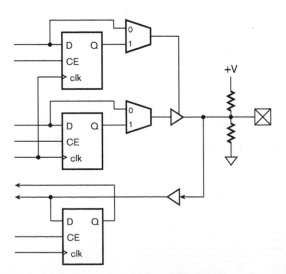

FIGURE 6.16   Typical organization of an FPGA I/O block.

accommodate the different drivers and loads to which different pins of an FPGA may be connected.

The RAM blocks in an FPGA provide for storage of information to be processed by the FPGA circuitry. As we shall see when we consider embedded computer systems in more detail, many applications require data to be input in blocks or streams, and for "chunks" of data to be processed at once. The RAM blocks can be used to store such chunks between processing steps. Also, when an embedded processor is implemented within an FPGA, RAM blocks provide a place to store the processor's instructions and the data upon which it operates. Typical modern FPGAs provide synchronous static RAM (SSRAM) blocks that can be programmed to be flow-through or pipelined, and that have two access ports that can be programmed to be read-only or read-write. The RAM blocks are each relatively small in capacity, but can be interconnected to form larger memories. Each block can be programmed to trade off the number of locations against the number of bits per location. For example, in a Xilinx Spartan-3 FPGA, each RAM block has a total of 18K bits of storage, which can be programmed to provide a $16K \times 1$-bit, $8K \times 2$-bit, $4K \times 4$-bit, $2K \times 9$-bit, $1K \times 18$-bit or 512×36-bit organization. (The 9-, 18- and 36-bit organizations can be used to provide a parity bit with each byte, or the extra bit per byte can be ignored.) The number of RAM blocks varies from 4 to 104 among different members of the Spartan-3 family. FPGAs from other vendors provide similar storage capacities and organizations.

Each of the various logic, I/O and RAM blocks on an FPGA connect to interconnection wires through programmable switches. The connections can be programmed so that a given input or output of a block can be connected (or not) to a wire that passes the block. The interconnections between logic blocks consist of a mix of short and long wires, and possibly wires of intermediate length, depending on the FPGA. Short wires connect nearby logic blocks, whereas long wires connect distant logic blocks or connect to a number of logic blocks distributed across the FPGA. It is the job of the place and route software to ensure that parts of the design are implemented in logic blocks in such a way that the interconnection resources can be programmed to "wire up" the design.

There are two forms of FPGA that differ in the way they are configured. The first form uses RAM cells to store the configuration information. The main advantage of this approach is that an FPGA can be programmed after the chip has been assembled into a system, without the need for any separate handling during manufacture. Furthermore, the system can be upgraded after delivery by storing new configuration information, rather than having to replace chips or other hardware. If the configuration is stored using volatile SRAM cells, it needs to be loaded each time power is applied to the system. Hence, the configuration needs to be stored in a separate nonvolatile memory, and additional circuits need to be included in the system to manage loading the configuration. The two main FPGA vendors, Xilinx and Altera, both use SRAM cells for their devices and provide specialized flash RAM devices for storing and configuring the FPGAs. Other vendors, such as Actel, provide FPGAs that use nonvolatile flash RAM cells for the configuration information. Such devices do not need the external components for storing or loading the configuration, thus reducing overall system complexity. However, the trade-off is a reduced maximum operating speed.

The second main form of FPGA uses *antifuses* to configure the device. An antifuse, as its name suggests, is a conductive connection that is formed during programming, as opposed to being blown. Since programming is done by forming a connection, no storage is needed, either inside the FPGA or externally. Moreover, the device is less susceptible to soft errors due to radiation (see Section 5.3). However, the device must be programmed separately before being installed in the final system. This requires additional manufacturing steps and handling, adding cost to the manufacturing process.

Platform FPGAs

It should be clear now that integrated circuit technology has developed continuously. This trend applies equally to FPGAs. As they have become denser and faster, it has become feasible to use them for applications requiring significant computational performance, such as audio and video processing and information encryption and decryption. In order to improve their usability for these kinds of applications, manufacturers have added specialized circuitry to the larger recent FPGAs, including processor cores, computer network transmitter/receivers and arithmetic circuits. Such FPGAs are often called *platform FGPAs*, meaning that the chip serves as a complete platform upon which a complex application can be implemented. Embedded software can run on the processor cores with instructions and data stored in block RAMs. The network connections can be used to communicate with other computers and devices, and the programmable logic and specialized arithmetic circuits can be used for high-performance data transformations required by the application. A minimal amount of circuitry is required externally to the FPGA, thus reducing the overall cost of the system.

Structured ASICs

Recently, manufacturers have developed a new kind of IC, called *structured ASICs*, that is midway between PLDs and standard-cell ASICs. A structured ASIC is an array of basic logic elements, like an FPGA. However, it is not programmable and omits the programmable interconnect. Moreover, the logic elements are generally very simple, comprising a collection of transistors that can be formed into logic gates and flip-flops. Whereas an FPGA is customized by loading a configuration program, a structured ASIC is customized by designing the top one or more layers of metal interconnection for the chip. Since the underlying logic elements and lower interconnection layers are fixed, the design effort and NRE cost for customization are much lower than those for a standard-cell ASIC. Further, since the structured ASIC is not programmable, just customized by a design and manufacturing process, the performance is potentially very close to that of a standard cell ASIC. Many observers expect that structured ASICs will become popular for complex medium- to high-volume applications over the next few years.

1. How does a programmable logic device differ from a fixed-function component?

2. What is a *fuse map*?

3. If crosses were drawn at the intersections (56, 28), (57, 0), (57, 7) and (58, 30) of the diagram in Figure 6.9, what logic function would be implemented?

4. Suppose the OLMC of Figure 6.12 is used for a state bit S2 of a finite-state machine. For each multiplexer, which input would be selected to make S2 available as an output and to feed it back for use in computing the next-state function?

5. What is the benefit of allowing a PLD in a system to be reprogrammed?

6. What are the purposes of logic blocks and I/O blocks in an FPGA?

7. What other blocks are included in an FPGA?

8. If an FPGA uses volatile SRAM cells to store configuration information, how is the configuration information stored and supplied to the FPGA?

9. What is an *antifuse*?

10. What distinguishes a platform FPGA from a simple FPGA?

KNOWLEDGE TEST QUIZ

6.3 PACKAGING AND CIRCUIT BOARDS

A single bare IC does not form a complete digital system. It needs to be packaged so that it can be connected to other ICs and components, including input and output displays for interacting with a user and connectors for cables for interacting with other systems. An IC is bonded into a package that serves several purposes. It protects the IC from moisture and airborne contaminants, it provides electrical connections, and it removes heat. There are numerous different kinds of IC package, each

FIGURE 6.17   An IC in a package with bond-wires connecting to the lead frame.

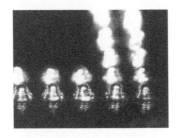

FIGURE 6.18   Connection bumps on a flip-chip IC.

FIGURE 6.19   A through-hole PCB.

with different physical, electrical and thermal properties. The choice of package depends on the number of connections required and the environment in which the product is to operate, among other factors.

Within a package, the IC is affixed to the bottom of a cavity. Fine gold wires are connected from pads on the edge of the IC to points on the package's lead frame (see Figure 6.17), which is the metal framework leading to the external package pins. The cavity is then sealed to protect the IC and the wires. As IC technology has developed, the maximum number of pins has increased, as have operating speeds. For a high pin-count, high performance IC, using bond wires introduces mechanical problems and delays and degrades signals. Recent packages for these ICs have adopted flip-chip technology. The connection pads on the IC are covered in conductive material forming bumps (Figure 6.18). The IC is then flipped over and affixed to the substrate of the package, with the bumps in direct contact with substrate connection points. The connection points lead to the external pins of the package.

The packaged ICs and other components in a system are assembled together on a *printed circuit board* (PCB). This consists of layers of fiberglass or other insulating material separating layers of metal wiring. The metal is deposited in a layer on a fiberglass sheet, and then etched using a photolithographic process, similar to that used in manufacturing ICs. Several layers are sandwiched together. Small holes are drilled through the layers and coated with metal to form connections, called *vias*, between the layers. The completed PCB contains all the circuit wiring needed for the product.

One form of PCB, a *through-hole* PCB (Figure 6.19), includes additional metal-coated holes into which IC package pins are inserted. Solder, a metal alloy with a low melting point, is melted into the holes to form electrical connections between the pins and the PCB wiring. Products using this form of manufacture need ICs in *insertion-type* packages, such as those shown in Figure 6.20. Dual in-line packages (DIPs) have two rows of pins with 0.1-inch spacing. These were among the first IC packages to be introduced, being used for SSI and MSI components, but are less common now. They are relatively large and are limited in the number of pins they can provide, with a 48-pin DIP being about the largest practical size. ICs requiring more pins can be packaged in a pin-grid array (PGA) package, having up to 400 or more pins. However, these have largely been replaced by newer forms of package, and are now mainly used for ICs such as computer CPUs that are to be mounted in sockets so that they can be removed. One of the advantages of through-hole PCBs is that they can be manually assembled, since the component sizes are manageable. This is good for low-volume products, since the cost of setting up a manufacturing run is less than that for automated assembly. However, the move to ICs with higher pin counts has reduced the applicability of this technology.

FIGURE 6.20   Insertion-type IC packages: DIPs (left) and PGAs (right).

The second form of PCB is a *surface-mount* PCB (Figure 6.21), so-called because components are mounted on the surface rather than being inserted in holes. This has the advantage of reduced manufacturing cost (for higher-volume products), finer feature sizes and increased circuit density. Surface mounting IC packages have pins or connection points that come into contact with a metal pad on the PCB. Solder paste is applied between each pin and pad and subsequently melted, forming the connection. There are numerous different surface mounting packages, some of which are shown in Figure 6.22. Quad flat-pack (QFP) packages have pins along all four sides, and are suitable for ICs with up to 200 or so pins. The spacing between pins varies from 1 mm for the packages with fewer pins, down to 0.65mm for the higher pin-count packages. Fine-pitch QFP packages allow increased pin count, up to nearly 400 pins, by reducing the pin spacing to 0.4mm. Given the delicacy of these pins, the packages are not suitable for manual handling and assembly. The most common package in use now for high pin-count ICs is the ball-grid array (BGA) package. Depending on the package size and the pin spacing, BGA

FIGURE 6.21   A surface-mount PCB.

FIGURE 6.22   Surface mounting IC packages: QFPs (left) and BGA (right).

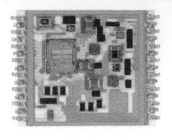

FIGURE 6.23 A multichip module.

packages can accommodate ICs with up to 1800 pins. Higher pin-count BGA packages are also being developed.

In recent times, high-density packaging techniques have been developed for use in products where space is constrained. A good example is a cell phone, in which small size and reduced weight are important marketing factors. Rather than placing each IC in a separate package and assembling several packages onto a PCB, *multichip modules* (MCMs) attach the bare chips to a ceramic substrate (see Figure 6.23). Interconnection wires and passive components (resistors and capacitors) are also printed or soldered onto the substrate. The complete module is then encapsulated with external connections made through package pins to a PCB. Even denser packaging can be achieved by building in three dimensions, rather than laying them out on a two-dimensional surface. For example, *chip stacking* involves placing two or more chips in a vertical stack. Connections can be made between adjacent chips by metal contacts, and between chips and the containing package by bond wires. Several flash memory manufacturers are using these techniques to provide high-capacity storage in very small packages. As demand for high-performance mobile devices increases, we can expect to see continued development of these high-density packaging techniques.

1.   How does flip-chip IC packaging differ from previous packaging technologies?

2.   What distinguishes surface-mount IC packages from insertion-type packages?

3.   What is a *via* in a PCB?

4.   For an IC with 1200 pins, what kind of package would most likely be used?

6.4   INTERCONNECTION AND SIGNAL INTEGRITY

When we introduced the digital abstraction in Chapter 1, we described signals as changing between low and high logic levels instantaneously. We emphasized, however, that this is an abstraction, and that real signals take time to change and time to propagate along signal wires. We have taken a relatively simple view of signal propagation between a source and a destination within a circuit. In practice, there are a number of complicating factors, particularly when the source and destination are in different ICs on a PCB. A signal change must propagate from the source driver, through the bond wire, package lead frame and pin of the source IC, along the PCB trace, through the pin, lead frame and bond wire of the destination IC, and into the receiver. Along this path, there are several influences that can cause distortion of the signal and introduce noise. The term *signal integrity* refers to the degree to which these effects are minimized. If we are using off-the-shelf ICs or PLDs, we do not have control over the path within the IC package. We must assume that the designers of the IC and package have done due diligence to maintain signal integrity. Alternatively, if we are implementing a design in an ASIC, we must take responsibility for signal integrity within the ASIC. Since this is a complex area, we largely defer it to an advanced reference on VLSI design, though many of the ideas that we discuss in this section do apply to ASIC design. In either case, using off-the-shelf parts or ASICs, we need to consider the effects of the PCB on signal integrity.

A change in a signal value causes a change in the current flowing through the PCB trace. This causes a change in the electric and magnetic fields around the trace. Propagation of those fields determines the speed of propagation of the signal change along the trace. In common PCB materials, the maximum propagation speed is approximately half the speed of light in a vacuum. Since the latter is $3 \times 10^8 \, \text{ms}^{-1}$, we can use 150mm per nanosecond as a good rule of thumb for signal propagation along a PCB trace. For low speed designs and small PCBs, this element of total path delay is insignificant. However, for high-speed designs, particularly for signals on critical timing paths, it is significant. Two cases in point are the routing of clock signals and parallel bus signals. If a clock signal is routed through paths of different lengths to different ICs, we may introduce clock skew, in much the same way that we described in Section 4.4. Similarly, if different signals within a parallel bus are routed along paths of different lengths, changes in elements of the bus may not arrive concurrently, and may be incorrectly sampled at the destination's receiver. In these cases, it may be necessary to tune the timing of the system by adding to the length of some PCB traces to match propagation delays. CAD tools used for PCB layout offer features to help designers perform such tuning semiautomatically.

A major signal integrity issue in PCB design is *ground bounce*, which arises when one or more output drivers switch logic levels. During switching, both of the transistors in the driver's output stage are momentarily on, and transient current flows from the power supply to ground. Ideally, the power supply can source the transient current without distortion. In reality, however, there is inductance in both the power and the ground connections, as shown in Figure 6.24. The inductance causes voltage spikes in the power supply and ground on the IC. This can cause voltage spikes on other output drivers, possibly causing false transitions in the receivers to which they are connected. It can also cause transient shifting of the threshold voltage of receivers on the IC, causing false transitions at those receivers. The effect is particularly pronounced when multiple drivers switch concurrently, for example, when the value on a parallel bus changes, since the transient current is much greater.

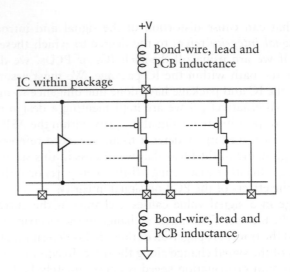

FIGURE 6.24 Inductance in the bond-wires, package leads and PCB connections for power and ground.

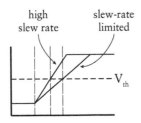

FIGURE 6.25 The effect of slew-rate limiting a signal. The signal takes longer to reach the threshold voltage V_{th}.

In order to reduce the effects of ground bounce, we can take a number of important measures. First, we can place *bypass capacitors* between power and ground at strategic places around a PCB. These capacitors hold a reserve of charge that can quickly supply the needs of switching drivers. A common rule of thumb is to place a capacitor close to each IC package. Values of 0.01μF to 0.1μF are common. Second, we can use separate PCB layers for the ground and power supply (Figure 6.26). This gives a low-inductance path for the power supply current and its ground return. It also has other benefits, mentioned below. Third, we can limit the rate of voltage change (the *slew rate*) and limit the drive current of the output drivers. These actions limit the rate of change of current, and so limit the inductive effect of the change. Components such as modern FPGAs have programmable output drivers that allow selection of slew rate and drive current limits. Of course, reducing the slew rate means that a signal takes longer to change from one logic level to the other, as illustrated in Figure 6.25. Hence, limiting slew rate may increase propagation delay through circuits, consequently requiring a reduction in clock rate. This is a case where a trade-off between speed of operation and noise immunity may be required. Finally, we can use differential signaling, discussed in Section 6.4.1, as a means of making the system more immune to noise induced by ground bounce.

FIGURE 6.26 Cross section of a multilayer PCB with ground and power planes.

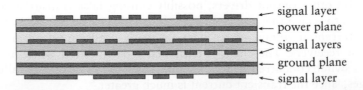

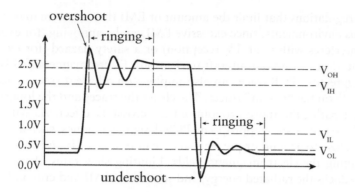

FIGURE 6.27 Overshoot, undershoot and ringing transmission-line effects.

Another signal integrity issue for high-slew rate signals is noise due to transmission-line effects. When the time for a transition between logic levels is similar to or shorter than the propagation delay along a signal path, the transition is affected by reflections at the driving and receiving ends of the path. A full analysis of the effects requires knowledge of the characteristic impedance of the path, as well as the source impedance of the driver and the terminating impedance of the receiver. Depending on the relationships between these values, the signal may suffer from partial transitions, overshoot, undershoot and ringing (Figure 6.27). The situation is made more complex if the signal wire is not a simple driver-to-receiver connection, but has multiple receivers along the path. PCB layout artifacts, such as vias and branching paths, also introduce further effects.

The main design techniques for managing transmission-line effects involve appropriate layout and proper termination of PCB traces. By running a trace of specific dimensions at a controlled distance between two ground or power planes in the PCB, we create a *stripline* transmission line with a controlled characteristic impedance. Where the transmission line effects are less critical, we can run a trace over just one plane, creating a *microstrip* transmission line. For critical signals, we can adopt circuit designs and layouts that avoid placing receivers along the PCB trace, or that group them together at the receiving end. Finally, we can include termination resistors to ensure proper matching of drivers and receivers to the characteristic impedance of the transmission line. In high-performance modern components, including FPGAs, the drivers include termination resistors on the IC. In other cases, we may need to include resistors as discrete components adjacent to IC pins.

As we mentioned earlier, transitions between logic levels on a signal cause electromagnetic fields to propagate around the PCB trace. Some of the field energy is radiated out from the system, and may impinge on other electronic systems, where it induces noise. This form of unwanted coupling is called *electromagnetic interference* (EMI). There are government and other regulations that limit the amount of EMI that a system may emit in various environments, since excessive EMI can be annoying (for example, if it interferes with your TV reception) or a safety hazard (for example, if it interferes with your aircraft navigation). Electromagnetic fields from an "aggressor" PCB trace can also impinge on adjacent traces, inducing *crosstalk* on the "victim" traces. The closer the traces and the longer their parallel paths, the more pronounced the crosstalk effect. As with other signal integrity issues, appropriate PCB design techniques, such as routing traces close to ground or power planes, can reduce EMI and crosstalk by containing the electromagnetic fields. Limiting slew rates of transitions also reduces the radiated energy, and so reduces EMI and crosstalk.

6.4.1 DIFFERENTIAL SIGNALING

The techniques for maintaining signal integrity that we have discussed so far are based on reducing the amount of interference induced on signal wires. Another technique, use of *differential signaling*, is based on the idea of reducing a system's susceptibility to interference. Rather than transmitting a bit of information as a single signal S, we transmit both the positive signal S_P and its negation S_N. At the receiving end, we sense the voltage difference between the two signals. If S_P − S_N is a positive voltage, then S is received as the value 1; if S_P − S_N is a negative voltage, then S is received as 0. This arrangement is illustrated in Figure 6.28. The assumption behind the differential signaling approach is that noise is induced equally on the wires for both S_P and S_N. Such common-mode noise is cancelled out when we sense the voltage difference. To show this, suppose a noise voltage V_N is induced equally on the two wires. At the receiver, we sense the voltage

$$(S\_P + V_N) - (S\_N + V_N) = S\_P + V_N - S\_N - V_N = S\_P - S\_N$$

For the assumption of common-mode noise induction to hold, differential signals must be routed along parallel paths on a PCB. While this might suggest a problem with crosstalk between the two traces, the fact that the signals are inverses of each other means that they both change at the same time, and crosstalk effects cancel out.

As well as rejecting common-mode noise, differential signaling also has the advantage that reduced voltage swings are needed for a given noise margin. Even though each of S_P and S_N switches between V_{OL} and V_{OH}, the differential swing at the receiver is between $V_{OL} - V_{OH}$ and $V_{OH} - V_{OL}$, that is, twice the swing of each individual signal. Reducing the voltage swing has multiple follow-on effects, including reduced switching current, reduced ground bounce, reduced EMI, and reduced crosstalk with other signals. Thus, use of differential signals can be very beneficial in high-speed designs.

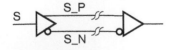

FIGURE 6.28 A differential driver and receiver.

1. What is meant by the term *signal integrity*?

2. How fast does a signal change propagate along a typical PCB trace?

3. What causes ground bounce in digital systems?

4. Where should bypass capacitors be placed on a PCB?

5. How does limiting the slew rate of an output driver improve signal integrity?

6. What design techniques can be used to mitigate transmission-line effects, such as overshoot, undershoot and ringing?

7. What are *EMI* and *crosstalk*?

8. How does differential signaling improve noise immunity?

9. For a 2.5V low-voltage differential signaling (LVDS) output, the nominal V_{OL} and V_{OH} voltages are 1.075V and 1.425V, respectively. What differential voltage swing is seen at the receiver?

KNOWLEDGE
TEST QUIZ

6.5 CHAPTER SUMMARY

▶ Improvements in IC manufacturing processes, especially in photo-lithography, have led to increased circuit speed and density. These trends are continuing.

▶ Yield, the proportion of manufactured ICs that work, is a significant determinant of cost. Reducing IC area reduces the chance of a defect on the IC.

▶ The 74xx00 families of small-scale integrated (SSI) and medium-scale integrated (MSI) circuits were the main components used in early and legacy digital systems. There are several 74xx00 families, varying in speed, power and logic thresholds. For new designs, 74xx00 components are largely superseded by programmable logic devices.

▶ Application-specific integrated circuits (ASICs) are ICs designed for particular applications. Application-specific standard products (ASSPs) are ASICs designed for particular market segments. In both cases, high non-recurring engineering (NRE) cost limits their use to high-volume applications, or applications that demand very high performance.

▶ In a full-custom ASIC, the IC circuitry is custom designed in detail. In a standard-cell ASIC, the circuit is implemented by a synthesis tool using predesigned gate and flip-flop cells from a library, thus reducing NRE cost.

▶ Programmable logic devices (PLDs) are standard parts that are programmed after manufacture to implement a circuit function.

▶ Programmable array logic (PAL) components are simple PLDs that implement simple combinational or sequential functions. Generic array logic (GAL) components include programmable macrocells instead of fixed-function output logic.

▶ Complex PLDs consist of a number of PAL structures and an inter-connection network integrated onto a single IC. They are useful for larger combinational or sequential designs.

▶ Field-programmable gate arrays (FPGAs) consist of an array of logic blocks, memory blocks and I/O blocks, and a programmable interconnection network. The logic blocks implement simple com-binational and sequential functions. Platform FPGAs also incorpo-rate processor cores, arithmetic circuits and other complex function blocks.

▶ ICs are embedded in packages and assembled onto printed circuit boards (PCBs) to form a complete digital system. Different package types are used for through-hole and surface-mount PCBs.

▶ Signal integrity refers to the minimization of distortion of digital signals due to parasitic capacitance and inductance. Effects include signal skew, ground bounce, transmission line effects (overshoot, undershoot and ringing), electromagnetic interference (EMI), and crosstalk. Effects are mitigated by careful PCB design.

▶ Differential signaling involves transmitting both a positive signal and its negation, and sensing the voltage difference between the two at a receiver. Differential signaling allows common-mode noise rejection and improved signal integrity.

6.6 FURTHER READING

Exponential Trends in the Integrated Circuit Industry, Scotten W. Jones, IC Knowledge LLC, 2004, http://www.icknowledge.com/trends/Exponential2.pdf. A good summary of the way in which a number of parameter values have changed exponentially over time, and the interrelationship among the parameters.

Introduction to Integrated Circuit Technology, 3rd Edition, Scotten W. Jones, IC Knowledge LLC, 2004, http://www.icknowledge.com/misc_technology/IntroToICTechRev3.pdf. Outlines the steps involved in manufacture, packaging and test of integrated circuits.

Digital Logic Pocket Data Book, Texas Instruments, Inc., 2002, http://focus.ti.com/lit/ug/scyd013/scyd013.pdf. Contains datasheets for 74*xx*00 family SSI, MSI and bus transceiver components.

CMOS VLSI Design: A Circuits and Systems Perspective, 3rd Edition, Neil H. E. Weste and David Harris, Addison-Wesley, 2005. An advanced reference on CMOS VLSI circuit design.

Signal Integrity Issues and Printed Circuit Board Design, Douglas Brooks, Prentice Hall, 2003. Introduces basic electrical characteristics of PCBs and components, and covers propagation delays, electromagnetic interference, transmission lines, crosstalk and power-supply decoupling. The author's website at www.ultracad.com provides numerous articles on these topics.

High-Speed Digital Design: A Handbook of Black Magic, Howard Johnson, Prentice Hall PTR, 1993. Covers aspects of analog circuit behavior that are relevant to high-speed digital circuit design, including signal integrity issues.

EXERCISES

EXERCISE 6.1 The 74LS85 component is a 4-bit cascadable magnitude comparator for unsigned binary integers. Details are provided in the Texas Instruments *Digital Logic Pocket Data Book* (see Section 6.6, Further Reading). Design a 16-bit magnitude comparator using four 74LS85 components and any additional gates required.

EXERCISE 6.2 Suppose a company is deciding between an ASIC and an FPGA implementation for a complex new design. Their estimates of NRE costs for the two alternatives are:

NRE COST COMPONENT	ASIC	FPGA
Staff	$2,500,000	$2,000,000
Infrastructure	$1,500,000	$1,000,000
Consumables and Services	$750,000	$100,000

The unit cost for manufacture of each ASIC will be $15.00, and the unit cost of purchasing and programming each FPGA will be $25.00. Which option is more cost-effective for production volumes of 100,000 units; 200,000 units; and 500,000 units? What is the production cost at which the two options are equally cost-effective?

EXERCISE 6.3 On a copy of Figure 6.9, draw the fuses that are required to implement the following VHDL assignments in a PAL16L8:

a) O8   <= not I9 or I10 when I1 else 'Z';

b) IO2   <= not (I1 and not I2) or (not I1 and I3 and not I8);

c) IO7   <= (I1 and not I2) or (not I3 and I10);

EXERCISE 6.4 Describe how an OLMC of a GAL22V10, shown in Figure 6.12, should be programmed to emulate the input/output circuit of:

a) IO2 of a PAL16L8

b) O1 of a PAL16L8

c) I1 of PAL16L8

d) an output of a PAL16R8

PROCESSOR BASICS

7

In this chapter we start our focus on embedded systems with an introduction to the kinds of processors that are used. We describe the way processors operate and give examples of the instructions that make up embedded software programs. We also describe the way instructions and data are encoded in binary and stored in memory. Finally, we examine ways of connecting the processor with memory components.

7.1 EMBEDDED COMPUTER ORGANIZATION

In Section 1.5.1, we introduced the idea of an embedded system, in which one or more computers form part of the system. The computers run programs that implement the functions required of the system. Unlike a general-purpose PC, a computer in an embedded system has just those resources required to support its specialized operation. In this section, we will describe some of the general properties of embedded systems and the processing elements they contain. We won't deal with how the processing elements are designed; that is a significant field of study in its own right. Instead, we will treat them as black-box circuit components that we can use to build a digital system.

A computer embedded in a digital system generally contains the elements shown in Figure 7.1. The *central processing unit* (CPU), often called a *processor core* when it is embedded as part of an IC, is the element that processes data according to a program. The kinds of processing it can perform include the arithmetic operations that we described in Chapter 3. It can also evaluate logical conditions and select among alternate operations based on the outcomes of the conditions. We will describe the way a program is formed in more detail in Section 7.2. Meanwhile, suffice it to say that the program is encoded in binary form and stored in the *instruction memory* shown in the figure. The data upon which the program operates are also encoded in binary form and stored in the *data*

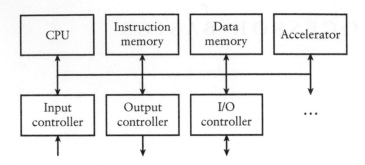

FIGURE 7.1 Elements of an embedded computer.

memory. In both cases, the memory is implemented using the kinds of memory components we described in Chapter 5. Whereas general purpose computers, such as PCs, usually store the instructions and data in the same memory, embedded computers typically separate the two. (This arrangement is often referred to as a *Harvard architecture*, named after the institution where the idea originated. The conventional approach with a single memory for instructions and data is called a *von Neumann architecture*, after the person who first described it.) The reason for the separation is that the instructions in an embedded computer are usually fixed during the manufacture of the system (or only occasionally upgraded in the field), and the amount of instruction memory required is known in advance. Hence, we usually store instructions in a ROM or flash memory component, and provide a RAM for the data memory. This differs from a general-purpose computer, in which one or more different programs need to be started at different times and run concurrently, and the amount of instruction memory is not known in advance.

The *input*, *output* and *input/output* (I/O) controllers in Figure 7.1 allow the computer to acquire data to be processed (input) and to deliver the results (output). In many embedded systems, the input data comes from sensors that sample physical properties, such as temperature, position, time, and so on. Similarly, the output data causes actuators to have a physical effect, such as moving a lever, turning a motor, heating some material, and so on. Input and output controllers can also deal with a user interface, consisting of switches, buttons and knobs for input and lights and LCD panels for outputs. For a complex user interfaces, devices such as a keyboard, mouse or display screen, as used in a general purpose computer, might also be employed. In all cases, the job of the input/output controller is to transform between a physical property or effect and a corresponding binary representation that can be processed by the CPU. We will describe how this can be done and how the CPU accesses the binary representation in Chapter 8.

The *accelerator* in Figure 7.1 is a specialized circuit designed to implement specific processing operations with higher performance than can be achieved using the CPU. Not all embedded systems include

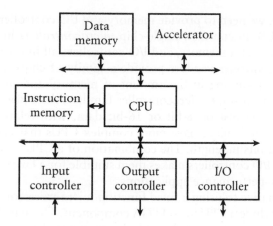

FIGURE 7.2 Organization of a high-performance embedded computer with multiple buses: one for the instruction memory, one for the data memory and an accelerator, and one for input/output controllers.

accelerators. The choice of whether to include an accelerator for any operation depends on the functional and performance requirements of the application, together with cost and other constraints that apply. We will discuss accelerators in more detail in Chapter 9, in which we include as an extended example an accelerator for detecting edges of objects in video images.

The final element in Figure 7.1 is the interconnection between the other elements. We use the term *bus* to refer to the collection of signals that form the interconnection. The figure shows just one bus connecting all of the elements. However, in more elaborate systems, there may be separate buses for connecting the memory and the input/output controllers with the CPU. There may even be separate buses for the instruction and data memories, since many high-performance processors can read further instructions concurrently with access to data by previous instructions. Accelerators, if included, might be connected to the CPU using the same bus as the memory, or using a separate dedicated bus. Figure 7.2 shows one possible organization for a high-performance embedded system with multiple buses. In this chapter, we will focus on the bus connecting the CPU and memory, and defer consideration of bus connections to input and output controllers and to accelerators until later chapters.

7.1.1 MICROCONTROLLERS AND PROCESSOR CORES

CPUs for embedded systems come in a range of sizes for different applications. Some are single-chip *microprocessors*, consisting of a CPU by itself in a package. Most CPUs used in general-purpose PCs are also available in versions suitable for embedded applications. Examples include Pentium family CPUs from Intel and the PowerPC from Freescale Semiconductor. Other microprocessors are designed specifically for embedded applications.

In both cases, we need to provide memory and I/O controllers as separate chips on a PCB. In contrast, single-chip *microcontrollers* include a CPU, instruction and data memory, and I/O controllers all in the one package. Many microcontroller vendors provide a family of chips, each with the same CPU, but varying in the amount of memory and the selection of I/O controllers. In some microcontroller families, the CPUs are relatively simple, operating just on 8-bit or 16-bit data, with relatively low performance. Other families have more complex CPUs that can operate on data up to 32 bits in length. The combination of a CPU with the on-chip memory and I/O controllers makes them suitable for a large range of cost-sensitive, low-performance applications.

An alternative to using a fixed function microprocessor or microcontroller is to include a CPU in an FPGA component. This has the advantage that the input/output controllers can be customized for an application, but still be included in the same package as the CPU. The CPU in the FPGA can be implemented as a fixed-function block embedded within the programmable fabric. The Virtex-II Pro and Virtex 4 FPGAs from Xilinx take this approach, and include one or more PowerPC processor cores. Alternatively, the CPU can be implemented as a *soft core* using the programmable resources of the FPGA. FPGA vendors provide soft core processor designs that users can include as part of their system. Examples include the MicroBlaze core from Xilinx, the Nios-II core from Altera, and the ARM core from Actel. These are all relatively high-performance CPUs that operate on data up to 32 or 64 bits in length. For simpler designs, a smaller soft core that operates on 8-bit data may suffice. It would take up less of the FPGA resources, and would fit in a smaller and cheaper FPGA component. The Xilinx PicoBlaze soft core is an example, as is the Gumnut core that we will introduce in Section 7.2.

If our design is implemented in an ASIC, we can also include a CPU and customized memory and input/output controllers. Several vendors provide processor core designs that can be included as blocks in ASICs. Among the most widely used are the ARM cores from ARM Ltd, the PowerPC cores from IBM, and the MIPS cores from MIPS Technologies. Given that we can customize the design on an ASIC, there is also opportunity to customize the CPU itself. Tensilica Inc. is a vendor that provides a customizable CPU based on the requirements of the program to be executed. Their approach involves analyzing the program and including only the CPU features needed to execute that program. They also allow extension of the CPU with customized hardware for specialized operations.

A final approach to mention is to include one or more *digital signal processors* (DSPs). These are specialized processing elements optimized for the kinds of operations involved in dealing with digitized signals, such as audio, video or other streams of data from sensors. Many signal processing applications require fixed-point or floating-point arithmetic operations to be performed at a high rate on large volumes of data. An ordinary CPU would not be able to meet the performance requirements. Nonetheless, such applications often need a conventional CPU to perform other operations, such as interacting with the user and overall coordination of system operation. Hence, DSPs are often combined with conventional CPUs in heterogeneous *multiprocessor* systems. Modern cell phones are good examples. Another approach to providing DSP functionality is to extend a conventional CPU with additional hardware and instructions for digital signal processing. Some processor cores from ARM and MIPS include such extensions, and Tensilica processor cores can be similarly customized. Since digital signal processing is an advanced topic, we will defer consideration of DSP cores and embedded multiprocessor systems to advanced reference books.

1. What are the main elements of an embedded computer?

2. Why do embedded computers usually have separate instruction and data memories?

3. What is the difference between a microprocessor and a microcontroller?

4. What is meant by a *soft core* processor in an FPGA?

KNOWLEDGE TEST QUIZ

7.2 INSTRUCTIONS AND DATA

The function performed by a CPU is specified by a program, which consists of a sequence of instructions. Each instruction specifies one simple step in the program, such as getting a piece of data from memory, or adding two numbers. The repertoire of instructions for a given CPU is called the *instruction set* of the CPU. We also use the term *instruction set architecture* (ISA) to refer to the combination of the instruction set and other aspects of the CPU that are visible to the programmer. CPUs from different vendors have quite significantly different instruction sets, so a sequence of instructions developed for one CPU will not work on a CPU from a different vendor. When we develop the program for an application, we usually use a *high-level language*, such as C, C++ or Ada, and use a software tool called a *compiler* to translate the program into a sequence of instructions that performs the same operations. Apart from allowing us to work at a higher level of abstraction, this has the advantage that the program can be ported to work on a CPU with a different instruction set simply by using a different translator. However, when we are developing an embedded system in which the CPU interacts with circuits that we design, we often need to monitor the instruction-by-instruction operation of the CPU as we test and debug the design. At this level, it is important to understand how a CPU represents and processes individual instructions. We will just describe CPU operation at this level, and defer a discussion of programming using high-level languages to other books.

The instructions of a program are encoded in binary and stored in successive locations of the instruction memory. The CPU *executes* the program by repeatedly following these steps:

1. *Fetch* the next instruction from the instruction memory.

2. *Decode* the instruction to determine the operation to perform.

3. *Execute* the operation.

In order to keep track of which instruction to fetch next, the CPU has a special register called the program counter (PC), in which the address of the next instruction is kept. In the fetch step, the CPU uses the contents of the PC to do a read access from the instruction memory, and then increments the PC value. In the decode step, the CPU determines the resources required to perform the operation specified by the instruction. In a simple CPU, the decode step is correspondingly simple. In a larger CPU, however, decoding may involve such actions as checking for resource conflicts and availability of data, and waiting until resources are free. In the execute step, the CPU activates the appropriate internal resources to perform the operation. This involves setting control signals to make multiplexers supply the required operands and arithmetic hardware perform the required operation, and enabling registers to receive results. In a simple CPU, these steps are performed in order, and when the execute step is finished, the CPU starts again with the fetch step. More complex, high performance CPUs, however, can overlap the steps, provided they produce the same outcome as if the steps were performed in order. Techniques used within CPUs to execute several instructions in parallel include *pipelining* and *superscalar* execution, described in the reference book on computer architecture (see Section 7.5).

The data on which instructions operate is encoded in binary in fixed-size quantities. The smallest data item is usually 8 bits, called a *byte*. It is often used to represent an unsigned or a signed integer, or a character. Simple CPUs can only operate on 8-bit data, so they are referred to as 8-bit CPUs. Larger CPUs can operate on 16-bit or 32-bit *words* of data, as well as on 8-bit data, so they are referred to as 16-bit or 32-bit CPUs, respectively.

Regardless of the sizes of data that can be operated upon, the data memory is usually organized with 8-bit locations, each separately addressed. 16-bit or 32-bit data is stored in two or four successive locations. The order of the bytes within a word varies between CPUs, as shown in Figure 7.3. *Little-endian* CPUs store the byte containing the least significant bits at the lower address and the byte containing the most significant bits at the higher address. In contrast, *big-endian* CPUs store the bytes in the opposite order. (The terms "little endian" and "big endian" originated in Jonathan Swift's *Gulliver's Travels*, in which the people of two countries fight over which end of their breakfast eggs should be cut open.

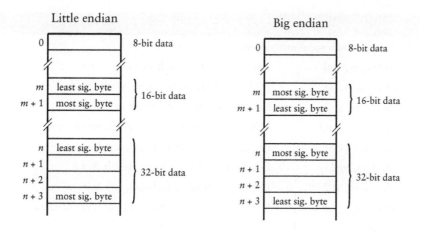

FIGURE 7.3   Little-endian (left) and big-endian (right) memory layout for data words.

The terms were adopted by Danny Cohen in an article, cited in Section 7.5, in which he argues that either byte ordering is acceptable, provided it is used consistently.) Some CPUs require that 16-bit data be stored at even addresses and that 32-bit data be stored at addresses that are a multiple of four. Others allow 16-bit and 32-bit data to be stored at any address.

7.2.1   THE GUMNUT INSTRUCTION SET

Rather than trying to describe the characteristics of the instruction sets of all CPUs, we will present one relatively simple example that embodies most of the important concepts. The CPU that we will describe is an 8-bit soft core called the *Gumnut*, developed by the author. (A gumnut is a small seedpod of an Australian eucalyptus tree. It is something small from which large things grow.) Further information and files are provided in the supplementary material for this book for use in FPGA designs. The complete Gumnut instruction set is listed in Table 7.1. We use a notation for instructions called *assembly code*. An assembly-code program can be translated by a software tool called an *assembler* into a sequence of binary-coded instructions to be loaded into the instruction memory.

The Gumnut has an instruction memory of up to 4096 instructions (using 12-bit addresses) and a data memory of 256 bytes (using 8-bit addresses). When the CPU is reset, it clears the PC to 0, and starts the fetch-decode-execute cycle, fetching the first program instruction from address 0 in the instruction memory. Within the CPU, there are eight general-purpose registers, named r0 through r7, that can hold data to be operated upon by instructions. Register r0 is special, in that it is hard-wired to have the value 0, and any updates to it are ignored. The CPU also has two single-bit *condition-code* registers called Z (zero) and

INSTRUCTION	DESCRIPTION
Arithmetic and logical instructions	
add *rd, rs, op2*	Add *rs* and *op2*, result in *rd*
addc *rd, rs, op2*	Add *rs* and *op2* with carry, result in *rd*
sub *rd, rs, op2*	Subtract *op2* from *rs*, result in *rd*
subc *rd, rs, op2*	Subtract *op2* from *rs* with carry, result in *rd*
and *rd, rs, op2*	Logical AND of *rs* and *op2*, result in *rd*
or *rd, rs, op2*	Logical OR of *rs* and *op2*, result in *rd*
xor *rd, rs, op2*	Logical XOR of *rs* and *op2*, result in *rd*
mask *rd, rs, op2*	Logical AND of *rs* and NOT *op2*, result in *rd*
Shift instructions	
shl *rd, rs, count*	Shift *rs* value left *count* places, result in *rd*
shr *rd, rs, count*	Shift *rs* value right *count* places, result in *rd*
rol *rd, rs, count*	Rotate *rs* value left *count* places, result in *rd*
ror *rd, rs, count*	Rotate *rs* value right *count* places, result in *rd*
Memory and I/O instructions	
ldm *rd, (rs) ± offset*	Load to *rd* from memory
stm *rd, (rs) ± offset*	Store to memory from *rd*
inp *rd, (rs) ± offset*	Input to *rd* from input controller register
out *rd, (rs) ± offset*	Output to output controller register from *rd*
Branch instructions	
bz ± *disp*	Branch if Z is set
bnz ± *disp*	Branch is Z is not set
bc ± *disp*	Branch if C is set
bnc ± *disp*	Branch if C is not set
Jump instructions	
jmp *addr*	Jump to *addr*
jsb *addr*	Jump to subroutine at *addr*
Miscellaneous instructions	
ret	Return from subroutine
reti	Return from interrupt
enai	Enable interrupts
disi	Disable interrupts
wait	Wait for interrupts
stby	Enter low-power standby mode

TABLE 7.1 The Gumnut instruction set. *rd* and *rs* are registers, *op2* is a register (*rs2*) or an immediate value (*immed*), *count* is count of number of places to shift or rotate, *disp* is a displacement from the next-instruction address, and *addr* is a jump target address.

C (carry). They are set to 1 or cleared to 0 depending on the result of certain instructions, and can be tested to decide among alternative courses of action in the program.

Arithmetic and Logical Instructions

The arithmetic and logical instructions operate on 8-bit data values stored in the CPU's general-purpose registers and store the result in the destination register, *rd*. For each instruction, one value is taken from a source register, *rs*. The other value, *op2*, either comes from a second source register (*rs2*) or is an *immediate value* (*immed*). An immediate value is a value that is specified as part of the instruction, rather than being stored in a register or in memory. For example, the instruction

```
add r3, r4, r1
```

adds the values currently in registers r4 and r1 and puts the result in r3. Similarly, the instruction

```
add r5, r1, 2
```

adds the immediate value 2 and the value currently in r1 and puts the result in r5. Note that the destination register can be the same as a source register. For example, the instruction

```
sub r4, r4, 1
```

updates register r4 by decrementing its value.

The addition and subtraction instructions treat the data values as 8-bit unsigned integers. The addc instruction includes the value of the C condition code as a carry-in bit, and the subc instruction includes the C value as a borrow-in bit. All of the instructions in this group modify the Z and the C bits. They set Z to 1 if the instruction result is 0, and they clear Z to 0 if the result is nonzero. The add and addc instructions set C to the carry-out bit of the addition, the sub and subc instruction set C to the borrow out of the subtraction, and the remaining logical instructions clear C to 0. We will see later in this section how the condition-code bits are used by branch instructions.

EXAMPLE 7.1   Write a sequence of instructions to evaluate the expression $2x + 1$, assuming the value of x is in register r3 and the result is to be put in r4.

SOLUTION   We can multiply x by 2 by adding it to itself. The required instructions are

```
add r4, r3, r3
add r4, r4, 1
```

EXAMPLE 7.2   Write a sequence of instructions that sets the Z bit to 1 if the least significant 4 bits of r2 have the value 0101.

SOLUTION   We can test whether a register value is equal to 0101 by subtracting 0101 from the value and putting the result in r0. The result value is ignored, but Z is set as a side-effect of the subtraction. However, the most significant 4 bits of r2 might contain 1s that we are not interested in, so we need to clear them to 0s before doing the subtraction. We can use an AND operation with the value 00001111 to clear the bit. The required instructions are:

```
and r1, r2, 0x0F
sub r0, r1, 0x05
```

The notation "0x" is a prefix for a hexadecimal value in the Gumnut assembly code notation. Thus, 0x0F is the value 00001111 and 0x05 is the value 00000101.

Shift Instructions

The shift instructions shift or rotate 8-bit values taken from the general purpose register *rs* and store the result in register *rd*. The number of places to shift or rotate is specified in the instruction as *count*. For example, the instruction

```
shl r4, r1, 3
```

reads the value currently in register r1, shifts it left by 3 places and puts the result in r4. The shift-left and shift-right instructions discard the bits shifted past the end of the 8-bit byte and fill the vacated bit positions with 0s. The rotate-left and rotate-right instructions copy the bits shifted past the end of the byte around to the other end. All of these instructions set Z to 1 if the instruction result is 0, and they clear Z to 0 if the result is nonzero. They set the C bit to the value of the last bit shifted past the end of the byte.

EXAMPLE 7.3   Write instructions that multiply the value in r4 by 8, ignoring the possibility of overflow.

SOLUTION   Recall from Section 3.1.2 that we can multiply an unsigned binary integer by 2^k by shifting k places to the left. Thus, since $8 = 2^3$, an instruction to multiply r4 by 8 is

```
shl r4, r4, 3
```

Memory and Input/Output Instructions

The Gumnut has separate instructions for accessing data memory and I/O controllers. We will discuss the operation of I/O controllers in detail in Chapter 8. For now, we simply point out that I/O controllers have registers that govern their operation, and that these registers can be read and written by the CPU. Just as locations in memory have addresses, each I/O controller register has an identifying address. The Gumnut uses 8-bit addresses for I/O controller registers, distinct from the 8-bit addresses it uses for locations in the data memory. We say that the Gumnut has separate *address spaces* for data memory and for I/O controller registers. This is in contrast to a number of other CPU instructions sets, in which I/O controller registers are part of the same address space as memory addresses. In those instruction sets, we say I/O registers are *memory mapped*.

For all of the Gumnut's memory and I/O instructions, the address to access is computed by adding the current value in *rs* and an offset value specified in the instruction. The load from memory instruction reads from the data memory at the computed address and puts the read value in register *rd*. The store to memory writes the value from register *rd* to the data memory at the computed address. The input and output instructions perform similar operations, but read or write to the I/O controller registers at the computed address. None of these instructions affect the values of the Z and C bits. As examples, the instruction

```
ldm r1, (r2)+5
```

calculates the memory address by adding the current value of r2 and the offset 5. It then reads from memory at that address and puts the read value in r1. Similarly, the instruction

```
stm r1, (r4)-2
```

stores the value from r1 into memory at the address 2 less than the current value of r4.

If we want to specify a particular address to access, we can use r0 as the register for *rs*. Recall that r0 always contains 0, so adding it to the offset value specified in the instruction just gives the offset value. In this case, we usually interpret the offset value as an unsigned 8-bit address. Our assembler tool allows us to imply the specification "(r0)" by omission and just write the address value, for example,

```
inp r3, 156
```

which reads from the I/O controller register at address 156 into r3. Similarly, if a register contains the address we want to access, we can use an offset of 0. Again, our assembler allows us to imply a 0 offset by omission, as in the instruction.

```
out r3, (r7)
```

EXAMPLE 7.4   Write instructions that increment a 16-bit unsigned integer stored in memory. The address of the least significant byte is in r2. The most significant byte is in the next memory location.

SOLUTION   Since the Gumnut arithmetic instructions only operate on 8-bit data, we need to do two adds, with the carry from the first used in the second. The instructions are

```
ldm r1, (r2)
add r1, r1, 1
stm r1, (r2)
ldm r1, (r2)+1
addc r1, r1, 0
stm r1, (r2)+1
```

Since the load and store instructions do not affect the C bit, the C result from the first addition is preserved and used in the addc instruction.

Branch Instructions

The branch instructions allow us to conditionally change the normal flow of execution. We mentioned earlier that the CPU follows a fetch-decode-execute loop to execute instructions at successive addresses in the instruction memory. It uses a program counter (PC) register to keep track of the next instruction address, and increments this register after fetching each instruction. The branch instructions modify the sequential flow of execution by changing the PC value. Each form of branch tests a condition, and if the condition is true, adds a signed 8-bit displacement value to the PC. The displacement, specified in the instruction, indicates how many locations forward or backward the next instruction to execute is from the current instruction. (A displacement of 0 refers to the instruction after the branch, since the PC has already been incremented after fetching the branch instruction.) If the condition is false, the PC is unchanged, and execution continues sequentially. The different branch instructions allow us to test each of the Z and C condition code bits for being set to 1 or not set to 1. Since these bits are affected by arithmetic, logical and shift instructions, we often deliberately precede a branch instruction with one of these instructions to compare data values. In other cases, the condition code setting occurs as a serendipitous side effect of data operations that we need to perform anyway.

EXAMPLE 7.5   Suppose the value in data memory location 100 represents the number of seconds elapsed in a time interval. Write instructions to increment the value, wrapping around to 0 when the value increments above 59.

SOLUTION   One possible sequence of instructions is

```
ldm r1, 100
add r1, r1, 1
sub r0, r1, 60
bnz +1
add r1, r0, 0
stm r1, 100
```

The first two instructions load the value into r1 and increment it. The sub instruction subtracts 60 from the new value and discards the result (by using r0 as the destination register). However, the Z condition code is updated as a side effect. If the new value is 60, the subtraction result is 0, so Z is set to 1; otherwise, it is cleared to 0. The branch instruction skips forward one instruction if Z is 0. The intervening add instruction, which is only executed when the incremented value was 60, overwrites the incremented value with 0. The final instruction, executed in all cases, stores the final value back to memory.

Jump and Miscellaneous Instructions

The first of the jump instructions, jmp, unconditionally breaks the sequential flow of execution by setting the PC to the address specified in the instruction.

EXAMPLE 7.6   Write instructions that test whether r1 is 0, and if so, clear the contents of memory location 100. If r1 is other than 0, the instructions should clear the contents of memory location 200 instead. Assume that the instructions start at address 10 in the instruction memory.

SOLUTION   In the required sequence of instructions we have two alternative actions to perform, depending on whether r1 is 0. Since instructions are laid out in linear order in the instruction memory, we need to put the instructions for the two alternatives one after the other. We need an unconditional jump at the end of the first alternative to bypass the instructions for the second alternative. The instructions are

```
10: sub r0, r1, 0
11: bnz +2
12: stm r0, 100
13: jmp 15
14: stm r0, 200
15: ...
```

The second of the jump instructions, jsb, is somewhat more involved than the simple jump instruction. It allows us to execute a *subroutine*, that is, a collection of instructions that perform some desired operations and that we can invoke from different parts of the program. Starting execution of a subroutine is referred to as *calling* the subroutine. The jsb instruction is used in tandem with the ret instruction, which returns from the subroutine to the place of the call. The sequence of instruction execution for a subroutine is shown in Figure 7.4. Execution proceeds sequentially until the jsb is encountered. The jsb saves

FIGURE 7.4   Flow of execution of subroutine calls. The subroutine is called from different places in the program, and in each case, returns to the instruction following the jsb.

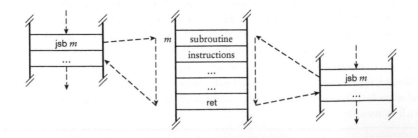

the incremented PC value (the return address) in an internal register and then updates the PC with the subroutine address specified in the instruction. This causes instructions in the subroutine to be executed. Eventually, the subroutine executes a ret instruction, which restores the saved return address to the PC. Thus, execution continues with the instruction after the jsb. The program can include several jsb instructions that all refer to the same subroutine. In each case, the return address saved is the address of the instruction after the jsb. This allows execution to return to the right place, regardless of where the subroutine was called from.

The instructions in the subroutine can include any in the CPU's instruction set. This raises the possibility that the subroutine might include a jsb to call a sub-subroutine. The sub-subroutine might include a further jsb to call a sub-sub-subroutine, and so on. When the sub-sub-subroutine returns, execution should continue just after the jsb in the sub-subroutine, and when it returns, execution should continue just after the jsb in the subroutine. In order to achieve this effect, the CPU needs more than just a single register to save return addresses. In fact, it needs a *push-down stack* of registers, as shown in Figure 7.5. Each time a jsb is executed, the return address for that jsb is pushed onto the stack. When a ret is executed, the return address used is the top entry on the stack, and that entry is popped from the stack. The Gumnut has a return-address stack that can hold up to eight entries, which is ample for most programs.

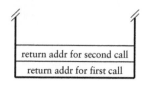

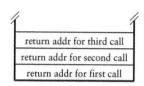

FIGURE 7.5   The push-down return-address stack after two nested calls (top) and a third nested call (bottom).

EXAMPLE 7.7   Suppose an application keeps track of a number of time intervals concurrently. Revise the sequence of instructions from Example 7.5 to form a subroutine that increments the number of seconds stored in the memory location whose address is in r2. Show how to call the subroutine to increment values in locations 100 and 102.

SOLUTION   We can rewrite the instructions to form a subroutine as follows:

```
ldm r1, (r2)
add r1, r1, 1
sub r0, r1, 60
bnz +1
add r1, r0, 0
stm r1, (r2)
ret
```

Assuming the first instruction in the subroutine is at location 20 in the instruction memory, the calling instructions are

```
add r2, r0, 100
jsb 20
add r2, r0, 102
jsb 20
```

The remaining miscellaneous instructions deal with *interrupts*, which are a way of responding to events signaled by I/O controllers. The enable-interrupt instruction allows the CPU to respond to interrupt events, and the disable-interrupt instruction prevents the CPU from responding. When the CPU responds to an interrupt event, it saves the address of the instruction it is about to execute and, instead, starts executing instructions in a special subroutine called an *interrupt handler*. The interrupt handler finishes with a return-from-interrupt (reti) instruction rather than a ret instruction. The wait instruction suspends execution until an interrupt occurs, and the stby instruction enters a low-power standby mode until an interrupt occurs. The difference is that the CPU would normally be able to respond to an interrupt immediately when suspended using a wait instruction, whereas it could take some time to power up from a stby instruction. We will describe interrupt processing in more detail as part of our discussion of input/output in Chapter 8.

7.2.2   THE GUMNUT ASSEMBLER

As we mentioned earlier, programs can be written in assembly language and translated into a sequence of binary-coded instructions by an assembler. The supplementary material for this book includes a simple assembler for the Gumnut, called *gasm*. The *gas User Guide*, also included in the supplementary material, provides a detailed description of the assembly language and how to use the assembler. We will describe a few key points here, illustrated by the program in Figure 7.6.

FIGURE 7.6 A Gumnut assembly language program to find the greater of two values.

```
; Program to determine greater of value_1 and value_2

                text
                org    0x000           ; start here on reset
                jmp    main

; Data memory layout

                data
value_1:        byte   10
value_2:        byte   20
result:         bss    1
```

(continued)

```
; Main program
                text
                org     0x010
main:           ldm     r1, value_1     ; load values
                ldm     r2, value_2
                sub     r0, r1, r2      ; compare values
                bc      value_2_greater
                stm     r1, result      ; value_1 is greater
                jmp     finish
value_2_greater: stm    r2, result      ; value_2 is greater

finish:         jmp     finish          ; idle loop
```

FIGURE 7.6   (*continued*)
A Gumnut assembly language
program to find the greater
of two values.

We have seen in VHDL models that we can include comments, starting with the characters "--", to describe parts of the model. We can also include comments in assembly language programs. In Figure 7.6, comments start with the ";" character and extend to the end of the line. Comments are especially important in assembly language programs, since each instruction performs only a single simple step. We use comments to describe the larger intent of a sequence of instructions.

The assembler lets us specify both the instructions to be included in the instruction memory and the contents of the data memory. We tell the assembler which memory we are specifying using the text (for instruction memory) and data (for data memory) *directives*. A directive does not represent a CPU instruction. Rather, it tells the assembler what to do when translating the program. Rather than requiring us to specify the address for each instruction and data item, the assembler adds instructions and data items at increasing addresses in each memory, starting at address 0. It automatically keeps track of where it is up to by using a *location counter* for each of the instruction and data memories. We can direct the assembler to change the location counter for the memory currently being filled by using an org (short for "origin") directive. For example, in Figure 7.6, the org 0x010 directive in the second text segment tells the assembler to continue placing instructions from location 010_{16}.

Within a data segment, we can include directives that specify the initial contents of data memory locations. The byte directive specifies the contents of an 8-bit location. The bss (short for "block starting with symbol") directive reserves a specified number of bytes of memory storage without initializing their content. We can precede each of these directives with a *label* that represents the starting address of the locations. The assembler works out the address for us. We can then refer to the label in instructions in the program. For example, the ldm instructions in Figure 7.6 refer to the labels value_1 and value_2 to load the initialized content of the data memory locations, and the stm instruction refers to the label result to store the greater value in the reserved location.

The advantage of using labels is that, when we revise the program, we don't need to revise the address values, since the assembler will work out new values when the program is reassembled.

Within a text segment, we include the instructions that form the program. Each instruction can be labeled, and the labels can be referenced in branch and jump instructions. Again, the assembler works out the instruction addresses represented by the labels, so that we don't have to work out branch displacements manually, or update references when we change the program.

One final point to note about the program in Figure 7.6 is that, once it completes its task, it doesn't stop executing. The Gumnut does not include any instructions for stopping. Instead, we include a *busy loop* at the end of the program. This just consists of an instruction that jumps back to itself, performing no useful work. Busy loops are common in embedded systems, since we usually do not want an embedded computer to stop (unless we turn the power off). An alternative is to have a CPU instruction or other facility that *suspends* operation until some activity is needed, such as responding to an I/O event. (On the Gumnut, we could use a wait or stby instruction.) This has the advantage that power consumption in the suspended state is typically much lower than in the active state. For this reason, suspending is preferred in battery-powered and other power-sensitive applications.

7.2.3   INSTRUCTION ENCODING

The instructions of a program are a form of information, and so, like any other information, can be encoded in binary. If we were to list all of the possible instructions, taking into account the operation to be performed and any registers, addresses, immediate values, and so on, we could devise an instruction coding taking up the smallest number of bits. However, decoding instructions would then be complex, leading to a large and slow decoder circuit within the CPU. Instead, instruction sets are usually encoded by separating a code word into distinct *fields*, each of which encodes one aspect of an instruction. The primary field is the *opcode*, short for operation code, that specifies the operation to be performed and, by implication, the layout of the remaining fields within the code word. By keeping the field layout simple and regular, we make the circuit for the instruction decoder simple and, hence, fast.

As an illustration, the instruction encoding for the Gumnut is shown in Figure 7.7. (The full details of the instruction encoding are described in Appendix D.) Each instruction code word is 18 bits long. The left-most bits, together with the function code (*fn*), form the opcode. Those instructions that specify register numbers have the numbers encoded in 3-bit binary form in separate fields of the instruction word. Similarly, instructions that specify immediate values, offsets, or displacements have those

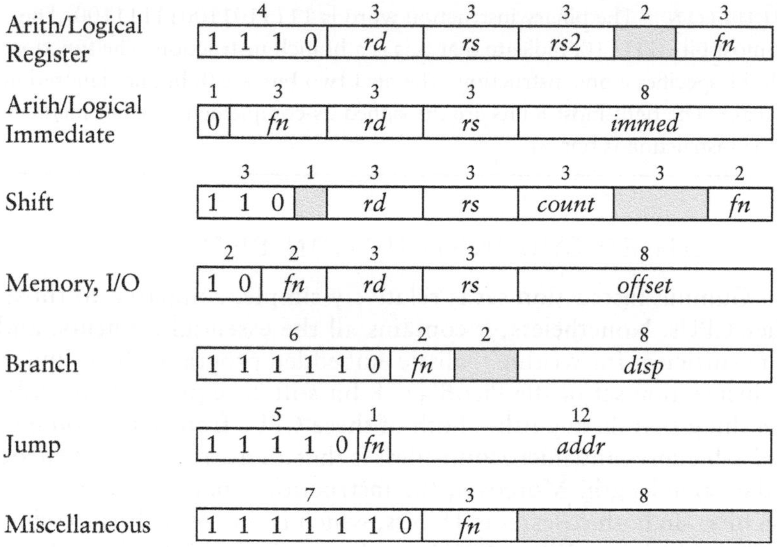

FIGURE 7.7 Instruction encoding for the Gumnut, showing the layout and size of fields within instructions.

values binary encoded in the right-most 8 bits of the instruction word. In several of the instruction formats, some bits remain unused. While this may waste some storage space within the instruction memory, the simplicity of encoding and the consequent simplicity of decoding is a trade-off worth making. As we mentioned earlier, it is the task of the assembler to translate instructions specified in textual assembly language into this binary encoding. Conversely, if we are testing a design that includes an embedded Gumnut, we may need to disassemble binary-coded instructions, that is, to determine the instructions corresponding to binary instruction code words processed by the embedded core.

EXAMPLE 7.8   Given that the function code for the addc operation is 001, what is the binary instruction word for the instruction

```
addc r3, r5, 24
```

SOLUTION   This is an arithmetic/logical immediate instruction, so the left-most bit is 0, and the function code is 001. The destination register r3 is encoded as 011, the source register number as 101, and the immediate value as 00011000. So the complete instruction word is 0 001 011 101 00011000, or, in hexadecimal, 05D18.

EXAMPLE 7.9   What instruction is represented by the hexadecimal instruction word 2ECFC?

SOLUTION   The binary instruction word is 111110110011111100. The left-most bits, 111110, indicate that this is a branch instruction. The function code 11 specifies a bnc instruction. The next two bits are 0, but are ignored in any case. The right-most 8 bits are the signed 2s-complement displacement –4. So the instruction is bnc –4.

7.2.4   OTHER CPU INSTRUCTION SETS

The Gumnut instruction set is relatively simple, compared to those of other CPUs. Nonetheless, it contains all the essential elements, and is quite sufficient for writing realistic embedded programs. It is similar to the instruction set of the PicoBlaze 8-bit soft core provided by Xilinx. One thing that distinguishes both of these CPUs from other commonly used 8-bit cores and microcontrollers is that all instructions are encoded in the same length. Moreover, the instruction length is not a multiple of 8 bits. (In both cases, it is 18 bits, which is one of widths to which a memory block in a Xilinx FPGA can be configured.) An example of an 8-bit microcontroller that takes a different approach is the 8051 from Intel and other vendors. It originated as a stand-alone microprocessor, and was subsequently released in microcontroller versions with various amounts of memory and I/O controllers included on chip. Its instruction set inherits from those of previous general purpose CPUs, in which a single memory address space was shared between instructions and data. Since locations in the 8051 memory are 8 bits wide, instructions are a multiple of 8-bit bytes. The opcode is included in the first byte. For some instructions the next one or two bytes contain further information to specify the instruction, such as an address and immediate data.

Another distinguishing characteristic of the 8051, compared to the Gumnut and PicoBlaze, is that the instruction set contains a much larger repertoire of operations. We call CPUs with instruction sets like this *complex instruction set computers* (CISCs), in contrast to the Gumnut and similar CPUs, which are *reduced instruction set computers* (RISCs). Many of the operations that can be expressed as one instruction on an 8051 would have to be implemented using a sequence of two or three instructions on a Gumnut. However, the complexity of the instruction set makes it much more difficult for the CPU to fetch and decode instructions. It also makes it difficult to implement a number of important CPU internal design techniques for increasing performance. For this and other reasons, RISC CPUs tend to dominate now.

The CPUs that we have mentioned thus far in this section are classified as 8-bit CPUs, as they operate only on 8-bit data. If the information to be represented in an embedded system is predominantly 16-bit, 32-bit or

64-bit data, using an 8-bit processor is very cumbersome. We may not be able to meet performance constraints, due to the number of instructions needed to implement 16-bit, 32-bit or 64-bit operations using 8-bit instructions. The alternative is to use a larger CPU whose instructions can operate on the larger data sizes directly. Most of the widely used processor cores for FPGAs and ASICs are 32-bit or 64-bit RISC CPUs. They have 32-bit or 64-bit registers and perform arithmetic and logical operations on data in those registers. They can load and store 8-bit, 16-bit, 32-bit and 64-bit data between registers and data memory. Instructions are encoded in fixed-length instruction words, usually 16 or 32 bits long. The larger, higher performance CPUs include instructions to operate on floating-point data as well as integers. Examples of this type of CPU include the PowerPC, ARM, MIPS and Tensilica cores that we mentioned earlier.

KNOWLEDGE
TEST QUIZ

1. What is meant by the *instruction set* of a CPU?

2. What three steps are repeatedly performed by a CPU to execute a program?

3. How does the CPU keep track of which instruction to execute next?

4. What is meant by the terms *little endian* and *big endian*?

5. What does an assembler do?

6. What does each of the following Gumnut instructions do?

```
addc r2, r3, 25
shr  r1, r1, 3
ldm  r5, (r1)+4
bnz  -7
jsb  do_op
ret
```

7. What is the binary instruction word for the following Gumnut instruction?

```
bnc +15
```

8. What Gumnut instruction is represented by the hexadecimal instruction word 05501?

7.3   INTERFACING WITH MEMORY

The way in which a CPU is connected to instruction and data memories depends on the implementation fabric used for both the CPU and the memories. In most embedded systems, the instruction memory is implemented with ROM, NOR flash memory, SRAM, or a combination of these. Including flash memory gives us the opportunity to upgrade the embedded software in the field. The data memory is usually implemented just with SRAM. Typically, the CPU and the memories each have a set of connection signals for the CPU/memory interface, and it is our job to join them together. If the two sets of signals are compatible, our job is relatively easy. Often, however, the sets of signals are designed in isolation, or according to different conventions. In such cases, we need to include glue logic to complete the interface.

One of the simplest cases of interfacing a CPU with memory is that of an embedded 8-bit core within an FPGA. The core includes interface signals that connect directly to those of the FPGA's memory blocks.

EXAMPLE 7.10   The memory interface signals of the Gumnut core are described in the following VHDL entity declaration:

```
library ieee;
use ieee.std_logic_1164.all, ieee.numeric_std.all;

entity gumnut is
  port ( clk_i     : in  std_logic;
         rst_i     : in  std_logic;
         inst_cyc_o : out std_logic;
         inst_stb_o : out std_logic;
         inst_ack_i : in  std_logic;
         inst_adr_o : out unsigned(11 downto 0);
         inst_dat_i : in  std_logic_vector(17 downto 0);
         data_cyc_o : out std_logic;
         data_stb_o : out std_logic;
         data_we_o  : out std_logic;
         data_ack_i : in  std_logic;
         data_adr_o : out unsigned(7 downto 0);
         data_dat_o : out std_logic_vector(7 downto 0);
         data_dat_i : in  std_logic_vector(7 downto 0);
         ... );
end entity gumnut;
```

Show how to include an instance of the Gumnut core in a VHDL model of an embedded system with a 2K × 18-bit instruction memory and a 256 × 8-bit data memory.

SOLUTION   The ports in the entity declaration can interface with the control signals of a flow-through SSRAM and a ROM implemented using FPGA SSRAM blocks, as described in Sections 5.2.2 and 5.2.5. In our architecture for our embedded system, we include the necessary signals to connect to an instance of the Gumnut entity, and use the signals in processes for the instruction and data memories. The architecture is

```vhdl
architecture rtl of embedded_gumnut is

  type ROM_2Kx18 is array (0 to 2047)
                      of std_logic_vector(17 downto 0);
  constant instr_ROM : ROM_2Kx18 := ( ... );

  type RAM_256x8 is array (0 to 255)
                      of std_logic_vector(7 downto 0);
  signal data_RAM : RAM_256x8;

  signal clk        : std_logic;
  signal rst        : std_logic;
  signal inst_cyc_o : std_logic;
  signal inst_stb_o : std_logic;
  signal inst_ack_i : std_logic;
  signal inst_adr_o : unsigned(11 downto 0);
  signal inst_dat_i : std_logic_vector(17 downto 0);
  signal data_cyc_o : std_logic;
  signal data_stb_o : std_logic;
  signal data_we_o  : std_logic;
  signal data_ack_i : std_logic;
  signal data_adr_o : unsigned(7 downto 0);
  signal data_dat_o : std_logic_vector(7 downto 0);
  signal data_dat_i : std_logic_vector(7 downto 0);
  ...

begin

  CPU : entity work.gumnut
    port map ( clk_i => clk,      rst_i       => rst,
      inst_cyc_o => inst_cyc_o, inst_stb_o => inst_stb_o,
      inst_ack_i => inst_ack_i,
      inst_adr_o => inst_adr_o, inst_dat_i => inst_dat_i,
      data_cyc_o => data_cyc_o, data_stb_o => data_stb_o,
      data_we_o  => data_we_o,  data_ack_i => data_ack_i,
      data_adr_o => data_adr_o,
      data_dat_o => data_dat_o, data_dat_i => data_dat_i,
      ... );

  IMem : process (clk) is
  begin
    if rising_edge(clk) then
```

(continued)

```
          if inst_cyc_o = '1' and inst_stb_o = '1' then
            inst_dat_i <=
              instr_ROM(to_integer(inst_adr_o(10 downto 0)));
            inst_ack_i <= '1';
          else
            inst_ack_i <= '0';
          end if;
        end if;
      end process IMem;

      DMem : process (clk) is
      begin
        if rising_edge(clk) then
          if data_cyc_o = '1' and data_stb_o = '1' then
            if data_we_o = '1' then
              data_RAM(to_integer(data_adr_o)) <= data_dat_o;
              data_dat_i <= data_dat_o;
              data_ack_i <= '1';
            else
              data_dat_i <= data_RAM(to_integer(data_adr_o));
              data_ack_i <= '1';
            end if;
          end if;
        end if;
      end process DMem;

      ...

    end architecture rtl;
```

Note that the instruction address port of the Gumnut core is 12 bits wide, whereas the 2K × 18-bit instruction memory uses an 11-bit-wide address. In this design, we simply leave the most significant address bit of the core unconnected. Each location in the instruction memory thus appears twice in the Gumnut's instruction address space: once at an address with the most significant bit 0, and once at an address with the most significant bit 1. We would normally just use one address for the location and ignore the other alias address.

Single-chip microcontrollers, such as those based on the 8051 described in Section 7.2.4, include a small amount of instruction and data memory on the microcontroller chip. However, many of them are able to address additional off-chip memory, using a number of the chip pins for the external memory interface signals. Since using the pins for this purpose reduces the number of pins available for inputs and outputs, the memory interface pins are often multiplexed to perform different functions at different times. This complicates the connection between the microcontroller and external memory.

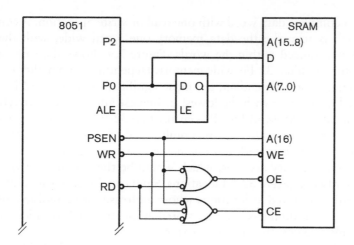

FIGURE 7.8 Connection between an 8051 microcontroller and an external combined instruction and data memory.

As an illustration, we will describe how to expand the memory of the 8051 microcontroller. The 8051 can access up to 64K bytes of instruction memory and 64K bytes of data memory, however, there are only 256 bytes of data memory and 4K to 16K bytes of instruction memory on the chip. The chip has two 8-bit input/output ports, P0 and P2, as well as a number of control signals, that can be used to connect to external memory. Figure 7.8 shows how they would be used to connect to an external 128K × 8-bit asynchronous SRAM, in which the lower 64K locations are used for instructions and the upper 64K locations for data. P2 provides the most significant address byte, and P0 is multiplexed with the least significant address byte and instruction and data bytes. Since information transfer on P0 is bidirectional, tristate drivers are used internally in the microcontroller and in the memory data pins.

The 8051 activates the address-latch enable (ALE) signal when it drives the least significant address bits on P0. We provide an 8-bit latch to hold these bits for the remainder of the memory access cycle. During an instruction read access, the 8051 activates the program-store enable (PSEN) signal, driving it to a low logic level. At other times, including data accesses, the signal is at a high logic level. Hence, we can use this signal directly as the most significant address bit to distinguish between instruction and data accesses to the external memory. The 8051 activates the RD signal during data read accesses and the WR signal during data write accesses. We use WR directly to control the memory's write enable (WE) signal. However, we need a small amount of glue logic to derive the chip enable (CE) and output enable (OE) signals. We could implement this glue logic, together with the address latch, in a small PAL component.

Microcontrollers and processor cores that access 16-bit, 32-bit or 64-bit data generally need data memories that are wider than 8 bits, even though addresses correspond to 8-bit locations. This allows the CPU to

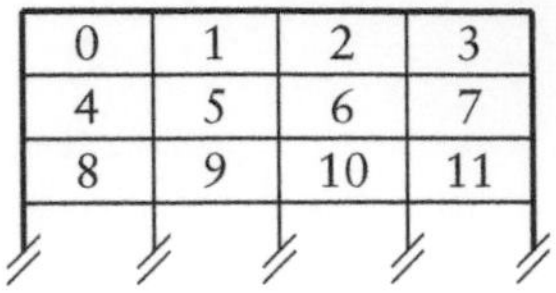

0	1	2	3
4	5	6	7
8	9	10	11

FIGURE 7.9 Arrangement of bytes within words in a 32-bit wide memory.

access a complete data word with one read or write operation. A common approach is to make the data memory one word wide, with the byte locations arranged within the words. Figure 7.9 shows the case of byte addressing within a 32-bit-wide memory. Depending on whether the CPU is big endian or little endian, the most significant byte of a 32-bit word is stored in the byte with the lowest or highest address, respectively, of a 32-bit location. Most 32-bit CPUs ensure that 32-bit data words are stored at locations whose addresses are a multiple of four. This allows the word to be read or written with just one memory access, rather than requiring two partial memory accesses, which would be the case if the word were split over two adjacent 32-bit locations. Similarly, CPUs ensure that 16-bit halfwords are stored at locations whose addresses are a multiple of two, and that 64-bit double-words are stored at locations whose addresses are a multiple of eight, for the same reason.

Reading from data memory is quite straightforward. A 32-bit CPU, for example, reads the whole 32-bit word containing the required data item. If the required item is only a 16-bit halfword or an 8-bit byte, the CPU usually extracts the item from the appropriate memory data signals and places it in a destination register. Writing a 32-bit word is similarly straightforward. The CPU places the word on the 32 memory data signals, and the memory performs a write operation. Writing a 16-bit halfword or an 8-bit byte is more involved, since we must ensure that the other bytes in the corresponding 32-bit memory location are not affected. The CPU typically provides separate *byte write enable* control signals instead of (or in addition to) the overall write enable control signal. Alternatively, it might provide separate *byte enable* signals instead of an overall memory enable signal. To write an 8-bit byte, the CPU places the byte value on the eight memory data signals corresponding to the position of the byte within a 32-bit word and activates the associated byte enable signal. The memory then performs a write operation, updating only the enabled byte within the addressed word. Similarly, to write a 16-bit halfword, the CPU places the halfword value on the appropriate 16 memory data signals and activates the associated two byte enable signals. The memory then writes only those two bytes of the addressed word.

EXAMPLE 7.11 The Xilinx MicroBlaze 32-bit processor core has connections to a 32K × 32-bit data memory as shown in Figure 7.10. (AS stands for "address strobe." This signal is active for each memory access.) Describe how the following memory operations proceed: a word read from address 00F00; a byte read from address 00F13; a word write to address 1E010; a byte write to address 1E016; and a halfword write to address 1E020.

SOLUTION Word read from 00F00: The address is a multiple of four. Write_Strobe is 0, so all four memory components perform a read operation, providing the 32-bit data on the Data_Read signal.

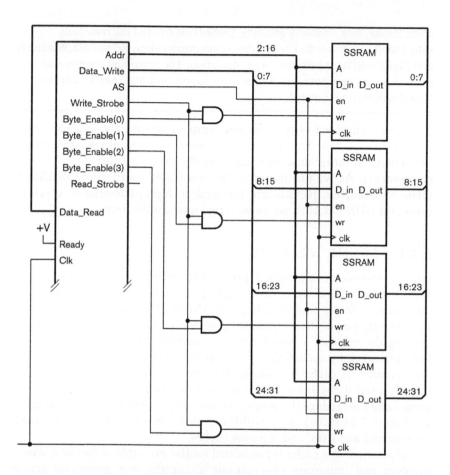

FIGURE 7.10 Connections from a Xilinx MicroBlaze core to a 32-bit data memory.

Byte read from 00F13: The address is 3 more than a multiple of four, so the byte is at offset 3 within a word. However, Write_Strobe is 0, so all four memory components perform a read operation, providing the 32-bit data on the Data_Read signal. The CPU extracts the required byte from Data_Read(24:31).

Word write to 1E010: The address is a multiple of four. Write_Strobe is 1 and all four Byte_Enable signals are 1, so all four memory components perform a write operation, taking the 32-bit data from the Data_Write signal.

Byte write to 1E016: The address is 2 more than a multiple of four, so the byte is at offset 2 within a word. The CPU provides the byte data on Data_Write(16:23). Write_Strobe and Byte_Enable(2) are 1, and the remaining Byte_Enable signals are 0. The memory component connected to Data_Write(16:23) performs a write operation. The remaining components perform a read operation, but the data they supply on Data_Read(0:7), Data_Read(8:15) and Data_Read(24:31) is ignored.

Halfword write to 1E020: The address is a multiple of four, so the halfword is at offset 0 within a word. The CPU provides the halfword data on Data_Write(0:15).

Write_Strobe, Byte_Enable(0) and Byte_Enable(1) are 1, and the remaining Byte_Enable signals are 0. The memory components connected to Data_Write(0:7) and Data_Write(8:15) perform a write operation. The remaining components perform a read operation, but the data they supply on Data_Read(16:23) and Data_Read(24:31) is ignored.

Some embedded systems require memory storage for large amounts of data. In such systems, it may be more appropriate to use dynamic memory (DRAMs) rather than SRAMs, given the lower cost per bit of DRAM components. As we mentioned in Section 5.2.4, controlling DRAMs is relatively complex, particularly for modern high-performance synchronous and DDR DRAMs, so we won't go into details here.

7.3.1   CACHE MEMORY

High performance embedded processors need to access instructions and data at higher rates than simple processors. For such processors, the memory access time of a large SRAM or DRAM memory system is significantly longer than the clock cycle time of the processor, potentially making the memory a performance bottleneck. Many processors avoid the bottleneck by including a *cache* in the path between the processor and memory. A cache is a small, fast memory that stores the most frequently used items from the main memory. By making access to these items faster, we reduce the average access time experienced by the processor. Figure 7.11 shows two possible organizations: a single cache for both instructions and data, and separate caches.

Operation of a cache is predicated on the *principle of locality*, which involves two important observations about the way programs access memory. The first is that a small proportion of instructions and data account for the majority of memory accesses over a given interval of time. The second is that those items stored in locations adjacent to a recently accessed item are likely to be accessed next. To take advantage of these

FIGURE 7.11   Processors with cache memories: a unified instruction/data cache for a single memory bus system (left), and separate instruction and data caches for a dual bus system (right).

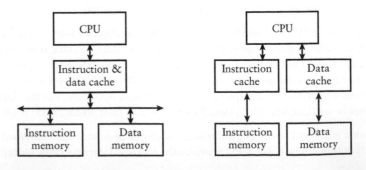

observations, we divide the collection of locations in main memory into fixed-sized blocks, often called *lines*, and copy whole lines at a time from main memory into the cache memory. When the processor requests access to a given memory location, the cache checks whether it already has a copy of the line containing the requested item. If so, the cache has a *hit*, and it can quickly satisfy the processor's request. If not, the cache has a *miss*, and must cause the processor to wait. The cache then copies the line containing the requested item from main memory into the cache memory. When the requested item is available in the cache, the processor can proceed with its requested access. The fact that neighboring items are also copied into the cache means that subsequent processor requests are likely to result in cache hits. As operation of the system proceeds, more and more lines are copied into the cache memory, resulting in a reduced miss rate. When the cache memory is full, some of the copied lines must be replaced by incoming lines. Ideally, the cache should replace the least recently used line. Since keeping track of usage history is complex, most caches use an approximation to determine which line to replace. In the steady state, caches can achieve miss rates of the order of 1% of processor requests. Thus, the average access time seen by the processor is very close to the access time of the cache memory.

For a system with cache memory, most of accesses to main memory are to entire lines, rather than to single locations. Since the processor is kept waiting during a main-memory operation, it is desirable to reduce the waiting time by making cache-line accesses as fast as possible. There are a number of advanced techniques that we can use to enable a higher rate of data transfer, or *memory bandwidth*. These include:

▶ Wide memory: Sufficient memory chips are used so that an entire cache line can be accessed at once. The line can then be transferred back to the cache on a wide bus in one clock cycle, or over a narrower bus in several clock cycles.

▶ Burst transfers: The CPU issues the first address of a line to be accessed in memory. The memory then performs a sequence of accesses at successive locations, starting from the first address. This technique obviates the time required to transfer the address for locations other than the first.

▶ Pipelining: The memory system is organized as a pipeline so that steps of different memory operations can be overlapped. For example, the pipeline steps might be address transfer, memory access, and returning read data to the CPU. Thus, the memory system could have three memory operations in progress concurrently, with one operation completed per clock cycle.

▶ Double data rate (DDR) operation: Rather than transferring data items only on rising clock edges, data can be transferred on both rising and falling clock edges. This doubles the rate at which data is transferred, hence the name.

These and a number of other techniques can be used in combination to form a memory system with sufficient bandwidth to allow the processor and cache to operate with minimal waiting time. A detailed discussion is beyond the scope of this book. The topic is addressed in books on computer organization and computer architecture (see Section 7.5).

KNOWLEDGE TEST QUIZ

1. When might we need glue logic to connect a memory to a CPU?

2. In the 8051 microcontroller, why are data signals and the least significant eight address signals multiplexed onto the same set of pins?

3. How many bits wide would the data memory for a 32-bit CPU typically be?

4. Why does a 32-bit CPU provide separate byte-enable signals for its data memory?

5. What two observations about the way programs access memory define the principle of locality?

6. What is meant by the terms cache hit and cache miss?

7. During a cache miss, what happens?

8. What is meant by the term *memory bandwidth*?

7.4 CHAPTER SUMMARY

▶ A computer system generally contains a central processing unit (CPU), instruction and data memory, input and output (I/O) controllers, and possibly special-purpose accelerators. The elements are interconnected by one or more buses.

▶ A microprocessor is a single-chip CPU that can be used in a general purpose computer or an embedded computer. A microcontroller is a single-chip computer incorporating a CPU, memory and I/O controllers. A digital signal processor (DSP) is a CPU specialized for processing streams of data from digitized signals.

▶ Microprocessors and CPUs in microcontrollers range in scale from simple 8-bit versions to complex 32-bit and 64-bit versions, referring to the size of data that can be processed in a single operation.

▶ CPUs can be implemented as predesigned cores and as soft cores.

▶ The instruction set of a CPU is its repertoire of instructions, usually including arithmetic and logical instructions, memory and I/O instructions, branch and jump instructions, and other miscellaneous instructions.

▶ Little-endian CPUs store multi-byte data with the least significant byte at the lowest address and the most significant byte at the highest address. Big-endian CPUs store the bytes in the opposite order.

▶ Instructions are encoded in binary. However, we usually develop programs using assembly language or a high-level language and use a translator (an assembler or compiler) to translate into binary-coded instructions.

▶ Instruction and data memories are usually connected directly to the CPU using memory-interface signals. Memories for 8-bit, 16-bit and 32-bit CPUs are commonly 8, 16 and 32 bits wide, respectively.

▶ Memories for high-performance CPUs can use a number of techniques for improving the memory bandwidth, including burst transfers, pipelining and double data rate (DDR) operation.

7.5 FURTHER READING

On Holy Wars and a Plea for Peace, Danny Cohen, Internet Engineering Note 137, 1980, available at http://www.rdrop.com/~cary/html/ endian_faq.html. This is the paper that originally adopted the terms "little endian" and "big endian" to refer to byte order.

Computer Architecture: A Quantitative Approach, 4th Edition, John L. Hennessy and David A. Patterson, Morgan Kaufmann Publishers, 2007. Includes a discussion of advanced memory system organization. The book also describes techniques, such as caches, used within high-performance CPUs to avoid delays due to memory accesses.

Computers as Components: Principles of Embedded Computing System Design, Wayne Wolf, Morgan Kaufmann Publishers, 2005. A more advanced reference on embedded systems design, covering CPU and DSP instruction sets, embedded systems platforms, and embedded software design.

Multiprocessor Systems-on-Chips, Ahmed Jerraya and Wayne Wolf, Morgan Kaufmann Publishers, 2004. Describes hardware, software and design methodologies for embedded systems containing multiple processor cores.

Engineering the Complex SOC: Fast, Flexible Design with Configurable Processors, Chris Rowen, Prentice Hall, 2004. Describes an approach to system-on-chip design based on extensible processors, using the Tensilica processor as an example.

ARM System-on-Chip Architecture, 2nd Edition, Steve Furber, Addison-Wesley, 2000. Describes the ARM instruction set, a number of ARM processor cores, and some examples of embedded applications using ARM cores.

Power Architecture Technology, IBM, http://www.ibm.com/developerworks/power. Resources describing the PowerPC architecture and processor cores.

See MIPS Run, 2nd Edition, Dominic Sweetman, Morgan Kaufmann Publishers, 2006. Describes the MIPS architecture, instructions set, and programming.

EXERCISES

EXERCISE 7.1   Suppose an embedded system includes two processor cores with a 32-bit wide dual-port memory for sharing data between the processors. Processor 1 is little endian, and processor 2 is big endian. Use the hexadecimal values 1234 (16 bits) and 12345678 (32 bits) to show how data is not shared correctly. How might the problem be remedied?

EXERCISE 7.2   Write Gumnut instructions to evaluate the expression $2(x + 1)$, assuming the value of x is in register r2 and the result is to be put in r7.

EXERCISE 7.3   Write Gumnut instructions to evaluate the expression $3(x - 1)$, assuming the value of x is in register r2 and the result is to be put in r7.

EXERCISE 7.4   Write Gumnut instructions to clear bits 0 and 1 of the value in register r1, leaving other bits unchanged, and to put the result in r2.

EXERCISE 7.5   Write Gumnut instructions to multiply the value in r4 by 18, ignoring the possibility of overflow. Hint: $18 = 16 + 2 = 2^4 + 2^1$.

EXERCISE 7.6   Write Gumnut instructions to increment the value in r3 modulo 60. If the result is 0, the value in r4 is to be incremented modulo 24.

EXERCISE 7.7   Write Gumnut instructions to test whether the 8-bit value in memory location 10 is equal to 99. If so, location 11 is to be set to 1; otherwise, location 11 is to be cleared to 0.

EXERCISE 7.8   Write Gumnut instructions to test whether r3 is 1 and input register 7 is also 1. If so, output register 8 is to be set to the hexadecimal value 3C.

EXERCISE 7.9   Write a Gumnut subroutine to clear a number of consecutive locations in memory to 0. The first address is provided in register r2 and the number of locations is provided in r3. Show a call to the subroutine to clear 10 locations starting from address 196.

EXERCISE 7.10   Write a complete Gumnut program to find the average of a sequence of eight 8-bit numbers stored in memory, and to write the result into a location in memory. Initialize the eight numbers to be the integers 2, 4, 6, ..., 16. Use a 16-bit sum to calculate the average, and shift instructions to divide by 8.

EXERCISE 7.11   Write a complete Gumnut program that monitors the value of input controller register 10. When the value changes from 0 to a nonzero value, the program increments a 16-bit counter and writes the counter value to output controller registers 12 (least significant byte) and 13 (most significant byte). The program should not terminate.

EXERCISE 7.12   Using the information in Appendix D, determine the encoding for the following Gumnut instructions:

a)   sub r3, r1, r0

b)   and r7, r7, 0x20

c)   ror r1, r1, 3

d)   ldm r4, (r3)+1

e)   out r4, 10

f)   bz +3

g)   jsb 0x68

EXERCISE 7.13 What Gumnut instructions are encoded by the following 18-bit hexadecimal values?

a)   009C0

b)   38227

c)   3353D

d)   24AFD

e)   3EA02

f)   3C580

g)   3F401

EXERCISE 7.14 Modify the design in Figure 7.8 to provide separate instruction and data memories for the 8051: a 64K × 8-bit ROM for the instruction memory and a 64K × 8-bit asynchronous SRAM for the data memory. The ROM has the same control signals as the SRAM except for the $\overline{\text{WE}}$ signal.

EXERCISE 7.15 Suppose a cache can satisfy a processor request in 5ns if it has a hit; otherwise the memory access time of 20ns must be added to the hit time. What is the average access time seen by the processor core for instructions for miss rates of 5%, 2% and 1%?

EXERCISE 7.16 Suppose a CPU with 32-bit instructions has an instruction cache with 16-byte lines. Addresses refer to bytes in memory. The cache is initially empty. Instructions are then fetched from the following addresses in order: 0, 4, 8, 92, 96, 100, 4, 8, 12, 16. For each fetch, determine whether the cache hits or misses. Assume no lines are replaced during execution of the sequence.

I/O INTERFACING

8

In the previous chapter, we introduced the notion of input/output (I/O) controllers that connect an embedded computer system with devices that sense and affect real-world physical properties. In this chapter, we will describe a range of devices that are used in embedded systems and show how they are accessed by an embedded processor and by embedded software.

8.1 I/O DEVICES

Digital systems with embedded computers are pervasive in our lives. We interact with many of them directly. Some are tools that we use in activities such as communication, entertainment, and information processing. These digital systems must incorporate human interface devices to allow us to control their operation and to receive responses. Other digital systems operate autonomously or under indirect control from us. Examples of such systems include industrial control systems, remote sensing devices and telecommunications infrastructure. These systems must incorporate devices to sense and affect the state of the physical world, as well as devices to communicate with one another, with controlling computers and with human interface devices.

Digital systems interact with the real world with *transducers*. An input transducer, or *sensor*, senses some physical property and generates an electrical signal that corresponds to the property. If the property is continuous in nature, such as temperature or pressure, the transducer may provide an analog signal that bears a continuous relationship with the physical property. Since digital systems deal with discrete representations of information, we need to convert the signal from analog to encoded digital form using a circuit called an *analog-to-digital converter*. Other forms of input transducer for continuous properties may provide discrete digital signals directly. An example is the shaft encoder for rotational position that we described in Section 3.1.3.

An output transducer, on the other hand, uses an electrical signal to cause a physical effect. Some transducers use an analog electrical signal to affect a physical property that is continuous in nature. An example is a loudspeaker that causes a continuously varying air pressure that we hear as a sound. To use such transducers in digital systems, we need a *digital-to-analog converter* circuit to convert from encoded digital form to an analog signal. Other forms of output transducer can use digital signals directly. Such transducers typically take a single-bit digital signal and cause a physical property to assume one of two values. For example, a transducer may cause a mechanical component to move to one position or another. Electromechanical transducers like this are often called *actuators*.

In the remainder of this section, we will describe a number of input and output devices that may be encountered in embedded systems. Then, in the next section, we will show how these devices can be connected to an embedded computer using input and output controllers.

8.1.1   INPUT DEVICES

Many digital systems include mechanically operated switches of various forms as input devices. These include push-button and toggle switches operated by human users, and microswitches operated by physical movement of mechanical or other objects. An example of the latter is a microswitch used to detect the presence of paper in a printer. In Section 4.4.1, we discussed ways in which switches can be connected as inputs to digital systems, and focused particularly on the problem of mechanical contact bounce and how to deal with it.

Keypads and Keyboards

Push-button switches are also used in keypads, for example, in phones, security system consoles, automatic teller machines, and other applications. In principle, we could treat each key in a keypad as a distinct push-button switch and connect it to the digital system as we have previously described. However, that would require a large number of signals and debouncing circuits, particularly for a large keypad. A more common technique is to arrange the key switches into a matrix, as shown in Figure 8.1, and to scan the matrix for closed contacts. When all of the key switches are open, all column lines (c1 through c3) are pulled high by the resistors. When a key switch is closed, one column line is connected to one row line (r1 through r4). We scan the matrix by driving one row line low at a time, leaving the rest of the row lines pulled high, and seeing if any of the column lines become low. For example, if the 8 key is pressed, c2 is pulled low when r3 is driven low. If more than one key in a given row is pressed at the same time, all of the corresponding column lines will be

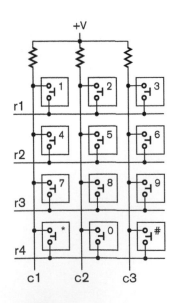

FIGURE 8.1   Keypad switches arranged in a scanned matrix.

pulled low when the row line is driven low. Thus, we are able to determine the same information about which keys are pressed as we would had we used individual connections for each key switch.

This raises the question of how the row lines are driven low. We could use a counter, together with circuitry that stores the count value and the column-line values for access by the embedded software. However, that would require synchronizing the processor with changes in count value so that the software could read the values at the appropriate times. A simpler approach is to provide a register into which the processor can write values to be driven on the row lines and another register for the processor to read the values of the column lines. This is shown in Figure 8.2. (We consider how the registers are attached to the processor in Section 8.2.) Since each of the key switches is a mechanical switch, it is subject to contact bounce. Thus we need to apply techniques for debouncing similar to those that we described for individual switches. The embedded software running on the processor needs to scan the matrix repeatedly. When it detects a key closure, it must check that the same key is still closed some time (say, 10 ms) later. Similarly, when it detects a key release, it must check that the same key remains released some time later. The scan must be repeated sufficiently often to debounce key presses without introducing a perceptible delay in response to key presses.

In a small digital system with a small keypad, the processing load to detect and debounce key presses would not be a significant part of the overall function of the system. The task of managing the keypad may safely be included as part of the main (or only) processor's workload. In other systems, it may be more appropriate to delegate the task, and possibly other I/O tasks, to subordinate embedded processors. The logical extension of this idea is illustrated by a keyboard for a general purpose computer. It has between 80 and 100 key switches arranged in a scanned matrix. Most keyboards include separate embedded processors whose entire workload consists of detecting key presses and dealing with roll-over (pressing a new key before the previously pressed key has been released), and communicating the information to the computer to which it is connected.

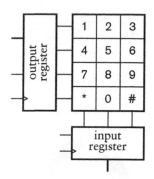

FIGURE 8.2 A keypad matrix with an output register for driving row lines and an input register for sensing column lines.

Knobs and Position Encoders

Historically, rotating knobs have been used in the user interfaces of electronic equipment to allow the user to provide information of a continuous nature. A common example is the volume control knob on audio equipment, or the brightness control on a light dimmer. In analog electronic circuits, the knob usually controls a variable resistor or potentiometer. With the introduction of digital systems, knobs were replaced by switches in many applications. For example, the volume control on audio equipment was replaced with two buttons, one to increase the volume and another

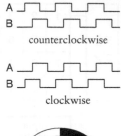

FIGURE 8.3 Operation of an incremental encoder: quadrature signals output from the encoder (top); an optical encoder disk (middle); and the disk and optical sensors attached to a shaft (bottom).

to decrease the volume. However, that form of control is not as intuitive or easy to use as a knob, so a digital form of knob is now used in many applications.

One form of digital knob input uses a shaft encoder, as we discussed in Section 3.1.3. This form has the advantage that the absolute position of the knob is provided as an input to the system. However, a simpler form of input device uses an *incremental encoder* to determine direction and speed of rotation. If the starting position or absolute position is not important, an incremental encoder is a good choice. An incremental encoder can also be used for a rotational position input in applications other than user interfaces, provided absolute positioning is not required. It can also be used for rotational speed input.

An incremental encoder operates by generating two square-wave signals that are 90° out of phase, as shown at the top of Figure 8.3. The signals can be generated either using electromechanical contacts, or using an optical encoder disk with LEDs and photo-sensitive transistors, as shown in the middle and at the bottom of Figure 8.3. As the shaft rotates counterclockwise, the A output signal leads the B output signal by 90°. For clockwise rotation, A lags B by 90°. The frequency of changes between low and high on each signal indicates the speed of rotation of the shaft.

A simple approach to using a knob attached to an incremental encoder involves detecting rising edges on one of the signals. Suppose we assume the knob is at a given position when the system starts operation. For example, we might assume a knob used as the volume control for a stereo is at the same setting as when the stereo was last used. (This would, of course, require the stereo to store the setting in a nonvolatile memory.) When we detect a rising edge on the A signal, we examine the state of the B signal. If B is low, the knob has been turned counterclockwise, so we decrement the stored value representing the knob's position. If, on the other hand, B is high, the knob has been turned clockwise, so we increment the stored value representing the knob's position. Using an incremental encoder instead of an absolute encoder in this application makes sense, since the volume might also be changed by a remote control. It is a change in the knob's position that determines the volume, not the absolute position of the knob.

Analog Inputs

Sensors for continuous physical quantities vary greatly, but they all rely on some physical effect that produces an electrical signal that depends on the physical quantity of interest. In most sensors, the signal level is small and needs to be amplified before being converted to digital form. Some sensors and the effects they rely on include:

▶ Microphones. These are among the most common sensors in our everyday lives, and are included in digital systems such as telephones, voice recorders and cameras. A microphone has a diaphragm that is displaced by sound pressure waves. In an electret microphone, for example, the diaphragm forms one plate of a capacitor. The other plate is fixed and has a permanent charge embedded on it during manufacture. The movement of the plates together and apart in response to sound pressure creates a detectable voltage across the plates that varies with the sound pressure. The voltage is amplified to form the analog input signal.

▶ Accelerometers for measuring acceleration and deceleration. A common form of accelerometer used in automobile air bag controllers, for example, has a microscopic cantilevered beam manufactured on a silicon chip. The beam and the surface over which it is suspended form the two plates of a capacitor. As the chip accelerates (or, more important, in the air bag application, decelerates), the beam bends closer to or farther from the surface. The corresponding change in capacitance is used to derive an analog signal.

▶ Fluid flow sensors. There are numerous forms of sensor that rely on different effects to sense flow. One form uses temperature-dependent resistors. Two matched resistors are self heated using an electric current. One of the resistors is placed into the fluid stream which cools it by an amount dependent on the flow rate. Since the resistance depends on the temperature, the difference in resistance between the two resistors depends on the flow rate. The resistance difference is detected to derive an analog input signal. Other forms of flow-rate sensor use rotating vanes, pressure sensing in venturi restrictions, and doppler shift of ultrasonic echoes from impurities. Different forms of sensor are appropriate for different applications.

▶ Gas detection sensors. Again, there are numerous forms that use different effects and are appropriate for different applications. As an example, a photo-ionizing detector uses ultraviolet light to ionize a sample of atmosphere. Gas ions are attracted to plates that are held at a potential difference. A circuit path is provided for charge to flow between the plates. The current in the path depends on the concentration of the gas in the atmospheric sample. The current is sensed and amplified to form the analog input signal.

Analog-to-Digital Converters

We mentioned earlier that analog input signals from sensors need to be converted into digital form so that they can be processed by digital circuits and embedded software. The basic element of an analog-to-digital converter (ADC) is a comparator, shown in Figure 8.4, which simply senses whether an input voltage (the + terminal) is above or below a reference voltage (the − terminal) and outputs a 1 or 0 accordingly.

FIGURE 8.4   A symbol for a comparator.

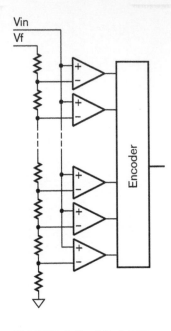

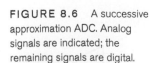

FIGURE 8.5   A flash ADC.

The simplest form of ADC is a *flash ADC*, illustrated in Figure 8.5. A converter with n output bits consists of a bank of $2^n - 1$ comparators that compare the input voltage with reference voltages derived from a voltage divider. For a given input voltage $V_{in} = kV_f$, where V_f is the full-scale voltage and k is a fraction between 0.0 and 1.0, a proportion k of the comparators have their reference voltage above V_{in} and so output 1, and the remaining comparators have their reference voltage lower than V_{in} and so output 0. The comparator outputs drive the encoder circuit that generates the fixed-point binary code for k. Flash ADCs have the advantage that they convert an input voltage to digital form very quickly. High-speed flash ADCs can perform tens or hundreds of millions of samples per second, and so are suitable for converting high bandwidth signals such as those from high-definition video cameras, radio receivers, radars, and so on. Their disadvantage is that they need large numbers of comparators. Hence, they are only practical for ADCs that encode the converted data using a relatively small number of bits. Common flash ADCs generate 8 bits of output data. We say they have a *resolution* of 8 bits, corresponding to the precision of the fixed-point format with which they represent the converted signal.

For signals that change more slowly, we can use a *successive approximation ADC*, shown in Figure 8.6. It uses a digital-to-analog converter (DAC) internally to make successively closer approximations to the input signal over several clock periods. To illustrate how the ADC works, consider a converter that produces an 8-bit output. When start input is activated, the successive approximation register (SAR) is initialized to the binary value 01111111. This value is provided to the DAC, which produces the first approximation, just less than half of the full-scale voltage. The comparator compares this approximation with the input voltage. If the input voltage is higher, the comparator output is 1, indicating that a better approximation would be above the DAC output. If the input voltage is lower, the comparator output is 0, indicating that a better

FIGURE 8.6   A successive approximation ADC. Analog signals are indicated; the remaining signals are digital.

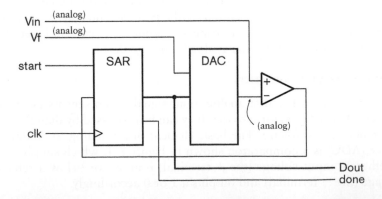

approximation would be below the DAC output. The comparator output is stored as the most significant bit in the SAR, and remaining bits are shifted down one place. This gives the next approximation, $d_70111111$, which is either one-quarter or three-quarters of the full-scale voltage, depending on d_7. During the next clock period, this next approximation is converted by the DAC and compared with the input voltage to yield the next most significant bit of the result and a refined approximation, $d_7d_6011111$. The process repeats over successive clock cycles, refining the approximation by one bit each cycle. When all bits of the result are determined, the SAR activates the done output, indicating that the complete result can be read.

The advantage of a successive approximation ADC over a flash ADC is that it requires significantly fewer analog components: just one comparator and a DAC. These components can be made to high precision, giving a high-precision ADC. 12-bit successive approximation ADCs, for example, are commonly available. The disadvantage, however, is that more time is required to convert a value. If the input signal changes by more than the precision of the ADC while the ADC is making successive approximation, we need to *sample and hold* the input. This requires a circuit that charges a capacitor to match the input voltage during a brief sampling interval, and then maintain the voltage on the capacitor while it is being converted. Another disadvantage of the successive approximation ADC is the amount of digital circuitry required to implement the SAR. However, that function could be implemented on an embedded processor, requiring just an output register to drive the DAC and an input bit from the comparator. The sequencing of successive approximations would then form part of the embedded software.

There are other forms of ADC apart from flash and successive approximation ADCs, each with advantages and disadvantages. Choice among them depends on the resolution, conversion speed and other factors dictated by the application. In practice, there is often a need to filter the analog input signal to ensure correct conversion to digital form. These considerations are beyond the scope of this book. More details can be found in books on digital signal processing mentioned in the Further Reading section.

8.1.2 OUTPUT DEVICES

Among the most common output devices are indicator lights that display on/off or true/false information. For example, an indicator might show whether a mode or operation is active, whether the system is busy, or whether an error condition has occurred. The simplest form of indicator is a single light-emitting diode (LED). It is low in cost, highly reliable, and easy to drive from a digital circuit, as Figure 8.7 shows. When the output from the driver is a low voltage, current flows through the LED,

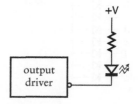

FIGURE 8.7 Output circuit for an LED indicator.

causing it to turn on. The resistor limits the current so as not to overload the output driver or the LED. We choose the resistance value to determine the current, and hence the brightness of the LED. When the output from the driver is a high voltage (near the supply voltage), the voltage drop across the LED is less than its threshold voltage, so no current flows; hence, the LED is turned off. We could, alternatively, connect the LED and resistor to ground, allowing a high output voltage to turn on the LED and a low output voltage to turn it off. However, output circuits designed to drive TTL logic levels are better able to sink current in the low state than to source current in the high state. Thus, it is more common to connect an LED as shown in Figure 8.7.

EXAMPLE 8.1   Determine the resistance for an LED pull-up resistor connected to a 3.3V power supply. The LED has a forward-biased voltage drop of 1.9V, and is sufficiently bright with a current of 2mA.

SOLUTION   Assuming the output driver low voltage is close to 0V, the voltage drop across the resistor must be $3.3V - 1.9V = 1.4V$. Using Ohm's Law with a current of 2mA means the resistance must be $1.4/0.002 = 700\Omega$. The closest standard value is 680Ω.

Displays

In Section 2.3.1, we introduced 7-segment displays and showed how we could decode a BCD value to drive the seven segments of a digit. In many applications, we have several digits to display. For example, an alarm clock typically has four digits for the hours and minutes of the time. While we could decode and drive each digit individually, that would require numerous output drivers, package pins and signals for the interconnections. Usually, it is more cost effective to connect the anodes or the cathodes of the LEDs for each digit in common, and to scan the digits. The connections for the LEDs in each digit, in this case, with common anodes, are shown in Figure 8.8. In addition to the seven LEDs for the segments, there is an LED for a decimal point (dp). The output connections for four digits are shown in Figure 8.9. Each of the outputs $\overline{A0}$ through $\overline{A3}$, when pulled low, turns on the transistor that enables a digit. We usually need these external transistors, since IC outputs cannot source enough current to drive up to eight LEDs directly.

To display four digits, we pull each of $\overline{A0}$ through $\overline{A3}$ low in turn. When $\overline{A0}$ is low, enabling the least significant digit, we drive the segment lines, $\overline{a}$ though $\overline{g}$ and $\overline{dp}$, low or high as required for the segment pattern for that digit. When $\overline{A1}$ is low, we drive the segment lines for the next digit, and so on. After driving the most significant digit, we cycle back to the least significant digit. If we cycle through the digits fast enough, our

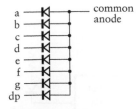

FIGURE 8.8   Connection of segment LEDs in a common anode 7-segment display.

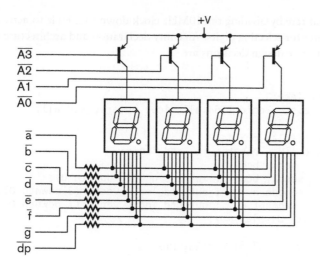

FIGURE 8.9 Connection of four 7-segment display digits.

eyes' persistence of vision smooths out any flickering due to each digit only being active 25% of the time.

The advantage of this scanned scheme is that we only need one signal for each digit plus one for each segment of a digit. For example, to drive four digits, we need 12 signals, compared with the 32 signals we would need had we driven segments individually. Depending on our application, we might use a counter or a shift register to drive the digit enable outputs and an 8-bit-wide multiplexer to select the values to drive onto the segment outputs. Often, however, the display is controlled by an embedded processor. In that case, we can simply provide output registers for the digit and segment outputs and let the embedded software manage the sequencing of output values.

EXAMPLE 8.2   Develop a VHDL model of a display multiplexer and decoder for the 4-digit 7-segment display shown in Figure 8.9. The circuit has four BCD inputs. The decimal point for the left-most digit should be lit, and the remaining decimal points not lit. The system clock has a frequency of 10MHz.

SOLUTION   The entity for the circuit has ports for the clock, reset and BCD inputs and for the the segment and anode outputs. Element 7 of the segment output drives the decimal point segment, and elements 6 down to 0 drive segments g through a, respectively. The outputs all use active-low logic. The circuit must include a multiplexer that selects each of the BCD inputs in turn. It decodes it to drive the 7-segment cathodes at the same time as activating the anode for the selected digit. Since we are relying on persistence of vision to avoid perceptible flicker, we need to cycle through the digits so that each is activated sufficiently frequently. A 50Hz cycle rate is acceptable. We can

achieve that rate by dividing the 10MHz clock down to 200Hz to activate a
2-bit counter for selecting digits. An entity declaration and architecture body to
implement these design decisions are

```vhdl
library ieee;
use ieee.std_logic_1164.all, ieee.numeric_std.all;

entity display_mux is
  port ( clk,  reset : in  std_logic;
         bcd0, bcd1,
         bcd2, bcd3  : in  unsigned(3 downto 0);
         anode_n     : out std_logic_vector(3 downto 0);
         segment_n   : out std_logic_vector(7 downto 0) );
end entity display_mux;

architecture rtl of display_mux is

  constant clk_freq      : natural : = 10000000;
  constant scan_clk_freq : natural : = 200;
  constant clk_divisor : natural : = clk_freq / scan_clk_freq;

  signal scan_clk  : std_logic;
  signal digit_sel : unsigned(1 downto 0);
  signal bcd       : unsigned(3 downto 0);
  signal segment   : std_logic_vector(7 downto 0);

begin

  -- Divide master clock to get scan clock
  scan_clk_gen : process (clk) is
    variable count : natural range 0 to clk_divisor - 1;
  begin
    if rising_edge(clk) then
      if reset = '1' then
        count := 0;
        scan_clk <= '0';
      elsif count = clk_divisor - 1 then
        count := 0;
        scan_clk <= '1';
      else
        count := count + 1;
        scan_clk <= '0';
      end if;
    end if;
  end process scan_clk_gen;

  -- increment digit counter once per scan clock cycle
  digit_counter : process (clk) is
```

(continued)

```vhdl
    begin
      if rising_edge(clk) then
        if reset = '1' then
          digit_sel <= "00";
        elsif scan_clk = '1' then
          digit_sel <= digit_sel + 1;
        end if;
      end if;
    end process digit_counter;

    -- multiplexer to select a BCD digit
    with digit_sel select
      bcd <= bcd0 when "00",
             bcd1 when "01",
             bcd2 when "10",
             bcd3 when others;

    -- activate selected digit's anode
    with digit_sel select
      anode_n <= "1110" when "00",
                 "1101" when "01",
                 "1011" when "10",
                 "0111" when others;

    -- 7-segment decoder for selected digit
    with bcd select
      segment(6 downto 0) <= "0111111" when "0000",    -- 0
                             "0000110" when "0001",    -- 1
                             "1011011" when "0010",    -- 2
                             "1001111" when "0011",    -- 3
                             "1100110" when "0100",    -- 4
                             "1101101" when "0101",    -- 5
                             "1111101" when "0110",    -- 6
                             "0000111" when "0111",    -- 7
                             "1111111" when "1000",    -- 8
                             "1101111" when "1001",    -- 9
                             "1000000" when others;    -- "-"

    -- decimal point is only active for digit 3
    segment(7) <= '1' when digit_sel = "11" else '0';

    -- segment outputs are negative logic
    segment_n <= not segment;

  end architecture rtl;
```

The scan_clk_gen process is the clock divider that generates the 200Hz clock for selecting digits. It sets the signal scan_clk to '1' for one master clock cycle at a 200Hz rate. The digit_counter process implements the 2-bit counter, incrementing the digit_sel signal each time scan_clk is '1'. The next two selected assignments use the digit_sel signal to select the BCD digit and to activate the corresponding anode. The remaining assignments decode the selected digit to drive the segment cathodes.

As an alternative to using LEDs for displays, some systems use liquid crystal displays (LCDs). Each segment of an LCD consists of liquid crystal material between two optical polarizing filters. The liquid crystal also polarizes light, and, depending on the angle of polarization, can allow light to pass or be blocked by the filters. The liquid crystal is forced to twist or untwist, thus changing its axis of polarization, by application of a voltage to electrodes in front of and behind the segment. By varying the voltage, we can make the segment appear transparent or opaque. Thus, LCDs require ambient light to be visible. In low light conditions, a back light is needed, which is one of their main disadvantages. The other disadvantages include their mechanical fragility and the smaller range of temperatures over which they can operate. They have several advantages over LEDs, including readability in bright ambient light conditions, very low power consumption, and the fact that custom display shapes can readily be manufactured.

Seven-segment displays are useful for applications that must display a small amount of numeric information. However, more complex applications often need to display alphanumeric or graphical information, and so may use LCD display panels. These can range from small panels that can display a few characters of text, to larger panels that can display text or images up to 320×240 dots, called *pixels* (short for picture elements). Beyond that size, systems would typically use the same kinds of display panels that are used in general purpose PCs. Since output for display panels is much more involved than output for simple segment-based displays, more complex control circuits are needed. We will return to control of display panels in Section 8.2.

Electromechanical Actuators and Valves

One of the simplest forms of actuator used to cause mechanical effects is a *solenoid*, shown in Figure 8.10. With no current flowing through the coil, the spring holds the steel armature out from the coil. When current flows, the coil acts as an electromagnet and draws the armature in against the spring. In a digital system, we can control the current in a small solenoid with a transistor driven by a digital output signal, as shown in Figure 8.11. The diode is required to absorb the voltage spikes that arise when the current through the inductive load is turned off.

The direct mechanical effect of activating a solenoid is a small linear movement of the armature. We can translate this into a variety of other effects by attaching rods and levers to the armature, allowing us to control the operation of mechanical systems. Hence, digitally controlled solenoids are widely used in manufacturing and other industrial applications.

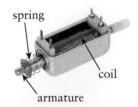

FIGURE 8.10   A solenoid actuator.

There are two important classes of devices based on solenoids, the first being solenoid valves. We can attach the armature of a solenoid to a valve mechanism, allowing the solenoid to open and close the valve, thus regulating the flow of a fluid or gas. This gives us a means of controlling chemical processes and other fluid or gas based processes. Importantly, a hydraulic solenoid valve (controlling flow of hydraulic fluid) or a pneumatic solenoid valve (controlling flow of compressed air) can be used to indirectly control hydraulic or pneumatic machinery. Such machines can operate with much greater force and power than electrical machines. So solenoid valves are important components in the interface between the disparate low-power digital electronic domain and the high-power mechanical domain.

The second class of device based on solenoids is relays. In these devices, the armature is attached to a set of electrical contacts. This allows us to open or close an external circuit under digital control. The reasons for using a relay are twofold. First, the external circuit can operate with voltages and currents that exceed those of the digital domain. For example, a home automation system might use a relay to activate mains power to a mains powered appliance. Second, a relay provides electrical isolation between the controlling and the controlled circuit. This can be useful if the controlled circuit operates with a different ground potential, or is subject to significant induced noise.

Motors

Whereas solenoids allow us to control a mechanical effect with two states, many applications require mechanical movement over a range of positions and at varying speeds. For these applications, we can use electric motors of various kinds, including stepper motors and servo motors. Both can be used to drive shafts to controlled positions or speeds. The rotational position or motion can be converted to linear position or motion using gears, screws, and similar mechanical components.

A stepper motor is the simpler of the two kinds of motors that we can control with a digital system. Its operation is shown in simplified form in Figure 8.12. The motor consists of a permanent magnet rotor mounted on the shaft. Surrounding the rotor is a stator with a number of coils that can be energized to form electromagnetic poles. The figure shows that, as coils are energized in sequence, the rotor is attracted to successive angular

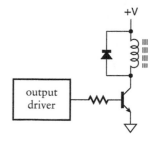

FIGURE 8.11 Solenoid controlled by a digital output.

FIGURE 8.12 Operation of a stepper motor.

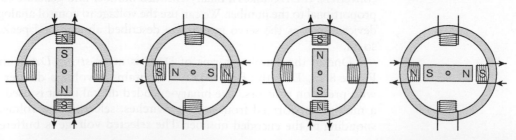

positions, stepping around through one rotation. The magnetic attraction holds the rotor in position, provided there is not too much opposing torque from the load connected to the motor shaft. The order and rate in which the coils are energized determines the direction and speed of rotation.

Practical stepper motors have more poles around the stator, allowing the motor to step with finer angular resolution. They also have varying arrangements of coil connections, allowing finer control over stepping. In practical applications, current through the coils is switched in either direction using transistors controlled by digital circuit outputs. The fact that the motor is activated by the on/off switching of current makes stepper motors ideal for digital control.

A servo-motor, unlike a stepper motor, provides continuous rotation. The motor itself can be a simple DC motor, in which the applied voltage determines the motor's speed, and the polarity of the applied voltage determines the direction of rotation. The "servo" function of the motor involves the use of feedback to control the position or speed of the motor. If we are interested in controlling position, we can attach a position sensor to the motor shaft. We then use a servo controller circuit that compares the actual and desired positions, yielding a drive voltage for the motor that depends on the difference between the positions. If we are interested in controlling the speed, we can attach a tachometer (a speed sensor) to the shaft, and again use a comparator to compare actual and desired speed to yield the motor drive voltage. In both cases, we can implement the servo controller as a digital circuit or using an embedded processor. We need a digital-to-analog converter to generate the drive voltage for the motor. We can use various position or speed sensors, including the position encoders we discussed in Sections 3.1.3 and 8.1.1.

Realistic servo control involves fairly complex computations to compensate for the nonideal characteristics of the motor and any gearbox and other mechanical components, as well as dealing with the effects of the mechanical load on the system. We won't go into any detail of those effects in this book.

Digital-to-Analog Converters

Digital-to-analog converters (DACs) are the complement of analog-to-digital converters. A DAC takes a binary-encoded number and generates a voltage proportional to the number. We can use the voltage to control analog output devices, such as the servo motors we described above, loudspeakers, and so on.

One of the simplest forms of DAC is an *R-string DAC*, shown in Figure 8.13. Like the flash ADC, it contains a voltage divider formed with precision resistors. The binary-encoded digital input is used to drive a multiplexer formed from analog switches, selecting the voltage corresponding to the encoded number. The selected voltage is buffered using

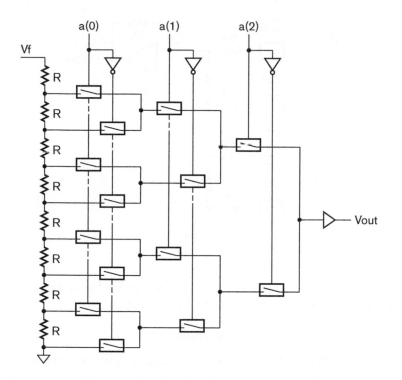

FIGURE 8.13   An R-string DAC.

a unity-gain analog amplifier to drive the final output voltage. This form of DAC works well for a small number of input bits, since it is possible to match the resistances to achieve good linearity. However, for a larger number of input bits, we require an exponentially larger number of resistors and switches. This scheme becomes impractical for DACs with more than eight to ten input bits.

An alternative scheme is based on summing of currents in resistor networks. One way of doing this is shown in Figure 8.14, sometimes called an *R/2R ladder DAC*. Each of the switches connected to the input bits connects the 2R resistance to the reference voltage *Vf* if the input is 1, or to ground if the input is 0. While the analysis is beyond the scope of this book, it can be shown that the currents sourced into the input node of the op-amp when the switches are in the 1 position are binary weighted. Those switches in the 0 position source no current. The superposition of the sourced currents means that the total current is proportional to the binary coded input. The op-amp voltage is thus also proportional to the binary coded input, in order to maintain the virtual ground at the op-amp input.

Just as there are numerous forms of analog-to-digital converter with various advantages and disadvantages, there are similarly numerous forms of digital-to-analog converter. We would choose an appropriate converter to meet the cost, performance and other constraints that apply

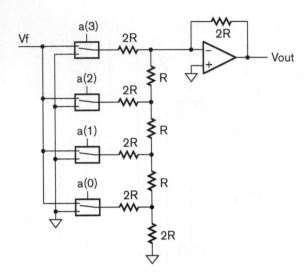

FIGURE 8.14   An R/2R ladder DAC.

to each application. More detail can be found in books on digital signal processing mentioned in the Further Reading section.

KNOWLEDGE TEST QUIZ

1. What is a sensor? What is an actuator?

2. Why would a digital system require a digital-to-analog converter?

3. How would we tell whether the 6 key in the keypad of Figure 8.1 is pressed?

4. Given the incremental encoder of Figure 8.3, if B is 1 when a 0 to 1 transition occurs on A, in which direction is the shaft rotated?

5. How many comparators are required in a flash ADC with a resolution of 8 bits?

6. How can we reduce the number of connections required for a multidigit 7-segment LED display?

7. What is the difference between a solenoid and a relay?

8. Identify two kinds of motor that we might control with a digital system.

9. If an application requires a 12-bit digital-to-analog converter (DAC), would we choose an R-string DAC or an R/2R ladder DAC? Why?

8.2 I/O CONTROLLERS

Given transducers, analog-to-digital converters and digital-to-analog converters, we can construct digital systems that include circuits to process the converted input information in digital form to yield output information. However, for an embedded computer to make use of the information, we need to include components that allow the embedded software to read input information and to write output information. For dealing with input, we can provide an *input register* whose content can be loaded from the digital input data and that can be read in the same way that the processor reads a memory location. For dealing with output, we can provide an *output register* that can be written by the processor in the same way that it writes to a memory location. The output signals of the register provide the digital information to be used by the output transducer. Many embedded processors refer to input and output registers as *ports*. Since it is such a commonly used term, we will make use of it, and take care to avoid confusion with ports of VHDL entities.

In practice, both input and output registers are parts of input and output controllers that govern other aspects of dealing with transducers under software control. We will start our discussion of I/O controllers in this section with some simple controllers that just include input and output registers for transferring data. We will then move on to consider more advanced controllers.

8.2.1 SIMPLE I/O CONTROLLERS

The simplest form of controller consists just of an input register that captures the data from an input device, or just an output register to provide data to a device. Usually, there are several I/O registers, so we need to select which register to read from or write to. This is similar to selecting which memory location to access, and is solved in the same way, namely by providing each register with an address. When the embedded processor needs to access an input or output register, it provides the address of the required register. We decode the address to select the register, and only enable reading or writing of that register.

As we mentioned in Chapter 7, some processors use memory mapped I/O; that is, they just use certain memory addresses to refer to I/O registers and use the same load and store instructions for accessing both memory location and I/O registers. We can use address decoding circuits connected to the processor to identify whether memory or I/O registers are being accessed, and enable the memory chips or the appropriate register as required. Other processors, like the Gumnut that we described in Chapter 7, have separate address spaces for memory and I/O registers, and include special instructions for reading and writing I/O registers. They provide control signals that distinguish between memory and I/O register access.

EXAMPLE 8.3   The signals provided by the Gumnut core for connecting to I/O registers are described in the following VHDL entity declaration:

```vhdl
library ieee;
use ieee.std_logic_1164.all, ieee.numeric_std.all;

entity gumnut is
  port ( clk_i      : in std_logic;
         rst_i      : in std_logic;
         ...
         port_cyc_o : out std_logic;
         port_stb_o : out std_logic;
         port_we_o  : out std_logic;
         port_ack_i : in  std_logic;
         port_adr_o : out unsigned(7 downto 0);
         port_dat_o : out std_logic_vector(7 downto 0);
         port_dat_i : in  std_logic_vector(7 downto 0);
         ... );
end entity gumnut;
```

The output port_adr_o is the port address, port_dat_o is the data written by an out instruction, port_dat_i is the data read by an inp instruction, port_cyc_o and port_stb_o indicate that a port read or write operation is to be performed, port_we_o indicates that the operation is a write, and port_ack_i indicates that the selected port is ready and has acknowledged completion of the read or write operation.

Develop a controller for the keypad matrix shown in Figure 8.2, and show how to connect the controller to a Gumnut core. Use output port address 4 for the matrix row output register and input port address 4 for the matrix column input register.

SOLUTION   The controller connects to the Gumnut I/O signals on one side and to the keypad row and column signals on the other side, as shown in Figure 8.15. We decode the port address from the Gumnut core externally to the controller to derive the strobe control signal (stb_i) for the controller.

FIGURE 8.15   Connection of a Gumnut core to a keypad controller.

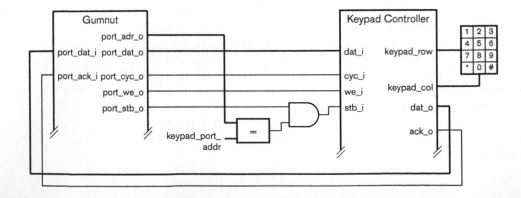

The VHDL entity declaration and architecture for the controller is

```vhdl
library ieee;
use ieee.std_logic_1164.all;

entity keypad_controller is
  port ( clk_i      : in  std_logic;
         cyc_i      : in  std_logic;
         stb_i      : in  std_logic;
         we_i       : in  std_logic;
         ack_o      : out std_logic;
         dat_i      : in  std_logic_vector(7 downto 0);
         dat_o      : out std_logic_vector(7 downto 0);
         keypad_row : out std_logic_vector(3 downto 0);
         keypad_col : in  std_logic_vector(2 downto 0) );
end entity keypad_controller;

architecture rtl of keypad_controller is

  signal col_synch : std_logic_vector(2 downto 0);

begin

  row_reg : process (clk_i) is
  begin
    if rising_edge(clk_i) then
      if cyc_i = '1' and stb_i = '1' and we_i = '1' then
        keypad_row <= dat_i(3 downto 0);
      end if;
    end if;
  end process row_reg;

  col_synchronizer : process (clk_i) is
  begin
    if rising_edge(clk_i) then
      dat_o <= "00000" & col_synch;
      col_synch <= keypad_col;
    end if;
  end process col_synchronizer;

  ack_o <= cyc_i and stb_i;

end architecture rtl;
```

The row_reg process represents the keypad row output register, storing the value to drive on the keypad row outputs. The col_synchronizer process represents the keypad column input register. Since the key switches may change at any time, we need to synchronize the input with the clock to avoid metastability failures. (We discussed this issue in Section 4.4.1.) In this design, we assume the keypad controller is the only thing driving the port_dat_o outputs, so we can assign directly to them regardless of the state of the control inputs. We will return to the topic of connecting multiple controllers in Section 8.3. The final assignment in the architecture body activates the port_ack_o output immediately on any port read or write operation, since there is no need to make the processor wait.

The controller is connected to a Gumnut core in an embedded system as shown in the following architecture body outline:

```
architecture rtl of embedded_system is

  signal ...

  constant keypad_port_addr : unsigned : = "04";
  signal keypad_stb_o : std_logic;

begin

  processor_core : entity work.gumnut(rtl)
    port map ( clk_i  => clk, rst_i => rst, ...,
             port_cyc_o => port_cyc_o, port_stb_o => port_stb_o,
             port_we_o  => port_we_o,  port_ack_i => port_ack_i,
             port_adr_o => port_adr_o, port_dat_o => port_dat_o,
                port_dat_i => port_dat_i, ... );

  keypad_stb_o <= '1' when port_adr_o = keypad_port_addr
                            and port_stb_o = '1' else '0';

  keypad : entity work.keypad_controller(rtl)
    port map ( clk_i => clk,
               cyc_i => port_cyc_o, stb_i => keypad_stb_o,
               we_i  => port_we_o,  ack_o => port_ack_i,
               dat_i => port_dat_o, dat_o => port_dat_i,
               keypad_row => keypad_row,
               keypad_col => keypad_col );

end architecture rtl;
```

The assignment to keypad_stb_o compares the Gumnut I/O port address with the value allocated for the keypad controller registers to derive the strobe signal for the keypad controller. The data input and output signals and the other control signals connect directly between the core and the controller.

While a simple I/O controller just has registers for input and output of data, more involved I/O controllers also have registers to allow the embedded processor to manage operation of the controller. Such registers might include *control registers*, to which a processor writes parameters governing the way transducers operate, and *status registers*, from which the processor reads the state of the controller. We often require such registers for controllers whose operation is sequential, since we need to synchronize controller operation with execution of the embedded software. As a consequence, we may have a combination of readable and writable registers used to control an input-only device or an output-only device.

EXAMPLE 8.4   In Section 8.1.1, we described a successive approximation analog-to-digital converter. It produces a binary-coded value representing the input voltage as a proportion of the full-scale reference voltage, V_f. We also mentioned that a sample-and-hold circuit can be used on the analog input if the voltage can change during the conversion process. Design a controller for a successive approximation ADC to connect to the Gumnut processor core. The controller has a control register whose contents govern operation of the converter. Bits 0 and 1 select among four alternate full-scale reference voltages. When a 1 is written to bit 2, the analog voltage is held and a conversion is started; when a 0 is written to the bit, the analog voltage is tracked. The controller also has a status register and an input data register. Bit 0 of the status register is 1 when a conversion is complete, and 0 otherwise. Other bits of the register are read as 0. The input data register contains the converted data.

SOLUTION   The controller circuit is shown in Figure 8.16. The control register is enabled when the least significant port address bit is 1 during a port write operation. The remaining port address bits are not decoded. Bits 0 and 1 of the register are decoded to control four analog switches that select the reference

FIGURE 8.16   Circuit for a controller for a successive approximation ADC.

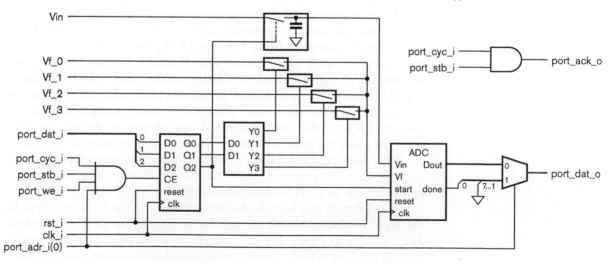

voltage. Bit 2 of the register controls the sample-and-hold component and the start signal of the ADC. The least significant port address bit is also used to select between the ADC data value and the ADC done status signal. Thus, when the processor performs a port read at address 0, it reads the ADC data, and when it performs a port read at address 1, it reads the done status.

8.2.2 AUTONOMOUS I/O CONTROLLERS

The simple I/O controllers in the previous section either involve no sequencing of operations, or just simple sequencing in response to accesses by a processor. More complex I/O controllers, on the other hand, operate autonomously to control the operation of an input or output device. For example, a servo-motor controller, given the desired position in an output register, might independently compute the difference between desired and actual position, compensate for mechanical lead and lag, and drive the motor accordingly. Interaction with the processor might only occur through the processor updating the desired position in the output register and monitoring the position difference by reading an input register. In some cases, if an autonomous controller detects an event of interest to the embedded software, for example, an error condition, the controller must notify the processor. We will discuss interrupts as a means of doing this in Section 8.5.2.

One reason for providing autonomy in the controller is that it allows the processor to perform other tasks concurrently. This increases the overall performance of the system, though at the cost of the additional circuitry required for the controller. Another reason is to ensure that control operations are performed fast enough for the device. If the device needs to transfer data at high rates, or needs control operations to be performed without delay, a small embedded processor may not be able to keep up. Making the I/O controller more capable may be a better trade-off than increasing the performance or responsiveness of the processor.

As an illustration of an autonomous controller, let us return to the LCD display panels that we mentioned in Section 8.1.2 as a form of output device for complex digital systems. LCD panels consist of a rectangular array of liquid crystal pixels. The electrodes are connected in rows on one side of the panel and in columns on the other side. A voltage is applied to one row at a time, and the column electrodes are variously set to the same or a complementary voltage to activate pixels in the selected row. In this way, the panel is scanned row by row to refresh the pixel states, in much the same way that a dynamic memory must be refreshed.

Since managing and refreshing an LCD panel requires a lot of activity, manufacturers of panels typically combine a display controller with a panel to form an LCD module. The display controller is an autonomous digital subsystem that includes memory for storing the information to be displayed on the panel and circuitry for refreshing the panel. An embedded computer treats the display controller as a specialized output controller, and provides it with updates to the stored information. In a graphical LCD module, the stored information consists of the image to be displayed, represented with one bit per pixel. In a character LCD module, the stored information consists of the binary code words for the characters. The display controller is responsible for decoding the character code words and rendering the image corresponding to the characters.

A specific example of an LCD module is the ASI-D-1006A-DB-_S/W module from All Shore Industries, Inc., a 100×60 pixel LCD panel that includes an SED1560 controller chip from Seiko Epson Corp. The module is designed to connect to 8-bit microcontrollers, such as the 8051 that we mentioned in Chapter 7. Figure 8.17 shows how this might be done. The controller chip has an internal memory for storing the image to be displayed on the LCD panel. The chip provides a control register to which the microcontroller can write encoded commands, a status register, and a data input/output register for access to the display memory. The microcontroller issues commands to the chip to configure the display and to load pixel data into the memory. Thereafter, the chip autonomously manages scanning the display using the pixel data in its memory, leaving the microcontroller free to perform other tasks.

As we mentioned above, the use of an autonomous controller may be appropriate for a device that must transfer input or output data at high rates. Often, such data must be written to memory (in the case of input data) or read from memory (in the case of output data). If the data transfer were done by a program copying data between memory and controller registers, that activity would consume much of the processor's time. An alternative, commonly adopted in high-speed autonomous controllers, is to use *direct memory access* (DMA), in which the controller reads data from memory or writes data to memory without intervention by the processor. The processor provides the starting memory address to the controller (by writing the address to a control register), and the controller then performs the data transfer autonomously. We can think of a controller that operates in this way as an accelerator for input/output operations. Since other forms of accelerator also use DMA for data transfer, we will defer a more detailed description of DMA until Chapter 9.

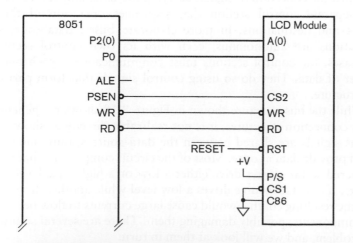

FIGURE 8.17 An LCD module connected to an 8051 microcontroller.

KNOWLEDGE TEST QUIZ

1. What is the purpose of an input register in an I/O controller? What is the purpose of an output register?

2. What is the purpose of a control register in an I/O controller? What is the purpose of a status register?

3. If an embedded processor uses memory mapped I/O, how do we distinguish accesses to memory from accesses to I/O registers?

4. Why might a controller for an input device have registers to which a processor can write?

5. What advantages do autonomous I/O controllers have over simple controllers?

8.3   PARALLEL BUSES

As we have seen, digital circuits consist of various interconnected components. Each component performs some operation or stores data. The interconnections are used to move data between the components. Where the data is binary coded, several signals are connected in parallel, one per bit of the encoding. Many of the interconnections we have seen thus far have been simple point-to-point connections, with one component as the source of data and a single separate component as the destination. In other cases, connections fan out from a single source to multiple destinations, allowing each of the destination components to receive data from the source.

In some systems, especially embedded systems containing processor cores, parallel connections carry encoded data from multiple sources to several alternate destinations. Such connection structures, shown conceptually in Figure 8.18, are called *buses*. In the simplest case, a bus is just the collection of signals carrying the data, and control remains in a separate control section that sequences operation of the data sources and destinations. In more elaborate buses, data sources and destinations are autonomous, each with its own control section. In such cases, the control sections must communicate to synchronize the transfer of data. They do so using control signals that form part of the bus structure.

While the bus structure shown in Figure 8.18 shows the general idea of bus connection structures, it is not realizable directly as shown. Since the bus signals are shared between the data sources, only one of them should provide data at once. Most of the circuit components that we have considered so far always drive either a low or a high logic level at their outputs. If one data source drives a low level while another drives a high level, the resulting conflict would cause large currents to flow between the two components, possibly damaging them. There are several solutions to this problem, and we will look at them in turn.

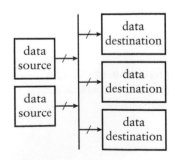

FIGURE 8.18 Conceptual connection structure for a bus.

8.3.1 MULTIPLEXED BUSES

One solution is to use a multiplexer to select among the data sources, as shown in Figure 8.19. The multiplexer selects the value to drive the bus signals based on a control signal generated by a control section. If the bus has n data sources, an n-input multiplexer is required for each bit of the encoded data transmitted over the bus. Depending on the number of sources and the arrangement of the components and signals on the integrated circuit chip, the multiplexer may be implemented as a single n-input multiplexer, or it may be subdivided into sections distributed around the chip. For example, if a bus has five data sources, two of which are on one side of a chip and the remaining three are on the other side, the bus wiring may be simplified by using a 2-input multiplexer adjacent to the two data sources and a 3-input multiplexer adjacent to the three data sources. The outputs of the multiplexers would then be connected to a 2-input multiplexer adjacent to the data destinations.

One extreme form of subdivision of bus multiplexers is the fully distributed structure shown in Figure 8.20. The data signals are connected in a chain going past all of the sources and then routed to the destinations. Each multiplexer either connects its associated data source to the chain (when the multiplexer's select input is 1) or forwards data from a preceding source (when the select input is 0). The advantage of this form of distributed multiplexer is the reduction in wiring complexity. It is often easier to route a set of signals in a chain past circuit blocks rather than trying to connect several data sources to a central hub.

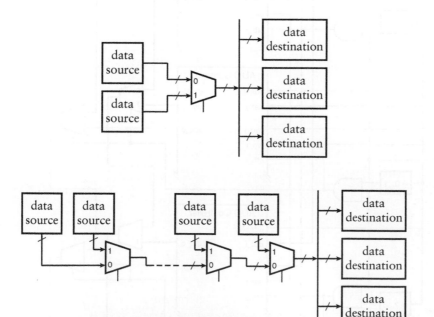

FIGURE 8.19 A bus using a multiplexer to select among data sources.

FIGURE 8.20 A distributed-multiplexer bus structure.

One example of a bus designed to use multiplexers is the Wishbone bus. The signals in the bus and their timing are specified in a standard document, referenced in the Further Reading section. The Gumnut core uses a simple form of Wishbone bus for each of the instruction, data and I/O port connections. The signals with a "_o" suffix are outputs from a component, and the signals with a "_i" suffix are inputs. Where multiple "_o" signals are to be connected to a "_i" signal, a multiplexer is required.

EXAMPLE 8.5   Show how, in an embedded system using a Gumnut core, the keypad controller of Example 8.3 and two instances of the ADC controller of Example 8.4, the components are interconnected using distributed multiplexers.

SOLUTION   The Gumnut core is the single source for the port address and control signals and for the output data signals, so no multiplexer is needed for those signals. The controllers each provide input data and ack signals, so distributed multiplexers are needed for them. We can decode the port address to derive the controller strobe signals and multiplexer select signals. We choose the first ADC controller when the port address is 0 or 1, the second ADC when the port address is 2 or 3, and the keypad controller when the port address is 4. The connections are shown in Figure 8.21.

FIGURE 8.21   Connection of two ADC controllers and a keypad controller to a Gumnut core using distributed multiplexers.

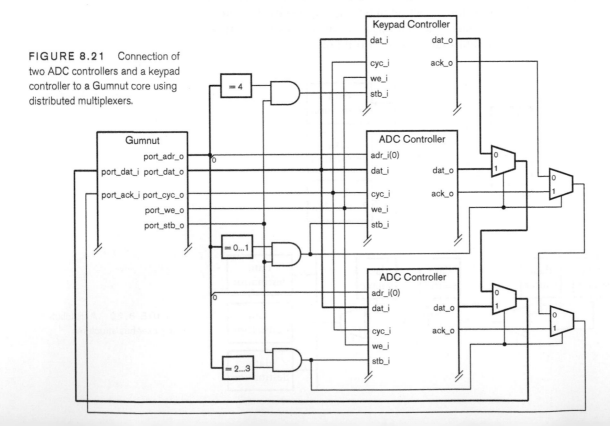

EXAMPLE 8.6   Develop a VHDL model for the embedded system of
Example 8.5.

SOLUTION   The architecture body is

```
architecture ADC_keypad of embedded_system is

  signal ...

  constant ADC0_port_addr   : unsigned : = X"00";
  constant ADC1_port_addr   : unsigned : = X"02";
  constant keypad_port_addr : unsigned : = X"04";

  signal ADC0_stb_o, ADC1_stb_o, keypad_stb_o : std_logic;
  signal ADC0_dat_o,
         ADC1_dat_o,
         keypad_dat_o,
         ADC0_dat_fwd,
         ADC1_dat_fwd : std_logic_vector(7 downto 0);
  signal ADC0_ack_o,
         ADC1_ack_o,
         keypad_ack_o,
         ADC0_ack_fwd,
         ADC1_ack_fwd : std_logic;
begin

  processor_core : entity work.gumnut(rtl)
    port map ( clk_i      => clk,
               rst_i      => rst, ...,
               port_cyc_o => port_cyc_o,
               port_stb_o => port_stb_o,
               port_we_o  => port_we_o,
               port_ack_i => ADC1_ack_fwd,
               port_adr_o => port_adr_o,
               port_dat_o => port_dat_o,
               port_dat_i => ADC1_dat_fwd, … );

  ADC0_stb_o <= '1' when
                    (port_adr_o and X"FE") = ADC0_port_addr
                    and port_stb_o = '1' else '0';
  ADC1_stb_o <= '1' when
                    (port_adr_o and X"FE") = ADC1_port_addr
                    and port_stb_o = '1' else '0';
  keypad_stb_o <= '1' when port_adr_o = keypad_port_addr
                       and port_stb_o = '1' else '0';

  keypad : entity work.keypad_controller(rtl)
    port map ( clk_i => clk,
```

(continued)

```
                    cyc_i => port_cyc_o,
                    stb_i => keypad_stb_o,
                    we_i  => port_we_o,
                    ack_o => keypad_ack_o,
                    dat_i => port_dat_o,
                    dat_o => keypad_dat_o, ... );

  ADC0 : entity work.ADC_controller(rtl)
    port map ( clk_i    => clk,
               rst_i    => rst,
               cyc_i    => port_cyc_o,
               stb_i    => ADC0_stb_o,
               we_i     => port_we_o,
               ack_o    => ADC0_ack_o,
               adr_i(0) => port_adr_o(0),
               dat_i    => port_dat_o,
               dat_o    => ADC0_dat_o, ... );

  ADC0_dat_mux :
    ADC0_dat_fwd <= ADC0_dat_o when ADC0_stb_o = '1' else
                    keypad_dat_o;

  ADC0_ack_mux :
    ADC0_ack_fwd <= ADC0_ack_o when ADC0_stb_o = '1' else
                    keypad_ack_o;

  ADC1 : entity work.ADC_controller(rtl)
    port map ( clk_i    => clk,
               rst_i    => rst,
               cyc_i    => port_cyc_o,
               stb_i    => ADC1_stb_o,
               we_i     => port_we_o,
               ack_o    => ADC1_ack_o,
               adr_i(0) => port_adr_o(0),
               dat_i    => port_dat_o,
               dat_o    => ADC1_dat_o, ... );

  ADC1_dat_mux :
    ADC1_dat_fwd <= ADC1_dat_o when ADC1_stb_o = '1' else
                    ADC0_dat_fwd;

  ADC1_ack_mux :
    ADC1_ack_fwd <= ADC1_ack_o when ADC1_stb_o = '1' else
                    ADC0_ack_fwd;

end architecture ADC_keypad;
```

The first group of assignments, after the Gumnut core instance, represent the port address decoders. They compare the port address from the processor core with the base addresses of the ADC controllers and the keypad controllers. For the ADC controllers, the port address is ANDed with the hexadecimal value FE to clear the least significant bit.

The instances of the ADC controllers are followed by assignments that represent the distributed multiplexers. The outputs of the multiplexers for the second ADC connect back to the Gumnut core port_dat_i and port_ack_i inputs.

8.3.2 TRISTATE BUSES

A second solution to avoiding contention on a bus is to use *tristate* bus drivers. We introduced tristate drivers in Chapter 5 as part of our discussion of connecting multiple memory components. We said that the outputs of a tristate driver can be turned off by placing it in a *high-impedance*, or *hi-Z*, state. The symbol for a tristate driver is shown in Figure 8.22. When the enable input is 1, the driver behaves like an ordinary output, driving either a low or a high logic level on the output. When the enable input is 0, the driver enters the high-impedance state by turning its output-stage transistors off.

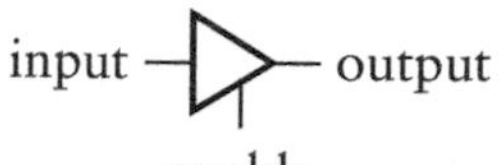

We can implement a bus with multiple data sources by using tristate drivers on the outputs of each data source. We use one driver for each bit of encoded data provided by the source, and connect the enable inputs of the drivers for a given source together, as shown in Figure 8.23. That way, a source either drives a data value onto the bus, or has all bits in the high-impedance state. The control section selects a particular source to provide data by setting the enable input of that source's drivers to 1, and all other enable inputs to 0.

One of the main advantages of tristate buses is the reduction in wiring that they afford. For each bit of the encoded data on the bus, one signal wire is connected between all of the data sources and destinations. However, there are some issues to consider. First, since bus wires connect all of the sources and destinations, they are generally long and heavily loaded with the capacitance of the drivers and inputs. As a consequence, the wire delay may be large, making high-speed data transfer difficult. Moreover, the large capacitance means we need more powerful output-stage circuits, increasing the area and power consumption of the chip.

A second issue is difficulty in designing the control that selects among data sources. The control section must ensure that one source's drivers are disabled before any other source's drivers are enabled. When we design the control section, we need to take into account the timing involved in disabling and enabling drivers. This is shown in Figure 8.24. When the

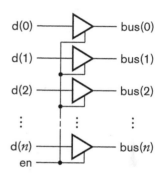

FIGURE 8.22 Symbol for a tristate driver.

FIGURE 8.23 Parallel connection of tristate drivers.

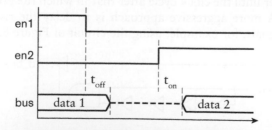

FIGURE 8.24 Tristate disable and enable timing.

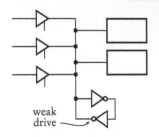

FIGURE 8.25 A bus keeper for maintaining valid logic levels.

enable input of a driver changes to 0, there is a delay, t_{off}, before the driver disconnects from the bus. Similarly, when the enable input changes to 1, there is a delay, t_{on}, before the driver delivers a valid low or high logic level on the bus. In the intervening time, the bus *floats*, indicated on the timing diagram by a dashed line midway between the low and high logic levels. Since there is no output driving a low or high logic level on the bus signals, each signal drifts to an unspecified voltage.

Letting the bus float to an unspecified logic level can cause switching problems in some designs. The bus signal might float to a voltage around the switching threshold of the bus destination inputs. Small amounts of noise voltage induced onto the bus wire can cause the inputs to switch state frequently, causing spurious data changes within the data destination and consuming power unnecessarily. We can avoid floating logic levels on the bus signals by attaching a *weak keeper* to the signal, as shown in Figure 8.25. The keeper consists of two inverters providing positive feedback to the bus signal. When the bus is forced to a low or high logic level by a bus driver, the positive feedback keeps it at that level, even if the forcing driver is disabled. The transistors in the output circuit of the inverter driving the bus are small, with relatively high on-state resistance, and so cannot source or sink much current. They are easily overridden by the output stages of the bus drivers.

When we need to change from one data source to another, it might seem reasonable to disable one driver at the same time as enabling the next driver. However, this can cause driver contention. If the t_{off} delay of the disabled driver is at the maximum end of its range and the t_{on} delay of the enabled driver is at the minimum end, there will be a period of overlap where some bits of the enabled driver may be driving opposite logic levels to those of the disabled driver. The overlap will be short-lived and is unlikely to destroy the circuit. However, it does contribute extra power consumption and heat dissipation and ultimately will reduce the operating life of the circuit. The overlap effect can be exacerbated by clock skew in the control section. If the flip-flop that generates the enabling signal receives its clock earlier than the flip-flop that generates the disabling signal, there will be an increased chance of overlap, even if the on and off delays of the tristate drivers are near their nominal values. Given these considerations, the safest approach when designing control for tristate buses is to include a margin of dead time between different data sources driving the bus. A conservative approach is to defer enabling the next driver until the clock cycle after that in which the previous driver is disabled. A more aggressive approach is to delay the rising edges of the enable signals, for example, using the circuit of Figure 8.26, to avoid

FIGURE 8.26 A circuit to delay the rising edge of a bus enable signal.

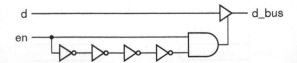

overlap between drivers. As many pairs of inverters are included as give the required delay. However, this approach requires very careful attention to timing analysis to ensure that it works effectively across the expected range of operating conditions.

A third issue relating to design of tristate buses is the support provided by CAD tools. Not all physical design tools provide the kinds of timing and static loading analyses needed to design tristate buses effectively. Similarly, tools that automatically incorporate circuit structures to enable testing of circuits after their manufacture don't always deal with tristate buses correctly. If the tools we use don't support tristate buses, we must resort to manual methods to complete and verify our design.

A final issue is that not all implementation fabrics provide tristate drivers. For example, many FPGA devices do not provide tristate drivers for internal connections, and only provide them for external connections with other chips. If we want to design a circuit that can be implemented in different fabrics with minimal change, it is best to avoid tristate buses.

In summary, tristate buses allow us to trade off significantly reduced wiring complexity against performance and design complexity, provided that our chosen implementation fabric allows tristate drivers and our CAD tool suite supports design and analysis of tristate buses. For designs that don't have stringent performance requirements, tristate buses can be a good choice. In the case of bus connections between chips on a printed circuit board, tristate buses are usually preferred. For that reason, fabrics such as FPGAs provide tristate drivers that can be used to drive output pins.

Modeling Tristate Drivers in VHDL

There are two aspects to modeling tristate drivers: representing the high-impedance state, and representing the enabling and disabling of drivers. In previous chapters, we have used the standard type std_logic to represent single-bit logic levels. The type contains the values '0' and '1' for representing 0 and 1 bit values, respectively. The type also contains the value 'Z' for representing the high-impedance state. In a VHDL model for a circuit, we can assign 'Z' to an output to represent disabling the output. Subsequently, assigning '0' or '1' to the output represents enabling it again.

EXAMPLE 8.7   Write a VHDL statement to model a tristate driver for an output signal d_out. The driver is controlled by a signal d_en, and when enabled, drives the value of an input d_in onto the output signal.

SOLUTION   We can use a conditional assignment statement, as follows:

```
d_out <= d_in when d_en = '1' else 'Z';
```

For multibit buses, we can use vector types whose elements are std_logic values. Examples include the types std_logic_vector from the std_logic_1164 package, as well as the vector types representing numeric values that we introduced in Chapter 3. While we can assign '0', '1' and 'Z' values individually to elements of vector signals, we usually assign either a string containing just '0' and '1' values to represent an enabled driver or a string of all 'Z' values to represent a disabled driver.

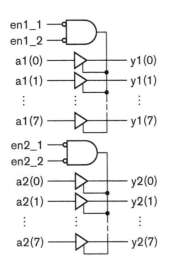

FIGURE 8.27   Internal circuit of the 16541 component.

EXAMPLE 8.8   The SN74x16541 component manufactured by Texas Instruments is a dual 8-bit bus buffer/driver in a package for use in a printed circuit board system. The internal circuit of the component is shown in Figure 8.27. Develop a VHDL model of the component.

SOLUTION   We can use ports of type std_logic_vector for each of the 8-bit inputs and outputs, and ports of type std_logic for the enable inputs. The entity declaration and architecture are

```
library ieee; use ieee.std_logic_1164.all;

entity sn74x16541 is
  port ( en1_1, en1_2,
         en2_1, en2_2 : in  std_logic;
         a1, a2        : in  std_logic_vector(7 downto 0);
         y1, y2        : out std_logic_vector(7 downto 0) );
end entity sn74x16541;

architecture rtl of sn74x16541 is
begin
  y1 <= a1 when en1_1 = '0' and en1_2 = '0' else (others => 'Z');
  y2 <= a2 when en2_1 = '0' and en2_2 = '0' else (others => 'Z');
end architecture rtl;
```

Each assignment within the architecture represents one of the 8-bit sections of the component. The condition after the when keyword determines whether the 8-bit tristate driver is enabled or disabled. The driver is disabled by assigning a vector value consisting of all 'Z' elements. We use the others notation to refer to all elements.

When we have multiple data sources for a tristate bus, our VHDL model includes multiple assignment statements that assign values to the bus signal. VHDL requires a signal with multiple drivers to be a *resolved signal*, meaning that it must *resolve* the values contributed by the separate assignments to determine the final value for the bus signal. Signals that we declare to be of any of the types std_logic, std_logic_vector, unsigned, signed, ufixed, sfixed, or float are automatically treated as resolved signals. If one assignment contributes '0' or '1' to a signal and all of the others contribute 'Z', the '0' or '1' value overrides the others and becomes the signal value. This corresponds to the normal case of one driver being enabled and the rest disabled. If one assignment contributes '0' and another contributes '1', we have a conflict. VHDL then uses the special value 'X', called *unknown*, as the final signal value, since it is unknown whether a real circuit would produce a low, high or invalid logic level on the signal. Depending on how the VHDL model of a data destination receiving an 'X' value is written, it might propagate the unknown value to its outputs, or produce arbitrary '0' or '1' values. Ideally, it would include an assertion statement that would detect unknown input values. If all assignments to a bus signal contribute 'Z', the final signal value is 'Z'. This corresponds to the bus signal floating. Again, since this does not represent a valid logic level, a VHDL model of a data destination receiving a 'Z' input should propagate an 'X' output and detect the error condition.

An important point to realize about the 'Z' and 'X' values is that they do not represent real logic levels in a physical circuit. Rather, assignment of 'Z' to an output is a notational device interpreted by synthesis CAD tools as implying a tristate driver for the output. Assignment of 'X' to an output is a notational device used in simulation to propagate error conditions in cases where we cannot determine a valid output value. We can write VHDL statements that test whether a signal has the value 'Z' or 'X', but it only makes sense to do so in testbench models, for example, in an assertion to verify that all drivers of a bus signal have been disabled or that there is no bus conflict. Since, according to our digital abstraction, signals in a physical circuit are only ever 0 or 1, a real digital component cannot sense any other level.

EXAMPLE 8.9 Suppose a VHDL architecture includes the following declarations and assignments

```
signal data_1, data_2, data_bus : unsigned(11 downto 0);
signal sel_1, sel_2             : std_logic;
...
data_bus <= data_1 when sel_1 = '1' else (others => 'Z');
data_bus <= data_2 when sel_2 = '1' else (others => 'Z');
```

Write an assertion to verify that the values of all elements of the bus signal are all valid logic levels, or that all drivers are disabled.

SOLUTION   We can make use of the is_X operation defined in the std_logic_ 1164 package. It tests whether any element of a std_logic_vector value is other than a valid logic level. Since the bus signal is of type unsigned, we need to convert the type of the signal value. The assertion is

```
assert not is_X(std_logic_vector(data_bus)) or
       (data_bus = (others => 'Z'));
```

Note that the is_X operation includes 'Z' elements as invalid logic levels, so we need to check for all elements being 'Z' separately.

8.3.3   OPEN-DRAIN BUSES

A third solution to avoid bus contention is to use *open-drain* drivers, as shown in Figure 8.28. Each driver connects the drain terminal of a transistor to the bus signal. When any of the transistors is turned on, it pulls the bus signal to a low logic level. When all of the transistors are turned off, the termination resistor pulls the bus signal up to a high logic level. If multiple drivers try to drive a low logic level, their transistors simply share the current load. If there is a conflict, with one or more drivers trying to drive a low level and others letting the bus be pulled up, the low-level drivers win. Sometimes, this kind of bus is called a *wired-AND* bus, since the bus signal is only 1 if all of the drivers output 1. If any driver outputs 0, the bus signal goes to 0. The AND function arises from the wiring together of the transistor drains. We can also use this form of bus with drivers that use bipolar transistors instead of MOSFET transistors. In that case, we connect the collector terminal of a transistor to the bus signal, as shown in Figure 8.29. Such a driver is called an *open-collector* driver.

Given the need for a pull-up resistor on each bus signal, open-drain or open-collector buses are usually found outside integrated circuits. For

FIGURE 8.28 Open-drain bus structure.

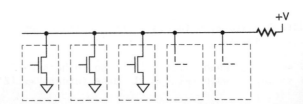

example, they may be used for a bus that connects a number of integrated circuits together, or for the signals in a backplane bus that connects a number of printed circuit boards together. Implementing pull-up resistors within an integrated circuit takes up significant area and consumes power. Hence, we usually use multiplexed or tristate buses within an integrated circuit chip. If we need the AND function that would be formed by open-drain connection, we can implement it with active gates.

Modeling Open-Drain and Open-Collector Connections in VHDL

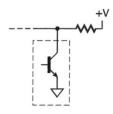

FIGURE 8.29   Open-collector bus driver.

We can model open-drain and open-collector drivers using the std_logic values '0' and 'Z'. We assign '0' to a signal to represent a driver whose output transistor is turned on, pulling the signal low. We assign 'Z' to the signal to represent a driver whose output transistor is turned off, since the transistor then acts as a switch that is opened. All that remains is to represent the pull-up resistor, which acts as a weak connection to a '1' level. The std_logic type includes two additional values: 'H' represents a weakly driven logic-high level, and 'L' represents a weakly driven logic-low level. We can represent a pull-up resistor with a constant assignment of 'H' to the bus signal, for example:

```
bus_sig <= 'H'; -- pull-up resistance
```

When a signal with open-drain or open-collector drivers is resolved, any '0' or 'X' values override the 'H' value. However, if all of the drivers are turned off, contributing 'Z' values, the 'H' value overrides them, giving a value of 'H' for the signal. If we simply use the value of the signal in logical operations, an 'H' value is treated the same as a '1' value. For example, given the following assignment:

```
gated_sig <= bus_sig and sig_en;
```

the result of the AND operation is either '0' or '1' (assuming the operands are not 'X'). If, however, we need to compare the bus signal with a '1' value, we should use the *strength stripping* operation To_X01, as in the following:

```
mux_out <= a1 when To_X01(bus_sig) = '1' else
           a0 when To_X01(bus_sig) = '0' else
           "XXXXXXX";
```

To_X01 converts '0' and 'L' to '0', '1' and 'H' to '1', and everything else to 'X'.

8.3.4 BUS PROTOCOLS

In most design projects, subsystems are often designed by different team members. Some subsystems may also be procured from external providers, or be implemented using off-the-shelf components. If the subsystems are to be interconnected using buses, it would be preferable for them to use the same bus signals with the same timing requirements; otherwise, interface glue logic is required. In order to facilitate connection of separately designed components, a number of common *bus protocols* have been specified. Some of the specifications are embodied in industry and international standards, whereas others are simply specifications agreed upon or promoted by component vendors. The specification of a bus protocol includes a list of the signals that interconnect compliant components, and a description of the sequences and timing of values on the signals to implement various bus operations.

Bus specifications and protocols vary, depending on their intended use. Some, intended for connecting separate chips on a circuit board or separate boards in a system, use tristate drivers for signals that have multiple data sources. Examples include the PCI bus used to connect add-on cards to personal computer systems, and the VXI bus used to connect measurement instruments to controlling computers. Others are intended for connecting subsystems within an IC. They have separate input and output signals, allowing for connection using multiplexers or switching circuits. Examples include the AMBA buses specified by ARM, the CoreConnect buses specified by IBM, and the Wishbone bus specified by the OpenCores Organization. Buses also vary in the number of parallel signals for transferring addresses and data, and in the speed of operation. Some, intended for high-speed data transfer, provide for the kinds of techniques we mention in Chapter 7, such as burst transfers and pipelining.

In this section, we will describe the relatively simple I/O bus protocol used by the Gumnut core. We have already introduced several aspects of the bus specification in preceding examples in this chapter. We will draw all of the aspects of the specification together here.

The Wishbone I/O bus signals for the Gumnut are described in the VHDL entity declaration in Example 8.3 and are shown as part of the Gumnut schematic symbol in Figure 8.21. To summarize, the signals are:

▸ port_cyc_o: a "cycle" control signal that indicates that a sequence of I/O port operations is in progress.

▸ port_stb_o: a "strobe" control signal that indicates an I/O port operation is in progress.

▸ port_we_o: a "write enable" control signal that indicates the operation is an I/O port write.

▶ port_ack_i: a status signal that indicates that the I/O port acknowledges completion of the operation.

▶ port_adr_o: the 8-bit I/O port address.

▶ port_dat_o: The 8-bit data written to the addressed I/O port by an out instruction.

▶ port_dat_i: the 8-bit data read from the addressed I/O port by an inp instruction.

When the Gumnut core executes an out instruction, it performs a port write operation. The timing of the operation is shown in Figure 8.30. Transitions are synchronized by the system clock. The Gumnut starts a write operation by driving the port_adr_o signals with the address computed by the out instruction and the port_dat_o signals with the data from the source register of the out instruction. It sets the port_cyc_o, port_stb_o and port_we_o control signals to 1 to indicate commencement of the write operation. The system in which the Gumnut is embedded decodes the port address to select an I/O controller and to enable the addressed output register to store the data. If the addressed controller is able to update the register within the first clock cycle, it sets the port_ack_i signal to 1 in that cycle, as shown in Figure 8.30(a). On the next rising clock edge, the Gumnut sees port_ack_i at 1 and completes the operation by driving port_cyc_o, port_stb_o and port_we_o back to 0. If, on the other hand, the addressed controller is slow and is not able to update the output register within the cycle, it leaves port_ack_i at 0, as shown in Figure 8.30(b). The Gumnut sees port_ack_i at 0 on the rising clock edge, and extends the operation for a further cycle. The controller can keep port_ack_i at 0 for as long as it needs to update the register. Eventually, when it is ready, it drives

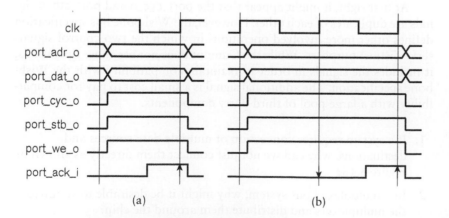

FIGURE 8.30 Timing for Gumnut I/O write operations: without wait cycles (a), and with one wait cycle (b).

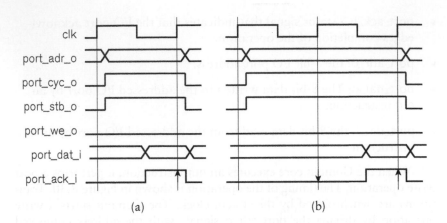

FIGURE 8.31 Timing for Gumnut I/O read operations: without wait cycles (a), and with one wait cycle (b).

port_ack_i to 1 to complete the operation. This form of synchronization, involving strobe and acknowledgment signals, is often called *handshaking*.

The Gumnut performs a port read operation when it executes an inp instruction. The timing for the operation, shown in Figure 8.31, is similar to that for a port write. The Gumnut starts the port read operation by driving the port_adr_o signals with the computed address, driving the port_cyc_o and port_stb_o signals to 1, and leaving port_we_o at 0. Again, the system decodes the address to select an I/O controller and enable the addressed input register onto the port_dat_i signals. The controller drives the port_ack_i signal to 1 as soon as it has supplied the data, either during the first cycle, as in Figure 8.31(a), or in a subsequent cycle, as in Figure 8.31(b). On seeing port_ack_i at 1, the Gumnut transfers the data from the port_dat_i signals to the destination register identified in the inp instruction. It then completes the port read operation by driving port_cyc_o and port_stb_o back to 0.

At first sight, it might appear that the port_cyc_o and port_stb_o signals are duplicates of each other. However, the Wishbone bus specification defines other more involved operations in which the two control signals serve distinct purposes. While the Gumnut does not use those operations, it includes the signals in order to maintain compatibility with the Wishbone specification. The additional signal is a small cost to pay for compatibility with a large pool of third-party components.

KNOWLEDGE TEST QUIZ

1. If a system requires connection of multiple data sources and destinations, why can we not just connect them directly as shown in Figure 8.18?

2. In a multiplexed bus system, why might it be desirable to subdivide the multiplexers and distribute them around the chip?

3. How does a tristate bus avoid logic-level contention on bus signals?

4. Why should we avoid floating bus signals?

5. What is a weak keeper?

6. What problems can arise if we disable one tristate bus driver at the same time as enabling the next driver? How can we avoid the problems?

7. Write a VHDL assignment that represents a tri-state bus driver for an 8-bit bus.

8. What value results on a VHDL std_logic signal when two tristate drivers are enabled and driving opposite logic levels?

9. Why is a signal connecting several open-drain drivers called a wired-AND connection?

10. Write a VHDL statement that represents a pull-up resistor for an open-drain bus.

11. What is the result of the strength stripping operation To_X01 applied to the values, '0', 'H', 'X', and 'Z'?

12. What is a bus protocol?

8.4 SERIAL TRANSMISSION

Throughout this book, we have described transfer of binary-encoded data using *parallel transmission*, in which we dedicate one signal wire per bit of encoded data. While this might appear to give us the fastest possible rate of data transfer, there are some disadvantages. The most obvious is that we require one signal wire per bit. For wide encodings, the wiring takes up significant circuit area, and makes layout and routing of the circuit more complex. For connections that extend between chips, parallel transmission requires more pad drivers and receivers, more pins, and more PCB traces. These all add cost to the system. Moreover, there are secondary effects, such as increased delay due to the extra space required for the connections, problems with crosstalk between wires routed in parallel, and problems with skew between signals. Dealing with these problems adds cost and complexity to the system. In this section, we will describe an alternative scheme for transferring binary-encoded data. The scheme is called *serial transmission*, since bits are transmitted one bit at a time in series over a single signal wire.

8.4.1 SERIAL TRANSMISSION TECHNIQUES

In order to transform data between parallel and serial form, we can use shift registers, introduced in Section 4.1.2. At the transmitting end, we load the parallel data into a shift register and use the output bit at one end of the register to drive the signal. We shift the content of the register one place at a time to drive successive bits of data onto the signal. At the receiving end, as each bit value arrives on the signal, we shift it into a shift register. When all the bits have arrived, the complete data code word is available in parallel form in the shift register. We sometimes use the term *serializer/deserializer*, or *serdes*, for shift registers used in this way. The advantage of serial transmission is that we only need one signal wire to transfer the data. Thus, we reduce the circuit area and cost for the connection. Moreover, if necessary, we can afford to optimize the signal path so that bits can be transferred at a very high rate. Some serial transmission standards in use today allow for rates exceeding 10 gigabits per second.

EXAMPLE 8.10   Show how a 64-bit data word can be transmitted serially between two parts of a system. Assume that the transmitter and the receiver are both within the same clock domain, and that the signal start is set to 1 on a clock cycle in which data is ready to be transmitted.

SOLUTION   At the transmitting end, we need a 64-bit shift register with parallel load control and an output from the least significant bit. At the receiving end, we also need a 64-bit shift register, but with a single-bit input and parallel data output. The connections are shown in Figure 8.32. The figure also shows the control section that sequences the serial transmission. When a start pulse occurs, the control section activates the receiver clock enable, rx_ce, for 64 cycles to shift the serial data in. The control section then pulses rx_rdy to indicate that the received data is ready. A timing diagram for one transmission is shown in Figure 8.33. We can implement the control logic with a counter and a simple finite-state machine.

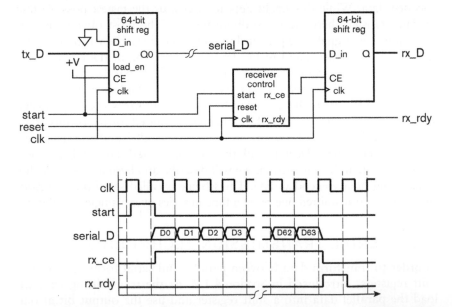

FIGURE 8.32   Serial transmission of 64-bit data within a clock domain.

FIGURE 8.33   Timing diagram for the serial receiver control.

One important issue that we need to address when transferring data serially is the order in which we transmit the bits. In principle, the order is arbitrary, so long as the transmitter and receiver agree. Otherwise, the receiver will end up with the bits in reverse order. In Example 8.10, we transmit the least significant bit first, and so shift bits into the receiver shift register at the most significant end, shifting them down to the least significant end. Fortunately, serial transmission in a system is often governed by a standard that specifies the order. This absolves us of the need to decide.

Another important issue is synchronization of the transmitter and the receiver. If we just drive the signal with the data bit values, there is no indication of when the time for one bit ends and the time for the next bit starts. This form of serial transmission is called *non-return to zero* (NRZ), and is illustrated in Figure 8.34, which shows the logic levels on a signal for NRZ serial transmission of the value 11001111, with the most significant bit being transmitted first. We assume in this case that the value on the signal when no bit is being transmitted is 0. In the figure, we have drawn a timescale showing the interval in which each bit occurs. However, that information is implicit, rather than being explicitly transmitted to the receiver along with the data. If the receiver, for some reason, assumed intervals twice as long for each bit, it would receive the value 10110000. To avoid this problem, we need to synchronize the transmitter and receiver, so that the receiver samples each bit value on the signal at some time during the interval when the transmitter drives the signal with the bit value.

There are three basic ways in which we can synchronize the transmitter and receiver. The first is by transmitting a clock on a separate signal wire. We saw this scheme in Example 8.10. The second is by signaling the start of a serial code word and relying on the receiver to keep track of the individual bit intervals. A common way of doing this originated with teletypes, which were computer terminals consisting of a keyboard and a printer connected to a remote computer using serial transmission. A refined version of such serial transmission is still used to connect some devices to serial communications ports on modern PCs.

In this second scheme, the signal is held at a high logic level when there is no data to transmit. When data is ready to be transmitted, transmission proceeds as shown in Figure 8.35, again with the most significant bit transmitted first. The signal is brought to a low logic level for one bit time to indicate the start of transmission. We call this the start bit. After that, the bits of data are transmitted, each for one bit time. We might also transmit a parity bit after the data bits, in case the signal wire is subject to induced noise, though this is not shown in the figure. This would allow us to detect some errors that might occur during transmission. Finally, we drive the signal high for one further bit time to indicate the end of transmission of the data. We call this the stop bit. We can then transmit

FIGURE 8.34   Serial transmission of the value 11001111.

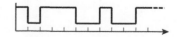

FIGURE 8.35   Serial transmission of the value 11100100 with start and stop bits.

the next piece of data, starting with a start bit, or leave the signal high if there is no data ready to transmit.

At the receiving end, the receiver monitors the logic level on the signal. While it remains at a high logic level, the receiver is idle. When the receiver detects a low logic level of the start bit, it prepares to receive the data. It waits until the middle of the first bit time and shifts the value on the signal into the receiving shift register. It then waits for further successive bit times, shifting each bit into the shift register. The complete data is available after the last bit is received. The receiver uses the stop-bit time to return to the idle state.

Note that the transmitter and the receiver must agree on the duration of the bit times on the signal. Usually, this is fixed in advance, either during manufacture or by programming. The transmitter and receiver typically have independent clocks, each several times faster than the serial bit rate. The sender uses its clock to transmit the data, and the receiver uses its clock to determine when to sense the data, synchronized by occurrence of the start bit. This is illustrated in Figure 8.36, in which the transmit clock and receive clock have slightly different frequencies and are not related in phase. Provided the difference is not too extreme, the drift from the nominal sampling time does not affect correct reception of the transmitted data.

Historically, computer component manufacturers provided a component called a *universal asynchronous receiver/transmitter*, or *UART*, for serial communications ports. The software on the computer could program the bit rate and other parameters. UARTs are still useful in some applications for connecting remote devices to digital systems via serial communications links. For example, an instrumentation system with remote sensors that transmit data at relatively low bit rates can use serial transmission managed by UARTs.

The third scheme for synchronizing a serial transmitter and receiver involves combining a clock with the data on the same signal wire. This avoids the need for tight clock synchronization, since there is an indication of when each bit arrives. As an example of such a scheme, we will describe *Manchester encoding*. As with NRZ transmission, Manchester encoding transmits each bit of data in a given interval. However, rather than representing each bit using one or other logic level, it represents a 0 with a transition from low to high in the middle of the bit interval, and a

FIGURE 8.36 Generation and sampling of serial data using transmitter and receiver clocks.

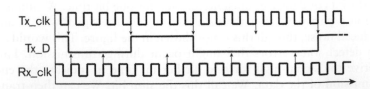

1 with a transition from high to low. (We could equally well choose the opposite assignment of transmissions, so long as transmitter and receiver agree.) At the beginning of the bit interval, a transition may be necessary to set the signal to the right logic level for the transition in the middle of the interval. Manchester encoding of the value 11100100 is shown in Figure 8.37, with the most significant bit transmitted first and with bit intervals defined by the transmitter's clock.

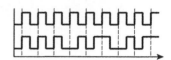

FIGURE 8.37 Manchester encoding of the value 11100100.

Since Manchester encoding of data is synchronized with the transmitter's clock and that clock is combined with the data, the receiver must be able to recover the transmitted clock and data from the signal. It does so using a circuit called a *phase-locked loop* (PLL), which is an oscillator whose phase can be adjusted to line up with a reference clock signal. A system using Manchester encoding usually transmits a continuous sequence of encoded 1 bits before transmitting one or more data words. The encoding of such a sequence gives a signal that matches the transmitter's clock. The receiver's PLL locks onto the signal to give a clock that can be used to determine the bit intervals for the transmitted data. This is shown in Figure 8.38.

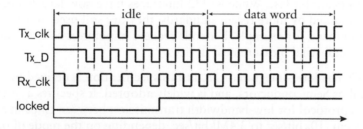

FIGURE 8.38 Synchronization of transmit and receive clocks by a PLL.

The main advantage of Manchester encoding over NRZ transmission is that it contains sufficient transitions to allow clock synchronization without the need for separate signal wires. The disadvantage is that the bandwidth of the transmission is double that of NRZ transmission. However, for many applications, that is not an overriding disadvantage. Manchester encoding has been used in numerous serial transmission standards, including the original Ethernet standard. Other serial encoding schemes that are similar in concept but more involved are now becoming widely used.

8.4.2 SERIAL INTERFACE STANDARDS

Given the advantages of serial transmission over parallel transmission for applications where distance and cost are significant considerations, numerous standards have been developed. These standards cover two broad areas of serial interfaces: connection of I/O devices to computers, and connection of computers together to form a network. Since most digital systems contain embedded computers, they can include standard interfaces for connecting components. The benefits of doing so include avoiding the need to design the connection from scratch, and being able to use off-the-shelf devices that adhere to standards. As a consequence, we can reduce the cost of developing and building systems, as well reducing the risk of designs not meeting requirements.

Some examples of serial interface standards for connecting I/O devices include:

▶ RS-232: This standard was originally defined in the 1960s for connecting teletype computer terminals with modems, devices for serial communication with remote computers via phone lines. Subsequently, the standard was adopted for direct connection of terminals to computers. Since most computers included RS232 connection ports, RS232 connections were incorporated in I/O devices other than terminals as a convenient way to connect to computers. Examples included user-interface devices such as mice, and various measurement devices. Serial transmission in RS232 interfaces uses NRZ encoding with start and stop bits for synchronization. Data is usually transmitted with the least significant bit first and most significant bit last. While RS232 interfaces have now largely been supplanted by more recent standards, they are still used in some equipment, for example, bar code readers in point-of-sale terminals, and industrial measurement devices.

▶ I^2C: The Inter-Integrated Circuit bus specification is defined by Philips Semiconductors, and is widely adopted. It specifies a serial bus protocol for low-bandwidth transmission between chips in a system (10kbit/sec to 3.4Mkbit/sec, depending on the mode of operation). It requires two signals, one for NRZ-coded serial data and the other for a clock. The signals are driven by open-drain drivers, allowing any of the chips connected to the bus to take charge by driving the clock and data signals. The specification defines particular sequences of logic levels to be driven on the signals to arbitrate to see which device takes charge and to perform various bus operations. The advantage of the I^2C bus is its simplicity and low implementation cost in applications that do not have high performance requirements. It is used in many off-the-shelf consumer and industrial control chips as the means for an embedded microcontroller to control operation of the chip. Philips Semiconductor has also developed a related bus specification, I^2S, or Inter-IC Sound, for serial transmission of digitally encoded audio signals between chips, for example, within a CD player.

▶ USB: The Universal Serial Bus is specified by the USB Implementers Forum, Inc., a nonprofit consortium of companies founded by the original developers of the bus specification. USB has become commonplace for connecting I/O devices to computers. It uses differential signaling (see Section 6.4.1) on a pair of wires, with a modified form of NRZ encoding. Different configurations support serial transfer at 1.5Mbit/sec, 12Mbit/sec or 480Mbit/sec. The USB specification defines a rich set of features for devices to communicate with host controllers. Since there is such a diversity of devices with USB interfaces, application-specific digital systems can benefit from inclusion of a USB host controller to enable connection of off-the-shelf devices. USB interface designs for inclusion in ASIC and FPGA designs are available in component libraries from vendors.

▶ FireWire: This is another high-speed bus defined by IEEE Standard 1394. Whereas USB was originally developed for lower bandwidth devices and subsequently revised to provide higher bandwidth, FireWire started out as a high-speed (400Mbit/sec) bus. There is also a revision of the standard defining transfer at rates up to 3.2Gbit/sec. FireWire connections use two differential signaling pairs, one for data and the other for synchronization. As with USB, there is a rich set of bus operations that can be performed to transmit information among devices on the bus. FireWire assumes that any device connected to the bus can take charge of operation, whereas USB requires a single host controller. Thus, there are some differences in the operations provided by FireWire and USB, and some differences in the applications for which they are suitable. FireWire has been most successful in applications requiring high-speed transfer of bulk data, for example, digital video streams from cameras.

EXAMPLE 8.11   Design an interface to connect an embedded Gumnut core to a remote temperature sensor. The temperature sensor is an Analog Devices AD7414 with an I^2C connection and an alert output that can be connected to a warning indicator.

SOLUTION   The OpenCores repository (see Section 8.7, Further Reading) contains an I^2C controller component that is Wishbone compliant. We can use it rather than designing a new I^2C controller from scratch. We connect the controller to the Gumnut core's Wishbone I/O bus, and provide pad connections to an external I^2C bus for connecting the temperature sensor. We connect the alert output of the sensor to an LED indicator. The sensor allows the embedded software to program threshold temperatures, beyond which the alert indicator is activated. The system design is shown in Figure 8.39. The use of the serial I^2C bus allows connection to the temperature sensor with only two wires, resulting in a substantial reduction in system cost compared to connection using a parallel bus.

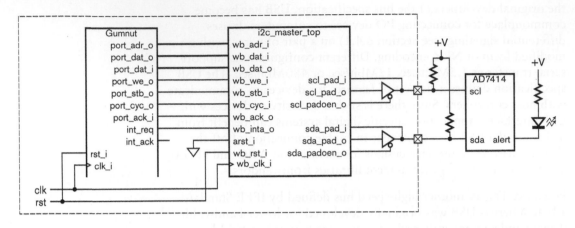

FIGURE 8.39 A temperature sensing system using an I²C serial bus.

KNOWLEDGE TEST QUIZ

1. What advantages does serial transmission of data have over parallel transmission?

2. How do we convert between parallel and serial form for serial data transmission or reception?

3. What determines the order in which we transmit bits of data?

4. What is meant by non-return to zero (NRZ) transmission?

5. What is the purpose of a start bit and a stop bit in serial transmission?

6. How does Manchester encoding represent 0 and 1 bits?

7. Why would we adopt a standard serial interface specification rather than developing a custom interface?

8. Which of I²C or FireWire would be most appropriate for connecting a motor controller and a digital video camera, respectively, to an embedded system?

8.5 I/O SOFTWARE

Now that we have described the hardware aspects of input and output, we turn our attention to the corresponding embedded software. We have seen that an out instruction in the Gumnut core invokes a port write operation to update an output register in an I/O controller, and an inp instruction invokes a port read operation to get the value from an input register. The embedded software running on the core needs to use out and inp instructions as part of the task of managing input and output devices to implement the functionality required of the system.

Since I/O devices interact with the real physical world, the embedded software needs to be able to respond to events when they occur, or to cause events at the right time. Dealing with *real time* behavior is one of the main differences between embedded software and programs for general purpose computers. Embedded software needs to be able to detect when events occur so that it can react. It also needs to be able to keep track of time so that it can perform actions at specific times or at regular intervals. In this section, we will introduce the basic mechanisms for synchronizing embedded software with I/O events.

8.5.1 POLLING

The simplest I/O synchronization mechanism is called *polling*. It involves the software repeatedly checking a status input from a controller to see if an event has occurred. If it has, the software performs the necessary task to react to the event. If there are multiple controllers, or multiple events to which the software must respond, the software checks each of the status inputs in turn, reacting to events as they occur, as part of a busy loop.

EXAMPLE 8.12   A factory automation system includes a safety monitoring subsystem based on an embedded Gumnut core. The core has alarm inputs from a number of machines that indicate various abnormal operating conditions. These are connected through a controller that has two input registers at addresses 16 and 17. Each bit of each register represents one alarm input, with the bit being 0 for normal operation and 1 for an alarm condition. The core also has a temperature sensor connected to an ADC. The converted value is available in an input register at address 20, represented as an 8-bit unsigned integer in °C. A temperature above 50°C is abnormal. The core has an output register at address 40. Writing a 1 to the least significant bit of the output register activates an alarm bell, and writing 0 deactivates it. Develop a polling loop for the embedded software to monitor the inputs and activate the alarm bell when any abnormal condition arises.

SOLUTION   The polling loop must repeatedly read the input registers. If any alarm input bit is 1, or if the temperature value is greater than 50°C, the alarm bell output bit must be set to 1; otherwise, it must be cleared to 0. The code is

```
alarm_in_1: equ 16   ; address of alarm_in_1 input register
alarm_in_2: equ 17   ; address of alarm_in_2 input register
temp_in:    equ 20   ; address of temp_in input register
alarm_out:  equ 40   ; address of alarm_out output register
```

(continued)

```
max_temp:    equ 50   ; maximum permissible temperature

poll_loop:   inp  r1, alarm_in_1
             sub  r0, r1, 0
             bnz  set_alarm ; one or more alarm_in_1 bits set
             inp  r1, alarm_in_2
             sub  r0, r1, 0
             bnz  set_alarm ; one or more alarm_in_2 bits set
             inp  r1, temp_in
             sub  r0, r1, max_temp
             bnc  set_alarm ; temp_in > max_temp
             out  r0, alarm_out ; clear alarm_out
             jmp  poll_loop
set_alarm:   add  r1, r0, 1
             out  r1, alarm_out ; set alarm_out bit 1 to 1
             jmp  poll_loop
```

Polling has the advantage that it is very simple to implement, and requires no additional circuitry beyond the input and output registers of the I/O controllers. However, it requires that the processor core be continually active, consuming power even when there is no event to react to. It also prevents the processor from reacting immediately to one event if it is busy dealing with another event. For these reasons, polling is usually only used in very simple control applications where there is no need for fast reaction times.

8.5.2 INTERRUPTS

Probably the most common way to synchronize embedded software with I/O events is through use of *interrupts*. The processor executes some background tasks, and when an event occurs, the I/O controller that detects the event interrupts the processor. The processor then stops what it was doing, saving the program counter so that it can resume later, and starts executing an *interrupt handler*, or *interrupt service routine*, to respond to the event. When it has completed the handler, it restores the saved program counter and resumes the interrupted program. In some systems, if there is no background task to run, the processor may enter a low-power standby state from which it emerges in response to an interrupt. This has the benefit of avoiding power consumption due to busy-waiting, though it may add delay to the interrupt response time if the processor requires some time to resume full-power operation.

Different processors provide different mechanisms for I/O controllers to request an interrupt. Some provide very simple mechanisms, such as that of the Gumnut core that we will describe shortly. Others provide more elaborate mechanisms, for example, allowing different controllers to be assigned different priorities, so that a higher-priority event can interrupt service of a lower-priority event, but not *vice versa*. Some provide a way for the controller to select the interrupt handler to be executed by the processor. However, there are some aspects that are common to most systems.

First, the processor must have an input signal to which controllers can connect to request interrupts. For older microprocessors and microcontrollers, the interrupt request signal is often an active-low signal pulled up with an external resistor. Each controller connects to the signal with an open-drain or open-collector driver, pulling the signal low to request an interrupt. Thus, the signal value is a wired-OR function of the individual controllers' requests. For processor cores that are designed to connect to on-chip I/O controllers, the interrupt request input is typically driven by active gates forming the OR of the controllers' requests.

Second, the processor must be able to prevent interruption while it is executing certain sequences of instructions, often called *critical regions*. Examples are instructions that update information shared between an interrupt handler and other parts of the embedded software. If the processor is part way through updating such information and is interrupted, the interrupt handler will see the partially updated information, which may not correctly represent a valid value. So processors generally have instructions or other means of *disabling interrupts* and *enabling interrupts*.

Third, the processor must be able to save sufficient information about the program it was executing when interrupted so that it can resume the program on completion of the interrupt handler. At the least, this includes saving the program counter value. Since the processor responds to an interrupt after completing one instruction and before starting the next, the program counter contains the address of the next instruction in the program. That is the instruction to be resumed after the interrupt handler. The processor must provide a register or some other storage in which to save the program counter. If there is other state information in the processor that might be modified by the interrupt handler, such as condition code bits, they must also be saved and restored.

Fourth, when the processor responds to an interrupt, it must disable further interrupts. Since response to an interrupt involves saving the interrupted program's state in registers, if the interrupt handler is itself interrupted, the saved state would be overwritten. Thus, the handler needs to prevent interruption, at least during the initial stages of responding to an interrupt.

Some processors allow the storage containing the saved state information to be read by a program. That allows a handler to copy the saved state into memory. The handler can then re-enable interrupts, allowing the interrupt handler itself to be interrupted to deal with another event. We call this *nested interrupt* handing. The handler must disable interrupts again when it has completed its operation so that it can restore the saved state before resuming the interrupted program.

Fifth, the processor must be able to locate the first instruction of the interrupt handler. The simplest way of doing this is for the handler to start at a fixed or predetermined address in the instruction memory. Alternative schemes involve the interrupting controller providing a *vector*: either a value used to form the address of the handler, or an index into a table of addresses in memory.

Finally, the processor needs an instruction for the interrupt handler to return to the interrupted program. Such a return from interrupt instruction restores the saved program counter and any other saved state.

The Gumnut processor core has all of these features, with the exception of nested interrupt handing. It has an input signal, int_req, that controllers can drive to 1 to request an interrupt. It includes two instructions in its instruction set: disi, for disabling interrupts; and enai, for enabling interrupts. When the core responds to an interrupt, it saves the program counter and the values of the Z and C condition codes in special internal registers, and disables further interrupts. The first instruction of the interrupt handler is located at address 1 in the instruction memory, so the processor simply loads that address into the program counter to start executing the handler. Finally, the Gumnut instruction set includes the reti instruction to return from an interrupt handler. It restores the saved values to the program counter and the Z and C condition code bits, and re-enables interrupts. Program execution then resumes from where it left off.

There are also requirements on I/O controllers that make interrupt requests. When an event occurs, the controller must activate the processor's interrupt request signal. However, the processor may not respond immediately. The requesting controller must keep the request signal active, otherwise the request may go unnoticed. Failure to respond to an event may be a critical error in some systems. Processors typically have a mechanism to *acknowledge* an interrupt request, that is, to indicate that the event has been noticed and that the interrupt handler as been activated. If there are multiple I/O controllers that can request interrupts, the processor needs to acknowledge each request individually, so that none are overlooked. Once a request has been acknowledged, the controller must deactivate the interrupt request signal. Otherwise, multiple responses might occur for the one event. In some cases, that can be as bad as missing an event.

The Gumnut core provides a simple interrupt acknowledgment mechanism. It has an output signal, int_ack, that it drives to 1 for one cycle when it responds to an interrupt request. If there is only one controller that can request interrupts in a Gumnut system, the controller can use the int_ack signal to clear its interrupt request state.

EXAMPLE 8.13   Design an input controller that has 8-bit binary-coded input from a sensor. The value can be read from an 8-bit input register. The controller should interrupt the embedded Gumnut core when the input value changes. The controller is the only interrupt source in the system.

SOLUTION   The controller contains a register for the input value. Since we need to detect changes in the value, we also need a register for the previous value, that is, the value on the previous clock cycle. When the current and previous values change, we set an interrupt-request state bit. Since there is only one interrupt source, we can use the int_ack signal from the processor core to clear the state bit. The controller circuit is shown in Figure 8.40.

FIGURE 8.40   Circuit for an input controller with interrupt request logic.

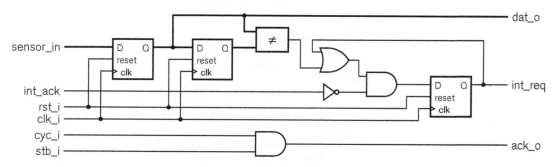

EXAMPLE 8.14   Develop a VHDL model of the input controller of Example 8.13.

SOLUTION   The entity declaration includes ports for the I/O bus, plus the interrupt request and acknowledge connections:

```
library ieee;
use ieee.std_logic_1164.all;

entity sensor_controller is
  port ( clk_i, rst_i : in  std_logic;
         cyc_i, stb_i : in  std_logic;
         ack_o        : out std_logic;
         dat_o        : out std_logic_vector(7 downto 0);
         int_req      : out std_logic;
         int_ack      : in  std_logic;
         sensor_in    : in  std_logic_vector(7 downto 0) );
end entity sensor_controller;
```

The architecture body is

```vhdl
architecture rtl of sensor_controller is

  signal prev_data,
         current_data    : std_logic_vector(7 downto 0);
  signal current_int_req : std_logic;

begin

  data_regs : process (clk_i) is
  begin
    if rising_edge(clk_i) then
      if rst_i = '1' then
        prev_data <= "00000000";
        current_data <= "00000000";
      else
        prev_data <= current_data;
        current_data <= sensor_in;
      end if;
    end if;
  end process data_regs;

  int_state : process (clk_i) is
  begin
    if rising_edge(clk_i) then
      if rst_i = '1' then
        current_int_req <= '0';
      else
        case current_int_req is
          when '0' =>
            if current_data /= prev_data then
              current_int_req <= '1';
            end if;
          when others =>
            if int_ack = '1' then
              current_int_req <= '0';
            end if;
        end case;
      end if;
    end if;
  end process int_state;

  dat_o <= current_data;
  int_req <= current_int_req;

  ack_o <= cyc_i and stb_i;

end architecture rtl;
```

The data_regs process represents the two data registers, one for the current sensor data value and one for the previous value. The int_state process represents the interrupt request and acknowledge logic. It is essentially a small finite state machine, with current_int_req encoding the state. In the state where current_int_req is '0', there is no interrupt request. However, if the current value changes from the previous value, current_int_req is set to '1'. The value of this output is used as the interrupt request signal to the processor. It stays '1', even when the current value and the previous value no longer differ. Eventually, when the processor responds to the interrupt and sets int_ack to '1', current_int_req is cleared back to '0'.

EXAMPLE 8.15   Show the Gumnut assembler code for the interrupt handler for the sensor controller interrupt. Assume the data register is read at port address 0.

SOLUTION   The interrupt handler is

```
              data
saved_r1:     bss     1

              text
sensor_data:  equ     0          ; address of sensor data
                                 ; input register

              org     1
              stm     r1, saved_r1
              inp     r1, sensor_data
              ...                ; process the data
              ldm     r1, saved_r1
              reti
```

Since the handler needs to use processor register r1, it must save whatever value is in there from the interrupted program. The data memory location saved_r1 is reserved for that purpose. The interrupt handler must be located at address 1 in the instruction memory. We use an org directive to ensure this. The instructions in the handler first save the contents of r1, then read the new value from the controller's input register. The handler then executes instructions that deal with the data. Finally, the handler restores the saved value to r1 and uses a reti instruction to resume the interrupted program.

If, in a Gumnut-based system, there are several controllers that can request an interrupt, the interrupt handler must be able to determine which controller requested an interrupt so that it can execute the appropriate response. In such a system, each controller must provide status information in a status register that indicates whether it has requested an interrupt. Furthermore, the int_ack signal is not sufficient to distinguish which request is acknowledged. Instead, the processor must perform some other action to acknowledge the interrupt. We could acknowledge and clear a controller's interrupt request as a side-effect of its status register being read. Alternatively, we could require a write operation to a control register to acknowledge the request.

8.5.3 TIMERS

As we mentioned earlier, many real-time embedded systems must perform actions at specific times or at periodic intervals. For these systems, we need to include some form of timer. We showed in Chapter 4 that we can use a counter to derive a periodic signal from the system clock. We can use such a signal as a time base: each cycle represents one unit of time in the embedded system. We also showed how we can use a loadable down counter as an interval timer. A common use for interval timers in real-time embedded systems is to generate an interrupt for the processor at some programmable multiple of a time base. The interval timer acts as an I/O controller, often called a *real-time clock*, with an output register for programming the time interval. The interrupt handler for the timer can then perform any required periodic actions.

EXAMPLE 8.16   Develop a VHDL model for a real-time clock controller for the Gumnut processor. The controller has a 10µs time base derived from a 50MHz system clock, and an 8-bit output register for the value to load into the counter. A write operation to the output register causes the counter to be loaded. After the counter reaches 0, it reloads the value from the output register and requests an interrupt. The controller has an input register for reading the current count value. The counter also has a 1-bit control output register. When bit 0 of the register is 0, interrupts from the controller are masked, and when it is 1, they are enabled. The counter has a status register, in which bit 0 is 1 when the counter has reached 0 and been reloaded, or 0 otherwise. Other bits of the register are read as 0. Reading the register has the side effect of acknowledging a requested interrupt and clearing bit 0. The counter output and input registers are located at the base port address, and the control and status registers are at offset 1 from the base port address.

SOLUTION   The entity declaration for the controller has ports for the I/O bus, and uses the stb_i port for the decoded base port address:

```vhdl
library ieee;
use ieee.std_logic_1164.all, ieee.numeric_std.all;

entity real_time_clock is
  port ( clk_i, rst_i      : in std_logic;   -- 50MHz clock
         cyc_i, stb_i, we_i : in std_logic;
         ack_o             : out std_logic;
         adr_i             : in std_logic;
         dat_i             : in unsigned(7 downto 0);
         dat_o             : out unsigned(7 downto 0);
         int_req           : out std_logic );
end entity real_time_clock;
```

The architecture body for the real-time clock controller is

```vhdl
architecture rtl of real_time_clock is

  constant clk_freq        : natural := 50000000;
  constant timebase_freq   : natural := 100000;
  constant timebase_divisor : natural :=
                            clk_freq / timebase_freq;

  signal count_value       : unsigned(7 downto 0);
  signal trigger_interrupt : std_logic;
  signal int_enabled,
         int_triggered     : std_logic;

begin

  counter : process (clk_i) is
    variable timebase_count : natural
                            range 0 to timebase_divisor - 1;
    variable count_start_value : unsigned(7 downto 0);
  begin
    if rising_edge(clk_i) then
      if rst_i = '1' then
        timebase_count := 0;
        count_start_value := "00000000";
        count_value <= "00000000";
        trigger_interrupt <= '0';
      elsif cyc_i = '1' and stb_i = '1' and adr_i = '0'
            and we_i = '1' then
        timebase_count := 0;
        count_start_value := dat_i;
        count_value <= dat_i;
        trigger_interrupt <= '0';
      elsif timebase_count = timebase_divisor - 1 then
        timebase_count := 0;
        if count_value = "00000000" then
        count_value <= count_start_value;
        trigger_interrupt <= '1';
      else
        count_value <= count_value - 1;
        trigger_interrupt <= '0';
      end if;
    else
      timebase_count := timebase_count + 1;
      trigger_interrupt <= '0';
    end if;
  end if;
end process counter;

control_reg : process (clk_i) is
```

(continued)

```
begin
  if rising_edge(clk_i) then
    if rst_i = '1' then
      int_enabled <= '0';
    elsif cyc_i = '1' and stb_i = '1' and adr_i = '1'
          and we_i = '1' then
      int_enabled <= dat_i(0);
    end if;
  end if;
end process control_reg;

int_reg : process (clk_i) is
begin
  if rising_edge(clk_i) then
    if rst_i = '1'
      or (cyc_i = '1' and stb_i = '1' and adr_i = '1'
          and we_i = '0') then
      int_triggered <= '0';
    elsif trigger_interrupt = '1' then
      int_triggered <= '1';
    end if;
  end if;
end process int_reg;

dat_o <= count_value when adr_i = '0' else
         "0000000" & int_triggered;

int_req <= int_triggered and int_enabled;

ack_o <= cyc_i and stb_i;

end architecture rtl;
```

The counter process represents the time-base divider, interval counter and counter output register. The variable timebase_count is used to divide the 50MHz clock to derive the 100kHz time base, and the variable count_start_value stores the value for the counter output register. The count value is represented by the signal count_value. The signal trigger_interrupt is an internal control signal used to manage interrupt requests. On reset, the variables and signals are cleared to zeros. When a port write operation is performed with the least significant address bit being '0', the written data is used to update count_start_value, and the counters are cleared to zeros again. On other clock cycles, the counters are incremented. When the time base counter reaches its terminal count, it wraps to zero, and count_value is decremented. When count_value reaches zero, it is reloaded from count_start_value, and the trigger_interrupt signal is set to '1'.

The control_reg process represents the control register, containing the interrupt-enable bit. On reset, the bit is cleared to '0'. Otherwise, when a write operation is performed with the least significant address bit being '1', the bit is updated with the written port data.

The int_reg process represents the one-bit state register that determines when an interrupt event has occurred. The signal int_triggered is set to '1' when the trigger_interrupt signal is '1', that is, when count_value is reloaded after having reached zero. The signal is cleared to '0' on reset, and also on a port read operation that reads the status register.

The remaining assignments implement the rest of the required functionality. The assignment to dat_o selects the value provided for a port read operation: either the count value or the interrupt status bit. The assignment to int_req causes an interrupt request when the triggering event has occurred and interrupt requests are enabled. The assignment to ack_o implements the controller's response to bus operations, indicating that the controller is ready without delay.

EXAMPLE 8.17   Suppose a Gumnut system includes the real-time clock controller of Example 8.16 with the registers located at base port address 16. Develop Gumnut code that calls the subroutine task_2ms every 2ms. In between activations, the program stands by in low-power mode. The subroutine should not be called as part of the interrupt handler, since other interrupts should be permitted during execution of the subroutine.

SOLUTION   The code is

```
;;; ------------------------------------------------
;;; Program reset: jump to main program

            text
            org     0
            jmp     main


;;; ------------------------------------------------
;;; Port addresses
rtc_start_count:        equ     16 ; data output register
rtc_count_value:        equ     16 ; data input register
rtc_int_enable:         equ     17 ; control output register
rtc_int_status:         equ     17 ; status input register


;;; ------------------------------------------------
;;; Interrupt handler

            data
int_r1:     bss     1       ; save location for
                            ; handler registers

            text
            org     1
```

(continued)

```
int_handler:          stm    r1, int_r1 ; save registers
check_rtc:            inp    r1, rtc_status ; check for
                                            ; RTC interrupt
                      sub    r0, r1, 0
                      bz     check_next
                      add    r1, r0, 1
                      stm    r1, rtc_int_flag ; tell main
                                             ; program
check_next:          ...

int_end:             ldm r1, int_r1 ; restore registers
                     reti


;;; -------------------------------------------------------
;;; init_interrupts:    Initialize 2ms periodic interrupt, etc.

                     data
rtc_divisor:         equ    199   ; divide 100kHz down
                                  ; to 500Hz
rtc_int_flag:        bss    1

                     text
init_interrupts:     add    r1, r0, rtc_divisor
                     out    r1, rtc_start_count
                     add    r1, r0, 1
                     out    r1, rtc_int_enable
                     stm    r0, rtc_int_flag
                     ...            ; other initializations
                     ret


;;; -------------------------------------------------------
;;; main program

                     text
main:                jsb    init_interrupts
                     enai
main_loop:           stby
                     ldm    r1, rtc_int_flag
                     sub    r0, r1, 1
                     bnz    main_next
                     jsb    task_2ms
                     stm    r0, rtc_int_flag
main_next:           ...
                     jmp    main_loop
```

The code is structured into separate sections and subroutines, each dealing with one part of the program. The first section deals with starting the main program when the system is reset. The instructions are located at address 0, and simply jump to the main program. The second section defines symbolic labels for the real-time clock controller registers. Reference to these labels makes the code easier to understand.

The subroutine init_interrupts initializes the real-time clock controller. It loads the value 199 into the controller's output register. This makes the controller count down from 199 to 0 and then restart from 199; thus, it divides the time base by 200 to give a 2ms period. The subroutine also sets the controller's interrupt-enable bit by writing 1 to the control register, and clears the rtc_int_flag location in memory. This location is used by the interrupt handler to indicate to the main program that a 2ms interrupt has occurred. The subroutine then proceeds with other initializations before returning to the caller.

The interrupt handler is located at instruction address 1. On responding to an interrupt, it checks the controllers in the system to determine the interrupt source, starting with the real-time clock controller. If the controller's status register is nonzero, the handler sets rtc_int_flag to 1, indicating to the main program that it should perform the 2ms task. The handler then proceeds to check for other interrupt sources before returning to the interrupted program.

The main program starts by calling the subroutine to initialize controllers and interrupts, then enables receipt of interrupts. It then stands by in low-power mode until an interrupt occurs. On return from the interrupt handler, the main program checks the rtc_int_flag location. If it is 1, a real-time clock interrupt has occurred, so the main program calls the task_2ms subroutine, as required, and then clears rtc_int_flag. The main program then performs any processing required for other interrupts that might have occurred. When that it done, it loops back and stands by for the next interrupt.

The code in Example 8.17 is a basic form of *real-time executive*, that is, a control program that schedules execution of tasks in response to interrupts and timer events. Vendors of microprocessors, microcontrollers and embedded processor cores generally provide more sophisticated *real-time operating systems* (RTOSs) for their products. There are also a number of third-party vendors who provide RTOSs that run on various processors. An RTOS generally includes an executive, together with software components to manage other resources, such as storage, input/output, communication and specialized processing resources. The advantage of using a real-time executive or an RTOS is that we can focus our software development effort on the aspects of our system that are different from other systems, and reuse proven code that deals with common embedded software mechanisms. We won't go into any further detail of real-time programming in this book. Instead, we refer to sources on the topic listed in the Further Reading section.

KNOWLEDGE
TEST QUIZ

1. In dealing with real-time behavior, what does embedded software need to do?

2. How does polling synchronize embedded software with I/O events?

3. Identify an advantage and a disadvantage of polling compared to other I/O synchronization schemes.

4. What action does a processor perform upon receiving an interrupt?

5. How does a processor prevent interruption while it is executing a critical region?

6. How does the processor determine where to resume program execution on completion of handling an interrupt?

7. What is an interrupt vector?

8. Why must a controller deactivate the interrupt request signal when its interrupt is acknowledged?

9. What purpose does a real-time clock serve in an embedded system?

10. What operations are performed by a real-time executive?

8.6 CHAPTER SUMMARY

▸ Transducers allow a digital system to interact with the physical world. Sensors generate an electrical representation of a physical property. Output transducers, including actuators, cause a physical effect.

▸ Input devices include switches, keypads, knobs, position encoders, and analog sensors.

▸ An analog-to-digital converter (ADC) produces a binary coded representation of an analog signal. ADCs include flash and successive-approximation ADCs.

▸ Output devices include indicator lights, 7-segment LED and LCD displays, electromechanical actuators and valves, motors, and analog output devices.

▸ A digital-to-analog converter (DAC) produces an analog signal proportional to a binary coded input. DACs include R-string and R/2R ladder DACs.

▸ An I/O controller includes input and output registers that provide an embedded processor with access to I/O data. It may also include control and status registers for managing operation of the controller.

▸ An autonomous controller may perform I/O operations while a processor performs other tasks concurrently.

▸ Buses connect multiple data sources and destinations. Parallel buses use one signal wire per bit of encoded data.

▸ Multiplexed buses use multiplexers to select data from one source at a time. Multiplexers can be centralized or distributed, depending on the wiring complexity of the system.

▸ Tristate buses allow direct connection of sources to destinations, using a high-impedance driver state to avoid contention. Tristate buses are not generally used within chips. The high-impedance state is modeled in VHDL using the 'Z' std_logic value.

▸ Open-drain and open-collector drivers allow wired-AND connections. Resistive pull-ups are modeled in VHDL using the 'H' std_logic value.

▸ Bus protocols specify the signals used and the sequences and timing of values to implement bus operations.

▸ Serial buses transmit bits in sequence over one wire. Shift registers are used to convert between parallel and serial transmission.

▶ Serial transmission requires synchronization between transmitter and receiver to determine the interval during which each bit is transmitted.

▶ Real-time software on an embedded processor must be able to react to I/O events and to keep track of time so that it can perform scheduled or periodic operations.

▶ Software can poll I/O controllers to determine when events occur.

▶ Interrupts are a mechanism for a controller to notify a processor of an event. The processor executes an interrupt handler to respond to the event, then resumes its interrupted task. The processor includes instructions for managing interrupts.

▶ Timers, or real-time clocks, issue periodic interrupts, allowing an embedded system to perform scheduled and periodic tasks.

8.7 FURTHER READING

Industrial Electronics: Applications for Programmable Controllers, Instrumentation and Process Control, and Electrical Machines and Motor Controls, 3rd Edition, Thomas E. Kissell, Prentice Hall, 2003. This is a comprehensive reference describing the kinds of input and output devices encountered in industrial settings, and the transducers and electronic circuits used to interface them to digital control systems.

Standard LCD Graphic Modules, Allshore Industries, www.allshore .com/lcd_displays/lcd_graphic_modules.asp. Provides data sheets on the ASI-D-1006A-DB-_S/W LCD module and the Seiko Epson SED1560 controller IC described in Section 8.2.2.

Understanding Digital Signal Processing, Richard G. Lyons, Prentice Hall, 2001. An introduction to the theory of digital signal processing (DSP).

WISHBONE System-on-Chip (SoC) Interconnection Architecture for Portable IP Cores, Revision B.3, OpenCores Organization, 2002, www.opencores.org/projects.cgi/web/wishbone/wbspec_b3.pdf. This is the specification document for the Wishbone bus used in this book.

OpenCores, www.opencores.org. From the website's FAQ, "OpenCores is a loose collection of people who are interested in developing hardware, with a similar ethos to the free software movement." The website hosts a repository of freely reusable core designs, many of which are compatible with the Wishbone bus.

Real-Time Concepts for Embedded Systems, Qing Li, Caroline Yao,
   CMP, 2003. A practical introduction to real-time programming for
   embedded systems.

EXERCISE 8.1   A calculator has keys arranged as shown in Figure 8.41.
Show how the key switches can be arranged in a scanned matrix.

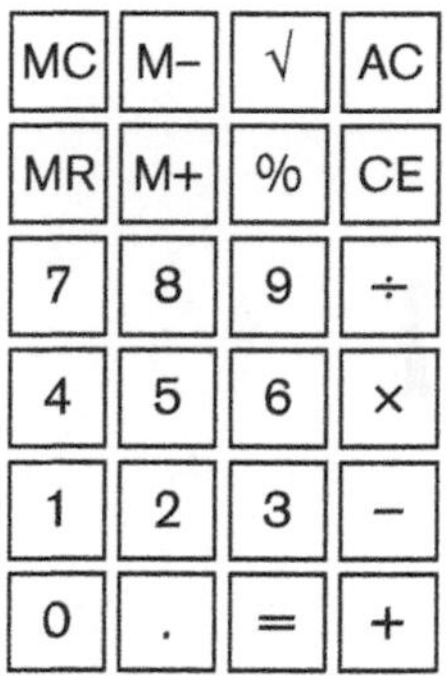

EXERCISE 8.2   Design a keypad controller to connect a Gumnut core to
the keypad described in Exercise 8.1. The controller should include an output
register for driving row lines and an input register for sensing column lines.

EXERCISE 8.3   Develop a Gumnut program that uses the keypad controller
described in Exercise 8.2 to scan the calculator keypad. When a key is pressed,
the program should call a subroutine labeled do_key to respond to the key press.
(Just include the subroutine call, not the instructions in the subroutine.) Assume
the output register is at port address 0 and the input register is at port address 1,
and omit switch debouncing.

EXERCISE 8.4   Show how the input controller described in Example 8.13
on page 377 can be used for a volume control knob with an incremental encoder.

EXERCISE 8.5   Develop a Gumnut interrupt handler that responds to
interrupts from the incremental encoder input of Exercise 8.4. The handler
should increment or decrement a value stored in memory as the knob is turned
clockwise or counterclockwise, respectively. The value should be limited to the
range 0 to 100.

EXERCISE 8.6   Develop a VHDL model of an 8-bit successive approximation
register (SAR) for use in an ADC (see Figure 8.6 on page 332).

EXERCISE 8.7   Develop a Gumnut subroutine to perform an analog-to-
digital conversion using successive approximation, returning an 8-bit result in
register r1. The Gumnut is connected to an output data register, an input status
register, an 8-bit DAC and a comparator as shown in Figure 8.42. The output
data register is written at port address 8. The input status register is read at port
address 8, and provides the value of the comparator output in the least significant bit, with other bits hardwired to 0.

EXERCISE 8.8   Some digital audio applications use an LED bar display,
consisting of a row of LED indicators to display the volume level of the audio
signal. Assuming that the loudness is proportional to the logarithm of the signal
amplitude, we can work out which LEDs to light by finding the left-most 1 bit in
the unsigned binary number representing the amplitude. Design a circuit to drive
an 8-LED common-anode bar display, given an 8-bit unsigned binary amplitude
value.

EXERCISES

MC	M−	√	AC
MR	M+	%	CE
7	8	9	÷
4	5	6	×
1	2	3	−
0	.	=	+

FIGURE 8.41

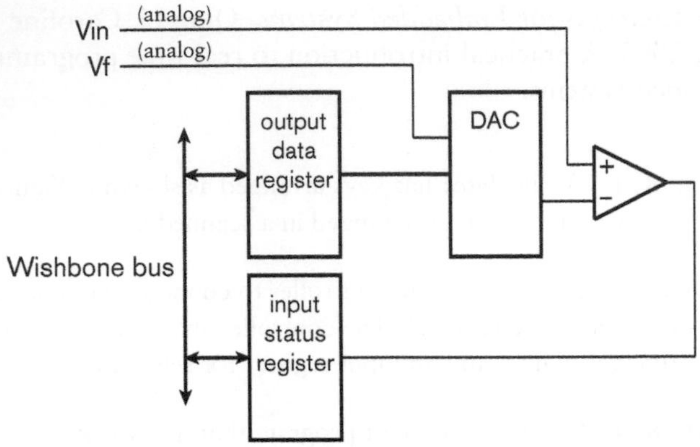

Vin —— (analog)
Vf —— (analog)

Wishbone bus

output data register

input status register

DAC

FIGURE 8.42

EXERCISE 8.9 Write a Gumnut subroutine that performs the function of the circuit described in Exercise 8.8. The subroutine takes an 8-bit unsigned binary amplitude value in r2 and outputs a corresponding value to an 8-bit register at port address 28, connected to the cathodes of an 8-LED common-anode bar display.

EXERCISE 8.10 Draw a schematic of a circuit corresponding to the display multiplexer of Example 8.2 on page 335.

EXERCISE 8.11 A 16-segment LED display, shown in Figure 8.43, can display alphabetic and numeric characters. Develop a circuit schematic and a VHDL model of a display decoder and driver to drive a 16-segment common anode LED display, given a 6-bit character-code input. Use a 64×16-bit ROM to decode the input. You needn't determine the ROM content for this exercise.

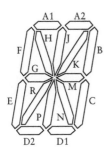

FIGURE 8.43

EXERCISE 8.12 Modify the display multiplexer/decoder design of Example 8.2 on page 335 to provide an 8-character alphanumeric scanned display, with eight 6-bit character code inputs. Use the ROM described in Exercise 8.11 to decode the character codes.

EXERCISE 8.13 Design an output controller to drive eight solenoids. The controller should have an 8-bit output register, and should connect to the Wishbone bus used by the Gumnut core.

EXERCISE 8.14 The ST Microelectronics L298 IC is a dual full-bridge driver that can be used to drive the kind of stepper motor shown in Figure 8.12 on page 339. The connections between the L298 and the motor (in simplified form) are shown in Figure 8.44. Determine the sequences of values on the inputs to the L298 to drive the stepper motor clockwise and counterclockwise.

EXERCISE 8.15 Assume the stepper motor driver described in Exercise 8.14 is connected to a Gumnut core through a 6-bit output register at port address 8, with bits 0 to 5 of the register controlling signals in1, in2, en_a, in3, in4 and en_b, respectively. Write a Gumnut subroutine to step the motor one-quarter turn, either clockwise, if r2 is 0, or counterclockwise, if r2 is 1. Hint: The subroutine

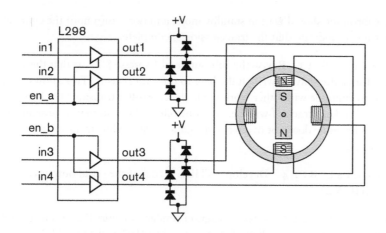

FIGURE 8.44

will need to keep track of the current state of the stepper motor control signals. Use a location in memory to save the state.

EXERCISE 8.16   Draw a diagram showing how the following components might be used to construct a handheld voice recorder: microphone, microphone amplifier, loudspeaker, loudspeaker amplifier, ADC, DAC, processor core, instruction memory, data memory, push-button switches. The recorder has buttons to record, play/pause, stop, skip forward, and skip backward.

EXERCISE 8.17   Draw a diagram similar to Figures 8.19 and 8.20 on page 351 showing multiplexed bus connection of two data sources, two data destinations, and two components that are both sources and destinations.

EXERCISE 8.18   Revise Figure 8.21 on page 352 to omit the second ADC controller.

EXERCISE 8.19   Revise the VHDL model in Example 8.6 on page 353 to omit the second ADC controller.

EXERCISE 8.20   Revise the VHDL model of Example 8.8 on page 358 to output 'X' values if the enable inputs are 'Z' or 'X'.

EXERCISE 8.21   Design a serial output controller for connection to the Gumnut core using the Wishbone bus. The controller should transmit each 8-bit data byte written to a data register using NRZ encoding with one start bit and one stop bit, as shown in Figure 8.35 on page 367. Transmission should occur at 9600 bits per second, with a transmit timing derived from a system clock with frequency 39.321600MHz ($= 9600 \times 4096$). When the stop bit has been transmitted, the controller should set an interrupt request output. The interrupt request output should be reset when the Gumnut int_ack signal is 1.

EXERCISE 8.22   Write a Gumnut subroutine to transmit a byte of data using the serial output controller of Exercise 8.21. Assume the data register is a port address 24 and that there are no other interrupt sources in the system.

The subroutine should wait in standby mode and not return until the controller interrupts to indicate that the transmission is complete.

EXERCISE 8.23   Revise the subroutine of Exercise 8.22 so that the subroutine returns after having written the byte to the data register. This allows the processor to continue with other work while the controller transmits the byte. You will need to keep track of whether the controller is busy so that a subsequent call to the subroutine does not overwrite the data register while transmission is still in progress.

EXERCISE 8.24   Develop a VHDL model of the serial output controller of Exercise 8.21.

EXERCISE 8.25   Draw a diagram similar to Figure 8.37 on page 369 showing Manchester encoding of the values 01100101 and 11110000.

EXERCISE 8.26   Design a circuit that has, as input, a transmit clock and an NRZ serial data signal (as in Figure 8.33 on page 366), and that generates a Manchester encoded serial data signal as output.

EXERCISE 8.27   Show how the system described in Example 8.11 on page 371 would be extended to connect to four AD7414 sensors.

EXERCISE 8.28   A Gumnut system includes a 4-digit 7-segment display, connected as shown in Figure 8.45. The anode data register is at port address 128, and the cathode data register is at port address 129. Write Gumnut assembly code for the task_2ms subroutine described in Example 8.17 on page 383 to scan the display. The BCD digits to display are stored in four bytes of memory labeled display_data. The subroutine should select one digit to drive each time it is called. Thus, four successive calls are required for a complete scan.

FIGURE 8.45

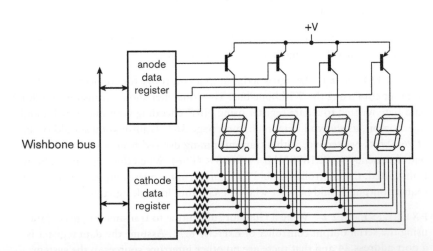

ACCELERATORS

9

In Section 7.1, as part of our introduction to embedded computer organization, we mentioned accelerators as optional components in embedded systems. If the system must perform some operation faster than is possible with embedded software running on a processor core, we can design custom hardware to perform the operation at the required speed. In this chapter, we will examine accelerators in more details and identify how they interact with an embedded processor.

9.1 GENERAL CONCEPTS

Many operations performed by digital systems consist of a number of steps. If a simple embedded processor core performs an operation, it performs the steps in sequence, with each step using one or more processor instructions. The rate at which the processor can execute instructions places a lower bound on the time it takes to perform the operation. The key to accelerating performance is *parallelism*: performing multiple steps at the same time, thus taking less time overall to complete the operation. The cost of parallelism is the additional components needed to perform the steps in parallel, since each component can only perform one step at a time. However, if sequential execution does not meet performance requirements, parallel hardware may be a higher-performance and lower-power alternative to using a faster (and more expensive) processor.

One place in which we can add hardware to achieve parallelism is within the processor core itself. As we saw in Chapter 7, a processor repeatedly fetches, decodes and executes instructions. Many processor cores use various techniques to perform these steps in parallel. For example, a processor might fetch a new instruction while decoding the preceding instruction and executing the instruction before that. A higher performance processor might fetch several instructions at once, decode them together, and use multiple function units to execute as many of them in parallel as it can. These and other techniques for achieving *instruction-level parallelism* are described in textbooks on computer architecture (see Section 9.5, Further Reading). While they can achieve performance improvements ranging from 2 times to perhaps 20 times over a simple processor core, the improvement comes at the cost of significantly increased complexity, area and power consumption. If an application requires much greater performance, or cannot afford the area and power consumption of a high-performance processor, a custom hardware accelerator may be a better option.

The extent to which we can improve performance depends on the amount of parallelism we can achieve, that is, on the number of steps we can perform at once. Many applications involve operations on data that has a regular, repetitive structure, and in which computation steps can be performed independently. For example, data from an audio source is a regular sequence of sample values. An operation that implements a volume control simply involves multiplying each sample value by the gain value. If several sample values are available at once, they can all be multiplied by the gain value in parallel. Similarly, video data from a camera consists of a sequence of frames, each of which is a rectangular array of picture elements (pixels). Many video processing operations can be performed within a frame in parallel across multiple pixels. Applications that involve less regularly structured data, or data that arrives at irregular intervals, are much harder to accelerate.

The amount of parallelism in some operations is limited only by the amount of data available at a given time. This applies to operations where each element of data can be processed independently of the others. Audio volume control is such a case. Other operations, however, involve *dependencies* that constrain parallelism. For example, some signal processing operations on audio streams involve combining successive sample values to produce values in a result stream. Filtering, as a case in point, involves combining several successive sample values to yield a single value in the output stream. Thus, we can't complete the processing for a given output sample until all of the required input values are available. Moreover, there are intermediate results that must be computed as part of the process, and the final result cannot be computed until all of the intermediate results have been computed.

In summary, we can accelerate performance of an operation by replication of hardware resources to perform steps in parallel, up to the limits on parallelism implied by the data dependencies and the availability of data. Practical design of accelerators involves applying enough parallelism to meet performance requirements, but not more, since that would increase cost and power unnecessarily.

In order to identify opportunities for parallelism, we would typically start with an abstract description of the processing operations to be performed by the system. This might take the form of an *algorithm* description expressed in a high-level language, such as a computer programming language or some other formal notation. The description identifies the data to be processed, how it is organized, and the sequence of processing steps to be performed. We then need to identify a *kernel* of the algorithm, that is, a part that involves the most intensive repetitive processing steps that take the most time. Such a kernel is a good candidate for an accelerator, since improving performance of the most time-consuming part of the algorithm gives the most payback. The remainder of the algorithm can then be implemented in embedded software.

We can quantify the performance gain achieved by accelerating a kernel of an algorithm. Suppose a system takes some amount of time, t, to execute the algorithm, and that a fraction, f, of that time is spent in executing the kernel. The remaining fraction, $1 - f$, is spent executing code other than the kernel. Thus,

$$t = ft + (1 - f)t$$

If our accelerator speeds up execution of the kernel by a factor s, the time spent in the kernel is divided by s, but the remaining time is unaffected. Thus the total execution time for the algorithm is reduced to

$$t' = \frac{ft}{s} + (1 - f)t$$

The overall speedup is the ratio of the original time to the reduced time:

$$s' = \frac{t}{t'} = \frac{ft + (1 - f)t}{\frac{ft}{s} + (1 - f)t} = \frac{1}{\frac{f}{s} + (1 - f)}$$

This formula expresses Amdahl's Law, named after Gene Amdahl, one of the pioneers of parallel computing. It indicates that the overall effect of speeding up a kernel depends strongly on the fraction of the original time taken up in executing the kernel. If that fraction is small, even a large speedup has little overall effect, since the nonaccelerated part dominates. On the other hand, if the fraction is large, accelerating the kernel has significant overall effect.

EXAMPLE 9.1   Suppose execution time is estimated for the various parts of an algorithm on an embedded processor. The algorithm has two kernels, one that consumes 80% of the execution time and another that consumes 15%. Using a hardware accelerator, we could speed up execution of the first kernel by a factor of 10 or the second kernel by a factor of 100. Which accelerator gives the best overall performance improvement?

SOLUTION   The overall speedup from accelerating the first kernel is

$$\frac{1}{\frac{0.8}{10} + (1 - 0.8)} = \frac{1}{0.08 + 0.2} = 3.57$$

Accelerating the second kernel gives an overall speedup of

$$\frac{1}{\frac{0.15}{100} + (1 - 0.15)} = \frac{1}{0.0015 + 0.85} = 1.17$$

Thus, even though the speedup for the second kernel is ten times that for the first kernel, the lower fraction of the original execution time for the second kernel means acceleration gives less overall improvement. Accelerating the first kernel is more effective.

Within the kernel, we need to identify an order in which to perform the computational steps. We need to ensure that data can be made available to be processed in order, and that intermediate results are computed before they are needed for subsequent steps. Other than those constraints, steps can potentially be performed in parallel. We finally need to determine which steps will actually be performed in parallel to meet the performance requirements. That then leads to an *architecture* for an accelerator, that is, a description of the processing blocks and the data flow between them.

There are two main schemes for implementing parallelism in accelerators. The first of these is simply to replicate components that perform a given step so that they operate on different elements of data. The speedup achieved through replication, compared to using just a single component, is ideally equal to the number of times the component is replicated. This scheme suits applications in which steps can be performed independently on the different data elements.

The second scheme for implementing parallelism is to break a larger computational step into a sequence of simpler steps, and to perform the sequence in a pipeline, as shown in Figure 9.1. (We introduced the concept of pipelining earlier in Section 4.1.1.) The pipeline stages perform their simple steps in parallel, each operating on a different data element or an intermediate result produced by the preceding stages. The overall computation by the pipeline for a given data element takes approximately the same time as a nonpipelined chain of components. However, provided we can supply data to the pipeline input and accept data at the pipeline output on every clock cycle, the pipeline completes one computation every cycle. Thus, the speedup compared to the nonpipelined chain is ideally equal to the number of stages. This scheme suits applications that involve complex processing steps that can be broken down into simpler sequences with each step depending only on the results of earlier steps. In

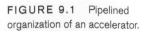

FIGURE 9.1   Pipelined organization of an accelerator.

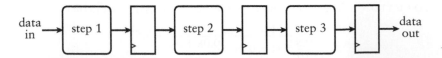

some applications involving independent complex computations, we can have replicated pipelines, giving the benefit of both schemes.

The analysis of systems, from algorithm description to accelerator architecture, is done early in the system design flow. It is often performed by expert system designers, drawing on their creativity and experience with previous systems. Automating this form of analysis has proven to be an extremely challenging problem, and early high-level synthesis tools have not been successful, except within very narrow application domains. More recently, a new generation of tools is starting to emerge and is showing promise in a wider range of applications, especially in audio, video and other signal-processing applications. As this technology matures, we should expect to see wider adoption in design methodologies. We will return to the topic of architecture analysis and its place in the design flow in our methodology discussion in Chapter 10.

The data for many systems involving accelerators is input or output data. In such systems, the I/O controller must transfer data between a device and the embedded system's memory at very high rates. Once the data is in memory, it can be processed by an accelerator, with the results also stored in memory. If these data memory accesses were mediated by a processor, copying data between memory and registers under software control, the rate of data transfer may be too slow. Instead, we can allow the controller and the accelerator to perform *direct memory access* (DMA), that is, to transfer data to and from memory autonomously. Instead of the processor initiating a memory access, the I/O controller or accelerator initiates an access, providing the required address and activating the memory control signals.

Since the processor and any subsystems that perform DMA must share access to the memory, and since the memory can only perform one access at a time, we need to ensure that processor and DMA accesses are interleaved. We must include an *arbiter* in the system, illustrated in Figure 9.2, that makes sure subsystems take turns to access the memory. Each *master* (the I/O controller, accelerator and processor) activates a

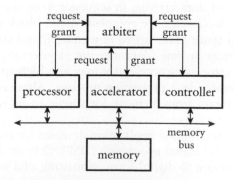

FIGURE 9.2 A multimaster system with an arbiter for the memory bus.

request signal to the arbiter when it needs to access the memory. The arbiter decides among them, based on a predetermined policy, and activates a grant signal for one of the subsystems. That subsystem then proceeds with its access, with the memory responding as a *slave*. Any other master with an active request must wait. When the granted master has completed its memory access, it releases its request. The arbiter can then activate another master's grant. Different applications may use different policies for deciding among competing requests, depending on whether a master can wait and for how long. Some applications use a round-robin policy, in which masters are granted access in strict turn. Other systems may require some masters to have priority over others in order to meet requirements for processing rates.

In many applications, the data to be processed by an accelerator is arranged in a regular pattern in memory, occupying *blocks* of adjacent or regularly spaced locations. The job of the accelerator is to process the data block by block. While it is processing one or more blocks, other parts of the system may be working on other blocks. As an example, several algorithms for processing still and video images divide each image into blocks of 8×8 or 16×16 pixels and process each block independently. Similarly, the MP3 format commonly used to encode audio data represents intervals of sound in frames that can be processed independently.

The datapath for a block-processing accelerator needs two main parts. The first part performs DMA to read and write data in memory. It includes circuits for generating addresses, using the starting addresses provided in registers by the processor and counters for keeping track of progress. The second part of the data path performs the required computation on the data. The control section for the accelerator sequences operation of the data path and synchronizes operation with the processor. Depending on the complexity of the operation and the bus protocol, sequencing might be done with one finite-state machine or with separate interacting machines for each activity.

Whereas a block-processing accelerator deals with blocks of data stored in contiguous memory locations, other forms of accelerators deal with *streams* of data arriving in sequence from some source. Thus, the two forms of accelerator are complementary: block processing deals with sequences in space (data stored in memory), and stream processing deals with sequences in time (data arriving at intervals). The source of data for a stream-processing accelerator may be a high-speed input device or another accelerator in a processing pipeline. Alternatively, data may be fetched in a stream from memory for supply to an output device or another accelerator.

One of the most common application domains for stream-processing accelerators is *digital signal processing* (DSP). One or more signals are converted from analog to digital form, consisting of a stream of sample values at periodic intervals. Processing operations include filtering, mixing, applying gain or attenuation, and conversion between time and frequency domains. Some application areas include audio and video processing, radio and radar signal processing, and analysis of data from sensors. For details of the mathematical basis for digital signal processing and the computational techniques used, refer to Section 9.5, Further Reading.

Having provided a means for an accelerator to access data, either in memory or through a stream connection, we also need to provide a way for embedded software to control operation of the accelerator. This may include providing data, such as parameters to be used in computations. It also includes synchronizing operation of the accelerator with other activities in the system, such as arrival of data from I/O controllers or other I/O events. Generally, this is done using input and output registers within the accelerator. Embedded software can then interact with the accelerator in much the same way as it interacts with autonomous I/O controllers. For example, an accelerator might include registers for the address and length of data in memory, for control of the operation to be performed and for status. Embedded software could write to the registers to initiate an operation, and rely on an interrupt from the accelerator when the operation is complete.

In some applications, it may be possible for the processor and an accelerator to operate with less strict synchronization. For example, the processor might generate units of work for the accelerator to perform and add information describing each unit to a first-in, first-out (FIFO) queue, like that described in Section 5.2.3. The accelerator can then accept each work unit when it is ready by reading the description from the head of the FIFO queue. FIFO queues can also be used for communication between multiple processors in a large-scale embedded system.

1. How does parallelism improve performance?

2. What factors constrain the amount of parallelism that can be achieved?

3. What aspects are described by an algorithm?

4. Why is it best to accelerate a kernel of an algorithm?

5. If a pipeline has four stages and accepts new input data on every clock cycle, what is the speedup compared to a nonpipelined chain of components?

6. What is direct memory access (DMA)?

7. What is the task of an arbiter in a multimaster system?

KNOWLEDGE
TEST QUIZ

8. What is the distinction between a block-processing accelerator and a stream-processing accelerator?

9. How does embedded software interact with an accelerator?

9.2 CASE STUDY: VIDEO EDGE-DETECTION

In this section, we will illustrate several aspects of accelerator design using, as an example, an accelerator for edge-detection in video images. This is somewhat of a compromise between what a real-world accelerator might do and what can be included here without overwhelming detail. Edge-detection is an important part of analyzing a scene in a video image, and has application in many areas such as security monitoring and computer vision. It involves identifying places in an image where there is an abrupt change in intensity. Those places usually occur at the boundaries of objects. Subsequent analysis of the edges can be used for recognizing what the objects are.

For this example, we will assume monochrome images of 640×480 pixels, each of 8 bits, stored row-by-row in memory with successive pixels, left to right in a row, at successive addresses. Pixel values are interpreted as unsigned integers ranging from 0 (black) to 255 (white). We will use a relatively simple algorithm, called the Sobel edge detector. It works by computing the derivatives of the intensity signal in each of the x and y directions and looking for maxima and minima in the derivatives. These are the places where the intensity is changing most rapidly. The Sobel method approximates the derivative in each direction for each pixel by a process called *convolution*. This involves adding the pixel and its eight nearest neighbors, each multiplied by a coefficient. The coefficients are often represented in a 3×3 *convolution mask*. The Sobel convolution masks, G_x and G_y, for the derivatives in the x and y directions, respectively, are shown in Figure 9.3. We can think of the derivative image being computed by centering each of the convolution masks over successive pixels in the original image. We multiply the coefficient in each mask by the intensity value of the underlying pixel and sum the nine products together to form two partial derivatives for the derivative image, D_x and D_y. Ideally, we would then compute the magnitude of the derivative image pixel as

$$|D| = \sqrt{D_x^2 + D_y^2}$$

However, since we are just interested in finding the maxima and minima in the magnitude, a sufficient approximation is

$$|D| = |D_x| + |D_y|$$

This approximation works, because the square-root and square functions are both monotonic (that is, they increase as the operand increases and decrease as the operand decreases). Hence, the maxima and minima in

−1	0	+1
−2	0	+2
−1	0	+1

G_x

+1	+2	+1
0	0	0
−1	−2	−1

G_y

FIGURE 9.3 Sobel convolution masks.

the true magnitude and the approximate magnitude occur at the same places in the image. Computing the approximation involves much less hardware than computing the square and square-root functions. We repeat the computation of the approximate magnitude for each pixel position in the image. Note that the pixels around the edge of the image do not have a complete set of neighboring pixels, so we need to treat them separately. The simplest approach is to set the value of $|D|$ for the edge pixels of the derivative image to 0. Since that is a relatively straightforward process and is not time consuming, we can implement it in software.

EXAMPLE 9.2   Express the Sobel edge-detection algorithm more formally in a *pseudo-code* notation, that is, a notation like a computer programming language.

SOLUTION   We will use a pseudo-code notation like VHDL. Let O(row, col) denote pixels in the original image, and D(row, col) denote pixels in the derivative image, where row ranges from 0 to 479 and col ranges from 0 to 639. Also, let Gx(i, j) and Gy(i, j) denote the convolution masks, where i and j range from −1 to +1. The algorithm is

```
for row in 1 to 478 loop
  for col in 1 to 638 loop
    sumx := 0; sumy := 0;
    for i in -1 to +1 loop
      for j in -1 to +1 loop
        sumx := sumx + O(row+i, col+j) * Gx(i, j);
        sumy := sumy + O(row+i, col+j) * Gy(i, j);
      end loop
    end loop
    D(row, col) := abs(sumx) + abs(sumy)
  end loop
end loop
```

EXAMPLE 9.3   Calculate the number of bits required to represent intermediate and final values for pixels in the Sobel convolution.

SOLUTION   Each pixel is represented as an 8-bit unsigned number. Given the coefficient values in the convolution masks, the partial products range from −510 to +510. Thus, the partial products should be represented using 10-bit signed numbers. There are nine partial products to add to form each of D_x and D_y. However, the coefficient values are such that the result values range from −1020 to +1020, which can be represented using 11 bits. We then need to add the two absolute values, giving a range of 0 to +2040 for $|D|$, which can also be represented in 11 bits. Since subsequent steps of the edge-detection operation involve determining which derivative pixels are above a certain threshold, we don't need to maintain 11 bits of accuracy for the results. Instead, it is more convenient to scale the results back to 8-bit values, since they can be packed back into memory in the same format as the original image.

EXAMPLE 9.4   Assuming a video frame rate of 30 frames per second, calculate the rate at which computations must be performed.

SOLUTION   Each frame consists of $640 \times 480 = 307{,}200$ pixels. Since there are 30 frames per second, pixels must be processed at a rate of $307{,}200 \times 30 = 9{,}216{,}000$ per second, that is, approximately 10 million per second.

EXAMPLE 9.5   Identify the parallelism that can be exploited to obtain the required performance.

SOLUTION   The computations required for all of the derivative pixels are independent of one another, since they only require values of the original image pixels. Thus, we could perform computations for as many derivative pixels in parallel as required. For computation of each derivative pixel, the *data dependency graph* is shown in Figure 9.4. This diagram shows the data required for each operation, starting with the pixels from the original image at the top, with intermediate results feeding through to dependent operations, yielding the derivative pixel at the bottom. We've elided partial products in which the coefficient is 0, since they don't contribute to the result. Inspection of the diagram shows that we can compute all of the partial products in parallel, since each partial product depends only on an original pixel value and a constant coefficient. We can then sum the two groups of six partial products in parallel, then compute the two absolute values in parallel, before summing them to produce the derivative pixel value.

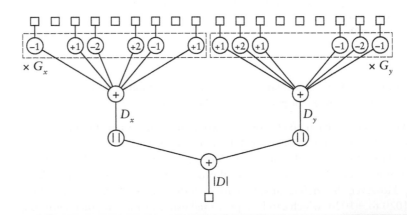

FIGURE 9.4   Data dependency graph for computation of a derivative pixel.

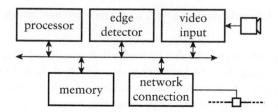

FIGURE 9.5   A video system incorporating an accelerator for edge detection.

The top-level view of the video system including the edge-detection accelerator is shown in Figure 9.5. Video input comes from an I/O controller for a video camera, which stores successive video frames in memory. Software on the processor directs the accelerator to operate on a given frame to produce the corresponding derivative image.

EXAMPLE 9.6   Suppose the memory in which the original and derivative images are stored is 32 bits wide, and that each 8-bit byte is individually addressed. Video frames are stored with one byte per pixel. The pixels of a row in a frame are stored from left to right at successive addresses, and rows are stored top to bottom, one after another in memory. Each memory read or write access takes 20ns, consisting of two cycles of a 100MHz system clock. Can the memory access data fast enough?

SOLUTION   Our earlier analysis showed that pixels arrive from the camera at a rate of approximately 10 million per second, or one every 100ns. If the video input controller stored each pixel to memory with a separate write access, it would consume 20% of the available memory bandwidth. A better alternative would be for the controller to aggregate four pixels and store them with a single write access, reducing its share of the memory bandwidth to 5%.

The edge-detection accelerator needs to produce a derivative pixel at the same rate at which input pixels arrive, that is, one every 100ns. Thus, writing the computed derivative pixels would consume a further 5% of the memory bandwidth, assuming groups of four derivative pixels are aggregated. Each pixel computation requires access to eight pixel values from the original image. A naive approach would involve reading each pixel with a separate read operation, and re-reading it when subsequently required to compute another derivative pixel. This approach would require eight reads per computed pixel, requiring 160% of the memory bandwidth. Clearly this is not possible.

Since each 32-bit word of memory contains four adjacent pixels in a row, we can reduce the bandwidth required for reading by using as many pixels as we can from each 32-bit read. For half the pixel positions, only three reads are needed (when the three pixels in each of the rows fall in the same word), and for the other half of the pixel positions, six reads are needed (when the three pixels in each of the rows cross word boundaries). So on average, each pixel computation would require 4.5 reads, requiring 90% of the memory bandwidth. This is still not feasible.

A further reduction can be afforded by noting that an original image pixel, once read, is used to compute three derivative pixels in each of the following, same, and preceding columns. So rather than re-reading it for those pixels, we can store it within the accelerator for use in computing multiple derivative pixels. We can save it just for computing the pixels to the left, in the same column, and to the right. We only need to read three words for every fourth pixel being computed, requiring 15% of the memory bandwidth. This, together with the 5% for video input and 5% for writing derivative pixels, is feasible, provided the remaining 75% of the bandwidth is sufficient for other operations to be performed by the system.

If we need to further reduce the bandwidth consumed by the edge detector, we could include small memories in the accelerator to store complete rows read from the main memory. This would allow each pixel to be read only once, reducing the bandwidth required for reading pixels to just 5%. The total for video input and edge-detection would then be 15% of the available bandwidth.

In our development of the edge-detector example, we will adopt the approach of reading three rows of four adjacent pixels from the original image and storing them in registers, rather than including memories for whole rows. We will design the accelerator to process blocks of data, where a block consists of the three complete rows of the original image used to form a complete row of the derivative image. As we will see, processing a block involves a start-up phase, a repetitive sequence of computation, and a completion phase. These phases are repeated for each derivative image row.

The architecture for the Sobel accelerator datapath is shown in Figure 9.6. It is essentially a pipeline, with pixel data read from the original image entering into the registers at the top right, flowing through the 3×3 multiplier array on the left, then down through the adders to the Dx and Dy registers, then through the absolute value circuits and adder to the |D| register, and finally into the register at the bottom left. The resulting derivative pixels are then written from that register to memory. (While a right-to-left data flow is opposite to usual practice, in this case, it has the advantage of preserving the same arrangement of pixels as that in an image.) We will describe the operation of the pipeline assuming initially that it is full of data. We will then discuss how to deal with starting it up at the beginning of an image row and draining it at the end of the row.

The pipeline generates the derivative pixels for a given row in groups of four. The accelerator reads four pixels from each of the preceding, current, and next rows in memory into the three 32-bit registers at the top right of the figure. Each register consist of four 8-bit pixel registers. Over

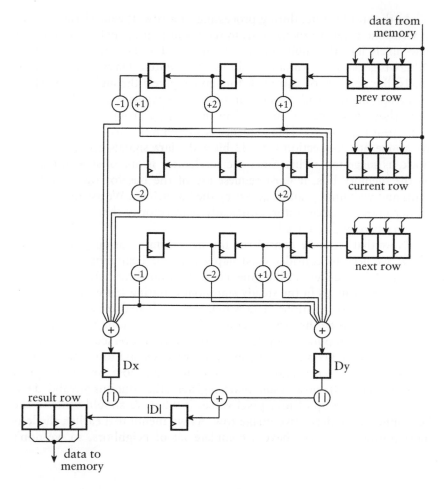

FIGURE 9.6 Architecture for the Sobel accelerator datapath.

the four subsequent clock cycles, pixels are shifted out to the left, one pixel at a time, into the multiplier array. Each cell in the array contains a pixel register and one or two circuits that multiply the stored pixel by a constant coefficient value. Since the coefficients are all -1, $+1$, -2, or $+2$, the circuits are not full-blown multipliers. Instead, multiplying by -1 is simply a negator, multiplying by $+1$ is a through connection with no circuitry, multiplying by -2 is a left shift of the result of a negator, and multiplying by $+2$ is simply a left shift. On each clock cycle, the array provides the partial products for a single derivative pixel, and the partial products are added and stored in the Dx and Dy registers. Also, on each clock cycle, the Dx and Dy values for the preceding pixel have their absolute values computed and added and stored in the |D| register. The resulting derivative pixel values are shifted into the result row register. When four result pixels are ready in the register, they are subsequently written to memory.

In the steady state, during processing of a row, the accelerator needs to write the pixels to memory from the result register before it can shift new pixels into the multiplier array and the Dx, Dy and |D| registers. Otherwise, the result values would be overwritten. Having written four pixels, the accelerator can push four more pixels through the pipeline, thus emptying the read registers and filling the result register. It can then write those result pixels and read in three more groups of four pixels, and repeat the process. This sequence is shown in Figure 9.7, assuming a Wishbone bus connection with 32-bit-wide data signals and a 100MHz clock, as suggested earlier. Since the accelerator is one of several masters on the memory bus, it must request use of the bus for the writes and reads and wait until granted access by the bus arbiter. We assume that the arbiter gives the accelerator sufficiently high priority that it can use the memory bandwidth it needs.

Now that we have considered the steady state during processing of a row, we need to consider what happens at the beginning of a row. In that case, the registers in the pipeline contain no valid data. So we start processing a row as in the steady state, but omitting the write operation for the first two iterations. Thereafter, the result register contains valid data, so we include the write operation in each iteration. Note that after the first four computation cycles, valid data has progressed into the pipeline as far as the Dx and Dy registers. After the second four computation cycles, valid data has progressed as far as the right-most three result pixel registers. The left-most result pixel register still contains invalid data. However, this group of four pixel values is what we should write to the beginning of the derivative image row. As we mentioned earlier, the left-most position does not have a complete set of neighbors, so we don't

FIGURE 9.7   Timing of pixel write and read operations and computation in the pipeline.

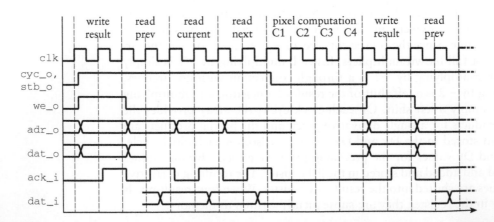

compute a value for it. We will rely on the embedded software to clear that pixel value to 0 subsequently.

When we reach the end of a row, we need to drain the pipeline. Since the number of pixels in a row is a multiple of four ($640 = 160 \times 4$), we can always read complete groups of four pixels each. After reading the last group, we perform four computation cycles normally. This gives us four result pixels to write, plus three remaining pixel values in the pipeline. We finish the row by writing the four result pixels, omitting the reads, performing four further computation cycles to drain the pipeline and shift the last pixel values into the required positions in the result register, and performing a final write. Note that this places an invalid value in the right-most result pixel register. This corresponds to the right-most pixel of a row, which does not have a complete set of neighbors. Again, we will rely on the embedded software to clear that pixel value to 0.

EXAMPLE 9.7   Develop VHDL RTL code to describe the datapath of Figure 9.6.

SOLUTION   The code in the architecture body for the Sobel accelerator is

```
-- Computation datapath signals

type pixel_array is array(-1 to +1,  -1 to  +1)
                         of unsigned(7 downto 0);

signal prev_row,
       curr_row,
       next_row   : unsigned(31 downto 0);
signal 0          : pixel_array;
signal Dx, Dy     : signed(10 downto 0);
signal abs_D      : unsigned(7 downto 0);
signal result_row : unsigned(31 downto 0);
...

-- Computational datapath

prev_row_reg : process (clk_i) is
begin
  if rising_edge(clk_i) then
    if prev_row_load = '1' then
      prev_row <= unsigned(dat_i);
    elsif shift_en = '1' then
      prev_row(31 downto 8) <= prev_row(23 downto 0);
    end if;
  end if;
end process prev_row_reg;
```

(continued)

```
curr_row_reg : process (clk_i) is
begin
  if rising_edge(clk_i) then
    if curr_row_load = '1' then
      curr_row <= unsigned(dat_i);
    elsif shift_en = '1' then
      curr_row(31 downto 8) <= curr_row(23 downto 0);
    end if;
  end if;
end process curr_row_reg;

next_row_reg : process (clk_i) is
begin
  if rising_edge(clk_i) then
    if next_row_load = '1' then
      next_row <= unsigned(dat_i);
    elsif shift_en = '1' then
      next_row(31 downto 8) <= next_row(23 downto 0);
    end if;
  end if;
end process next_row_reg;

pipeline : process (clk_i) is
begin
  if rising_edge(clk_i) then
    if shift_en = '1' then
    abs_D <= resize( (unsigned(abs Dx)
                        + unsigned(abs Dy)) srl 3, 8 );
  Dx <= - signed(resize(O(-1, -1), 11)) -- - 1 * O(-1, -1)
        + signed(resize(O(-1, +1), 11)) -- + 1 * O(-1, +1)
        - (signed(resize(O( 0, -1), 11)) -- - 2 * O(0, -1)
              sll 1)
        + (signed(resize(O( 0, +1), 11)) -- + 2 * O( 0, +1)
              sll 1)
        - signed(resize(O(+1, -1), 11)) -- - 1 * O(+1, -1)
        + signed(resize(O(+1, +1), 11)); -- + 1 * O(+1, +1)
  Dy <=   signed(resize(O(-1, -1), 11)) -- + 1 * O(-1, -1)
        + (signed(resize(O(-1, 0), 11)) -- + 2 * O(-1,  0)
              sll 1)
        + signed(resize(O(-1, +1), 11)) -- + 1 * O(-1, +1)
        - signed(resize(O(+1, -1), 11)) -- - 1 * O(+1, -1)
        - (signed(resize(O(+1, 0), 11)) -- - 2 * O(+1,  0)
              sll 1)
        - signed(resize(O(+1, +1), 11)); -- - 1 * O(+1, +1)
    O(-1, -1) <= O(-1, 0);
    O(-1,  0) <= O(-1, +1);
    O(-1, +1) <= prev_row(31 downto 24);
    O( 0, -1) <= O(0, 0);
    O( 0,  0) <= O(0, +1);
    O( 0, +1) <= curr_row(31 downto 24);
```

(continued)

```
      O(+1, -1) <= O(+1, 0);
      O(+1, 0)  <= O(+1, +1);
      O(+1, +1) <= next_row(31 downto 24);
    end if;
  end if;
end process pipeline;

result_row_reg : process (clk_i) is
begin
  if rising_edge(clk_i) then
    if shift_en = '1' then
      result_row <= result_row(23 downto 0) & abs_D;
    end if;
  end if;
end process result_row_reg;
```

The first three processes in the architecture represent the three registers into which groups of four pixels are read from memory. Each process has a separate control signal governing loading, since the registers are loaded in successive memory read operations. They share a control signal for shifting, since they all shift a pixel out into the pipeline in parallel.

The pipeline process, as its name suggests, represents the computational pipeline of the accelerator. The signals to which the process assigns, governed by the shift_en control signal, represent the pipeline registers. The signal O is a 3×3 array of pixel values, with indices corresponding to the difference in row and column numbers from those of the derivative pixel computed from the register values. For example, the element with indices $(-1, +1)$ contains the pixel in the previous row and next column from the pixel being computed. Values are shifted into this array leftward from the left-most 8 bits of each of the input registers. The Dx and Dy values are computed from the array element values. In each case, the values are resized to 11 bits and converted to signed numbers, as we discussed earlier in our analysis of the precision requirements for the computation. Multiplying by 2 is performed with a logical shift left by one position, and multiplying by a negative coefficient is implemented by subtraction instead of addition. The absolute values of the Dx and Dy values are converted back to unsigned representation, added, and then scaled back from 11 to 8 bits to yield the final derivative pixel value.

The result_reg process represents the register that accumulates groups of four derivative pixels for writing to memory. Pixels are shifted into this register under control of the shift_en signal.

We mentioned earlier that a block-processing accelerator needs circuits for address generation, as well as for processing the data. Our Sobel accelerator needs circuits to compute the addresses for reading pixels from the original image and for writing pixels to the derivative image. We will provide a register into which the embedded software can write the base addresses for the original image and the derivative image in memory. The address generator needs to determine pixel addresses using the base addresses. We will require that all addresses are word aligned, that is, that they are all multiples of four. This means the two least significant address bits are always 00, and so do not need to be computed or explicitly stored.

EXAMPLE 9.8   Given a base address B for an image in memory, derive equations for computing the address of a pixel in row r and column c of the image. Rows and columns are numbered from 0.

SOLUTION   The image size is 480 rows of 640 pixels per row. The starting address of row r is

$$B + r \times 640$$

The pixel in column c in that row is then located at address

$$B + r \times 640 + c$$

We can treat the expression $r \times 640 + c$ as an address offset from the base address.

EXAMPLE 9.9   Design the address generator datapath for the Sobel accelerator. Assume main memory is 4Mbytes in size, organized as $1M \times 32$ bits.

SOLUTION   The address generator needs two base address registers: O_base, for the original image, and D_base, for the derivative image. Since pixels are processed in groups of four, the least significant two address bits are always 0, and so do not need to be explicitly stored in the address registers.

There are several alternatives for deriving the read and write addresses, including maintaining counters for the image rows and columns. However, we can avoid the need to multiply by 640 by counting pixel offsets from the base addresses, as shown in Figure 9.8. In the case of the original image, we start counting from an offset of 0 and increment by 1 for each group of four pixels read from memory. We add the offset to the base address to form the pixel-group address for the previous row. We add 640/4 to that to form the read address for the current row, and add 1280/4 to form the read address for the next row (assuming 00 for the least significant bits in both cases). In the case of the derivative image, we start counting from an offset of 640/4 and increment by 1 for each memory write. The multiplexer in the figure selects the appropriate computed address to drive the memory address bus.

EXAMPLE 9.10   Develop VHDL RTL code to describe the address generator of Figure 9.8.

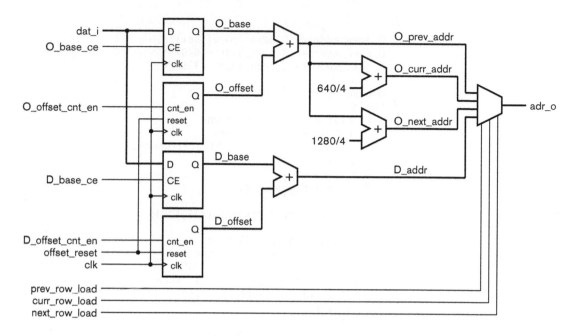

FIGURE 9.8   Datapath for the address generator.

SOLUTION   The code in the architecture body for the Sobel accelerator is

```
-- Address generator

O_base_reg : process (clk_i) is
begin
  if rising_edge(clk_i) then
    if O_base_ce = '1' then
      O_base <= unsigned(dat_i(21 downto 2));
    end if;
  end if;
end process O_base_reg;

O_offset_counter : process (clk_i) is
begin
  if rising_edge(clk_i) then
    if offset_reset = '1' then
      O_offset <= (others => '0');
    elsif O_offset_cnt_en = '1' then
      O_offset <= O_offset + 1;
    end if;
  end if;
end process O_offset_counter;
```

(continued)

```
O_prev_addr <= O_base + O_offset;
O_curr_addr <= O_prev_addr + 640/4;
O_next_addr <= O_prev_addr + 1280/4;

D_base_reg : process (clk_i) is
begin
  if rising_edge(clk_i) then
    if D_base_ce = '1' then
      D_base <= unsigned(dat_i(21 downto 2));
    end if;
  end if;
end process D_base_reg;

D_offset_counter : process (clk_i) is
begin
  if rising_edge(clk_i) then
    if offset_reset = '1' then
      D_offset <= (others => '0');
    elsif D_offset_cnt_en = '1' then
      D_offset <= D_offset + 1;
    end if;
  end if;
end process D_offset_counter;

D_addr <= D_base + D_offset;

adr_o(21 downto 2) <=
  O_prev_addr when prev_row_load = '1' else
  O_curr_addr when curr_row_load = '1' else
  O_next_addr when next_row_load = '1' else
  D_addr;
```

The processes O_base_reg and D_base_reg represent the base address registers for the original and derivative images, respectively. The processes O_offset_counter and D_offset_counter represent the counters for pixel groups read and written, respectively. The registers and counters are governed by control signals generated by the accelerator's control section. The adders are represented by the combinational assignments to the four address signals O_prev_addr, O_curr_addr, O_next_addr and D_addr. The assignment to the bus address signal adr_o represents the multiplexer that chooses among the generated addresses for memory read and write operations.

The remaining aspect of the Sobel accelerator design is control sequencing. We have touched on the sequence needed for computation of the derivative image, row-by-row and pixel-group at a time. This includes sequencing of write and read operations with the accelerator as a bus master. We also need to sequence the accelerator's response as a bus slave when the embedded software writes to the base address registers. Finally, we need to provide for synchronization with the embedded software controlling the accelerator. That requires some additional control and status registers, as follows:

▸ A control register that, when written to, causes the accelerator to start processing an image. The value written is ignored.

▸ A control register with an interrupt enable bit in bit 0.

▸ A status register in which bit 0 is the done bit, set to 1 when the processor has completed processing an image. Other bits are read as 0. When the done bit is 1 and the interrupt enable bit is 1, the accelerator requests an interrupt. Reading the done bit has the side effect of acknowledging the interrupt and clearing the bit.

To keep the bus interface simple, we will map each of these registers at 32-bit aligned addresses. The complete register map is shown in Table 9.1.

REGISTER	OFFSET	READ/WRITE
Interrupt control	0	write-only
Start	4	write-only
Original image base address	8	write-only
Derivative image base address	12	write-only
Status	0	read-only

TABLE 9.1   Register map for the Sobel accelerator.

EXAMPLE 9.11   Develop VHDL model code for the accelerator's bus slave interface.

SOLUTION   The timing for the bus slave operations is shown in Figure 9.9. Both write and read operations are initiated in a cycle where cyc_i and stb_i are 1. In each case, the accelerator can respond by setting ack_o to 1 in the next

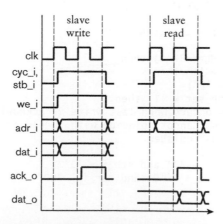

FIGURE 9.9   Timing for slave bus write and read operations.

cycle, then back to 0 in the following cycle. We need to decode the bus address input to derive a select signal for the accelerator, and use the less significant address bits to determine which register to read or write. For write operations, we generate clock-enable signals using combinational logic. In the case of a write to the start-register address, since there is no real register, we derive a control signal, start, that will be used by the accelerator control section to initiate a computation sequence. For read operations, we form the data value to be returned to the processor. The only real register is the status register, for which we return the value of the done bit, zero extended to 32 bits wide. For other register offsets, we just return all zeros. The read value is multiplexed with the value of the result row register to drive the accelerator's data output bus, dat_o. The model code describing these aspects is

```
-- Wishbone slave interface

  start <= '1' when cyc_i = '1' and stb_i = '1'
                    and we_i = '1' and adr_i = "01" else '0';

  O_base_ce <= '1' when cyc_i = '1' and stb_i = '1'
                    and we_i = '1' and adr_i = "10" else '0';

  D_base_ce <= '1' when cyc_i = '1' and stb_i = '1'
                    and we_i = '1' and adr_i = "11" else '0';

  int_reg : process (clk_i) is
  begin
    if rising_edge(clk_i) then
      if rst_i = '1' then
        int_en <= '0';
      elsif cyc_i = '1' and stb_i = '1'
            and we_i = '1' and adr_i = "00" then
        int_en <= dat_i(0);
      end if;
    end if;
  end process int_reg;

  status_reg : process (clk_i) is
  begin
    if rising_edge(clk_i) then
      if rst_i = '1' then
        done <= '0';
      elsif done_set = '1' then
        -- This occurs when last write is acknowledged,
        -- and so cannot coincide with a read of the
        -- status register.
        done <= '1';
      elsif cyc_i = '1' and stb_i = '1'
            and we_i = '0' and adr_i = "00"
```

(continued)

```
             and ack_o_tmp = '1' then
       done <= '0';
     end if;
   end if;
 end process status_reg;

int_req <= int_en and done;

ack_gen : process (clk_i) is
begin
  if rising_edge(clk_i) then
    ack_o_tmp <= cyc_i and stb_i and not ack_o_tmp;
  end if;
end process ack_gen;

ack_o <= ack_o_tmp;

-- Wishbone data output multiplexer

dat_o <= (31 downto 1 => '0') & done -- status register read
         when cyc_i = '1' and stb_i = '1' and we_i = '0'
             and adr_i = "00" else
       (others => '0')        -- other registers read as 0
         when cyc_i = '1' and stb_i = '1' and we_i = '0'
             and adr_i /= "00" else
       std_logic_vector(result_row);   -- for master write
```

EXAMPLE 9.12   Develop the control section to sequence computation of the derivative image.

SOLUTION   We can use a finite-state machine to sequence the computation. Since much of the sequence is repetitive, we can use counters to keep track of progress. We will use one counter to keep track of how many rows have been computed, starting from 0 and incrementing up to 477. We will use a second counter to keep track of iterations across the columns, starting from 0 and incrementing up to 159. The state transition diagram for the FSM is shown in Figure 9.10. We have only shown the states and the transition conditions to avoid cluttering the diagram. Also, we have not shown transitions from a state back to itself. We assume that if a transition condition from a given state is false, the FSM stays in that state for the next cycle.

The FSM is initially in the idle state. When the start signal is activated by a write to the start register, the FSM starts the initial sequence of reads and computations for the first row. This consists of reading the first three groups of original image pixels and then performing four computation cycles. After that, the FSM enters a loop in which it reads three more groups of original image pixels, performs four computation cycles, and then writes a group of result pixels. As

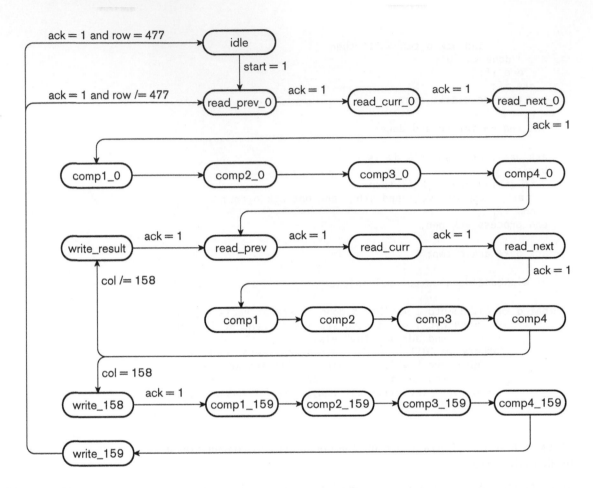

FIGURE 9.10 State transition diagram for the Sobel accelerator control section.

we will see when we look at the output function of the FSM, the column counter is incremented after each write. At the end of the last computation cycle, the FSM either continues with the loop (if the column counter is not 158) or goes to a state to start draining the pipeline (if the column counter is 158). Draining the pipeline involves one state for writing the penultimate result group, four cycles of computation, and one last state for writing the final result group. The row counter is incremented after this final write. The FSM then goes back either to the initial sequence for the next row (if the row counter is not 477) or to the idle state (if the row counter is 477, the terminal count).

The output functions for the FSM are shown in Tables 9.2 and 9.3. To make the tables a little easier to read, we have left entries blank where the control outputs are 0, and only shown the cases where they are 1. Some of the control signals are Moore outputs, depending on the current state only. They are shown in Table 9.2. Other control signals are Mealy outputs. For these, in Table 9.3, we have shown the input conditions that, along with the current state, determine their values. As in the state transition diagram, we have omitted the complementary conditions. In those cases, the Mealy outputs remain 0.

CURRENT STATE	offset_reset	row_reset	col_reset	prev_row_load	curr_row_load	next_row_load	shift_en	cyc_o	we_o
idle	1	1	1						
read_prev_0			1	1				1	
read_curr_0					1			1	
read_next_0						1		1	
comp1_0							1		
comp2_0							1		
comp3_0							1		
comp4_0							1		
read_prev				1				1	
read_curr					1			1	
read_next						1		1	
comp1							1		
comp2							1		
comp3							1		
comp4							1		
write_result								1	1
write_158								1	1
comp1_159							1		
comp2_159							1		
comp3_159							1		
comp4_159							1		
write_159								1	1

TABLE 9.2 Output functions for the Moore control outputs of the FSM.

CURRENT STATE	CONDITION	row_cnt_en	col_cnt_en	O_offset_cnt_en	D_offset_cnt_en	done_set
idle	start = 1					
read_prev_0	ack_i = 1					
read_curr_0	ack_i = 1					
read_next_0	ack_i = 1			1		
comp1_0	–					
comp2_0	–					
comp3_0	–					
comp4_0	–					
read_prev	ack_i = 1					
read_curr	ack_i = 1					
read_next	ack_i = 1			1		
comp1	–					
comp2	–					
comp3	–					
comp4	col /= 158					
comp4	col = 158					
write_result	ack_i = 1			1	1	
write_158	ack_i = 1			1	1	
comp1_159	–					
comp2_159	–					
comp3_159	–					
comp4_159	–					
write_159	ack_i = 1 and row /= 477	1			1	
write_159	ack_i = 1 and row = 477				1	1

TABLE 9.3   Output functions for the Mealy control outputs of the FSM.

EXAMPLE 9.13   Develop VHDL model code for the control section.

SOLUTION   The control-section code includes declarations of internal signals for the control FSM, the row and column counters, and the control signals:

```
type state_type is (idle,
                    read_prev_0,   read_curr_0,  read_next_0,
                    comp1_0,       comp2_0,
                    comp3_0,       comp4_0,
                    read_prev,     read_curr,    read_next,
                    comp1,         comp2,
                    comp3,         comp4,
                    write_result,  write_158,
                    comp1_159,     comp2_159,
                    comp3_159,     comp4_159,
                    write_159);

signal current_state, next_state : state_type;
signal row : natural range 0 to 477;
signal col : natural range 0 to 159;

signal O_base_ce, D_base_ce : std_logic;

signal start : std_logic;
signal offset_reset, row_reset, col_reset : std_logic;

signal prev_row_load, curr_row_load,
       next_row_load                      : std_logic;
signal shift_en                           : std_logic;
signal row_cnt_en, col_cnt_en             : std_logic;
signal O_offset_cnt_en, D_offset_cnt_en   : std_logic;
signal int_en, done_set, done             : std_logic;
signal cyc_o_tmp, ack_o_tmp               : std_logic;
```

The two counters used by the control section to keep track of progress through rows and columns, respectively, are represented by the row_counter and col_counter processes:

```
row_counter : process (clk_i) is
begin
  if rising_edge(clk_i) then
    if row_reset = '1' then
      row <= 0;
    elsif row_cnt_en = '1' then
```

(continued)

```
          row <= row + 1;
        end if;
      end if;
    end process row_counter;

    col_counter : process (clk_i) is
    begin
      if rising_edge(clk_i) then
        if col_reset = '1' then
          col <= 0;
        elsif col_cnt_en = '1' then
          col <= col + 1;
        end if;
      end if;
    end process col_counter;
```

Next, the model includes processes representing the finite-state machine using the techniques we have described in previous chapters. The state register is represented by the state_reg process:

```
    state_reg : process (clk_i) is
    begin
      if rising_edge(clk_i) then
        if rst_i = '1' then
          current_state <= idle;
        else
          current_state <= next_state;
        end if;
      end if;
    end process state_reg;
```

The fsm_logic process combines both the state transition function and the output function into the one process. The process also includes expressions comparing the row and column counter values with their terminal count values, rather than performing the comparisons in separate combinational statements. Combining these aspects into a single process makes the VHDL model somewhat more compact and simpler to understand, since the FSM is somewhat larger than those we have previously described.

```
    fsm_logic : process (current_state, start, ack_i,
                         row, col) is
    begin
      offset_reset    <= '0';  row_reset      <= '0';
```

(continued)

```
col_reset         <= '0';
row_cnt_en        <= '0';  col_cnt_en       <= '0';
O_offset_cnt_en <= '0';  D_offset_cnt_en <= '0';
prev_row_load    <= '0';  curr_row_load    <= '0';
next_row_load    <= '0';
shift_en          <= '0';  cyc_o_tmp        <= '0';
we_o              <= '0';  done_set         <= '0';
case current_state is
  when idle =>
    offset_reset <= '1'; row_reset <= '1'; col_reset <= '1';
    if start = '1' then
      next_state <= read_prev_0;
    else
      next_state <= idle;
    end if;

  when read_prev_0 =>
  col_reset <= '1'; prev_row_load <= '1'; cyc_o_tmp <= '1';
    if ack_i = '1' then
      next_state <= read_curr_0;
    else
      next_state <= read_prev_0;
    end if;

  when read_curr_0 =>
    curr_row_load <= '1'; cyc_o_tmp <= '1';
    if ack_i = '1' then
      next_state <= read_next_0;
    else
      next_state <= read_curr_0;
    end if;

  when read_next_0 =>
    next_row_load  <= '1'; cyc_o_tmp <= '1';
    if ack_i = '1' then
      O_offset_cnt_en <= '1';
      next_state <= comp1_0;
    else
      next_state <= read_next_0;
    end if;

  when comp1_0 =>
    shift_en   <= '1';
    next_state <= comp2_0;
  ...
  when comp4 =>
    shift_en <= '1';
    if col = 158 then
```

(continued)

```
              next_state <= write_158;
            else
              next_state <= write_result;
            end if;

        when write_result =>
          cyc_o_tmp <= '1'; we_o <= '1';
          if ack_i = '1' then
            col_cnt_en <= '1'; D_offset_cnt_en <= '1';
            next_state <= read_prev;
          else
            next_state <= write_result;
          end if;

        when write_158 =>
          cyc_o_tmp <= '1'; we_o <= '1';
          if ack_i = '1' then
            col_cnt_en <= '1'; D_offset_cnt_en <= '1';
            next_state <= comp1_159;
          else
            next_state <= write_158;
          end if;
        ...
        when write_159 =>
          cyc_o_tmp <= '1'; we_o <= '1';
          if ack_i = '1' then
            D_offset_cnt_en <= '1';
            if row = 477 then
              done_set <= '1';
              next_state <= idle;
            else
              row_cnt_en <= '1';
              next_state <= read_prev_0;
            end if;
          else
            next_state <= write_159;
          end if;
    end case;
  end process fsm_logic;

  cyc_o <= cyc_o_tmp;
  stb_o <= cyc_o_tmp;
```

Now that we have developed all of the hardware required for the Sobel accelerator, the remaining part is the embedded software that controls its operation. As we mentioned when we introduced this example, video edge-detection is used in a range of application areas. So rather than redesigning the control software for each application, it makes better sense to develop a software component that can be reused from one application to another. We can do this by developing a driver that provides a set of operations that gives application software an abstract view of the accelerator. Each application can then use the driver as one part of a collection of software components that implements the required functionality. For example, an application that recognizes objects in video images might apply edge-detection to each image in a video stream, followed by grouping of edges and matching against a database of edge patterns. Such software development is just as important as the hardware development in a complete application. A more complete treatment can be found in books on embedded system software development (see Section 9.5, Further Reading).

1. If image pixels were represented using only 6 bits instead of 8, how many bits would be required for the values of Dx, Dy and |D|?

2. Can the value of |D| for a given derivative-image pixel be computed in parallel with the values of Dx and Dy? Why, or why not?

3. If the memory read and write time is increased from two cycles to four, would there be sufficient memory bandwidth for video input and edge-detection?

4. Why do we not compute values for the left-most and right-most pixels in each row of the derivative image?

5. How does the embedded software initiate processing of an image? How does it determine when processing is complete?

6. What would happen if the software attempted to initiate processing when processing of a previous image was not yet complete?

7. Is the FSM that sequences computation a Mealy, Moore, or hybrid FSM?

9.3   VERIFYING AN ACCELERATOR

Throughout this book, we have stressed the importance of verification as part of our design methodology. It is particularly important when designing accelerators, given their relative complexity. We need to ensure that the design will operate correctly with all legal data values, and that it will interact with the embedded processor correctly. Since the space of all possible data values and operational sequences is astronomically large, it is not feasible to test the design exhaustively. Rather, we need to develop a verification plan that covers a variety of operating conditions. We will return to this in more detail in our methodology discussion in Chapter 10. Meanwhile, we will illustrate a simpler approach to simulation-based verification of the Sobel accelerator described in Section 9.2.

One way to approach verification of a complex accelerator is to verify the different aspects of its operation independently. For example, we might verify the following aspects of the Sobel accelerator one by one, adopting a "divide and conquer" approach:

▶ Slave bus operations

▶ Computation sequencing

▶ Master bus operations

▶ Address generation

▶ Pixel computation

Clearly all of these aspects of the accelerator must work correctly for the accelerator as a whole to work. However, verifying each in turn is much simpler than trying to verify all aspects at once. Having verified that the slave bus operations function correctly, we can then use them to initiate computation. Then we can check that computation follows the intended sequence of steps, with master bus operations proceeding correctly, ignoring the actual addresses and pixel values. We can then make sure addresses are being generated correctly, and finally check that pixel values are computed correctly. Verifying a stream-processing accelerator would proceed similarly, but we would additionally need to verify that the accelerator interacts correctly with the source of data being processed.

For this verification process, we need to construct a testbench that mimics the behavior of the embedded system containing the accelerator. If we have a verified model of the embedded processor, we can include it in the testbench and write small test programs to run on it. The test programs write to accelerator registers to set up and initiate operations. On the other hand, if no processor model is available, we can write a *bus functional model* of the processor, that is, a model that performs a predetermined sequence of bus operations without actually executing any processor instructions. Our testbench also needs to include a memory model and bus arbiter. The memory, like the processor, need not be a fully functional model. Instead, it might simply engage in write and read operations on the bus, generating read data according to a predetermined rule and discarding write data. These simplifications allow us to focus our verification effort on the accelerator, and to create test cases in a controlled manner.

EXAMPLE 9.14   Develop a testbench for the Sobel accelerator that includes a bus functional processor model. The processor should program the accelerator to operate on an original image at address 008000_{16} to generate a derivative image at 053000_{16}. It should then read the status register once every $10\,\mu s$ until the done bit is set. The testbench should also include a bus arbiter that gives the accelerator priority, and a bus functional memory that returns 0 for reads and discards data from writes.

SOLUTION   Our testbench is modeled after the general system organization shown in Figure 9.2. The accelerator is the design under verification, and the arbiter and bus functional processor and memory form the remainder of the testbench. We also include a clock and reset generator. The outline of the testbench architecture body is

```vhdl
architecture sobel_test of testbench is

  constant t_c : time := 10ns;
  constant mem_base : unsigned(22 downto 0)
    := "000" & X"00000";

  constant sobel_reg_base : unsigned(22 downto 0)
    := "100" & X"00000";
  constant sobel_int_reg_offset    : natural := 0;
  constant sobel_start_reg_offset  : natural := 4;
  constant sobel_O_base_reg_offset : natural := 8;
  constant sobel_D_base_reg_offset : natural := 12;
  constant sobel_status_reg_offset : natural := 0;

  signal clk, rst : std_logic;

  signal bus_cyc, bus_stb, bus_we : std_logic;
  signal bus_sel : std_logic_vector(3 downto 0);
  signal bus_adr : unsigned(22 downto 0);
  signal bus_ack : std_logic;
  signal bus_dat : std_logic_vector(31 downto 0);
  signal int_req : std_logic;

  signal sobel_cyc_o, sobel_stb_o, sobel_we_o : std_logic;
  signal sobel_adr_o : unsigned(21 downto 0);
  signal sobel_ack_i : std_logic;
  signal sobel_stb_i : std_logic;
  signal sobel_ack_o : std_logic;
  signal sobel_dat_o : std_logic_vector(31 downto 0);
  ...

begin

  clk_gen : process is
  begin
    clk <= '1'; wait for t_c / 2;
    clk <= '0'; wait for t_c / 2;
  end process clk_gen;

  rst <= '1', '0' after 2.5 * t_c;
```

(continued)

```
duv : entity work.sobel(rtl)
  port map ( clk_i => clk,            rst_i => rst,
             cyc_o => sobel_cyc_o,    stb_o => sobel_stb_o,
             we_o  => sobel_we_o,
             adr_o => sobel_adr_o,    ack_i => sobel_ack_i,
             cyc_i => bus_cyc,        stb_i => sobel_stb_i,
             we_i  => bus_we,
             adr_i => bus_adr(3 downto 2),
             ack_o => sobel_ack_o,
             dat_o => sobel_dat_o,    dat_i => bus_dat,
             int_req => int_req );
  ...

end architecture sobel_test;
```

The clock generator process uses the constant t_c for the clock cycle time, giving a clock frequency of 100 MHz. The constants mem_base and sobel_base define the base addresses of the memory (000000_{16}) and the Sobel accelerator registers (400000_{16}). Additional constants define the offsets from the base address for the control and status registers. Next, the testbench includes signals for the bus address, data and control signals. As we will see shortly, these are multiplexed from the various sources in the system. The testbench also declares signals for connection specifically to the Sobel accelerator. Within the architecture body, the accelerator is instantiated as the design under verification (duv) and connected to the signals.

The testbench code for the processor bus functional model is

```
signal cpu_cyc_o, cpu_stb_o, cpu_we_o : std_logic;
signal cpu_sel_o : std_logic_vector(3 downto 0);
signal cpu_adr_o : unsigned(22 downto 0);
signal cpu_ack_i : std_logic;
signal cpu_dat_o : std_logic_vector(31 downto 0);
signal cpu_dat_i : std_logic_vector(31 downto 0);
...

processor_bfm : process is

  procedure bus_write ( adr : in unsigned(22 downto 0);
                        dat : in std_logic_vector(31 downto 0) ) is
  begin
    cpu_adr_o <= adr;
    cpu_sel_o <= "1111";
```

(continued)

```
        cpu_dat_o <= dat;
        cpu_cyc_o <= '1'; cpu_stb_o <= '1'; cpu_we_o <= '1';
        wait until rising_edge(clk) and cpu_ack_i = '1';
    end procedure bus_write;

begin
    cpu_adr_o <= (others => '0');
    cpu_sel_o <= "0000";
    cpu_dat_o <= (others => '0');
    cpu_cyc_o <= '0'; cpu_stb_o <= '0'; cpu_we_o <= '0';
    wait until rising_edge(clk) and rst = '0';
    -- Write 008000 (hex) to O_base_addr register
    bus_write(sobel_reg_base
                + sobel_O_base_reg_offset, X"00008000");
    -- Write 053000 + 280 (hex) to D_base_addr register
    bus_write(sobel_reg_base
                + sobel_D_base_reg_offset, X"00053280");
    -- Write 1 to interrupt control register (enable interrupt)
    bus_write(sobel_reg_base
                + sobel_int_reg_offset, X"00000001");
    -- Write to start register (data value ignored)
    bus_write(sobel_reg_base
                + sobel_start_reg_offset, X"00000000");
    -- End of write operations
    cpu_cyc_o <= '0'; cpu_stb_o <= '0'; cpu_we_o <= '0';
    loop
        wait for 10 us;
        wait until rising_edge(clk);
        -- Read status register
        cpu_adr_o <= sobel_reg_base + sobel_status_reg_offset;
        cpu_sel_o <= "1111";
        cpu_cyc_o <= '1'; cpu_stb_o <= '1'; cpu_we_o <= '0';
        wait until rising_edge(clk) and cpu_ack_i = '1';
        cpu_cyc_o <= '0'; cpu_stb_o <= '0'; cpu_we_o <= '0';
        exit when cpu_dat_i(0) = '1';
    end loop;
    wait;
end process processor_bfm;
```

The processor waits for completion of system reset, then performs the required sequence of bus write operations to initialize the accelerator. For each bus operation, described by the bus_write procedure, the processor assigns the appropriate values to the address, data and control signals, then waits for the accelerator to acknowledge completion of the operation. After completion of the write to the start register, the processor enters a loop in which it waits for $10 \mu s$, resynchronizes with the clock, then reads the accelerator status register. When the accelerator acknowledges completion of the read operation, the processor checks whether the done bit is 1. If so, the processor exits the loop and waits indefinitely, completing the test.

The testbench code for the memory bus functional model is

```
signal mem_stb_i  : std_logic;
signal mem_sel_i  : std_logic_vector(3 downto 0);
signal mem_ack_o  : std_logic;
signal mem_dat_o  : std_logic_vector(31 downto 0);
...

mem : process is
begin
  mem_ack_o <= '0';
  mem_dat_o <= X"00000000";
  wait until rising_edge(clk)
             and bus_cyc = '1' and mem_stb_i = '1';
  if bus_we = '0' then
    mem_dat_o <= X"00000000"; -- in place of read data
  end if;
  mem_ack_o <= '1';
  wait until rising_edge(clk);
end process mem;
```

The memory repeatedly waits until the bus_cyc and mem_stb_i signals are both '1', indicating that a memory operation is required. If bus_we is '0', the operation is a read, so the memory provides zeros on the data outputs. In the case of a write operation, the memory does nothing with the input data. In either case, the memory sets the acknowledge signal to '1', and then on the next clock cycle clears the signal back to '0', completing the operation.

The arbiter for the testbench is somewhat more involved than the other testbench components. It uses the sobel_cyc_o and cpu_cyc_o signals as requests from the Sobel accelerator and the processor, respectively, and generates sobel_gnt and cpu_gnt grant signals. When either of the request signals is activated, the arbiter activates the corresponding grant. If both requests are activated in the same cycle, the arbiter gives preference to the accelerator, activating its grant and leaving the processor's grant inactive until the accelerator's request is deactivated. Since the grant outputs depend not only on the values of the request inputs, but also on the preceding history of request values, the arbiter must be implemented as a sequential circuit using an FSM. The state transition diagram is shown in Figure 9.11. The FSM is a Mealy machine, since that allows us to activate a grant signal in the same cycle in which the corresponding request is activated.

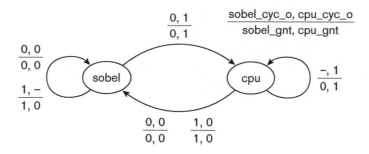

FIGURE 9.11 State transition diagram for the testbench arbiter.

The testbench code for the arbiter is

```
type arbiter_state_type is (sobel, cpu);
signal arbiter_current_state,
       arbiter_next_state : arbiter_state_type;

signal sobel_gnt, cpu_gnt : std_logic;
...

arbiter_fsm_reg : process (clk) is
begin
  if rising_edge(clk) then
    if rst = '1' then
      arbiter_current_state <= sobel;
    else
      arbiter_current_state <= arbiter_next_state;
    end if;
  end if;
end process arbiter_fsm_reg;

arbiter_logic : process (arbiter_current_state,
                         sobel_cyc_o, cpu_cyc_o) is
begin
  case arbiter_current_state is
    when sobel =>
      if sobel_cyc_o = '1' then
        sobel_gnt <= '1'; cpu_gnt <= '0';
        arbiter_next_state <= sobel;
      elsif sobel_cyc_o = '0' and cpu_cyc_o = '1' then
        sobel_gnt <= '0'; cpu_gnt <= '1';
        arbiter_next_state <= cpu;
      else
        sobel_gnt <= '0'; cpu_gnt <= '0';
        arbiter_next_state <= sobel;
      end if;
    when cpu =>
```

(continued)

```
        if cpu_cyc_o = '1' then
           sobel_gnt <= '0'; cpu_gnt <= '1';
           arbiter_next_state <= cpu;
        elsif sobel_cyc_o = '1' and cpu_cyc_o = '0' then
           sobel_gnt <= '1'; cpu_gnt <= '0';
           arbiter_next_state <= sobel;
        else
           sobel_gnt <= '0'; cpu_gnt <= '0';
           arbiter_next_state <= sobel;
        end if;
    end case;
end process arbiter_logic;
```

The rest of the testbench code represents the bus multiplexers and slave select logic:

```
signal sobel_sel, mem_sel : std_logic;
...

-- Bus master multiplexers and logic

bus_cyc <= sobel_cyc_o when sobel_gnt = '1' else cpu_cyc_o;
bus_stb <= sobel_stb_o when sobel_gnt = '1' else cpu_stb_o;
bus_we  <= sobel_we_o when sobel_gnt = '1' else cpu_we_o;

bus_sel <= "1111" when sobel_gnt = '1' else cpu_sel_o;

bus_adr <= '0' & sobel_adr_o when sobel_gnt = '1' else
           cpu_adr_o;

sobel_ack_i <= bus_ack and sobel_gnt;
cpu_ack_i   <= bus_ack and cpu_gnt;

-- Bus slave logic

sobel_sel <= '1' when (bus_adr and ("111" & X"FFFF0"))
                        = sobel_reg_base else
             '0';
mem_sel <= '1' when (bus_adr and "100" & X"00000")
                        = mem_base else '0';

sobel_stb_i <= bus_stb and sobel_sel;
mem_stb_i <= bus_stb and mem_sel;

bus_ack <= sobel_ack_o when sobel_sel = '1' else
           mem_ack_o   when mem_sel = '1' else
           '0';
```

(continued)

```
-- Bus data multiplexer

bus_dat <= sobel_dat_o when (sobel_gnt = '1' and bus_we = '1')
                        or (sobel_sel = '1'
                            and bus_we = '0') else
           cpu_dat_o when (cpu_gnt = '1' and bus_we = '1') else
           mem_dat_o;
```

The grant signals from the arbiter determine which source provides values for the bus control and address signals. They also gate the acknowledge signals back to the masters, so that a master that is waiting for the bus does not receive an acknowledgment from a slave for the active master's bus operation. The bus slave logic decodes addresses and determines which slave is selected. The select signals gate the strobe signal from the active master to the selected slave, and multiplex the selected slave's acknowledgment signal onto the bus_ack signal. The bus data multiplexer determines the source of data for the bus_dat signal, depending on which master is active, which slave is selected, and whether the bus operation is a read or a write.

We can simulate the testbench of Example 9.14 to verify that the Sobel accelerator correctly responds to slave bus operations and performs master bus operations with correct addresses. We need to observe the values of the bus control and address signals, as well as the internal signals of the accelerator. Figure 9.12 shows a simulation waveform display of the bus signals during initialization of the accelerator by the processor bus functional model. Figures 9.13 through 9.15 show the internal signals of the accelerator during the start of processing a row (Figure 9.13), during steady state processing (Figure 9.14), and upon completion of processing a row and commencement of the next row (Figure 9.15). Finally, Figure 9.16 shows the internal signals on completion of processing an entire image.

While the verification shown here might give us confidence that the design is correct, it is by no means complete. For example, it doesn't demonstrate that the computation produces correct values according to the specification of the algorithm, and it doesn't show that the control sequencing is correct for all possible interactions between the accelerator and other bus masters. Creating test cases for simulation-based verification to cover all of these aspects is infeasible, given the number of permutations of data values and ways in which components can interact. Instead, we need to turn to more sophisticated verification techniques, such as constrained random test generation, coverage analysis, and property-based formal verification. We will return to the topic of verification again in Chapter 10, but we also refer the interested reader to advanced books on verification listed in Section 9.5, Further Reading.

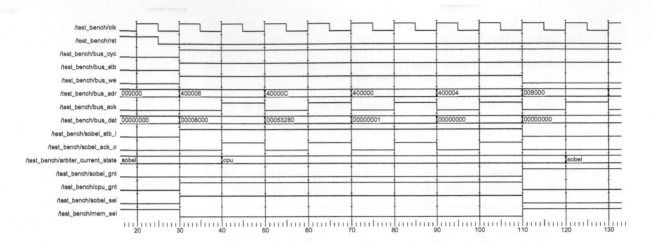

FIGURE 9.12 Waveform display of bus operations for initializing the Sobel accelerator.

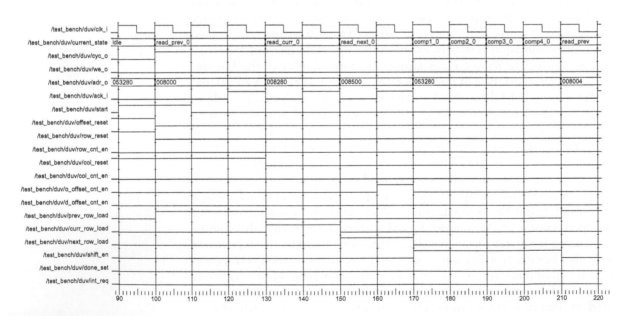

FIGURE 9.13 Waveform display of accelerator internal signals at the start of row processing.

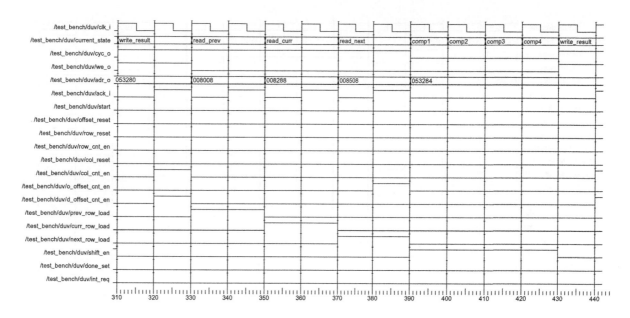

FIGURE 9.14 Waveform display showing row processing in the steady state.

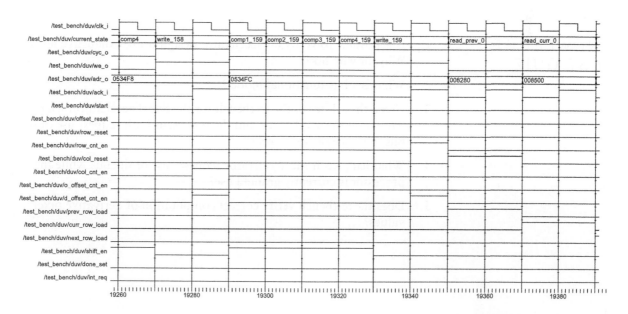

FIGURE 9.15 Waveform display showing completion of one row and commencement of the next.

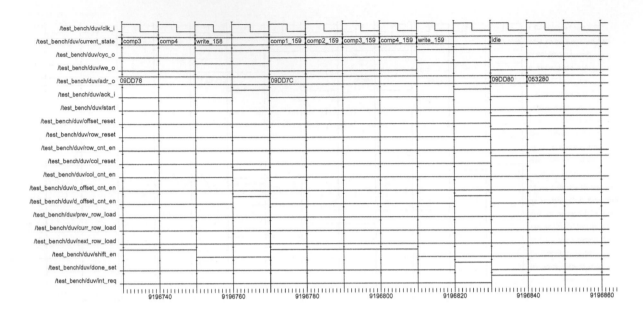

FIGURE 9.16 Waveform display showing completion of image processing.

KNOWLEDGE TEST QUIZ

1. Is it possible to verify an accelerator design using exhaustive testing? Why, or why not?

2. What is a bus functional model?

3. Given the arbiter in the testbench for the Sobel accelerator, what happens if the accelerator and the processor both request use of the bus in the same clock cycle?

4. What happens if the accelerator requests use of the bus while the processor is currently granted use?

5. Does the testbench verify correct computation of derivative pixel values?

9.4   CHAPTER SUMMARY

▶ Parallelism, performing multiple processing steps at once, allows accelerators to reduce the time required to complete an operation.

▶ An accelerator achieves parallelism by replicating hardware resources and by pipelining. This leads to cost/performance and power/performance trade-offs.

▶ The degree of achievable parallelism is constrained by data dependencies within a computation.

▶ Designing an accelerator involves analyzing an algorithm and identifying a kernel to be implemented in hardware. The remainder of the algorithm is implemented in embedded software.

▶ Amdahl's Law quantifies the overall speedup from accelerating a kernel of an algorithm.

▶ Accelerators and high-speed I/O controllers can use direct memory access (DMA) to transfer data to or from memory without processor intervention. An address generator in such a unit calculates memory addresses for DMA.

▶ An arbiter determines which of several bus masters can use the bus at any time to access bus slaves, such as memory and I/O controller registers.

▶ A block-processing accelerator processes blocks of data stored in memory. Many video and still-image processing applications are block oriented.

▶ A stream-processing accelerator processes data arriving from a source in a sequence of values. Digital-signal processing (DSP) is often stream oriented.

▶ Accelerators include control and status registers for use by embedded software.

▶ Verification of an accelerator using exhaustive simulation is generally not feasible. Aspects of operation can be verified independently, but a complete verification plan should include other forms of verification.

9.5   FURTHER READING

Computer Architecture: A Quantitative Approach, 4th Edition, John L. Hennessy and David A. Patterson, Morgan Kaufmann Publishers, 2007. An advanced textbook on computer architecture, covering instruction-level parallelism in depth.

Parallel Computer Architecture: A Hardware/Software Approach, David E. Culler and Jaswinder Pal Singh, Morgan Kaufmann Publishers, 1999. An in-depth treatment of parallel computing. While the book focuses on parallel computers, many of the principles can also be applied to architectures of hardware accelerators.

Understanding Digital Signal Processing, Richard G. Lyons, Prentice Hall, 2001. An introduction to the theory of digital signal processing (DSP).

Computers as Components: Principles of Embedded Computing System Design, Wayne Wolf, Morgan Kaufmann Publishers, 2005. Includes a discussion of accelerators in the context of embedded hardware and software design, with a video-processing accelerator as a case study.

Embedded Software Development with eCos, Anthony J. Massa, Prentice Hall, 2003. Describes the Embedded Configurable Operating System (eCos), including the hardware abstraction layer.

Comprehensive Functional Verification: The Complete Industry Cycle, Bruce Wile, John C. Goss and Wolfgang Roesner, Morgan Kaufmann Publishers, 2005. A detailed treatment of functional verification strategies and techniques.

EXERCISES

EXERCISE 9.1 In computer graphics applications, a three-dimensional vector representing a point's position in space can be transformed by multiplying by a 3×3 matrix:

$$\begin{bmatrix} P_x' \\ P_y' \\ P_z' \end{bmatrix} = \begin{bmatrix} a_{11} & a_{12} & a_{13} \\ a_{21} & a_{22} & a_{23} \\ a_{31} & a_{32} & a_{33} \end{bmatrix} \begin{bmatrix} P_x \\ P_y \\ P_z \end{bmatrix}$$

Determine the data dependencies in the computation and thus the maximum available parallelism.

EXERCISE 9.2 Devise a pipeline architecture that can perform the computation described in Exercise 9.1 using all the available parallelism. Assume a new input vector arrives and a result can be accepted on every clock cycle.

EXERCISE 9.3 If a kernel of an algorithm is accelerated by a factor of 100, and the kernel accounts for 90% of execution time before acceleration, what is the overall speedup?

EXERCISE 9.4 For a kernel that consumes 90% of execution time, what speedup for the kernel would be required to achieve an overall speedup of 5?

EXERCISE 9.5 Suppose there are two options for accelerating a kernel that consumes 80% of a system's execution time. Option 1 accelerates the kernel by a factor of 100 and increases the system's cost by a factor of 2. Option 2 accelerates the kernel by a factor of 200 and increases cost by a factor of 4. What is the ratio of cost to performance for each option, compared to the original system?

EXERCISE 9.6 Use the Sobel convolution masks to compute the approximate derivative pixel at the center of each of the 3 × 3 image blocks shown in Figure 9.17. The numbers represent the pixel intensities.

255	255	255
255	255	255
255	255	255

0	0	0
127	127	127
255	255	255

255	223	191
255	223	191
255	223	191

0	63	127
63	127	191
127	191	255

FIGURE 9.17

EXERCISE 9.7 Consider a lower-performance version of the Sobel accelerator, dealing with video frames of 320 × 240 8-bit pixels at a rate of 15 frames per second. Repeat the analysis of Example 9.6 to determine a suitable approach to memory accesses for this version of the accelerator.

EXERCISE 9.8 In Example 9.6, we mentioned that we could reduce the memory bandwidth consumed by the Sobel accelerator by storing rows of pixels in small memories, so that each pixel need only be read from memory once. Revise Figure 9.6 to show how this might be done using two row memories. Hint: It may simplify your design to assume that a memory can read and modify a location in a single cycle, with the read data provided on the data output and the new data taken from the data input. Memory components in some implementation fabrics can operate in this way.

EXERCISE 9.9 Write a pseudocode outline (not detailed code!) of embedded software for an edge-detection application using the Sobel accelerator. Assume the camera input controller has a register for the base address in memory for the next frame to be acquired, and interrupts the processor when the frame has been acquired. The software should then use the accelerator to perform edge detection. When the accelerator has completed its operation, the software should perform a subroutine for post-detection analysis. The software should maintain three images in memory: one being acquired from the camera, one being processed by the accelerator, and one undergoing post-detection analysis.

EXERCISE 9.10 Revise the memory bus functional model code to provide a synthetic image (that is, an artificially constructed image) when locations in the original image are read. The synthetic image should contain a 320 × 240 white rectangle centered on a black background. Hint: use the memory address during a read to determine whether the location being accessed is within the image. If so, then calculate the row and column numbers from the address. If the row and columns lie within the rectangle, the memory should return white pixel values; otherwise, it should return black pixel values.

EXERCISE 9.11   Calculate the values of the derivative pixels that should result from Sobel edge detection of the synthetic image described in Exercise 9.10.

EXERCISE 9.12   Write a checker process for inclusion in the testbench using the revised memory bus functional model of Exercise 9.10. The checker should verify that derivative pixels written to memory by the Sobel accelerator have the values determined in Exercise 9.11.

DESIGN METHODOLOGY 10

Now that we have completed our coverage of design techniques for digital systems, we return to the topic of design methodology that we introduced in Chapter 1. If you have undertaken lab projects in conjunction with studying this book, you will have put many of the ideas relating to design and verification into practice. In this chapter, we will expand upon those ideas and also consider the larger context in which digital systems are designed.

10.1 DESIGN FLOW

In our introductory discussion of design methodology, we commented that design of a real-world digital system is a complex undertaking, usually requiring a team of people. We stressed the importance of taking a systematic approach in order to manage both the complexity of the design itself and the many interactions between the participants. We introduced the notion of a design methodology, which codifies the process of design, verification and preparation for manufacture of a product. For the relatively small projects we have described in this book, a systematic design approach has been of some benefit. For large real-world projects, having a clearly specified methodology and sticking to it are indispensable. Many projects have failed, not through technical problems, but through lack of control of the design process itself.

While many aspects of electronics design are standardized across the industry, design methodology is not one of them. Indeed, it would be difficult to settle on a standardized methodology, since there is significant difference among design projects. Moreover, the suite of tools available to designers evolves quite rapidly. Hence, each organization typically defines its own methodology based on the kinds of the design projects it undertakes, and evolves the methodology from project to project.

Our prototypical design methodology, introduced in Chapter 1, divided the design flow into a number of stages: functional design, synthesis, and physical design. Each stage includes verification steps to ensure that the design meets its requirements and satisfies constraints. Figure 10.1 shows the elements of the design flow, including hierarchical hardware/software codesign, integrated into a single diagram. The product of the design process is a set of data files used in manufacturing the product. Each manufactured unit is then tested and delivered to the end customer or market. We also refined the design flow for embedded systems to include design and verification of the embedded software.

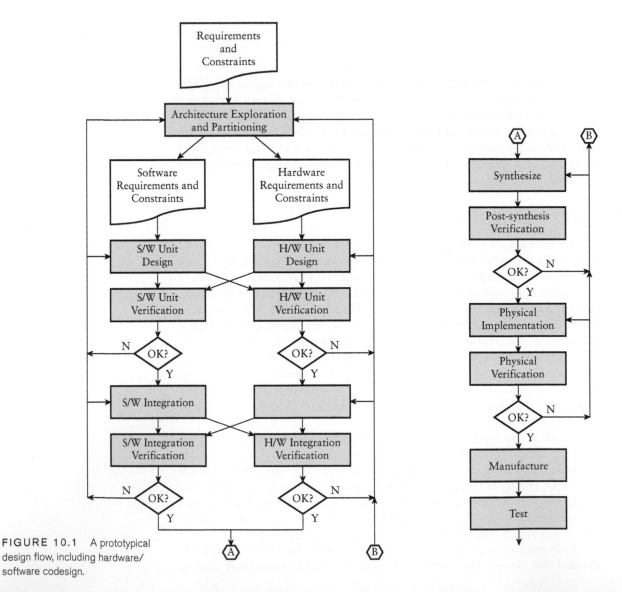

FIGURE 10.1 A prototypical design flow, including hardware/software codesign.

For many complex systems, the hardware is simply the platform upon which to deliver the software, with most of the functionality of the system implemented in the software. In such systems, developing the software is a major proportion of the system development effort.

A key part of a design methodology is the set of electronic design automation (EDA) tools used to support it. For nearly all designs, it is not feasible to make physical prototypes to ensure that the design is correct and meets constraints. Instead, we design models as *virtual prototypes* in forms that EDA tools can analyze and refine. In the remainder of this section, we will explore the stages of a design methodology in greater depth, identifying the models we create, the kinds of analysis we perform, and the EDA tools that we use. We will assume initially a linear flow, starting from a design concept and following through to physical implementation. In the subsequent section, we will consider how the process of design optimization can cause us to return to earlier stages in the flow, leading to a cyclic, evolutionary design flow.

10.1.1 ARCHITECTURE EXPLORATION

A digital system is designed and manufactured to meet some functional requirements, subject to various constraints. Unless the system is a minor modification to a previous design, there is a large space of possible designs that could meet the requirements. Of those possible designs, some would violate the constraints and so would not be feasible candidates. Others would satisfy the constraints, so we would choose from among them. Our choice is typically guided by one or more objective functions, such as cost, performance, power consumption, or reliability, that we seek to optimize. Of course, until we have undertaken some design effort, we have nothing to analyze to determine whether a candidate design meets requirements and satisfies constraints. Clearly we don't want to go through the complete design process for each candidate. Instead, we need to identify sufficient information about a candidate at an abstract level to be able to estimate the value of relevant properties of the system. The estimates don't need to be perfect, just sufficiently accurate to decide whether the candidate would meet requirements and constraints, and to allow comparison of candidates to select among them. We use the term *architecture exploration*, or alternatively, *design space exploration*, to refer to the task of abstract modeling and analysis of candidate designs. The term originates in the concept of exploring the space of possible system architectures.

One important aspect of architecture exploration is partitioning of operations among components of a system. Partitioning is essentially the application of a divide-and-conquer problem-solving strategy. If our system requirements involve a number of processing steps, we can divide our system into a number of components, each of which performs one of the processing steps. The components interact with one another to complete the overall task of the system. When working at the abstract level of architecture exploration, the components need not be physical parts of the system. Instead, we can think of *logical partitioning*, that is simply identifying parts of the system that will implement the various processing steps. This form of partitioning is also called *functional decomposition*. We can also think about the kinds of physical components that we might include in the system and how the logical partitions can be mapped to the physical partitions. The *physical partitions* can include processor cores, accelerators, memories and I/O controllers. Also important is hardware/software partitioning. We addressed this as part of our discussion of accelerator design in Section 9.1. A given logical component may be mapped to a specialized hardware component whose only task is to implement that logical component. Another logical component may be mapped to a software task run on a processor core under control of a real-time operating system.

As an example of system partitioning, consider a road transport monitoring system that checks whether freight trucks drive from one part of the country to another in too short an interval. (Such a system is deployed in the author's neighborhood in Australia.) Stations on freeways each have a video camera on a gantry over the road. The video images are analyzed to identify the license plate of each truck passing underneath, and the time and license number are logged. The information is transmitted to a central facility for recording and comparison with information from other stations. A hypothetical functional decomposition of the monitoring station is shown at the top of Figure 10.2. It includes logical components for input of video from a camera, filtering to remove noise, edge-detection,

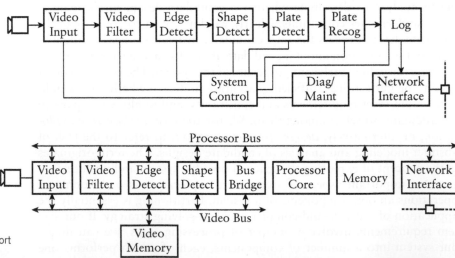

FIGURE 10.2 Logical partitioning (top) and physical partitioning (bottom) of a transport monitoring system.

shape detection, license plate detection, character recognition to identify the license number, logging, network interface, system control, and diagnostic and maintenance tasks. This logical structure can be mapped onto the physical structure shown at the bottom of Figure 10.2. In this case, the physical components comprise an embedded system with accelerators for video processing up to the shape-detection stage. License plate detection and recognition, logging, system control, and diagnostics and maintenance tasks are mapped onto software tasks running on the processor core.

As we mentioned in Section 9.1, architecture exploration and partitioning is often done by expert system designers. Decisions made in this early stage of the design flow have a major impact on the rest of the design. Unfortunately, it is very difficult to automate these tasks using EDA tools. Instead, system designers often rely on *ad hoc* system models, expressing algorithms in programming languages such as C or C++, and using spreadsheets and mathematical modeling tools to analyze system properties. By far the most valuable asset in this stage is the experience of the system designer. Lessons learned from previous projects can be brought to bear on new design projects. Nonetheless, design automation for architecture exploration and system-level modeling has been an active area of research for some time, and we should expect tool support to improve from its present immature state.

However architecture exploration design is done, whether by systematic or *ad hoc* means, the result is a high-level specification of the system. For each of the components in the system, the specification describes the function it is to perform, the connections to other components, and the constraints upon its implementation. The specification might be expressed in a language that can be executed or simulated, such as certain forms of the Unified Modeling Language (UML). Such an executable specification has the benefit of being more precise than a specification written in a natural language, such as English. Moreover, execution can answer questions of interpretation, such as what a component is supposed to do in certain circumstances. The specification is used as the input to the next stage of the design flow.

10.1.2 FUNCTIONAL DESIGN

The next stage to consider is functional design, which has been the main topic of this book. Our architecture specification has decomposed the system into physical components, each of which must implement one or more logical partitions. As we mentioned in Chapter 1, architectural design is the top level of a top-down design process. We can decompose each component into subcomponents, which we then design and verify as units. That might also involve further decomposition, until we reach a level of complexity that is manageable.

Before we embark on that division into subcomponents, we can develop a *behavioral model* of the component, expressing its functionality at an intermediate level of abstraction between system level and register-transfer level. The behavioral model might include a description of the algorithm to be implemented by the component without detailed cycle-by-cycle timing, or it might just be a bus functional model. The purpose of the behavioral model is to allow function verification of the component before proceeding to detailed implementation. Once the functionality is verified, the same testbenches can be used with the detailed implementation models to verify them under the same test cases. As an example, consider the Sobel edge-detection accelerator that we discussed in Chapter 9. We presented a pseudo-code description of the algorithm in Example 9.2 on page 401. We could develop a behavioral model of the edge-detector based on that algorithm. While the behavioral model might read and write pixels in a different order and with different timing, they should produce the same derivative image from a given original image.

In order to implement a given component, we can take several approaches. One approach is to design a new implementation by refining the higher-level model, using the design techniques we have discussed in previous chapters. An alternative, however, is to reuse a component from a previous system, from a library of components, or from a component vendor, if a suitable reusable component is available. We often use the term *intellectual property*, or *IP*, to refer to such reusable components, since they constitute a valuable intangible resource. The most obvious benefit of reuse is the saving in design time it affords. Moreover, if the IP has been specifically developed for reuse and has been thoroughly verified as a unit, then we can save effort in verification. Even if an IP block does not exactly meet the requirements for our system, we may be able to adapt it with less effort than would be required for a fresh start. If the IP performs the required function, but does not have quite the right interface connections or timing, we might be able to embed it in a *wrapper*, circuitry that deals with the differences. If the IP has almost the required function and the source code is available to us, we might be able to make minor changes to adapt its functionality to our needs.

In circumstances where there is no reusable IP available and we must implement a component from scratch, we should still think of reuse. We should consider whether the component might be reused in a subsequent system. In that case, we should spend extra effort to ensure that the component is well specified, its functionality is verified under all operating conditions, and its use and implementation are well documented. The effort spent doing that when developing the component will be recouped in the subsequent projects.

Another alternative for implementing a component may be to use a *core generator*, which is an EDA tool that generates a model of a component based on parameters that describe its function. Core generators are available for common kinds of functions, such as memories, arithmetic units, bus interfaces, digital signal processing, and finite-state machines.

Figure 10.3 is a screenshot of a typical core generator, in this case a tool provided by Xilinx for generating cores to be implemented in its FPGAs. (The Xilinx Core Generator is included as part of the Xilinx ISE tool suite linked from the companion website.) The screenshot shows the kinds of core function that can be generated. For each function, parameters controlling operation of the generated core can be specified, as illustrated in Figure 10.4 for a content-addressable memory core. The tool then automatically generates a suite of design files for the specified function, including HDL source code for behavioral simulation and net-list files for inclusion in the physical design. Core generators are available for ASIC-based designs as well as for FPGAs. In both cases, using a generated core can save a substantial amount of design and verification effort, and is well worth considering.

Throughout this book, we have shown how to design components using the VHDL hardware description language. VHDL has much in common with computer programming languages, and many of the same considerations apply to managing the design process. In particular, it is important to write VHDL models in such a way that they are clearly understandable and can be maintained throughout their life cycle. Many organizations adopt coding style rules to help ensure the quality of the model code. An example of such style rules is the *VHDL Modelling Guidelines* document published by the European Space Agency (see Section 10.7, Further Reading). Some EDA vendors also provide style checking tools, sometimes called *lint* tools (after the Unix *lint* program for checking C programs) that verify whether code meets a set of rules.

Another aspect of hardware model development in common with software development is the need for revision management, also referred to as source code control. Usually, there are multiple designers working on the model code for a component, and they make numerous revisions as they develop and verify the code. Revision management software helps coordinate their work by maintaining a repository of versions of the code. Typically, designers work with their own copy of the code. As they complete a change, they commit the revised code back to the repository. Other designers periodically update their copies with changes committed to the repository. Usually, changes made by different designers affect different parts of the code, so the changes can be merged automatically. Where changes conflict, the designers must manually reconcile them. Some vendors' EDA tool suites include proprietary revision management tools. Others use open source tools, such as the Concurrent Version System (CVS) or the more recent Subversion tool. The author has used both of these to good effect, for software development, digital design, and other projects.

10.1.3 FUNCTIONAL VERIFICATION

Throughout this book, we have stressed the importance of verification of digital designs, focusing mainly on simulation-based verification of

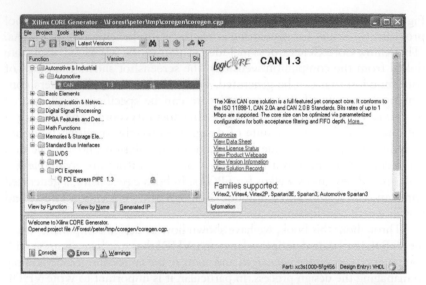

FIGURE 10.3 Xilinx Core Generator showing categories of core functions.

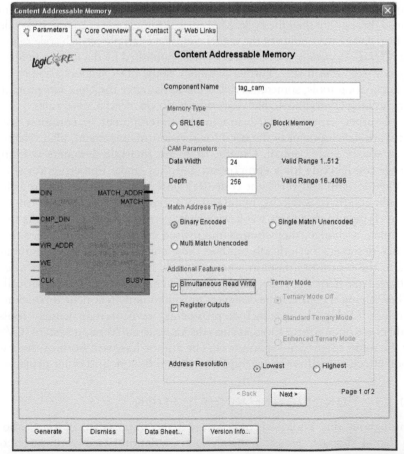

FIGURE 10.4 A dialog for specifying parameters for a generated core.

functionality. We have shown how to construct testbenches that stimulate a design under verification and that include checkers to monitor outputs of the design. Successful verification of a system requires a verification plan that identifies what parts of the design will be verified, the functionality that will be verified, and how verification will be performed. Without a verification plan, there are no criteria to determine whether verification is complete.

The first question, what parts to verify, can be answered by appealing to the hierarchical decomposition of the system. Since the system is composed of subsystems, each subsystem must be correct for the entire system to be correct. Thus, verifying each subsystem can be considered to be a prerequisite for verifying the entire system. This argument can be repeated recursively, leading to a bottom-up verification strategy. As each designer works on a bottom-level component, they verify that it meets its functional requirements. Those components can then be integrated into the next-level subsystem, which is then verified. The process is repeated, up to the top level of the system.

The second question, what functionality to verify, can be answered by appealing to the specification for each component. That, in itself, is good motivation for ensuring well-written specifications. Without a clear statement of what a component is supposed to do, we cannot verify with confidence that the component does what is required of it. At the lower levels of the design hierarchy, the functionality of each component is relatively simple, and so the component can be verified fairly completely. At higher levels of the design hierarchy, the functionality of subsystems and the complete system gets much more complex. Thus, it is much harder to verify that a subsystem or the system meets functional requirements under all circumstances. Instead, we might focus on the interactions among components, for example, checking for adherence to protocols.

We use the term *coverage* to refer to the proportion of functionality that is verified. Historically, *code coverage* has been used as a figure of merit. It refers to the proportion of lines of code that have been executed at least once during simulation of the design. The benefit of using code coverage is that it is easy to measure, but it does not give a reliable indication that all of the required functionality has been implemented and implemented correctly. Instead, we should use *functional coverage*, even though it is harder to quantify. Aspects of functional coverage include the distinct operations that have been verified, the range of data values that have been applied, the proportion of states of registers and state machines that have been visited, and the sequences of operations and values that have been applied. Coverage measurement tools are now available that allow monitoring of signal and storage values within a model during simulation and measurement of which ranges of values have been observed. These tools can help us to identify operating conditions that have and have not been verified.

The third question in the verification plan is how to verify. There are a number of techniques that we can apply. We have already illustrated the approach of *directed testing* using simulation in several examples throughout this book. Directed testing involves identifying particular test cases to apply to the DUV and checking the output for each test case. This approach is very effective for simpler components where there are only a small number of categories of stimulus. However, for more complex components, achieving significant function coverage is not feasible, and so we must complement directed testing with other techniques. Another approach that has gained acceptance is *constrained random testing*. This involves a test case generator randomly generating input data, subject to constraints on the ranges of values allowed for the inputs. Specialized verification languages, such as Vera and *e*, include features for specifying constraints and random generation of data values to be used as stimulus to a DUV. More recently, similar features have been included in System-Verilog, an extension to the Verilog hardware description language. Similar features are also planned for inclusion in a future revision of VHDL.

Both directed and constrained random testing require checkers that ensure that the DUV produces the correct outputs for each applied test case. If, as part of our top-down design process, we have developed a behavioral model of a component, we may be able to use it to simplify the checker for the register-transfer level implementation. We can create a comparison testbench, illustrated in Figure 10.5, that verifies that the implementation has the same functionality as the behavioral model. We use the same test-case generator to provide test cases to two instances of the design under verification: one an instance of the behavioral model, and the other an instance of the RTL implementation. The checker then compares the outputs of the two instances, making any necessary adjustments for timing differences.

Directed and constrained random testing are both simulation-based verification techniques. The problem inherent in simulation-based verification is that it is not feasible to attain 100% coverage. The number of possible input cases and sequences is too large for exhaustive simulation to be feasible. *Formal verification*, on the other hand, allows complete verification that a component meets a specification. The specification is embodied in one or more asserted properties, expressed in a *property specification language*, such as PSL (see Section 10.7, Further Reading). PSL can be used as an adjunct to a hardware description language such

FIGURE 10.5 A comparison testbench for comparing outputs of a behavioral model and its RTL refinement.

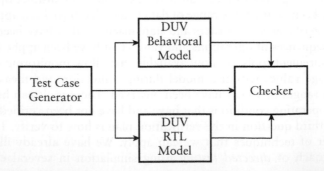

as VHDL or Verilog. The recent revision of VHDL also allows PSL to be embedded in a VHDL model, either as part of the design or in a testbench model. As an alternative, the SystemVerilog extension of Verilog includes features similar to those of PSL for expressing properties.

A property can be as simple as a Boolean expression relating the values of signals in the design. More commonly, it involves temporal expressions relating sequences of values over time. For example, a property might specify that activation of a select signal followed by an enable signal on the next clock cycle is followed by activation of an acknowledge signal within three clock cycles and deactivation of all three signals on the subsequent cycle. A formal verification tool performs *state-space exploration* to verify the asserted properties. It exhaustively examines all possible sequences of input values, determines the resulting values for signals in the design at all clock cycles, and checks that the asserted properties hold for all cycles. Where a property does not hold, the tool uses the sequence of input values to construct a counter-example leading to the failure. Properties can also be used to express assumptions about the inputs to a component, which help the formal verification tool limit the space of possible values that it has to explore.

The strength of formal verification is that it provides a rigorous proof that the assertions hold. However, its completeness is only as good as the properties that are verified. If those properties do not cover all of the functional requirements, then a formal verification does not achieve complete functional coverage. Moreover, writing properties that completely and accurately capture the intent of a specification is very difficult. In many organizations using formal verification, that task is left to expert verification engineers who work alongside design engineers. A further difficulty is that state-space exploration is a computationally intense problem, so verification of numerous complex assertions may be intractable. It may be necessary to limit the use of formal verification to parts of a design or to a subset of the complete functionality, and to use other techniques for other parts.

While properties are a necessary part of formal verification, they can also be used in simulation-based verification of a system. In particular, a property can be used to generate a checker that monitors the values of signals mentioned in the property during the course of a simulation and tests whether the property holds. One way to generate such a checker is to automatically translate the property into an HDL model that can be simulated along with the functional model. Alternatively, a simulator that supports the property language can directly interpret the property during simulation, in which case the checker is implied by the property. The advantage of this approach to verification using properties is that the properties are often a more abstract and succinct expression of the specification to be verified, compared to a checker expressed as an HDL model. Moreover, the properties can be re-used for formal verification after an initial phase of simulation-based verification. Thus, the specification is expressed in the single form of properties, rather than being replicated in potentially inconsistent forms.

Hardware/Software Co-Verification

In an embedded system, much of the system's functionality may be implemented in software that interacts with hardware. In order to verify functionality of the system, we need to verify the software and its interaction with the hardware. In principle, if we had hardware models of the processor and the instruction and data memories, we could verify the software by simulating its execution on the hardware models. We would load the instructions into the instruction memory and start the simulation. The operation of the processor, fetching and executing instructions and reading and writing I/O controller and accelerator registers, would then be simulated. The problem with this approach is that it is very slow, since simulation of each processor instruction involves simulation of much of the detail of hardware operation. While that may be necessary or useful for verification of some aspects of the system, for example, verifying detailed timing of interrupt request and service, it makes system-level verification difficult.

Fortunately, there are approaches we can take to make hardware/software co-verification much faster. One approach recognizes that software and hardware development are usually done by different people. Allowing the software development team to start software verification as early in the design process as possible reduces the overall time to complete the system. The key to this is to divide the software into two layers: a lower layer that depends on the hardware, and an application layer that is insulated from the hardware by the lower layer. The lower layer, sometimes called the *hardware abstraction layer* (HAL), or the *board support package* (BSP) for processors that are components on a printed circuit board, contains driver code and interrupt service routines for I/O controllers, memory management code, and so on. It provides an abstract interface that can be called by the application layer.

With this division of the embedded software, development and verification of the application layer code can proceed without waiting for the hardware design. Instead, a software verification tool can emulate the operations provided by the hardware abstraction layer. For example, where the final system includes an output display panel, the verification tool might provide a virtual panel that displays as a window on the software developer's computer screen. The software developer can write the software in a programming language, making calls to the emulated abstraction layer, and run the software on their host computer. In this way, the software can run at close to real time. The disadvantage is the lack of detailed interaction with hardware, which might mask timing problems inherent in the software.

For more detailed verification of embedded software, we can use an *instruction set simulator* (ISS). Rather than compiling and executing the software on the host computer, an ISS uses code compiled into the instructions of the target embedded processor. It then simulates execution of those instructions, but without simulating the detailed hardware operations of the target processor. Simulation of software on an ISS is intermediate in speed between native execution on a host computer and execution on a hardware model of the target processor. Moreover, because execution is simulated, the tool can perform detailed analysis and debugging of the software. Again, the platform support layer can be emulated, allowing verification before the hardware design is available.

Once the hardware design team has developed models of the hardware that interacts with the embedded software, we can perform *cosimulation* of the hardware and software. This usually involves a collaboration between an ISS and a simulator for the hardware model. The two simulators run concurrently, communicating when the processor performs bus read and write operations. Initially, at least, the hardware models need not be fully functional behavioral or RTL models. Bus functional models may be sufficient to verify that code in the hardware abstraction layer correctly reads and writes to registers. As more detailed implementation models of the hardware become available, they can be substituted for the bus functional models. Operation of the hardware design under software control can then be verified. Since cosimulation is much slower than executing the software on a real processor, we would typically run only small sections of the embedded software for this kind of verification. Ultimately, however, full start-up and operation of the embedded application software can be verified using cosimulation, albeit very slowly.

10.1.4   SYNTHESIS

Having performed functional design and verification of a digital system, the next stage in the design flow is synthesis, that is, the refinement of the functional design to a gate-level net list. For most designs, synthesis can be performed largely automatically using an RTL synthesis tool. Where a design is complex, has very high performance requirements, and is implemented as an ASIC, it may be necessary to custom design the circuitry of some subsystems. However, we won't go into that process in this book, referring the reader instead to books on ASIC design listed in Section 10.7, Further Reading. In this book, we will focus on automatic RTL synthesis, particularly as it is used for FPGA-based designs.

RTL synthesis, as the name suggests, starts with models of the design refined to the register-transfer level. This means that we cannot use all of the features of our hardware description language (VHDL or Verilog) arbitrarily. Many language features are only suitable for high-level behavioral modeling and for writing testbenches, and cannot be synthesized into equivalent gate-level circuits. Moreover, most RTL synthesis tools do not accept use of all language features that could, in principle, be synthesized. Instead, they require that RTL models be written using a subset of language features, and that code implying hardware be structured using various templates. For example, most synthesis tools require that sequential hardware be expressed using process statements. For edge-triggered registers, they require that the process statements be written using the template structures we have shown in this book, or using a process with exactly one wait statement, at the start of the process, that waits for a clock edge. They generally prohibit multiple wait statements occurring throughout a process, though in principle many such process statements imply reasonable hardware. The reasons for these restrictions are largely historical and financial, and are related to development of synthesis tool technology. Early synthesis tools performed relatively simple pattern recognition on the HDL source code to determine which hardware circuits were implied. Subsequent developments focused more on improving the quality and optimization of the synthesized hardware, rather than expanding the scope of acceptable input code. Moreover, there was little user demand for removing the restrictions, since designers could largely get their job done with the restrictions in place.

Since different synthesis tools accept different subsets of the input language, it can be difficult to develop RTL models that are portable across a range of tools. To help designers write interoperable models, the IEEE had defined two standard coding styles for synthesizable models, one for VHDL (IEEE Standard 1076.6) and the other for Verilog (IEEE Standard 1364.1). The initial version of each of these standards (1076.6-1999 and 1364.1-2002) defined a subset that was portable across a number of tools—essentially a "lowest common denominator" subset. The VHDL subset is summarized in Appendix C, and is largely what we have followed in RTL models in this book. A later revision of the VHDL synthesis standard (1076.6-2004) has extended the VHDL subset substantially to include more constructs that are "in principle" synthesizable. Most tool vendors accept portions of the extended subset, if not all of it. We should expect vendors to meet the extended standard as they develop their tools further. Meanwhile, we need to consult the documentation for each tool we use to determine what we can and can't write in synthesizable code.

A synthesis tool starts by analyzing the model, checking to make sure the code conforms to its style requirements. It also performs some design rule checks, such as checking for unconnected outputs, undriven inputs, and multiple drivers on nonresolved signals. The tool then *infers* hardware constructs for the model. This involves things like:

▶ Analyzing signal declarations to determine the encoding and the number of bits required to represent the data.

▶ Analyzing expressions and assignments to identify combinational circuit elements, such as adders and multiplexers, and to identify the input, output and intermediate signal connections.

▶ Analyzing process statements to identify the clock and control signals, and to select the appropriate kinds of flip-flops and registers to use.

For each of these inferred hardware elements, the synthesis tool determines an implementation using primitive circuit elements selected from a *technology library*. This is a collection of components that are available within the implementation fabric selected for the design. For ASICs, the technology library is usually provided as part of a larger design kit by the ASIC vendor, that is, the company that will ultimately manufacture the ASIC design. For FPGAs, the technology library is usually embedded in the tools provided by the FPGA vendor. Typical components in a technology library include inverting and noninverting gates with a small number (2 to 4) of inputs, small multiplexers, carry chain components, and flip-flops.

The process of translating the design into a circuit of library components is guided by *synthesis constraints* that we specify. Such constraints include bounds on clock periods and propagation delays. We will return to the topic of constraints in Section 10.2 as part of our discussion of design optimization. The synthesis tool uses the constraints to choose among alternative implementations. For example, two alternative circuit structures might implement the required functionality inferred from the RTL code: one with fewer gates connected in a deeper chain, and so with greater propagation delay; and the other with more gates connected less deeply, and so with less propagation delay. If our constraints specified minimal overall delay as the synthesis goal, the tool would choose the latter implementation. In making such timing-based choices, the tool uses a simple wire model to determine the wire delays, based on average wire lengths and loading, since at this stage the actual layout and wiring is yet to be done.

While we rely on the synthesis tool to determine the implementation for much of our design, there are cases where we may need to directly instantiate specific predetermined implementations. One such case is use of components created using the core generators that we discussed in Section 10.1.2. The core generator creates both a simulation model and an implementation optimized for the target implementation fabric. When we synthesize the design, we need to instantiate the version of the generated core that includes the optimized implementation. The way in which we do this varies from tool to tool, so as with many aspects of usage of specific tools, we would need to consult the documentation.

As we mentioned in Chapter 1, synthesis of a design is followed by a further verification step, in this case, to verify that the implementation produced by the synthesis tool meets timing constraints. We will return to the topic of timing analysis in Section 10.2. We also simulate the implemented design to ensure that it meets functional requirements. This might appear to be unnecessary, since we would expect the synthesis tool to faithfully implement the design described by the RTL model and verified using RTL simulation. However, there are two good reasons for performing gate-level simulation. First, the technology library includes estimates of timing parameters of the components used in the gate-level design. The gate-level simulation allows us to verify that the design still meets functional requirements, taking into account these timing parameters. It may expose subtle timing errors that were not evident in the RTL code. Second, there are ways in which we can write RTL model code that produce different behavior in the RTL simulation and the synthesized hardware. For example, if we omit an asynchronous control signal from the sensitivity list in a VHDL process, the simulated behavior of the process will differ from the behavior of the register produced by a synthesis tool. A synthesis tool should issue a warning in such cases, but in a complicated design with many messages issued by tools, such warnings may be overlooked. Simulating the synthesized design and making sure it behaves in the same way as the RTL design is a good check that we have used the tools correctly.

10.1.5   PHYSICAL DESIGN

The final stage in the design flow is physical design, in which we refine the gate-level design into an arrangement of circuit elements in an ASIC, or build the programming file that configures each element in an FPGA. While many of the same steps are used for the two kinds of implementation fabric, there are differences in the techniques used in the EDA tools for physical design of ASICs and FPGAs.

Physical design for ASICs, in its basic form, consists of *floorplanning*, *placement*, and *routing*. The first step, floorplanning, involves deciding where each of the blocks in the partitioned design is to be located on the chip. There are a number of factors that influence the floor plan. Blocks that have a large number of connections between them should be placed near each other, since that reduces wire length and wiring congestion. Similarly, blocks that are connected to external pins should be placed near the edge of the chip. The position of those blocks also determines the allocation and positioning of pins for external connections. The blocks should be arranged to make the chip as close to square as possible, since that influences the size of the package that can be used. Square chips are easier to package than rectangular chips. Floorplanning also involves the arrangement of power supply and ground pins and internal connections, and, importantly, the connection and distribution of clock signals across the chip. Finally, floorplanning also involves provision of channels for laying out interconnections between blocks. Devising a good floorplan for an ASIC can be quite challenging. EDA tools can assist by providing graphical tools to help us visualize floorplans and rearrange blocks, ensuring all the time that a floorplan is feasible, and by analyzing alternative floorplans to determine figures of merit.

Having determined a floorplan for an ASIC, we then proceed to placement and routing. This step involves positioning each cell in a synthesized design (placement) and finding a path for each connection (routing). The main goals are to position all cells and route all connections (not always achievable!), while minimizing area and delay of critical signals. Many of the signal integrity issues that we discussed in Chapter 6 also come into play in this stage. The result of placement and routing is a suite of files to send to the chip foundry for fabrication. We can also generate detailed timing information, based on the actual positions of components and wires, and use this in a more accurate simulation model of the gate-level design. This detailed timing simulation is a final check that our design meets its timing constraints. Given the overwhelming amount of detail involved, placement and routing are largely automated by EDA tools. Smith, in his book on ASICs, gives a good overview of the techniques used by tools (see Section 10.7, Further Reading).

In contrast, physical design for FPGAs involves deciding how to implement the synthesized design using the programmable resources of a prefabricated chip. The FPGA chip designers would have used ASIC design techniques for physical design of the FPGA chip, but that is a separate process, completed before we start our physical design targeted at the FPGA.

Physical design for FPGAs also starts with floorplanning, but in this case, the problem is much more constrained. A good arrangement of blocks in the FPGA fabric reduces the number of long-distance interconnects, giving a faster and more readily routed implementation. It also simplifies connections to I/O blocks and their associated pins. For many smaller FPGA-based applications, the floorplan generated automatically by the vendor's EDA tools is sufficient. However, for larger designs, if we have difficulty fitting a design into a given FPGA, we either attempt to improve the floorplan, or use a larger FPGA. Floorplanning tools for FPGAs provide many features similar to those of tools for ASICs, since many of the considerations are similar.

Before an FPGA design proceeds to placement and routing, we need to perform an intermediate step, *mapping*. This involves identifying the FPGA-specific resources to be used for each of the library components instantiated in the synthesized design. For example, instances of gates and multiplexers would be mapped to look-up tables, and instances of flip-flop components to the specific flip-flops provided in logic blocks.

Also, components representing specific FPGA resources, such as carry chain circuits and RAM blocks, would be mapped at this stage. The result is an implementation of the design using logic blocks, I/O blocks and FPGA-specific resources, as opposed to the library cells used by the synthesis tool.

Placement and routing of an FPGA seeks to achieve much the same goals as for ASICs, namely, identifying specific blocks and routing wires in the FPGA to use for the mapped blocks, while minimizing area and delay of critical signals. As with floorplanning, this step is best left to automatic tools provided by the FPGA vendor. If we need to improve placement or routing, we do so by specifying constraints on placement and timing, rather than by trying to intervene directly. The final result of the placement and routing step for an FPGA is a *bit file* specifying how the FPGA is to be configured. We can also generate detailed timing information for the design based on the internal timing parameters of the logic blocks and interconnect in the FPGA, and run final timing simulations to verify that the implemented design meets timing constraints.

In a simplified design flow, we leave physical design until after we have completed functional design and synthesis. In the more realistic flows used in industry, aspects of physical design are interwoven with earlier stages. For example, aspects of physical design are being increasingly used in synthesis tools, allowing a tool to choose among alternative implementations based on how the placement and routing affect area and timing of the circuits. Ultimately, if we understand how physical design issues affect the quality of the final implementation, we are in a better position to make good design trade-offs early in the design flow.

KNOWLEDGE TEST QUIZ

1. What is meant by the term architecture exploration?

2. What is the distinction between logical partitions and physical partitions of a system?

3. Identify the information described in a high-level specification of a system.

4. What is a behavioral model of a component? What is its purpose?

5. What are the benefits of reusing an IP block to implement a component?

6. Identify three kinds of function that can be implemented using a core generator.

7. If several designers are collaborating on development of model code, what tool can they use to coordinate their changes?

8. What aspects of the design flow does a verification plan cover?

9. Describe the difference between code coverage and functional coverage. Which is more important for ensuring correctness of a design?

10. Briefly outline how constrained random testing works.

11. Identify some advantages and disadvantages of formal verification over simulation-based testing.

12. What is a hardware abstraction layer for embedded software?

13. What is an instruction set simulator?

14. Why do RTL synthesis tools only accept a subset of a hardware description language's features?

15. What is a technology library used by a synthesis tool?

16. Why should we perform gate-level simulation of the circuit produced by a synthesis tool?

17. Briefly describe the purpose of floorplanning, placement, and routing.

10.2 DESIGN OPTIMIZATION

In the previous section, we described a design flow assuming that the design meets constraints at each stage. In most design projects, this ideal situation does not hold. Instead, we need to perform some optimization of the design, possibly making trade-offs of one property against another. Moreover, if we discover during some stage of the design flow that there is no feasible optimization, we need to revisit earlier stages to revise design choices we had previously made. Thus, realistic design flows are not linear, starting with design concept and leading directly to final implementation. Instead, they are cyclic, with the design evolving as more "back-end" implementation detail informs design choices made in "front-end" stages.

In this section, we will consider three main properties of a design that are usually constrained and that we often seek to optimize: area, timing, and power. At each stage of the design flow, we can make decisions that affect these properties. Many of the decisions also affect other aspects of the system, so we must make tradeoffs among the properties. Our decisions early in the flow, starting with architecture exploration and partitioning, generally have the greatest impact. The range of choices we might consider can quite easily involve an order of magnitude difference in a property. For example, if we compare an architecture that makes heavy use of parallelism with an alternative sequential architecture, we would expect the parallel version to have significantly higher performance, but at the cost of significantly greater area. If our sole concern was minimizing area, we would choose the sequential architecture; however, that might not meet performance constraints. Once we move to subsequent stages of the design flow, we are generally less able to affect the properties of the design to such a degree. If fine-tuning is insufficient, then we need to revisit earlier stages to make more substantial changes.

10.2.1   AREA OPTIMIZATION

As we have previously mentioned, the area of a circuit is a significant determinant of cost. For circuits implemented as chips, the cost of fabricating circuits on a wafer must be apportioned among the chips on that wafer. Larger chips thus take on a larger share of the wafer fabrication cost. Further, since chips are rectangular and wafers are circular, larger chips leave more wasted area near the edges of the wafer, so the proportion of wafer cost borne by each chip is not simply the ratio of chip area to wafer area. Instead, the relationship is nonlinear. A further nonlinearity arises from the fact that larger area increases the likelihood of a defect occurring on any given chip and causing the chip not to function. Since nonfunctional chips must be discarded after the wafer is fabricated, the remaining functional chips must bear the cost of fabricating and testing the nonfunctional chips. Putting all of the relevant effects together leads to final chip cost being approximately proportional to the square of the chip area. Next, the chip must be packaged, as we described in Section 6.3. A larger chip requires a larger and more costly package than a smaller chip. Also, since the chip is presumably larger because it has more transistors than a smaller chip, it consumes more power, and so the resulting heat must be dissipated. These effects also lead to package costs that are nonlinearly related to area. For circuits that are implemented using multiple packaged chips on a PCB, the cost of the system is likewise sensitive to area for many similar reasons. PCBs are also subject to defects during manufacture and assembly, and must be packaged in enclosures or cases. For both chip and PCB designs, the manufacturing, packaging and testing costs are largely beyond our control as designers. Thus, we can only affect cost indirectly through managing the area of our design.

In Section 10.1.5, we described the floorplanning step of physical design. One of the goals of floorplanning is to find an arrangement of blocks that minimizes area. We can also do some floorplanning, at least at a preliminary level, as part of the partitioning step of architecture exploration. Consideration of how the blocks of a partitioned design can be arranged on a chip may exclude some candidate architectures as infeasible, and may favor others that have less area. At the partitioning step, we can also estimate the number of pins that will be required for the chip, since that will influence the floorplan. In particular, if the pin count is large, the area required for the pad ring may constrain the overall area of the chip, leading to consideration of alternative architectures with reduced pin counts. By making these decisions early in the project, we can avoid wasting time on a design that we subsequently discover cannot be made to fit well on a chip. Of course, in order to do an early floorplan, we need to be able to estimate the area required for each block in the partitioned design. We can use those estimates as area budgets for subsequent design steps. When we get to physical design, we can finally validate the estimates. If they are significantly inaccurate, we would need to revisit the floorplan.

In the functional design stage of the design flow, we can influence circuit area through our choice of components, whether explicitly instantiated or implied by RTL model code. For example, as we discussed in Chapter 3, different forms of adders and multipliers have differing circuit complexity and hence circuit area, traded off against propagation delay or cycle count for performing an operation. If we directly instantiate such circuits, generated by a core generator, for example, we can influence overall circuit area. Also, choosing minimal bit-widths for data, as we did in the Sobel edge detector design in Chapter 9, helps to keep circuit area to a minimum, since components that process the data can then be of the minimal size, and the minimal amount of wiring is required between the components.

In the synthesis stage, we can influence the circuit area by specifying constraints to the synthesis tool. At a broad level, we can direct the tool to use a synthesis strategy that favors minimizing area instead of delay, or to use additional effort to optimize the design instead of reducing turnaround. In the case of hierarchically structured designs, we can direct the tool to try to optimize across block boundaries, possibly combining components from different blocks in order to reduce area. In cases where a tool does not automatically infer use of special resources within an implementation fabric, such as RAMs or ROMs, we might provide hints that specific parts of a design be implemented using specific resources. Synthesis tools vary in the details of how such directives and hints are specified. Most allow us to embed specifications in the RTL model code in the form of attributes, and to write separate constraint files containing specifications. The latter is generally preferred, since embedded attributes make the model code less reusable in other designs.

Finally, in the physical design stage, we can influence circuit area through intervention in the floorplanning, placement and routing of the circuit. At this level, however, we are just fine tuning. We cannot readily change the number or kind of components used or the amount or connectivity of the wiring between them. That is why decisions made earlier in the flow have more significant impact.

10.2.2   TIMING OPTIMIZATION

The aim of timing optimization is to ensure that a design meets performance constraints. Performance and timing are essentially the inverses of each other. We usually think of performance in terms of the number of operations completed per unit time. The inverse of this is the time taken to complete an operation. Our goal is to maximize the number of operations per second, or, conversely, to minimize the time per operation. In the architecture exploration stage of the design flow, we have the greatest impact on performance through application of parallelism, limited by the data dependencies involved. Clearly, increasing parallelism is in conflict with minimizing area and power, since the extra resources required to realize the parallelism take up area and consume power. As we have mentioned, we need to make trade-offs, applying just enough parallelism to reach performance requirements, but no more.

As we move through the design flow, our emphasis tends away from performance as the figure of merit and shifts more toward timing. This makes sense, since at lower levels of abstraction, we focus on design of individual blocks that perform operations. We generally try to find a circuit for the block that performs the required operation in the least amount of time, consistent with our other constraints.

As part of performance analysis of candidate architectures, we need to make estimates of the clock frequency that can be achieved. Alternatively, the clock frequency may be specified *a priori* due to external requirements on the system. Either way, the resulting clock period is a constraint that passes through to subsequent stages of the design flow. We have seen, during our discussion of sequential timing in Section 4.4, that the clock period constrains the propagation delay on combinational paths through the register-transfer-level circuit. That includes paths from block inputs through combination logic to register inputs, and paths from register outputs through combinational logic to block outputs. If blocks are designed by separate designers, we must ensure that the combined path from a register output in one block to a register input in another block meets the timing constraint. One way to do this is to allocate a *timing budget* each block, specifying maximum clock-to-output delays and input-to-clock setup times for each block. Any deviation from the budget must be specifically agreed between designers, documented, and carefully verified. A common instance of this approach is to require each block to have registered outputs, essentially limiting the clock-to-output budget to the register output delay and the input-to-clock budget to most of a clock cycle. In a large high-speed design, where inter-block wiring delay may be significant, it may also be appropriate to require registered inputs to blocks.

In the functional design stage of the design flow, we can influence timing through our choice of components. This is a similar argument to that for influencing area in the functional design stage, except that the two objectives may be in conflict with each other. More frequently, we use directives and hints to a synthesis tool to optimize timing of the detailed design, and then analyze the resulting synthesized circuit to verify that timing constraints are met. If they are not, we might revise the directives and hints and resynthesize. If we are unable to meet constraints through this iterative process, we need to revisit earlier stages of the design flow and make larger changes to the design at higher levels of abstraction.

Analyzing the synthesized design is typically done with a *static timing analysis* tool. The tool uses timing estimates for each of the components in the technology library, together with simple wire-load models. Since the design has not been placed and routed at this stage, the delays due to the lengths of wires can only be estimated. Moreover, the actual propagation delay of each library component and the load on each wire may vary as a result of mapping the design to technology-specific components. However, the estimates used are sufficient to guide optimization at this stage. The static timing analysis tool aggregates the delay through combinational circuits and wiring between successive registers, based on the timing model we described in Section 4.4. It thus identifies the critical path in the design and determines whether the clock period constraint is met.

In the physical design stage, we can fine tune timing by choice of placement of components and wires. However, since this process is very computationally intensive, it is unlikely that we could do a better job manually than an EDA tool can do automatically. What we can do is control how much computational effort is applied, at the expense of additional run-time to perform the placement and routing. Once the physical design has been performed, it is possible to extract accurate delay values for components and wiring. We can then repeat the static timing analysis using these values to verify whether timing constraints have been met. If they have not, we need to revisit earlier stages of the design flow to improve the timing of the circuit.

EXAMPLE 10.1   Synthesize and implement the Sobel accelerator design described in Section 9.2, targeting a Xilinx XC3S200-5 Spartan-3 FPGA with a clock frequency of 100MHz (that is, a clock period of 10ns).

SOLUTION   We will outline the process using the Xilinx ISE tool suite. Detailed information on using the tools is found in the documentation provided by the vendor. We first create a project, specifying the target device and include the VHDL entity and architecture source files. We then create a constraint file, using the constraints editor, containing the clock constraint, as follows:

```
NET "clk_i" TNM_NET = "clk_i";
TIMESPEC "TS_clk_i" = PERIOD "clk_i" 10 ns HIGH 50 %;
```

Next, we invoke the synthesis tool using the default options. The synthesis report produced by the tool includes a post-synthesis timing report, shown in Figure 10.6. The tool has only achieved a clock period of 14.174ns, so our timing constraint is not met. The report also indicates the path with the longest delay, starting at a flip-flop storing a bit of the array signal named O in the VHDL source and ending at a flip-flop for a bit of the Dx signal. The path passed through several LUTs and carry-chain components that implement the

FIGURE 10.6 Initial post-synthesis timing report for the Sobel accelerator.

```
Timing constraint: Default period analysis for Clock 'clk_i'
   Clock period: 14.174ns (frequency: 70.552MHz)
   Total number of paths / destination ports: 169373 / 623
-------------------------------------------------------------------------------
Delay:              14.174ns (Levels of Logic = 20)
   Source:          O<-1>_1_0 (FF)
   Destination:     Dx_10 (FF)
   Source Clock:    clk rising
   Destination Clock: clk rising

   Data Path: O<-1>_1_0 to Dx_10
                         Gate    Net
      Cell:in->out fanout Delay  Delay  Logical Name (Net Name)
      -------------------------------  -------------------------
       FDE:C->Q        3  0.626  1.066  O<-1>_1_0 (O<-1>_1_0)
       LUT2:I0->O      1  0.479  0.000  Msub__addsub0000_lut<0> (N68)
       MUXCY:S->O      1  0.435  0.000  Msub__addsub0000_cy<0> (Msub__addsub0000_cy<0>)
       MUXCY:CI->O     1  0.056  0.000  Msub__addsub0000_cy<1> (Msub__addsub0000_cy<1>)
       MUXCY:CI->O     1  0.056  0.000  Msub__addsub0000_cy<2> (Msub__addsub0000_cy<2>)
       MUXCY:CI->O     1  0.056  0.000  Msub__addsub0000_cy<3> (Msub__addsub0000_cy<3>)
       MUXCY:CI->O     1  0.056  0.000  Msub__addsub0000_cy<4> (Msub__addsub0000_cy<4>)
       XORCY:CI->O     1  0.786  0.851  Msub__addsub0000_xor<5> (_addsub0000<5>)
       LUT2:I1->O      1  0.479  0.000  Msub__addsub0001_lut<5> (N96)
       MUXCY:S->O      1  0.435  0.000  Msub__addsub0001_cy<5> (Msub__addsub0001_cy<5>)
       XORCY:CI->O     1  0.786  0.851  Msub__addsub0001_xor<6> (_addsub0001<6>)
       LUT2:I1->O      1  0.479  0.000  Madd__addsub0002_lut<6> (N117)
       MUXCY:S->O      1  0.435  0.000  Madd__addsub0002_cy<6> (Madd__addsub0002_cy<6>)
       XORCY:CI->O     1  0.786  0.851  Madd__addsub0002_xor<7> (_addsub0002<7>)
       LUT2:I1->O      1  0.479  0.000  Msub__addsub0003_lut<7> (N148)
       MUXCY:S->O      1  0.435  0.000  Msub__addsub0003_cy<7> (Msub__addsub0003_cy<7>)
       MUXCY:CI->O     1  0.056  0.000  Msub__addsub0003_cy<8> (Msub__addsub0003_cy<8>)
       XORCY:CI->O     1  0.786  0.976  Msub__addsub0003_xor<9> (_addsub0003<9>)
       LUT1:I0->O      1  0.479  0.000  _addsub0003<9>_rt (_addsub0003<9>_rt)
       MUXCY:S->O      0  0.435  0.000  Madd__add0001_cy<9> (Madd__add0001_cy<9>)
       XORCY:CI->O     1  0.786  0.000  Madd__add0001_xor<10> (_add0001<10>)
       FDE:D              0.176         Dx_10

      -------------------------------------------------
      Total               14.174ns (9.580ns logic, 4.594ns route)
                              (67.6% logic, 32.4% route)
```

adders and subtracters for original-image pixels to produce the Dx pixel. Given the number of additions and subtractions involved, it should not be surprising that this is the critical path.

We can proceed further along the design flow to placement and routing to see if the back-end tools can improve on the estimated timing. Using the default settings for the place-and-route (PAR) tool does indeed give some improvement, but not enough. The post-PAR static timing analysis report indicates a clock period of 12.865ns. The report also includes suggestions for achieving further improvement, including changing the synthesis and PAR settings to use greater effort and to apply multiple PAR passes to explore more alternative physical layouts. Even with these changes, the tools are only able to reduce the clock period to 12.052ns. So we must revisit an earlier stage of the design flow to achieve greater improvement.

Since the critical path is the section of the circuit from the O registers to the Dx register, we can try to reduce the delay there. The path to the Dy register is similar, so any change we make to the Dx path should be mirrored in the Dy path to avoid it becoming critical. Both paths consist of addition and subtraction of six operands (see Figure 9.6 on page 405). In the VHDL model shown in Example 9.7, we simply expressed each of these computations as a chain of additions and subtractions. If we review synthesis style guidelines for writing VHDL, they recommend grouping chained arithmetic operations. For example, given the expression

```
a + b + c + d
```

a synthesis tool might generate a chain of adders, as shown at the left of Figure 10.7. The propagation delay through the circuit is three adder delays. On the other hand, if the expression is parenthesized as

```
( a + b ) + ( c + d )
```

the tool might generate an adder tree, as shown at the right of Figure 10.7. This arrangement has a propagation delay of only two adders. If more operands are involved, the difference between the chain structure and the tree structure is even more significant.

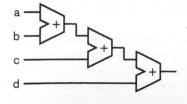 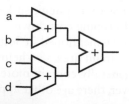

FIGURE 10.7 Adders connected in a chain (left) and a tree structure (right).

We can test whether the synthesis tool treats arithmetic expressions in this way by rewriting the statements in the pipeline process of the VHDL code as follows:

```
Dx <= ( - signed(resize(O(-1,  -1), 11))     -- - 1 * O(-1, -1)
           + signed(resize(O(-1,  +1), 11))     -- + 1 * O(-1, +1)
           -(signed(resize(O( 0,  -1), 11))     -- - 2 * O( 0, -1)
             sll 1) )
       + ((signed(resize(O( 0,  +1), 11))       -- + 2 * O(0,  +1)
             sll 1)
           - signed(resize(O(+1,  -1), 11))     -- - 1 * O(+1, -1)
           + signed(resize(O(+1,  +1), 11)) );-- + 1 * O(+1, +1)
Dy <= (    signed(resize(O(-1,  -1), 11))     -- + 1 * O(-1, -1)
           +(signed(resize(O(-1,   0), 11))     -- + 2 * O(-1,  0)
             sll 1)
           + signed(resize(O(-1,  +1), 11)) ) -- + 1 * O(-1, +1)
       - ( signed(resize(O(+1,  -1), 11))     -- - 1 * O(+1, -1)
           +(signed(resize(O(+1,   0), 11))     -- - 2 * O(+1,  0)
             sll 1)
           + signed(resize(O(+1,  +1), 11)) );-- - 1 * O(+1, +1)
```

With this change in place, and using maximum optimization options for the synthesis and PAR tools, we reduce the synthesis estimate of the clock period to 9.515ns and the post-PAR estimate to 9.864ns. This just satisfies our timing constraint, with a 1.4% margin. In practice, we would prefer to have a larger margin, and so might try further revision of the model code to reduce the path delay. If regrouping the arithmetic expression shown above is insufficient, we might be able to move part of the computation to the paths after the Dx and Dy registers, where there is more slack.

10.2.3   POWER OPTIMIZATION

As digital systems have become more complex, power consumption has become a more significant constraint in their design. This is particularly the case for mobile battery-operated devices, such as cell phones, PDAs, and portable media players. The amount of power consumed by the circuit directly affects how long the device functions on a single battery charge, or, alternatively, how large a battery is required. Even in fixed mains-powered systems, power consumption is important. Electrical power consumed by a circuit is turned into heat, which must be dissipated through the chip and system packaging. Dealing with additional heat dissipation adds cost to a system, so keeping power consumption to a minimum is part of keeping cost down.

As we mentioned in Section 10.2.1, many of the approaches to minimizing circuit area also help reduce power consumption, since larger circuits generally contain more transistors, each of which consumes power. However, there are other approaches we can consider that focus on power consumption. One such approach is to identify blocks of a system that remain idle for substantial periods during the system's operation, and to remove power from those blocks during idle periods. Some laptop computers take this approach, for example, by powering down a network card when the computer is not connected to a network cable. In some instrumentation applications, an embedded microcontroller only need be active for small periods of time to sample data inputs and determine control settings. During other times, the microcontroller can power down. Several commercially available microcontroller chips have a standby mode for this purpose, and can be "woken up" in response to activation of an input signal. Recent processor cores also include power management features, allowing the processor to operate at various power levels, including powering down completely, and to control power levels of other system components.

While powering down blocks of a system can reduce average power consumption significantly, it is not simple to implement. In particular, if other parts of the system to which the block is connected must continue operating, then the interface signals must be disabled to avoid spurious activation of the parts that remain active. Further, when power is restored to a block, it takes a significant number of clock cycles before the block can resume operation. This delay may affect performance, so the technique is only appropriate where the delay can be tolerated.

In CMOS circuits, the predominant technology used for digital systems, most of the power consumption occurs when transistors change between their on and off states. Moreover, the greater the fanout load connected to a circuit, the greater the power required to switch the load between high and low logic levels. We discussed these effects in Chapter 1, where we introduced the term *dynamic power* consumption. In a clocked synchronous digital system, we have many flip-flops, all of which are governed by a global clock signal. Each flip-flop contains several transistors that switch state on clock edges, even if the stored data does not change. These transitions consume power without affecting the computation performed by the circuit. If the performance requirements of a system are not constant, that is, if there are periods where high performance is required and other periods where lower performance is acceptable, we can reduce dynamic power consumption by reducing the clock frequency. This requires that the source of the clock signal be adjustable. Often, we would implement power management through clock frequency control within the real-time operating system of an embedded computer. The clock generator in such a system would need to be adjustable under program control.

Another common way of reducing power in CMOS systems is *clock gating*. This involves turning off the clock to parts of a circuit whose stored values do not need to change. We have seen how to use a clock-enable signal to control activity of a single flip-flop or register. However,

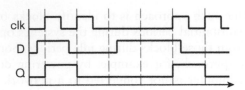

FIGURE 10.8 Timing diagram for a flip-flop with a gated clock.

such components are still affected by clock transitions, even when the clock-enable input is 0. With clock gating, the components see no clock transitions when the clock is turned off, as shown in Figure 10.8. Here, the clock is gated off for two cycles. During that interval, the component consumes no dynamic power.

Gating a clock is not as simple as inserting an AND gate in the clock signal. Given the delay in an AND gate, the resulting clock edges would be skewed from those of the ungated clock, making it difficult to meet timing constraints. Also, since the gating control signal is typically generated by a clocked control section, a naive approach can lead to glitches on the gated clock signal, as shown in Figure 10.9. The glitch may cause unreliable triggering of the components to which the gate clock is connected. Further, if the control signal has glitches, for example, due to differing delays in paths through combination logic that generates the signal, those glitches may be passed through to the gated clock. The solution to these problems is not to express the clock gating in the RTL model of the circuit. Rather, we should treat clock gating as a power optimization to be implemented by clock insertion tools during physical design. Several clock synthesis tools can perform such power optimization.

As with optimization of other parameters, we need to perform analysis of a circuit design to determine whether power constraints are met. in the final circuit, the actual power consumption will depend on the clock frequency and on the relative frequency of changes of signal values, compared to the clock frequency. A power analysis tool can estimate the circuit's power consumption based on estimates of those frequencies. A good way to make such estimates is to monitor the values of signals during simulation of a model of the circuit. The tool can then combine the acquired data with power consumption data from the technology library and load models for the interconnecting signals.

FIGURE 10.9 Glitch on a gated clock due to poor design.

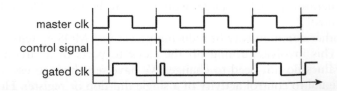

1. If we need to achieve a major improvement in system performance, should we focus effort in earlier or later stages of the design flow?

2. Why is cost of a chip nonlinearly related to circuit area?

3. How can we affect circuit area during the functional design stage of the design flow?

4. Identify a means of improving system performance that we might consider in the architecture exploration stage. What trade-offs arise from improving performance?

5. How does a timing budget help a design team to meet timing constraints?

6. What is the purpose of specifying timing constraints for synthesis?

7. How does a static timing analyzer verify timing for a synthesized design and for a placed and routed design?

8. If timing constraints are not met, what must we do?

9. Briefly describe two techniques for reducing power consumption.

10. Why should clock gating not be implemented in RTL model code? How is it better implemented?

10.3 DESIGN FOR TEST

The design flow that we have described starts with architecture exploration and proceeds through to delivery of design files for manufacture of ASICs or programming of FPGAs. Once the chips have been manufactured, they must be tested to ensure that they work correctly. There are several reasons why manufactured chips might not work, including problems arising during wafer manufacture and during packaging. Some problems cause whole batches of chips to fail, whereas others produce isolated failures. The aim of testing after manufacture is to identify the defective chips so that they can be discarded rather than supplied to customers. Ideally, the cause and location of faults can be recorded so that the manufacturing process can be adjusted to enhance yield.

In the case of ASICs, developing tests for the chip is part of the design process. In the case of FPGAs, test development is part of the process of developing the FPGA as an implementation fabric. The FPGA used as the implementation target of a final design is already tested as part of its manufacture. However, the FPGA must still be inserted into a PCB as part of a larger system, which itself must be tested as part of its manufacture. In many systems, faulty PCBs are not discarded. Instead, defective chips on the PCBs are replaced, and the repaired PCB retested and returned to service.

For a simple circuit, testing could involve applying test cases on the chip inputs and verifying that the chip produces the correct outputs. This is similar in concept to verification during the design stage. However, rather than verifying that the design meets functional requirements, the intention is to verify that the manufactured chip performs as designed. Nonetheless, the same question of feasibility arises. As the number of possible input values and input sequences increases, testing that the chip operates correctly in all cases becomes infeasible. The time available for testing is much less than for design verification, since thousands or millions of chips must be tested individually. Furthermore, testing requires use of test equipment to physically apply input values and measure output values. Such equipment is a costly resource, so its use must be minimized.

We can reduce the time and cost involved in testing a system by including additional circuitry to improve the system's testability. Such circuitry includes elements that make internal nodes observable, or that perform testing automatically as a special mode of system operation. We use the term *design for test* (DFT) to refer to design techniques that seek to improve testability. In the remainder of this section, we will describe the way in which we can develop test cases specifically aimed at locating faults. We will then look at two approaches to adding hardware to designs to improve testability.

10.3.1   FAULT MODELS AND FAULT SIMULATION

As part of the design flow, we need to develop a set of *test vectors*, or *test patterns*, that is, combinations of input values for the circuit that can be used to expose faults. The idea is that application of each test vector (or small sequence of test vectors) should cause the circuit to produce a given output. If the circuit produces a different output, the circuit is faulty. A good choice of vectors may also be able to locate the fault within the circuit.

In order to work out how to expose faults, we need to consider the kinds of faults that can occur in a circuit. We rely on *fault models* that are abstractions of the effects produced by faults. We use a *fault simulator* that simulates the operation of the circuit with a given fault injected at a given location. The simulator applies test vectors until an incorrect output results, indicating that the fault has been detected. If no incorrect output is produced for all of the test vectors, the fault remains undetected by that set of vectors. The simulator repeats the simulation for other faults and other locations in the circuit. Once all of the faults have been simulated, the *fault coverage* of the test vectors, that is, the proportion of faults detected, can be determined. Ideally, the fault coverage should be 100%, but for a large design, this may not be feasible. In choosing the test vectors, we can use an automatic test pattern generator (ATPG), an EDA tool that analyzes a circuit and seeks to create a minimal set of test vectors with as close to full coverage as possible.

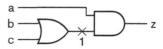

FIGURE 10.10   A circuit with a stuck-at-1 fault.

A simple and commonly used fault model is the *stuck-at* model, in which an input or output of a gate in a circuit can be stuck at 0 or stuck at 1, rather than being able to change between 0 and 1. Such a fault might be caused by a short circuit to the ground or power supply. This is illustrated in Figure 10.10, in which an input to the AND gate is stuck at 1. For some input combinations to the circuit (b = 1 or c = 1), the value at the stuck node would normally be the same as the stuck-at value; the circuit would produce the correct output, and we would not detect the fault. For other input combinations (b = 0 and c = 0), the value at the stuck node would be the opposite of the stuck-at value. Whether we could detect the fault would then depend on the remaining logic between the stuck node and the circuit's output. In this circuit, if a = 0, the output value is independent of the value at the stuck node, so the fault is masked. However, if a = 1, the value of the stuck node is propagated to the output, allowing us to detect the fault. In general, detecting the fault involves applying a combination of input values that *sensitizes* the path from the fault to the output and that drives the stuck node to the opposite of its stuck-at value. A node in a circuit is *observable* if a fault at the node can be made to result in an incorrect output value. The node is *controllable* if there are input combinations that cause the node to take on a given value. Observability and controllability of nodes in a circuit determine the testability of the circuit.

Other fault models consider the transistor-level circuits for gates, and involve transistors being stuck on or stuck off. This allows us to detect faults that are not adequately represented by the stuck-at fault model. For example, given the output driving circuit of a gate, shown in Figure 10.11, a fault might cause the upper transistor to be stuck on. When the gate should be driving a high logic level, its output is correct. However, when it should be driving a low logic level, both transistors are on. This creates a voltage divider, and the output logic level is an invalid level between the valid high and low levels. How this propagates through the rest of the circuit depends on the input thresholds of the gates connected to the driver, and may not be detectable simply by examining the output values. A testing approach used for such faults is to measure the steady-state current drawn from the power supply (I_{DDQ}) to detect the increase when both transistors are on. Further fault models include bridging faults, representing short-circuit connections between signal wires; delay faults, in which the propagation delay of a circuit is longer than normal; and faults in storage elements.

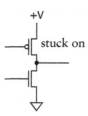

FIGURE 10.11   Output driver circuit of a gate.

10.3.2   SCAN DESIGN AND BOUNDARY SCAN

In this book, we have described the RTL view of digital systems, consisting of combinational circuits that transform and transfer data between registers. The fault models and fault detection techniques that we described in the preceding section work well for combinational circuits, but are difficult to adapt to detect faults in registers and other storage elements. The problems are compounded when the registers are buried deep within a datapath, since they are significantly more difficult to control and observe.

Scan design techniques address this problem by modifying the registers to allow them to be chained into a long shift register, called a *scan chain*, as shown in Figure 10.12. Test vectors can be shifted into the registers in the chain, under control of the *test mode* input, thus making them controllable. Stored values can also be shifted out of the registers, thus making them observable. Furthermore, the chain of registers allows us to control and observe the combinational blocks between registers. We can test each combinational block separately. This process consists of shifting test values into the register chain until the test vector for each block reaches the input registers for that block. We then run the system in its normal operational mode for one clock cycle, clocking the output of each block into the block's output registers. We also need to apply test vectors to the external inputs of the system and observe the external outputs of the system, in order to test any combinational input and output circuits. Finally, we shift the result values out through the register chain. The test equipment controlling the process compares the output values with the expected results to detect any faults. This sequence is repeated until all of the test vectors have been applied to all of the combinational blocks, or until a fault is detected.

This form of design for test has a number of advantages and disadvantages. Chief among its advantages is the increased controllability and observability provided. This makes achieving high fault coverage feasible, especially for large circuits. We can reduce the test generation problem to that of testing combinational circuits, which can largely be automated by

FIGURE 10.12   Connection of modified registers in a scan chain.

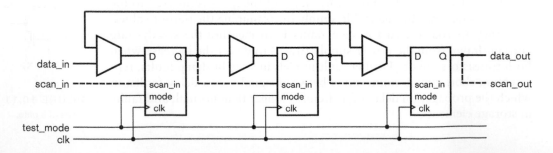

ATPG tools to achieve 100% fault coverage. Moreover, the modification of the registers to allow them to function as shift registers can also be automated. One approach is to design and synthesize the circuit normally, generating a gate-level circuit with flip-flops implementing the registers. Then as part of physical design, a tool can substitute modified flip-flops that have a shift mode, as shown in Figure 10.13. The circuit is placed normally. Finally, during the routing step, connections are made between adjacent flip-flops to form the shift-register chain. This minimizes the routing overhead and interference with other signal wires. The DFT tools can compensate for the resulting placement-dependent ordering of test vector input and output bits in the scan chain.

The main disadvantage of scan design is the overhead, both in circuit area and delay. The modified flip-flops have additional circuitry, including an input multiplexer to select between the normal input and the output of the previous flip-flop in the scan chain. The area overhead for scan design has been estimated at between 2% and 10%. The input multiplexer imposes additional delay in the combinational path leading to the flip-flop input. If the path is a critical timing path, performance of the whole system is affected. Another disadvantage of scan design, when compared to some other DFT techniques, is that the scan chain is very long. Shifting test vectors in and result vectors out takes a large fraction of test time, so the system cannot be tested at full operational speed. That overhead time can be reduced by dividing the scan chain into segments that can be shifted in parallel. However, each chain requires separate input and output pins, so we must compromise between test-time overhead and test-pin-count overhead.

One issue to consider in adding hardware to enhance testability is the possibility of faults within the test hardware. In the case of scan chains, faults could prevent the flip-flops from operating as shift registers correctly, and the wiring for the scan chains could be defective. Fortunately, we can test the scan chain itself quite readily. We simply need to insert a sequence of 0s and 1s into the chain and shift it through to the output of the chain. If we see the sequence unchanged at the expected time, we know the scan chain is defect free. We can then proceed to test the internal circuits of the system.

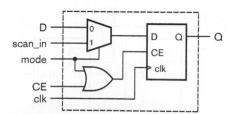

FIGURE 10.13   Modified flip-flop for use in a scan chain.

The concept of scan design can be extended for use in testing the connections between chips on a PCB, leading to a technique called *boundary scan*. The idea is to include scan-chain flip-flops on the external pins of each chip. To test the PCB, the test equipment shifts a test vector into the scan chain. When the chain is loaded, the vector is driven onto the external outputs of the chips. The scan-chain flip-flops then sample the external inputs, and the sampled values are shifted out to the test equipment. The test equipment can then verify that all of the connections between the chips, including the chip bonding wires, package pins and PCB traces, are intact. Various test vectors can be used to detect different kinds of faults, including broken connections, shorts to power or ground planes, and bridges between connections.

The success of boundary scan techniques led to the formation of the Joint Test Action Group (JTAG) in the 1980s for standardizing boundary scan components and protocols. The term JTAG has now become synonymous with boundary scan in its basic and extended forms, supporting automatic testing of individual chips and PCBs containing multiple chips. Standardization has been managed for some time by the IEEE as IEEE Standard 1149.1.

The JTAG standard specifies that each component have a *test access port* (TAP), consisting of the following connections:

▶ Test Clock (TCK): provides the clock signal for the test logic.

▶ Test Mode Select Input (TMS): controls test operation.

▶ Test Data Input (TDI): serial input for test data and instructions.

▶ Test Data Output (TDO): serial output for test data and instructions.

There is also an optional Test Reset Input (TRST), but it is not widely used in practice. Figure 10.14 shows a typical connection of automatic test equipment (ATE) to the TAPs of components on a PCB. Figure 10.15 shows the test logic within each component. The TAP controller governs operation of the test logic. There are a number of registers for test data and instructions, and a chain of boundary scan cells inserted between external pins and the component core. A typical boundary scan cell is

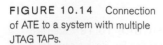

FIGURE 10.14   Connection of ATE to a system with multiple JTAG TAPs.

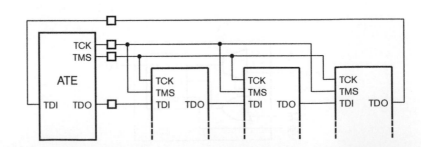

shown in Figure 10.16. Depending on the control inputs to the cell, data can flow straight through, input data can be captured, output data can be driven, and test data can be shifted through. Input and output pins of the component each require just one cell. Tristate output pins require two cells: one to control and observe the data, and the other to control and observe the output enable. Bidirectional pins require three cells, as they are a combination of a tristate output and an input.

The TAP Controller operates as a simple finite-state machine, changing between states depending on the value of the TMS input. Different states govern shifting of data into the Instruction Register or one of the data registers (including the scan chain). The JTAG standard defines a number of instructions formats for operations that select among data registers, control the mode of the scan chain, and so on. There are also instructions for component-specific extensions, including built-in self test modes that we will discuss in Section 10.3.3. The JTAG standard also defines the

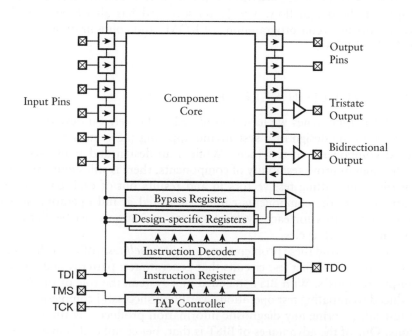

FIGURE 10.15 Architecture of a component with JTAG boundary scan.

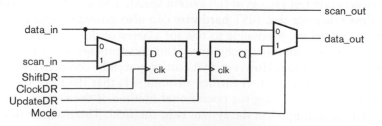

FIGURE 10.16 A JTAG boundary scan cell for an input or output pin.

Boundary Scan Description Language (BSDL), which is a subset of VHDL used to specify the pins, registers, and instructions implemented in the test logic of a component. We can use a BSDL description of a component, together with a set of test vectors, as input to ATE for testing the component and the board in which it is embedded.

While the boundary scan technique originated as a means of testing connections between components on a board, the JTAG boundary scan cells have been designed to allow testing of the component core also. The cells can be configured to isolate the component core's inputs from the package input pins. Test data can be shifted into the cells at the inputs and then driven onto the core's inputs. The core's outputs can be sampled into the cells at the output pins and then shifted out to the ATE. Thus, the JTAG architecture solves two problems: in-circuit testing of components in a system, and in-circuit testing of the connections between the components. This flexibility has led to the widespread use of the standard, with EDA tools available for insertion of test logic into designs during various stages of the design flow. The JTAG standard has also been extended to support in-circuit programming of ROMs and configuration of PLDs, including FPGAs.

10.3.3   BUILT-IN SELF TEST (BIST)

The DFT approaches we have considered so far rely on developing test vectors during design of a system and applying the vectors to manufactured components to test them. While scan-design and boundary scan techniques improve testability of components, there is still significant time overhead in shifting test vectors in and results out of each component. Furthermore, the components cannot be tested at full operating speed, since test vectors for each cycle of system operation must be shifted in over many clock cycles.

We can avoid these problems using *built-in self test* (BIST) techniques, which involve adding test circuits that generate test patterns and analyze output responses. With BIST included in a system, the role of the ATE is reduced to initiating test operations, verifying successful completion, or, if a test fails, storing any diagnostic information produced by the BIST circuits. One of the advantages of BIST is that, being embedded in a system, it can generate test vectors at full system speed. This significantly reduces the time taken for test. BIST hardware can also generate multi-cycle test sequences, making it possible to test for classes of defects that are difficult to expose with other techniques. The disadvantage, of course, is the larger area overhead. However, that cost may well be compensated for by the reduced testing cost. A further advantage is that the BIST hardware remains available during the operational lifetime of the system, and can be used for testing when the system is in the field. There are stories told of customers who were unaware of a fault in a system until a service engineer arrived to repair the system. A system with BIST and redundant components, capable of reporting faults to a service center over a network connection, makes such stories plausible.

There are two main aspects to consider when designing a BIST implementation: how to generate the test patterns, and how to analyze the output response to determine whether it is correct. The main problem is to devise circuits for these aspects that are not excessively large and that do not adversely affect normal system performance.

The most common means of generating test patterns is a *pseudo-random test pattern generator*. Unlike true random sequences, pseudo-random sequences can be repeated from a given starting point, called the *seed*. Nonetheless, they have similar statistical properties to true random sequences. Pseudo-random sequences can be readily generated with a simple hardware structure called a *linear-feedback shift register* (LFSR). Figure 10.17 shows an LFSR for generating sequences of 4-bit values. The sequence is initiated by presetting the flip-flops, generating the test value 1111 as the seed. On successive clock cycles, the LFSR generates values in the sequence shown in Figure 10.17. The sequence contains all possible 4-bit values except 0000. In most applications, it is desirable to include that value also. Fortunately, we can modify the LFSR to form a *complete feedback shift register* (CFSR), as shown in Figure 10.18, which generates all possible values. Similar circuits can be designed to generate pseudo-random test vectors of other lengths. Placement of the XOR gates within the LFSR is determined by the *characteristic polynomial* of the LFSR, referring to the mathematical theory underlying LFSR operation. Discussion of the theory is beyond the scope of this book.

Analyzing the output response of a circuit to the test patterns presents more of a problem. In most cases, it would be infeasible to store the correct output response for comparison with the circuit's output response,

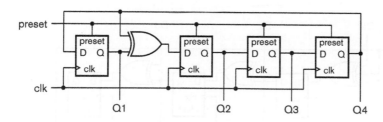

FIGURE 10.17 A 4-bit LFSR for generating pseudo-random test vectors.

$1111 \rightarrow 1011 \rightarrow 1001 \rightarrow 1000 \rightarrow 0100 \rightarrow 0010 \rightarrow 0001 \rightarrow 1100$
$0111 \leftarrow 1110 \leftarrow 0101 \leftarrow 1010 \leftarrow 1101 \leftarrow 0011 \leftarrow 0110$

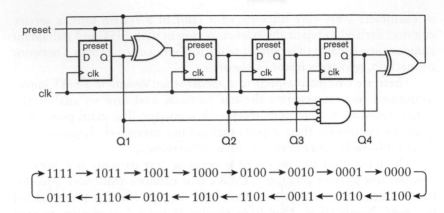

FIGURE 10.18 A 4-bit CFSR for generating pseudo-random test vectors.

$$1111 \rightarrow 1011 \rightarrow 1001 \rightarrow 1000 \rightarrow 0100 \rightarrow 0010 \rightarrow 0001 \rightarrow 0000$$
$$0111 \leftarrow 1110 \leftarrow 0101 \leftarrow 1010 \leftarrow 1101 \leftarrow 0011 \leftarrow 0110 \leftarrow 1100$$

since the storage required could well be larger than the circuit under test. Instead, we need to devise a way of compacting the expected output response and the circuit's output response. Doing so requires less storage, and comparison hardware, though at the cost of circuitry to compact the circuit's outputs. There are several schemes for output response compaction, but the most commonly used is *signature analysis*. This technique is closely related to use of LFSRs for test pattern generation, and the same mathematical theory underlies operation of both. A *signature register* forms a summary, called a *signature*, of a sequence of output responses. Two sequences that differ slightly are likely to have different signatures. Figure 10.19 shows an example of a *multiple-input signature register* (MISR), with four inputs from a circuit under test and a 4-bit signature.

Use of BIST using an LFSR for test-pattern generation and a signature register for response analysis requires us to perform a logic simulation of the circuit without faults. Since the sequence generated by the LFSR is determined by the seed, we can perform the simulation with that sequence

FIGURE 10.19 A 4-bit MISR with four inputs from a circuit under test.

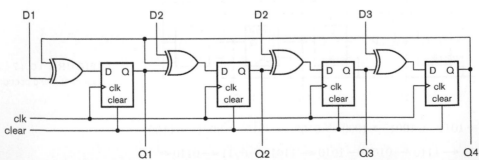

of input values. We use the output values from the simulation to compute the expected signature, and save the signature for use during test. When BIST of a circuit is initiated, either by ATE during manufacturing test or by an in-system test controller during system operation, the LFSR generates test patterns and the MISR computes the signature of the actual circuit's outputs. The ATE or test controller then shifts the computed signature out of the MISR and compares it with the expected signature. If they are the same, no fault is detected (though there is a chance that an actual fault remains undetected). If they differ, there is definitely a fault.

In this section, we have introduced some of the basic concepts of BIST. In practice, BIST and design for test techniques have been developed to much greater sophistication. There is a whole segment of the electronics design industry devoted to test, with vendors offering a range of software and test equipment. Design for test techniques sit at the interface between the design community and test community. In any significant electronics design project, it is essential that design and test engineers communicate throughout the design process to ensure that the system will be testable during manufacture, and, if required, during operation once deployed.

KNOWLEDGE
TEST QUIZ

1. What is meant by the term design for test?

2. What does a fault simulator do?

3. Describe the stuck-at fault model, and identify circuit defects that are represented by the model.

4. What is meant by a circuit node being controllable and observable?

5. How does I_{DDQ} testing detect transistor stuck-on faults?

6. What changes must be made to a circuit to create a scan chain?

7. Identify an advantage and a disadvantage of scan design over testing using external pins only.

8. How does boundary scan enhance testability of PCB-based systems?

9. Identify the signals required in a JTAG test access port.

10. What is the purpose of a JTAG TAP controller in a component?

11. Why does a bidirectional tristate pin require three boundary scan cells?

12. What circuits are added to a system for built-in self test?

13. Why is BIST useful after manufacturing test of a system?

14. What purposes do LFSRs and MISRs have in signature-based BIST?

10.4 NONTECHNICAL ISSUES

We have now covered all of the technology-related aspects of digital system design that we sought to cover in this book. However, they are not the only aspects we must think about. We finish by discussing a few nontechnical issues that bear upon design projects. In some cases, the best technical choices may not be the best choices when all things are considered.

Electronics products, like most products, go through *life cycles*. Product design is just one stage in the life cycle. It is preceded by market research and financial modeling. After design, manufacturing facilities and supply channels, and sales and distribution channels need to be established. Depending on the product, there may be a need for maintenance and repair, or for customer service. During the product's lifetime, it may be redesigned to meet changing needs, or may be reused in other products. Finally, the product becomes obsolete and is retired from production and support.

There are various financial models that can be applied to estimate revenue from a product over its life cycle. Generally, revenue from a product typically peaks early in the product's life cycle, and tails off until obsolescence. The non-recurring engineering (NRE) costs of developing the product, along with other up-front costs, must be met from the revenue stream. If the product is aimed at a competitive market, entering the market early has a critical impact on revenue. Late entry allows competitors to gain market share, reducing the revenue available for the late product, and possibly making it unprofitable. Hence, *time-to-market* is an important nontechnical measure for a design project. For many consumer products, such as cell phones and media players, product life cycles are very short, so there is only a short window of opportunity to gain sufficient revenue for profitability. Time-to-market pressures for design of such products are very intense.

In other industry segments, products have very long life cycles. Examples are military systems and telecommunications infrastructure. For such products, attributes such as reliability and maintainability are important. For example, considerations such as longevity of a vendor company may override technical considerations in choice of components for a system. Such long-lived products must typically be supported throughout their lifetime. Hence, the design phase will involve more than just the technical design of the circuit. It will also involve development of design documentation, and liaison with support service providers to develop support plans, procedures and documents.

Another important factor to consider in design of digital systems is that the implementation technology continues to evolve rapidly. Each generation of chip technology allows more transistors to be packed into a given chip area, more bits of storage per memory chip, and higher clock frequencies. If the design process for a complex system spans an 18-month period, a new technology generation is likely to be available when the product reaches the manufacturing stage. Designing using the previous generation may well lead to a product with lower performance or capacity than competitors' products. When we start a design project, we must be aware of technology trends and make projections to determine the appropriate technology for the future manufacture of our product.

As we have mentioned throughout this book, design of a digital system is a complex undertaking. For smaller systems, a small team of engineers can feasibly deal with product definition and specification, detailed design, verification, and manufacture. Even so, a systematic methodology reduces the risk of the product development project going off the rails. For larger systems, a larger development team is typically needed. Different team members bring expertise in different areas to the project. Indeed, larger teams are often structured with subteams being responsible for different aspects of the design methodology, such as architectural definition, detailed design, verification, test development, and liaison with the manufacturing facility. It is important for individual team members to understand the structure of the overall project and the context in which they are working. In particular, maintaining good communication and information flow within the project is critical. Good project management is essential to a successful outcome.

1. Identify some of the main stages in a product's life cycle.

2. Why is time-to-market critical for some products?

3. For products with long lifetimes, what additional activities are often required, beyond technical design of the product?

4. If a system is designed for a competitive market using technology that is current at the start of the design process, what risk does the product face?

KNOWLEDGE
TEST QUIZ

10.5 IN CONCLUSION

We have now completed our foundational study of digital system design. We started with the basic elements of digital logic, gates and flip-flops, and showed how they can be used in circuits that meet given functional requirements. Given the complexity of requirements for most modern systems, we appealed to the principle of abstraction as a means of managing complexity. In particular, we use hierarchical composition to build blocks from the primitive elements, and systems from those blocks. By this means, we were able to reach the level of complete embedded systems, comprising processors, memories, I/O controllers, and accelerators, without becoming overwhelmed by the detailed interactions of the millions of transistors involved. Throughout our study, we also paid attention to the real-world effects that arise in digital circuits and the constraints that they imply. We showed how a disciplined design methodology helps us meet functional requirements while satisfying constraints.

The study of digital systems in this book serves as a foundation for further studies in several areas. On the hardware side, VLSI design involves study of techniques for design of circuit elements and systems on silicon integrated circuits, typically starting with CMOS digital circuits, but then extending to analog, radio frequency (RF) and mixed analog/digital systems. Also, micro-electromechanical systems (MEMS) and microfluidics are becoming increasingly important, particularly for the interface between digital systems and the real world. At the interface between hardware and software, digital design leads to studies in computer organization and computer architecture, since a computer is just a specialized kind of digital system. Embedded systems design also lies on this interface, since the trade-off between hardware and software implementation is an important aspect. On the software side, studies in operating systems and compiler design benefit from an understanding of how computer hardware functions. Finally, the electronic design automation tools themselves make an interesting topic for advanced study, and are key to successful digital system design methodologies.

The integrated circuit technology on which digital systems are based has been continually developing since its inception. Moore's Law characterizes the development, describing how the number of transistors available on an IC increases exponentially. This rule has held solidly since the 1960s until now, despite several predictions during that time that one road block or another would halt progress. The semiconductor industry still holds to Moore's Law, with the International Technology Roadmap for Semiconductors (see www.itrs.net) mapping out several future generations of IC technology over the next fifteen years. Feature sizes for leading-edge processes are projected to be reduced from 65nm in 2007, through 45nm, 32nm, and 22nm, down to 16nm in 2019. Making use of the enormous number of transistors that will be available (and that will be required on a chip to make fabrication economically viable) will be one of the main challenges facing designers. Many commentators predict that ASICs will become increasingly uneconomical for all but the largest-volume applications. Instead, customizable platforms targeted at various classes of applications will become more common. Most designs will involve customization of such platforms, either through programmable fabric embedded in the platform, or through customization of the final layers of metal wiring on the chip. Of course, the further into the future we look, the harder it is to predict, as alternative evolutions diverge and the probability of some disruptive new technology emerging increases. Whatever happens, digital systems design promises to remain an exciting endeavor.

10.6 CHAPTER SUMMARY

▶ A design methodology codifies the process of design, verification and preparation for manufacture of a product. It involves development of virtual prototypes to support design analysis and refinement.

▶ Architecture exploration is the process of modeling and evaluating candidate designs at a high level of abstraction. A system is partitioned for subsequent refinement. Logical partitioning identifies functional components, whereas physical partitioning identifies physical hardware components. Logical functions are mapped onto physical partitions.

▶ Functional design refines partitions to a level from which implementations can be synthesized. Components may be implemented through IP reuse or by core generators.

▶ Functional verification ensures that the refined design meets functional requirements, and can be performed using simulation and formal verification. Functional coverage is the proportion of functionality verified.

▶ Hardware/software co-verification uses instruction-set simulators and hardware emulation to test software before hardware models are available. Software and hardware can be tested together using cosimulation.

▶ An RTL synthesis tool refines HDL models to gate-level circuits composed of components from a technology library, subject to timing and area constraints.

▶ Physical design involves floorplanning to arrange the blocks of a circuit, and placement and routing of the gate-level cells.

▶ A design can be optimized at various stages in the design flow. The main parameters we seek to optimize are area, timing, and power consumption. This usually involves making trade-offs.

▶ Design for test enhances testability of a product, thus reducing test cost. Testing involves applying test patterns to a circuit's inputs and verifying that the expected outputs are produced.

▶ Fault models represent the effects of defects in a circuit, and are used by a fault simulator to determine fault coverage of a set of test vectors.

▶ Testability of a circuit can be enhanced by adding test hardware. Scan design involves modifying registers to form shift registers for

▶ shifting test vectors into a circuit and output results out. Boundary scan, including that specified by the JTAG standard, supports in-circuit testing of components and PCBs.

▶ Built-in self test (BIST) adds autonomous test circuits to components for use in manufacturing and in-field testing.

▶ Various nontechnical issues affect the design process, including business and life-cycle considerations.

10.7 FURTHER READING

VHDL Modelling Guidelines, European Space Agency, 1994, available from the Hamburg VHDL Archive, http://tams-www.informatik. uni-hamburg.de/vhdl/. Description of coding style and modeling guidelines, with several examples.

A VHDL Modeling Guide, TP-804, Version 1.0, Naval Research Laboratory (NRL), the Naval Surface Warfare Center (NSWC) and the Naval Air Warfare Center, Aircraft Division (NAWC-AD), 1994, available from the Hamburg VHDL Archive, http://tams-www .informatik.uni-hamburg.de/vhdl/. Specification of coding style and modeling guidelines for component models and testbenches.

The ASIC Handbook, Nigel Horspool and Peter Gorman, Prentice Hall PTR, 2001. A detailed description of a design flow for ASICs, including discussion of both technical and nontechnical aspects.

Application-Specific Integrated Circuits, Michael John Sebastian Smith, Addison-Wesley Professional, 1997 (see also http://www-ee.eng .hawaii.edu/%7Emsmith/ASICs/HTML/ASICs.htm). A description of ASIC technology and design methodology.

Surviving the SOC Revolution: A Guide to Platform-Based Design, Henry Chang *et al.*, Kluwer Academic Publishers, 1999. Deals with the basic principles of a design methodology addressing a platform-based approach to the design of embedded systems.

Winning the SoC Revolution, Grant Martin and Henry Chang (editors), Kluwer/Springer, 2003. A followup to *Surviving the SOC Revolution*, including several case studies of design methodologies used for real commercial projects. Also includes a discussion of nontechnical aspects of design.

Handbook on Electronic Design Automation of Integrated Circuits, Louis Scheffer, Luciano Lavagno, and Grant Martin (Editors), CRC,

2006. A comprehensive overview of the design automation algorithms, tools, and methodologies used to design integrated circuits.

Reuse Methodology Manual for System-On-A-Chip Designs, 3rd Edition, Michael Keating, Russell John Rickford, and Pierre Bricaud, Springer, 2006. Describes a design methodology for creating reusable ASIC designs.

Comprehensive Functional Verification: The Complete Industry Cycle, Bruce Wile, John C. Goss, and Wolfgang Roesner, Morgan Kaufmann Publishers, 2005. A detailed treatment of functional verification strategies and techniques and their place in the design flow.

Writing Testbenches Using SystemVerilog, Janick Bergeron, Springer, 2006. Presents many of the functional verification features that were added to the Verilog language as part of SystemVerilog, and shows how they are used in the verification process.

Verification Methodology Manual for SystemVerilog, Janick Bergeron, Eduard Cerny, Alan Hunter, and Andy Nightingale, Springer, 2005. Describes a methodology for verification of complex digital systems using a layered approach.

UML for SoC Design, Grant Martin and Wolfgang Müller (editors), Springer, 2005. A collection of the main contributors of the UML and SoC workshop at the 2004 Design Automation Conference, presenting approaches to executable UML, UML translations for FPGA synthesis and SystemC simulation, as well as UML-specific SoC methodologies.

A Practical Introduction to PSL, Cindy Eisner, Dana Fisman, Springer, 2006. Describes the Property Specification Language PSL, including its use for simulation-based and formal verification, and touches on methodological issues.

The e-Hardware Verification Language, Sasan Iman, Sunita Joshi, Springer, 2004. Provides a detailed coverage of the *e*-hardware verification language (HVL), and its use in implementing a verification environment.

The Art of Verification with Vera, Faisal Haque, Jonathan Michelson, Khizar Khan, Verification Central, 2001. Covers the elements of the Vera testbench tool and the OpenVera language using examples to show how they can be used to verify different types of designs.

Assertion-Based Design, Harry D. Foster, Adam C. Krolnik, and David J. Lacey, Springer, 2004. Describes an approach to design that facilitates verification, based on assertions expressed using the Open Verification Library (OVL), Property Specification Language (PSL), and SystemVerilog.

A Designer's Guide to Built-In Self Test, Charles E. Stroud, Kluwer Academic Publishers, 2002. A comprehensive reference on the theory and practice of BIST, including fault models, test pattern generation, signature analysis, and scan-based design.

APPENDIX A

KNOWLEDGE TEST QUIZ ANSWERS

ANSWERS FOR CHAPTER 1

SECTION 1.2

1. The two values used in binary representation are 0 and 1.

2. If one input is 0 and the other is 1, the output is 0. If both inputs are 0, the output is 0. If both inputs are 1, the output is 1.

3. If one input is 0 and the other is 1, the output is 1. If both inputs are 0, the output is 0. If both inputs are 1, the output is 1.

4. A multiplexer selects between two inputs, based on the value of a select input. The output value is the same as the selected input value.

5. The outputs of a combinational circuit depend only on the current values of the inputs, whereas the outputs of a sequential circuit depend on the current and past values of inputs.

6. A flip-flop stores one bit of information.

7. The term rising edge refers to a transition of a clock signal from 0 to 1. The term falling edge refers to a transition from 1 to 0.

SECTION 1.3

1. The TTL output voltage levels are 0.4V maximum for logic low and 2.4V minimum for logic high. The input threshold voltages are 0.8V minimum for logic low and 2.0V maximum for logic high. The noise margins are 0.4V for both logic low and logic high.

2. The term fanout refers to the number of inputs driven by a given output.

3. The propagation delay of a component is the time for a change of logic level at an input to cause a corresponding change at the output.

4. Minimizing the fanout of a component reduces the capacitive loading on the output, thus reducing the propagation delay.

5. Yes, wires do contribute to delay in a circuit, since it takes a nonzero amount of time for a signal change to propagate along the wire from an output to an input.

6. The setup time is the interval for which a value to be stored must be present on the data input before a rising clock edge. The hold time is the interval for which the value must remain unchanged after the rising clock edge. The clock-to-output time is the interval from a rising clock edge until the stored data appears at the output.

7. Static power consumption arises from leakage current flowing through transistors that are turned off. Dynamic power consumption arises from the charging and discharging of load capacitance when outputs switch between logic levels.

8. No, the cost of an IC is disproportionately dependent on area.

SECTION 1.4

1. The two parts are the entity declaration and an architecture body.

2. For each port, the entity declaration specifies the name, whether the port is an input (in) or an output (out), and the type.

3. A structural model is one that describes a circuit as a collection of interconnected components. A behavioral model is one that describes the function performed by the circuit.

4. Functional verification involves ensuring that the design performs the required function. Timing verification involves ensuring that the design meets its timing constraints.

5. One approach is to interpret the model as an executable program using a simulator. Another approach is formal verification, in which properties of the design are proven mathematically.

6. Synthesis involves automatic refinement of a model at a higher level of abstraction into a structural model at a lower level of abstraction.

SECTION 1.5

1. A design methodology is the systematic process of design, verification and preparation for manufacture of a product. A design methodology specifies the tasks undertaken, the information required and produced by each task, the relationships between the tasks, including dependencies and sequencing, and the CAD tools used.

2. A design methodology makes the design process more reliable and predictable, thus reducing risk and cost.

3. We must revisit a previous task to correct the error.

4. Top-down design involves developing an overall organization of a digital system to meet requirements, then designing and verifying each of the subsystems and sub-subsystems, and finally integrating and verifying the entire system.

5. Two implementation fabrics are field-programmable gate arrays (FPGAs) and application-specific integrated circuits (ASICs).

6. An embedded system is a digital system in which one or more computers are embedded as part of the circuit and programmed to implement part of the required functionality.

7. Hardware/software codesign refers to the practice of designing the hardware and the software of an embedded system together.

ANSWERS FOR CHAPTER 2

SECTION 2.1

1. The truth table for $a \cdot \overline{b} + \overline{c}$ is

a	b	c	$a \cdot \overline{b} + \overline{c}$
0	0	0	1
0	0	1	0
0	1	0	1
0	1	1	0
1	0	0	1
1	0	1	1
1	1	0	1
1	1	1	0

2. Truth tables for the two expressions are

a	b	$\overline{a \cdot b}$	$\overline{a} + \overline{b}$
0	0	1	1
0	1	1	1
1	0	1	1
1	1	0	0

Since the expressions have the same value for all combinations of values for *a* and *b*, the expressions are equivalent.

3. An expression is in sum-of-products form if it is the logical OR of a number of logical AND terms of variables or negations of variables.

4. The truth table for the AND-OR-Invert gate is

a	b	c	d	$\overline{a \cdot b + c \cdot d}$
0	0	0	0	1
0	0	0	1	1
0	0	1	0	1
0	0	1	1	0
0	1	0	0	1
0	1	0	1	1
0	1	1	0	1
0	1	1	1	0
1	0	0	0	1
1	0	0	1	1
1	0	1	0	1
1	0	1	1	0
1	1	0	0	0
1	1	0	1	0
1	1	1	0	0
1	1	1	1	0

5. Buffers are used to reduce the load on an output that must be connected to many inputs.

6. A compacted truth table is

a	b	c	f_1
0	–	–	0
1	0	–	0
1	1	0	1
1	1	1	0

7. The benefit is that it gives us more scope for optimizing the circuit. We might be able to identify two candidate circuits that both produce the required outputs for the combinations we do care about, but that differ in their output for the "don't care" combinations. If one of the candidates better meets constraints than the other, we would choose it, accepting whatever result it yields for the "don't care" combinations.

8. The dual is $\overline{a \cdot (b + c)} = (\overline{a} + \overline{b}) \cdot (\overline{a} + \overline{c})$. We need to pay attention to the order of application of operators when forming the dual.

9. f <= (a and not b) or not c;

10. We should allow a CAD tool to synthesize and optimize a circuit based on constraints and our chosen implementation fabric, since a CAD tool can generally do a better job than we could do manually.

SECTION 2.2

1. A 5-bit code can have up to $2^5 = 32$ code words.

2. The minimum number of bits needed is $\lceil \log_2 12 \rceil = 4$ bits.

3. There are seven values to represent, so we need a 7-bit one-hot code. A possible code is Monday: (1, 0, 0, 0, 0, 0, 0), Tuesday: (0, 1, 0, 0, 0, 0, 0), Wednesday: (0, 0, 1, 0, 0, 0, 0), Thursday: (0, 0, 0, 1, 0, 0, 0), Friday: (0, 0, 0, 0, 1, 0, 0), Saturday: (0, 0, 0, 0, 0, 1, 0), Sunday: (0, 0, 0, 0, 0, 0, 1).

4. signal s : std_logic_vector(0 to 7);

   Alternatively, we could use a descending index range:

   ```
   signal s : std_logic_vector(7 downto 0);
   ```

5. s <= "00000000";

6. A single bit flip in a one-hot code always produces an invalid code. If a 0 bit flips to 1, the resulting word has two 1 bits, which is invalid for a one-hot code. If the single 1 bit flips to 0, the resulting word has no 1 bits, which is also invalid.

7. Every valid code word in the augmented code has an odd number of 1 bits. If a bit flip changes a 0 to a 1, the result has an additional 1 bit, giving an even number of 1 bits. Similarly, if a bit flip changes a 1 to a 0, the result has one fewer 1 bits, again giving an even number of 1 bits. In either case, the error can be detected by testing whether there is an even number of 1 bits.

8. No, parity cannot be used to correct a bit flip. While we can determine that a bit has flipped, we cannot determine which bit has flipped. So we cannot determine the original uncorrupted code word.

SECTION 2.3

1. $y_4 = a_2 \cdot \overline{a_1} \cdot \overline{a_0}$

2. intruder_zone(2) = '0', intruder_zone(1) = '1', intruder_zone(0) = '1'.

   This is the code 011, which represents Zone 4. Thus the output would be incorrect.

3. We would not be able to distinguish the case of an intrusion being detected in Zone 1 from the case of no intrusion being detected.

4. It ranks the inputs in priority order. If a given input is 1 and no higher priority inputs are 1, the encoder outputs the code corresponding to the given input, regardless of whether any lower priority inputs are 1.

5. The BCD code 0101 represents the decimal digit 5.

6. The 7-segment code corresponding to the BCD code 0011 is 1001111 (the digit 3).

7. A multiplexer allows us to select among two or more data inputs. The value of the output is the same as the value of the selected input. The selection is determined by a separate select input.

8. We need to encode the choice of which input to select. Since there are 6 possible choices, we need $\lceil \log_2 6 \rceil = 3$ select input bits.

9. We can use five single-bit-wide 2-to-1 multiplexers. Each multiplexer chooses between the corresponding bits of the encoded data inputs. The select input is connected in common.

10. The signal would be at a high logic level, indicating falsehood of the statement "the door is closed."

11. To turn the motor on, we need to assign a low logic level, represented in VHDL by the value '0'.

SECTION 2.4

1. The purpose of a testbench model is to provide input values to the design under verification (DUV) and to check that the output values are correct.

2. s <= "0101"; wait for 1 ms;

3. After executing the last statement, the process starts again from the first statement in the process body.

4. The outputs of the design under verification would not have responded to the change at the inputs. Instead, they would still reflect the output generated in response to the previous input values.

5. If the circuit is very simple, comprising just one or two gates, it may be appropriate to implement it using discrete gates in individual packages. An example is a circuit that deals with minor differences in signals connecting off-the-shelf ICs.

6. A PLD is a programmable logic device, that is, a circuit containing some number of gates whose interconnections can be programmed.

ANSWERS FOR CHAPTER 3

SECTION 3.1

1. A number x is represented using n bits $x_{n-1}, x_{n-2}, \ldots, x_0$, with

$$x = x_{n-1}2^{n-1} + x_{n-2}2^{n-2} + \cdots + x_0 2^0$$

2. An n-bit unsigned binary number can represent values from 0 to $2^n - 1$.

3. Since $8191 = 2^{13} - 1$, we need 13 bits in the vector for x. The declaration is

```
signal x : unsigned(12 downto 0);
```

4. In octal: $01\ 011\ 101 \Rightarrow 135_8$. In hexadecimal: $0101\ 1101 \Rightarrow 5D_{16}$.

5. $10010011 \Rightarrow 000010010011$. The 12-bit result represents the same value as the original.

   $10010011 \Rightarrow 010011$. The 6-bit result does not represent the same value, since a significant 1 bit was truncated.

6.
```
  0 1 0 0 1 0 1 0
  0 1 1 0 0 0 0 0
0 1 0 0 0 0 0 0 0
─────────────────
  1 0 1 0 1 0 1 0
```

The addition does not overflow, since the carry bit out of the most significant position is 0.

7. In a ripple-carry adder, the carry bit into each position is determined using the carry bit from the next less significant position. Thus, in the worst case, carries must propagate through all bit positions of the adder before the final result is determined. In a carry-lookahead adder, the carry bit into each position is determined using just the operand bits and the carry bit into the

adder. Thus, the delay before the final result is determined is less than the worst-case delay of the ripple-carry adder.

8. We need to declare a signal for a 17-bit intermediate result:

```
signal tmp_s3 : unsigned(16 downto 0);
```

The required assignments are

```
tmp_s3 <= ('0' & s1) + ('0' & s2);
c_out <= tmp_s3(16);
s3 <= tmp_s3(15 downto 0);
```

9.   0 1 0 0 1 0 1 0
   − 0 1 1 0 0 0 0 0
  1 1 1 0 0 0 0 0 0
   ─────────────────
    1 1 1 0 1 0 1 0

The subtraction does underflow, since the borrow bit out of the most-significant position is 1.

10. Use the control signal to complement the second operand by passing each operand bit through an XOR gate with the control signal as the other gate input. Also, connect the control signal to the carry input of the adder.

11. smaller <= '1' when a < b else '0';

12. Since $16 = 2^4$, multiply by performing a logical shift left by four places; that is, shift the bits of the number left by four places and append four 0 bits on the right. Divide by 16 by performing a logical shift right by four places; that is, shift the bits of the number four places to the right, truncating the four right-most bits and appending four 0 bits on the left.

13. The product of two n-bit unsigned binary numbers requires $2n$ bits.

14. Gray codes are often used to avoid incorrect position values arising when multiple bits of a code change at once. In a Gray code, only one bit changes between adjacent code words.

SECTION 3.2

1. The weight of the left-most bit is negative (-2^{n-1}) for 2s-complement representation, whereas it is positive $(+2^{n-1})$ for unsigned representation.

2. The range of values is -2^{11} to $+2^{11} - 1$, that is, -2048 to $+2047$.

3. Since $512 = 2^9$, we need 10 bits in the vector for the number. The declaration is

    ```
    signal x : signed(9 downto 0);
    ```

4. $01110001 \Rightarrow 000001110001$. The 12-bit result represents the same value as the original.

    $01110001 \Rightarrow 110001$. The 6-bit result does not represent the same value, since the truncated bits are different from the resulting sign bit.

    $11110011 \Rightarrow 111111110011$. The 12-bit result represents the same value as the original.

    $11110011 \Rightarrow 110011$. The 6-bit result represents the same value as the original, since the truncated bits are the same as the resulting sign bit.

5. $\overline{11110010} + 1 = 00001101 + 1 = 00001110$.

6. The operand to be subtracted is complemented before being input to the adder, and the carry input of the adder is set to 1.

7. Since $16 = 2^4$, multiply by performing a logical shift left by four places; that is, shift the bits of the number left by four places and append four 0 bits on the right. Divide by 16 by performing an arithmetic shift right by four places; that is, shift the bits of the number four places to the right, truncating the four rightmost bits and replicating the original sign bit four times on the left.

SECTION 3.3

1. The number x is represented by the bits $x_{m-1}, \ldots, x_0, x_{-1}, \ldots, x_{-f}$ as:

    $$x = x_{m-1}2^{m-1} + \cdots + x_0 2^0 + x_{-1}2^{-1} + \cdots + x_{-f}2^{-f}$$

2. The range is -2^{m-1} to $2^{m-1} - 2^{-f}$.

3. We need 9 pre-binary-point bits and 4 post-binary-point bits. So the declaration is

    ```
    signal x : ufixed(8 downto -4);
    ```

4. s3 <= resize(s1 – s2, s3);

5. The product requires 28 bits: 10 pre-binary-point bits and 18 post-binary-point bits.

SECTION 3.4

1. $4.5_{10} = 100.1_2 = 1.001_2 \times 2^2$. The exponent is represented as $2 + 2^4 - 1 = 17$, that is 10001. The mantissa is represented using just the post-binary-point bits. The number is positive, so the sign bit is 0. The floating-point representation is 010001001000000000.

2. 0000000000000000 represents $+0.0$.

   0111100000000000 represents $+\text{Infinity}$.

   0100010000000000 has a biased exponent of 8, so the actual exponent is $8 + 1 - 2^3 = 1$. The number is $+1.1_2 \times 2^1 = 11_2 = 3$.

3. We require the exponent size e to be the smallest such that $2^{2^{e-1}} > 100$; that is $2^{e-1} \geq 7$, or $e = 4$. For 4 decimal digits of precision, we need at least $4/0.3 = 14$ mantissa bits.

   ```
   signal x : float(4 downto -14);
   ```

4. The type float32 corresponds to IEEE single precision, with a range of approximately $\pm 1.2 \times 10^{-38}$ to $\pm 1.7 \times 10^{38}$ and a precision of approximately 7 decimal digits.

ANSWERS FOR CHAPTER 4

SECTION 4.1

1. The process is sensitive to a rising edge of the clock. On a rising edge, the process copies the value of the data input to the data output.

   ```
   reg: process (clk) is
   begin
     if rising_edge(clk) then
       q <= d;
     end if;
   end process reg;
   ```

2. We call such an arrangement a *pipeline*.

3. The clock-enable input controls when the register updates the stored value. The register only updates the stored value when the CE input is 1 at the time of a rising clock edge. If the CE input is 0 on a rising clock edge, the register maintains the stored value unchanged.

4. When an asynchronous reset becomes 1, the flip-flop or register is reset to zero immediately, whereas a synchronous reset is only acted upon at the time of a clock edge.

5. A shift register allows the stored data to be shifted by one position.

6. The term "transparent" refers to the fact that data is transmitted through to the output while the latch-enable input is 1.

7. The implication is that the output is to maintain its previous value, requiring a latch in the hardware. Thus, the circuit is sequential, not combinational.

SECTION 4.2

1.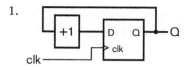

2. The maximum count value is $2^n - 1$. After that, the counter advances to 0.

3. The terminal count, $k - 1$, is decoded and fed back to the synchronous reset. Thus, the counter advances from $k - 1$ to 0.

4. A decade counter is a modulo 10 counter. It counts from 0 up to 9 and then advances to 0.

5. An interval timer is a counter whose clock input is a periodic signal with period t. The counter is loaded with a value k. The terminal count is reached after an interval of $k \times t$. The terminal-count output signal is used to trigger an activity after expiration of the time interval.

6. The accumulated delay may exceed the clock period. In that case, there will be clock cycles during which the counter outputs don't reach the correct value before the end of the cycle.

SECTION 4.3

1. The datapath contains the combinational circuits that implement the basic operations required of the digital system and the registers that store intermediate results.

2. The control section performs control sequencing, that is, ensuring that the datapath performs the required operations in the right order and at the right times.

3. Control signals govern the operation of the datapath elements: selecting operations to be performed and enabling registers. Status signals indicate whether certain conditions of interest are true, for example, whether data has certain values, or whether input data is available.

4. In a Mealy finite-state machine, the output function depends on both the current state and the values of the inputs. If the input values change during a clock cycle, the output values may change as a consequence. In a Moore

finite-state machine, on the other hand, the output function depends only on the current state, and not on the input values.

5. type states is (s0, s1, s2, s3);

6. Mealy-style output labels are attached to the arcs. Moore-style output labels are attached to the state bubbles.

SECTION 4.4

1. "Register transfer level" refers to a level of abstraction in our view of a digital system, in which we focus on transfer of data between registers through combinational subcircuits.

2. $t_{co} + t_{pd} + t_{su} < t_c$.

3. The critical path is the path from register output to register input with the longest propagation delay.

4. The critical path delay determines the shortest clock cycle time for the system. Since all operations are performed in times determined by the clock, the critical path determines the overall system performance.

5. We need to focus optimization effort on the critical path or paths to reduce their delay.

6. Clock skew refers to a situation where a clock edge arrives at different registers at different times, due to different wire delays in the wires that distribute the clock signal.

7. Use of registered inputs and outputs avoids the need for a clock period to accommodate both combinational subcircuit delays and delays though external pins and wiring.

8. Changes on asynchronous inputs around the time of a clock edge can induce metastability in input registers.

9. We must debounce an input from a switch to avoid spurious activation of the system's response to switch movements.

10. Whereas a combinational testbench simply applies each test case and check for the correct result, a sequential testbench must synchronize comparison of results with application of test cases. It must ensure that correct results occur at the correct times and in the correct order, and that no spurious results occur at other times.

11. Rather than having a single global clock signal for the entire chip or system, instead, the system is divided into several regions, each with its own local clock. Where signals connect from one region to another, they are treated as asynchronous inputs.

ANSWERS FOR CHAPTER 5

SECTION 5.1

1. The capacity of the memory is $4096 \times 24 = 98{,}304$ bits. The memory requires $\log_2 4096 = 12$ address bits.

2. The effect of a write operation is for the memory to write data present on the data inputs at the location whose address is present on the address inputs. The effect of a read operation is for the memory to read the content of the location whose address is present on the address inputs and drive the data value on the data outputs.

3. We would connect the address and control signals in common, and connect the data inputs and outputs of each component to separate groups of four data input and output signals, as shown in Figure A.1.

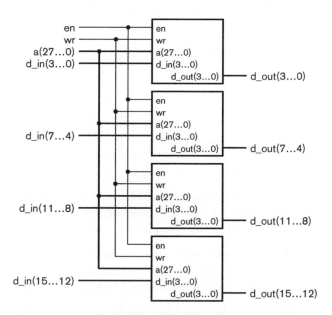

FIGURE A.1

4. We would use a 2-to-4 decoder to decode two address bits to select a memory, and a 4-to-1 multiplexer for the data outputs, shown in Figure A.2.

5. The location would reside in component 0.

6. The states of a tristate driver are logic low, logic high and high impedance.

7. We can omit the output multiplexer and simply connect the data outputs of the memory components together. Moreover, the separate data inputs and outputs can be combined into bidirectional data connections.

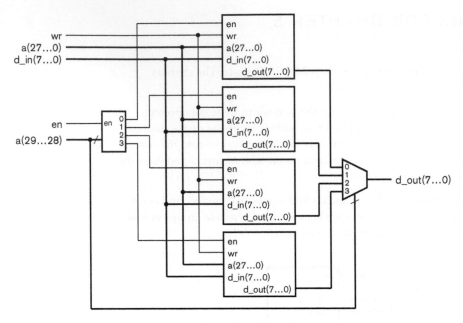

FIGURE A.2

SECTION 5.2

1. A RAM can perform both write and read operations, whereas a ROM can perform read operations only. The data in a ROM must either be placed in the memory when the device is manufactured or be programmed into the device after manufacture.

2. A volatile memory requires power to maintain the stored data and loses data if power is removed. A nonvolatile memory, on the other hand, maintains the stored data while power is removed.

3. In a static RAM, the stored data persists indefinitely so long as power is applied to the memory component. In a dynamic RAM, on the other hand, data decays unless it is periodically refreshed by being read and rewritten.

4. The access time of a RAM is the delay from the start of a read operation to having valid data at the outputs.

5. The need to set up and hold address and data values before and after activation of the control signals and to keep the values stable during the entire cycle means that we must either perform operations over multiple clock cycles, or use delay elements to ensure correct timing within a clock cycle.

6. Flow-through SSRAMs have registers on the inputs, but not on the data outputs. Pipelined SSRAMs, on the other hand, have registers both on the inputs and on the data outputs.

7. Memory storage is represented by a signal of an array type. Each array element represents a storage location.

8. A multiport memory can perform multiple memory access operations at the same time, giving a higher overall access rate, and allowing separate subsystems to access the memory independently without contention.

9. We should consult the data sheet for the memory component to understand the effect.

10. FIFO stands for "first-in, first-out," and refers to the order in which data is written to and read from the memory.

11. A FIFO allows us to smooth out the flow of data between the domains. Data arriving is written into the FIFO synchronously with the sending domain's clock, and the receiving domain reads data synchronously with its clock.

SECTION 5.3

1. Soft errors are transient, and involve a bit flip in a memory cell without a permanent effect on the cell's capacity to store data. Hard errors persist in a memory circuit. A memory cell or chip affected by a hard error is no longer able to store data.

2. In DRAMs, soft errors are typically caused by high-energy neutrons generated by collision of cosmic rays with atoms in the earth's atmosphere. The neutrons collide with silicon atoms in the DRAM chip, leaving a stream of charge that can disrupt the storage or reading of charge in a DRAM cell.

3. We are not able to take any action to correct the data, since parity does not allow us to identify which bit flipped causing the parity error.

4. Single-error correction and double-error detection of 4-bit data words requires $\log_2 4 + 2 = 4$ check bits per word.

ANSWERS FOR CHAPTER 6

SECTION 6.1

1. Photolithography is the use of a photographic process to draw on the surface to control which areas are affected by processing steps.

2. The larger the IC area, the greater the IC cost. The higher the defect density, the greater the cost, since defective ICs must be discarded and their cost amortized over the remaining good ICs.

3. "L" stands for low power, and "S" refers to the use of Schottky diodes to reduce switching delays.

4. The term glue logic refers to simple logic circuits for interconnecting LSI components.

5. ASIC stands for application-specific integrated circuit, and ASSP stands for application-specific standard product.

6. It would probably not make sense to design an ASIC for a customized system, since the production volume would be very low. The high NRE cost would be amortized over a small number of units, making the unit cost excessive. It would probably be better to develop an FPGA-based design.

7. It would most likely make sense to develop an ASIC for an engine control system, since the production volume would be very high. The ASIC would have lower manufacturing cost per unit than a programmable part. Amortizing the NRE for the ASIC over the large production run would result in a lower overall cost per unit than that for the FPGA.

SECTION 6.2

1. Whereas the function performed by a fixed-function component is determined by the logic circuit for the component, the function of a programmable logic device can be programmed after manufacture.

2. A fuse map is a file used by a programming instrument to determine which fuses to blow in a programmable array logic (PAL) device.

3. Output O8 implements the function $I2 \cdot \overline{\overline{IO2} + I10}$, with the output enabled by the condition I9.

4. The state bit S2 would be stored in the flip-flop. The 4-to-1 multiplexer would select input 3, taking the negated value of S2 from the flip-flop and negating it again to present S2 at the output. The 2-to-1 multiplexer would select input 0, feeding the negated value of S2 back to the AND array. The buffer/inverter provides both S2 and its negated form to the array.

5. The system can be upgraded after delivery by storing new configuration information, rather than having to replace chips or other hardware.

6. The logic blocks can be programmed to implement simple combinational or sequential logic functions. The I/O blocks can be programmed to be registered or nonregistered, as well as implementing various specifications for voltage levels, loading and timing.

7. FPGAs also contain embedded RAM blocks and a programmable interconnection network. The more recent FPGAs also include special circuits for clock generation and distribution.

8. The configuration needs to be stored in a separate nonvolatile memory, and additional circuits need to be included in the system to manage loading the configuration.

9. An antifuse is a conductive connection that is formed during programming, as opposed to being blown.

10. A platform FGPA serves as a complete platform upon which a complex application can be implemented. It includes specialized circuitry such as one or more processor cores, computer network transmitter/receivers and arithmetic circuits. In contrast, a simple FPGA includes only basic logic blocks, embedded memory and I/O blocks.

SECTION 6.3

1. In flip-chip packaging, the connection pads on the IC are covered in conductive material forming bumps. The IC is flipped over and affixed to the substrate of the package, with the bumps in direct contact with substrate connection points. The connection points lead to the external pins of the package. In contrast, previous packaging involved placing the IC circuit-side up in a package cavity, and joining bond wires from the IC's connection points to the lead frame in the package.

2. Insertion-type packages have pins that are designed to be inserted through plated holes in a printed circuit board (PCB). Solder is melted into the holes to form electrical connections between the pins and the PCB wiring. In contrast, surface-mount packages have pins or connection points that come into contact with a metal pad on the PCB. Solder paste is applied between each pin and pad and subsequently melted, forming the connection.

3. Vias are small holes drilled through the layers of a PCB and coated with metal to form connections between the layers.

4. A ball-grid array (BGA) package would mose likely be used.

SECTION 6.4

1. The term signal integrity refers to the degree to which distortion and noise effects upon a signal path are minimized.

2. A signal propagates at approximately 150mm per nanosecond along a typical PCB trace.

3. Ground bounce arises when one or more output drivers switch logic levels. During switching, both of the transistors in the driver's output stage are momentarily on, and transient current flows from the power supply to ground. The inductance in both the power and the ground connections causes voltage spikes in the power supply and ground on the IC.

4. Bypass capacitors should be placed between power and ground close to each IC package.

5. Limiting the slew rate limits the inductive effect resulting from the change in voltage levels.

6. Transmission-line effects can be mitigated by appropriate layout and proper termination of PCB traces. For example, we can design PCB traces to form stripline or microstrip transmission lines.

7. EMI is electromagnetic interference, causes by field energy radiated out from a system due to logic switching. Crosstalk is noise induced on a PCB trace by fields radiated from an adjacent trace.

8. When differential signaling is used, noise is induced equally on the wires for both the signal, S_P, and its negation, S_N. Such common-mode noise is cancelled out when the voltage difference is sensed by a receiver.

9. The differential voltage swings between $1.075V - 1.425V = -0.35V$ and $1.425V - 1.075V = +0.35V$, that is, a swing of 0.7V.

ANSWERS FOR CHAPTER 7

SECTION 7.1

1. The main elements are a central processing unit (CPU), often called a processor core); an instruction memory; a data memory; input, output, and input/output controllers; possibly one or more accelerators; and one or more buses to connect the elements together.

2. The instructions in an embedded computer are usually fixed during manufacture of the system (or only occasionally upgraded in the field), and the amount of instruction memory required is known in advance. Hence, we usually store instructions in a ROM or flash memory component, and provide a separate RAM for the data memory.

3. A microprocessor is a CPU in a package by itself, whereas a microcontroller includes a CPU, instruction and data memory, and I/O controllers all in the one package.

4. A soft core processor is a CPU implemented using the programmable resources of the FPGA.

SECTION 7.2

1. The instruction set of a CPU is its repertoire of instructions.

2. The steps are: fetching the next instruction from the instruction memory, decoding the instruction to determine the operation to perform, and executing the operation.

3. The CPU has a special register called the program counter (PC), in which the address of the next instruction is kept.

4. The terms refer to the order of the bytes within a word in memory. Little-endian CPUs store the byte containing the least significant bits at the lower address and the byte containing the most significant bits at the higher address. In contrast, big-endian CPUs store the bytes in the opposite order.

5. An assembler translates an assembly-code program into a sequence of binary-coded instructions to be loaded into the instruction memory.

6. addc r2, r3, 25: Adds the value in r3 and the immediate value 25, and puts the result in register r2.

   shr r1, r1, 3: Shifts the value in r1 right by three places and puts the result back in r1.

   ldm r5, (r1)+4: Loads a value from memory into r5. The address from which the value is loaded is calculated by adding the value in r1 and the offset value 4.

   bnz −7: If the Z condition code bit is 0, the displacement value −7 is added to the PC, causing a branch in the flow of program execution. Otherwise, execution continues sequentially.

   jsb do_op: The address of the next instruction in memory is saved by pushing it onto the return address stack. The address represented by the label do_op is then put in the PC, causing a transfer of control to that address.

   ret: The address at the top of the return address stack is copied to the PC, and the return address stack is popped by one entry. The effect is to return control from a subroutine.

7. The instruction bnc +15 is a branch instruction, so bits 17−12 are 111110. Bits 11−10 are 11, representing the branch condition. Bits 7−0 encode the displacement in 2s-complement form, namely, 00001111. Bits 9−8 are not used, so we don't care what their values are. The binary encoding for the instruction is thus 11111011−−00001111. Assuming the don't care bits are set to 0, the encoding in hexadecimal is 3EC0F.

8. The binary instruction word is 000101010100000001. Since bit 17 is 0, the instruction is an arithmetic/logical instruction with an immediate operand. Bits 16−14 are 001, encoding the function addc. Bits 13−11 are 010, encoding the destination register r2, and bits 10−8 are 101, encoding the source register r5. Bits 7−0 are 00000001, encoding the immediate value +1. The instruction is thus addc r2, r5, +1.

SECTION 7.3

1. If the CPU and memory have incompatible signals for interconnection, we need glue logic to complete the interface.

2. Multiplexing the data and address signals allows more package pins to be used for inputs and outputs.

3. The data memory would typically be 32 bits wide, allowing a complete data word to be accessed with one read or write operation.

4. The byte-enable signals are used to ensure that, when a byte within a 32-bit word is modified, the other bytes in the corresponding 32-bit memory location are not affected.

5. The first observation is that a small proportion of instructions and data account for the majority of memory accesses over a given interval of time. The second observation is that those items stored in locations adjacent to a recently accessed item are likely to accessed next.

6. When the processor requests access to a given memory location, the cache checks whether it already has a copy of the line containing the requested item. A cache hit refers to the case where the check succeeds, allowing the cache to satisfy the processor's request immediately. A cache miss refers to the case where the check fails, and the processor must wait.

7. During a cache miss, the cache copies the line containing the requested item from main memory into the cache memory. When the requested item is available in the cache, the processor can proceed with its requested access.

8. The term memory bandwidth refers to the rate of transfer of data to or from the memory.

ANSWERS FOR CHAPTER 8

SECTION 8.1

1. A sensor is an input transducer that senses some physical property and generates an electrical signal that corresponds to the property. An actuator is an electromechanical output transducer that causes a mechanical component to move to one position or another.

2. If the property sensed by an input transducer is continuous in nature, the transducer may provide an analog signal that bears a continuous relationship with the physical property. Since digital systems deal with discrete representations of information, we need to convert the signal from analog to encoded digital form.

3. We would drive the row line r2 low and sense the column line c3. If c3 is low, the 6 key must be pressed, connecting c3 to r2. Otherwise, the key is not pressed, leaving c3 pulled high by the pull-up resistor.

4. The shaft is rotated in a clockwise direction.

5. An 8-bit flash ACD requires 255 comparators.

6. We can connect the like terminals of all of the LEDs in each digit in common so that we can activate each digit in turn. We connect the other terminal of corresponding segment LEDs together. Activating a segment connection lights that segment of the activated digit. Thus, the number of connections is seven (for the segments) plus the number of digits (for the common anodes or cathodes).

7. A solenoid moves the armature for its mechanical effect. A relay moves the armature to open or close a set of electrical contacts, that is, to achieve an electrical effect on an external circuit.

8. Two kinds of motor are stepper motors, which provide rotation in a series of steps, and servo-motors, which provide continuous rotation.

9. We would use an R/2R ladder DAC, since the number of resistors required for an R-string DAC would be too great.

SECTION 8.2

1. An input register allows the embedded software to read input information from input devices, and an output register allows the embedded software to write output information to output devices.

2. A control register allows embedded software to provide parameters governing the way transducers operate, and a status register allows embedded software to read the state of the controller.

3. We can use address decoding circuits to identify whether memory or I/O registers are being accessed, and enable the memory chips or the appropriate register as required.

4. If the operation of the input controllers is sequential, we need to synchronize controller operation with execution of the embedded software. The controller might have control registers allowing the embedded software to initiate input sequences.

5. An autonomous I/O controller allows the processor to perform other tasks concurrently. This increases the overall performance of the system. Moreover, an autonomous I/O controller may be able to transfer data at higher rates than a processor using a simple controller, or to perform control operations with less delay.

SECTION 8.3

1. If one data source drives a low level while another drives a high level, the resulting conflict would cause large currents to flow between the two components, possibly damaging them.

2. Subdivision of the multiplexers may allow the chip wiring to be simplified.

3. Contention is avoided on a tristate bus by enabling at most one source at a time to drive the bus high or low. The remaining sources are disabled, with their drivers in the high-impedance state.

4. The bus signal might float to a voltage around the switching threshold of the bus destination inputs. Small amounts of noise voltage induced onto the bus wire can cause the inputs to switch state frequently, causing

spurious data changes within the data destination and consuming power unnecessarily.

5. A weak keeper consists of two inverters providing positive feedback to the bus signal. When the bus is forced to a low or high logic level by a bus driver, the positive feedback keeps it at that level, even if the forcing driver is disabled.

6. If the t_{off} delay of the disabled driver is at the maximum end of its range and the t_{on} delay of the enabled driver is at the minimum end, there will be a period of overlap where some bits of the enabled driver may be driving opposite logic levels to those of the disabled driver. While the overlap is unlikely to destroy the circuit, it does contribute extra power consumption and heat dissipation and ultimately will reduce the operating life of the circuit. We can avoid the problems by including a margin of dead time between different data sources driving the bus.

7. d_out <= d_in when d_en = '1' else "ZZZZZZZZ";

8. The resulting value is 'X', denoting an unknown logic level.

9. Such a signal is called a wired-AND connection since the bus signal is only 1 if all of the drivers output 1. If any driver outputs 0, the bus signal goes to 0.

10. bus_sig <= 'H';

11. The results are '0', '1', 'X', and 'X', respectively.

12. A bus protocol is a specification of the signals that interconnect components and the sequences and timing of values on the signals to implement various bus operations.

SECTION 8.4

1. Serial transmission uses fewer signal wires, drivers and receivers, thus reducing circuit area and simplifying layout and routing. For connections between chips, it also uses fewer pads, pins and PCB traces. These all lead to reduced cost. Secondary advantages include avoidance of crosstalk and skew.

2. We use a shift register, sometimes called a *serializer/deserializer*, or *serdes*, at each of the transmitting and receiving end.

3. In principle, the order is arbitrary, so long as the transmitter and receiver agree. Often, serial transmission in a system is governed by a standard that specifies the order.

4. In NRZ transmission, we drive the serial signal with the values of successive data bits. There is no indication of when the time for one bit ends and the time for the next bit starts.

5. The start bit indicates the start of transmission, allowing the receiver to synchronize with the transmitter. The stop bit indicates the end of transmission of the data.

6. Manchester encoding transmits each bit of data in a given interval. It represents a 0 with a transition from low to high in the middle of the bit interval, and a 1 with a transition from high to low. (Alternatively, the opposite assignment of transmissions could be used, so long as transmitter and receiver agree.) At the beginning of the bit interval, a transition may be necessary to set the signal to the right logic level for the transition in the middle of the interval.

7. We would adopt a standard serial interface specification to avoid the need to design the connection from scratch, and to be able to use off-the-shelf devices that adhere to standards. As a consequence, we would reduce the cost of developing and building a system, as well reducing the risk of the design not meeting requirements.

8. I^2C would be most appropriate for connecting a motor controller to an embedded system, since the application has a low bandwidth requirement. FireWire would be most appropriate for connecting a digital video camera, since the application has a high bandwidth requirement.

SECTION 8.5

1. Embedded software needs to be able to detect when events occur so that it can react. It also needs to be able to keep track of time so that it can perform actions at specific times or at regular intervals.

2. Polling involves the software repeatedly checking a status input from a controller to see if an event has occurred. If it has, the software performs the necessary task to react to the event.

3. Polling has the advantage that it is very simple to implement, and requires no additional circuitry beyond the input and output registers of the I/O controllers. Polling has the disadvantage that it requires that the processor core be continually active, consuming power even when there is no event to react to. It also prevents the processor from reacting immediately to one event if it is busy dealing with another event.

4. The processor stops what it was doing, saving the program counter so that it can resume later, and starts executing an interrupt handler, or interrupt service routine, to respond to the event.

5. Processors generally have instructions for disabling and enabling interrupts. A processor can execute the instruction to disable interrupts before entering the critical region, and execute the instruction to enable interrupts on completion of the critical region.

6. The processor must save the value of the program counter in a register or some other storage when an interrupt occurs. It can then use the saved value as the location at which to resume execution.

7. An interrupt vector is either a value used to form the address of the interrupt handler, or an index into a table of interrupt handler addresses in memory.

8. The controller must deactivate the interrupt request signal to avoid multiple responses for the one event.

9. A real-time clock generates an interrupt for the processor at some programmable multiple of a time base. The interrupt handler for the timer can then perform any required periodic actions.

10. A real-time executive schedules execution of tasks in response to interrupts and timer events.

ANSWERS FOR CHAPTER 9

SECTION 9.1

1. Parallelism involves performing multiple steps at the same time, thus taking less time overall to complete an operation.

2. Parallelism is constrained by data dependencies and the availability of data.

3. An algorithm describes the data to be processed, how it is organized, and the sequence of processing steps to be performed.

4. Since a kernel is the most time-consuming part of executing an algorithm, accelerating it gives the most payback, that is, the most significant reduction in execution time.

5. The speed up factor is four, since the pipeline completes an operation every clock cycle, whereas the nonpipelined chain only completes an operation once every four clock cycles.

6. DMA is the process whereby an I/O controller or accelerator transfers data to and from memory autonomously, rather than having the processor copy the data.

7. An arbiter makes sure that masters take turns to access the memory.

8. A block-processing accelerator processes the data arranged in blocks of adjacent or regularly spaced locations in memory. A stream-processing accelerator, on the other hand, processes streams of data arriving in sequence from some source.

9. Generally, embedded software interacts with an accelerator using input and output registers within the accelerator, in much the same way as interaction

with autonomous I/O controllers. Alternatively, the software may use FIFO queues to interact with an accelerator, particularly if less strict synchronization is required.

SECTION 9.2

1. With 6-bit pixels, the partial products range from -126 to $+126$. Thus, the partial products should be represented using 8-bit signed numbers. There are nine partial products to add to form each of Dx and Dy. However, the coefficient values are such that the result values range from -252 to $+252$, which can be represented using 9 bits. We then need to add the two absolute values, giving a range of 0 to $+504$ for |D|, which can also be represented in 9 bits.

2. The value of |D| cannot be computed in parallel with the values of Dx and Dy, since the latter two values are needed as inputs to the computation of |D|. In other words, there is a dependence of |D| on Dx and Dy.

3. Doubling the read and write times reduces the memory bandwidth to half of the original bandwidth. As a result, video input would consume 10% of the bandwidth, reading pixels into the accelerator would consume 30%, and writing derivative pixels would consume a further 10%, for a total of 50% of the bandwidth. Thus, there would be sufficient bandwidth for these operations. However, the remaining 50% may be insufficient for the embedded processor's requirements.

4. The left-most and right-most pixels in each row do not have a complete set of neighboring pixels; hence, we cannot compute the derivative for those pixels using the convolution masks.

5. The embedded software initiates processing of an image by writing to the Start register at offset 4 in the accelerator's I/O register address map. Processing is complete when the done bit (bit 0) in the Status register at offset 0 is 1. The embedded software can read this bit to determine when processing is complete. It can also enable an interrupt to be triggered when the done bit changes to 1.

6. When the software initiates processing by writing to the Start register, the start signal is set to 1 for the duration of the write access. That signal causes the control FSM, when in the idle state, to begin the control sequence. While an image is being processed, the FSM is not in the idle state, and so does not respond to the start signal. Hence, if the software writes to the Start register in that time, there is no effect. If the processor also writes to either of the base address registers, the newly written values would be used to generate addresses for the operation already in progress, thus corrupting the operation. For these reasons, the software must wait until one image is completely processed before initiating processing for another image.

7. The FSM is a hybrid FSM, since the outputs O_offset_cnt_en, D_offset_cnt_en, row_cnt_en and col_cnt_en depend on both the current state and the FSM inputs, whereas other outputs depend only on the current state.

SECTION 9.3

1. It is not feasible to test an accelerator design exhaustively, since the space of all possible data values and operational sequences is astronomically large.

2. A bus functional model is a model that engages in bus operations without actually performing internal operations such as executing processor instructions or performing memory reads and writes.

3. If both requests are activated in the same cycle, the arbiter gives preference to the accelerator, activating its grant and leaving the processor's grant inactive until the accelerator's request is deactivated.

4. If the accelerator requests use of the bus while the processor is currently granted use, the arbiter leaves the accelerator's grant inactive until the processor's request is deactivated. This allows the processor to complete its use of the bus without preemption.

5. No, the testbench only exercises the bus functionality of the accelerator.

ANSWERS FOR CHAPTER 10

SECTION 10.1

1. The term architecture exploration refers to the task of abstract modeling and analysis of candidate designs for a system.

2. Logical partitions are parts of the system that implement the various processing steps. They are subdivisions of the system functionality. Physical partitions are the physical components to which the logical partitions are mapped. The physical partitions can include processor cores, accelerators, memories and I/O controllers.

3. For each of the components in the system, a high-level specification describes the function it is to perform, the connections to other components, and the constraints upon its implementation.

4. A behavioral model of the component expresses its functionality at an intermediate level of abstraction between system level and register-transfer level. The behavioral model might include a description of the algorithm to be implemented by the component without detailed cycle-by-cycle timing, or it might just be a bus functional model. The purpose of the behavioral model is to allow functional verification of the component before proceeding to detailed implementation.

5. The benefits of reuse include savings in design time and verification effort.

6. The kinds of function that can be implemented using a core generator include memories, arithmetic units, bus interfaces, digital signal processing, and finite-state machines.

7. The designers can use a revision management tool, also referred to as source code control tool.

8. A verification plan identifies what parts of the design will be verified, the functionality that will be verified, and how verification will be performed.

9. Code coverage refers to the proportion of lines of code that have been executed at least once during simulation of the design, whereas functional coverage refers to the proportion of functionality that has been verified. Functional coverage includes aspects such as the distinct operations that have been verified, the range of data values that have been applied, the proportion of states of registers and state machines that have been visited, and the sequences of operations and values that have been applied. Functional coverage is more important than code coverage for ensuring correctness.

10. Constrained random testing involves a test case generator randomly generating input data, subject to constraints on the ranges of values allowed for the inputs.

11. Formal verification allows complete verification that a component meets a specification, since it provides a rigorous proof that the assertions embodying the specification hold. Contrast this with simulation-based verification, in which the number of possible input cases and sequences is too large for exhaustive simulation to be feasible. However, the completeness of formal verification depends on the properties that are verified. If those properties do not cover all of the functional requirements, then a formal verification does not achieve complete functional coverage. Moreover, writing properties that completely and accurately capture the intent of a specification is very difficult. A further difficulty is that state-space exploration is a computationally intense problem, so verification of numerous complex assertions may be intractable.

12. The hardware abstraction layer is the lower layer of embedded software that depends on the hardware. It contains driver code and interrupt service routines for I/O controllers, memory management code, and so on. It provides an abstract interface that can be called by the upper application layer.

13. An instruction set simulator simulates execution of instructions, but without simulating the detailed hardware operations of the target processor.

14. Many language features are only suitable for high-level behavioral modeling and for writing testbenches, and cannot be synthesized into

equivalent gate-level circuits. For those features that are, in principle, synthesizable, acceptance may also depend on the historical development of the synthesis tool.

15. A technology library is a collection of primitive components that are available within the implementation fabric selected for the design.

16. There are two reasons to performing gate-level simulation. First, it allows us to verify that the design still meets functional requirements, taking into account the timing estimates from the technology library. Second, there are ways in which we can write RTL model code that produce different behavior in the RTL simulation and the synthesized hardware. Simulating the synthesized design and making sure it behaves the same as the RTL design is a good check that we have used the tools correctly.

17. Floorplanning involves deciding where each of the blocks in the partitioned design is to be located on the chip. Placement involves positioning each cell in a synthesized design, and routing involves finding a path for each connection.

SECTION 10.2

1. We should focus effort in earlier stages to have the greatest impact.

2. First, since chips are rectangular and wafers are circular, larger chips leave more wasted area near the edges of the wafer, so the proportion of wafer cost borne by each chip is not simply the ratio of chip area to wafer area. Second, larger chip area increases the likelihood of a defect occurring on any given chip and causing the chip not to function. Since nonfunctional chips must be discarded after the wafer is fabricated, the remaining functional chips must bear the cost of fabricating and testing the nonfunctional chips. Third, a larger chip requries a larger and more costly package than a smaller chip. Fourth, since the chip is presumably larger because it has more transistors than a smaller chip, it consumes more power, and so the resulting heat must be dissipated, leading to increased package costs.

3. In the functional design stage of the design flow, we can influence circuit area through our choice of components. Also, choosing minimal bit-widths for data helps to keep circuit area to a minimum, since components that process the data can then be of the minimal size, and the minimal amount of wiring is required between the components.

4. In the architecture exploration stage, we can improve performance through application of parallelism, limited by the data dependencies involved. We may need to trade off parallelism against area and power, since the extra resources required to realize the parallelism take up area and consume power.

5. A timing budget specifies maximum clock-to-output delays and input-to-clock setup times for each block designed by members of the team. This helps to ensure that the combined path from a register output in one block to a register input in another block meets timing constraints.

6. Specifying timing constraints allows a synthesis tool to optimize timing of the detailed design, and then to analyze the resulting synthesized circuit to verify that timing constraints are met.

7. For the synthesized design, a static timing analyzer uses timing estimates for each of the components in the technology library, together with simple wire-load models. The tool aggregates the delay through combinational circuits and wiring between successive registers and thus identifies the critical path in the design to determine whether the clock period constraint is met. For a placed and routed design, the tool repeats the analysis using more accurate delay values for components and wiring extracted from the implemented circuit.

8. If timing constraints are not met, we need to revisit earlier stages of the design flow to improve the timing of the circuit.

9. One technique for reducing power concumption is to identify blocks of a system that remain idle for substantial periods during the system's operation, and to remove power from those blocks during idle periods. Another technique, usable if the performance requirements of a system are not constant, is to reduce dynamic power consumption by reducing the clock frequency. A third technique is clock gating, which involves turning off the clock to parts of a circuit whose stored values do not need to change.

10. Implementing clock gating in RTL model code can lead to clock skew and glitches being introduced by the synthesis tool. Clock gating is better implemented as a power optimization by clock insertion tools during physical design.

SECTION 10.3

1. Design for test refers to design techniques that seek to improve testability.

2. A fault simulator simulates the operation of a circuit with a given fault injected at a given location. The simulator applies test vectors until an incorrect output results, indicating that the fault has been detected. If no incorrect output is produced for all of the test vectors, the fault remains undetected by that set of vectors.

3. The stuck-at model represents a defect in which an input or output of a gate in a circuit is stuck at 0 or stuck at 1, rather than being able to change between 0 and 1. Such a fault might be caused by a short circuit to the ground or power supply.

4. A node is controllable if there are input combinations that cause the node to take on a given value. A node in a circuit is observable if a fault at the node can be made to result in an incorrect output value.

5. In CMOS circuits, transistors occur in complementary pairs, with one or other turned on at a time. If a transistor is stuck on, turning on the other transistor in the pair causes excess current to flow from the power supply (I_{DDQ}). I_{DDQ} testing identifies transistor stuck-on faults by detecting this excess current.

6. The registers in the circuit must be modified to allow them to be chained into a long shift register.

7. Advantages include increased controllability and observability, making high fault coverage feasible, and the ability to automate testability by modification of the registers to allow them to function as shift registers. The disadvantage is the overhead, both in circuit area and delay.

8. Boundary scan enhances testability of PCB-based systems by allowing testing of the connections between chips on a PCB. Boundary scan involves including scan-chain flip-flops on the external pins of each chip. The test equipment shifts a test vector into the scan chain and drives the vector onto the external outputs of the chips. The scan-chain flip-flops then sample the external inputs, and the sampled values are shifted out to the test equipment. The test equipment can then verify that all of the connections between the chips, including the chip bonding wires, package pins and PCB traces, are intact.

9. The signals in a JTAG test access port are Test Clock (TCK), provides the clock signal for the test logic; Test Mode Select Input (TMS), controls test operation; Test Data Input (TDI), serial input for test data and instructions; Test Data Output (TDO), serial output for test data and instructions; and optionally Test Reset Input (TRST).

10. The TAP controller governs operation of the test logic.

11. A bidirectional tristate pin requires one boundary scan cell for the data, another for the output enable, and a third for the input.

12. Built-in self test requires addition of test circuits that generate test patterns and analyze output responses.

13. The BIST hardware remains available during the operational lifetime of the system, and can be used for testing when the system is in the field.

14. A linear-feedback shift register (LFSR) is used to generate pseudo-random sequences of test patterns. A multiple-input signature register (MISR) forms a summary, called a signature, of a sequence of output responses.

SECTION 10.4

1. Some of the stages are market research and financial modeling, product design, manufacturing, sales and distribution, maintenance, and obsolesence.

2. Late entry allows competitors to gain market share, reducing the revenue available for the late product, and possibly making it unprofitable.

3. For such products, the design phase will also involve development of design documentation, and liaison with support service providers to develop support plans, procedures and documents.

4. Designing using technology that is current at the start of the design process may well lead to a product with lower performance or capacity than competitors' products.

APPENDIX B

INTRODUCTION TO ELECTRONIC CIRCUITS

In Chapter 1, we described the abstractions that underlie digital logic, namely, use of two discrete voltage levels and instantaneous switching between them. We also mentioned that these abstractions are only applicable if we adhere to certain design disciplines. Otherwise, the non-ideal nature of the components from which we construct digital circuits becomes significant, breaking assumptions that underpin the abstractions. When designing digital circuits, it is important to have an understanding of the nondigital, or *analog*, nature of components and circuits so that we avoid behavior that invalidates the digital abstractions. This Appendix summarizes the knowledge about analog electronic components and circuits assumed in our discussion of digital electronics. Our aim is to support digital design, so we do not go into the depth found in a course or book on circuits.

B.1 COMPONENTS

Electronic circuits and components can be described at a fundamental level in terms of electric and magnetic fields. Physicists use Maxwell's equations to characterize the properties of fields, but that is at too low a level of abstraction for our purposes. Instead, we start with voltage (V), measured in volts, and current (I), measured in amperes, as our basic quantities of interest. We will describe electronic circuits as assemblies of components interconnected by wires, and analyze the voltage and current at various places within each circuit. We will focus on the components in this section, and return to circuits in Section B.2.

We can think of a component as a physical entity with two or more *terminals*, or *pins*. Current can flow into or out of terminals, and there may be voltage differences between terminals. The various components that we will consider differ in the relationships between current and voltage at their terminals. Just as we can use the notion of models to think about systems in general, we can use it to think about components. We will use mathematical models to describe the relationship between current and voltage for each kind of component. We will use relatively simple models that capture properties of interest to digital designers and that ignore more detailed effects that are generally not relevant.

B.1.1   VOLTAGE SOURCES

One of the simplest components to describe is a constant voltage source, which we can represent graphically using the symbol at the left of Figure B.1. The voltage difference between its positive terminal (marked +) and its negative terminal (marked −) is a constant. Voltage sources are usually used to provide operating power to digital circuits, either in the form of batteries (whose symbol is shown in the center of Figure B.1), or power supply units connected to the mains supply. In this book, we use the symbols shown at the right of Figure B.1: the top symbol for a positive-voltage power supply and the bottom symbol for a 0V ground. Since voltages are all relative, labeling the ground as 0V is nominal. Other voltages are then measured relative to that reference.

An ideal voltage source maintains the constant voltage difference between its terminals regardless of the current flowing through them. Real voltage sources used as power supplies can only do that for a limited range of current flow. Within that range, the voltage is approximately constant within specified bounds. However, once the current exceeds a specified limit, the power supply may either reduce the voltage or fail. We would normally design a system so that the current drawn from a power supply remains within the specified limits, with some margin, to avoid the possibility of failure.

FIGURE B.1   Symbols for a voltage source (left), a battery (center), power supply (top right), and ground (bottom right).

B.1.2   RESISTORS

A resistor (Figure B.2) is a component whose voltage drop across its terminals is linearly related to the current flowing through the terminals. This property is expressed in Ohm's law:

$$V = IR \tag{B.1}$$

For an ideal resistor, the *resistance R* is constant. In practice, it is slightly dependent on temperature and other effects, but we can usually ignore those dependencies in digital circuit design. For practical resistors, the resistance, measured in ohms (Ω), can range from around 1Ω to $1M\Omega$.

Since there is a voltage drop across a resistor and current flowing through it, there is work done. A resistor converts electrical energy into heat, with the power being the product of voltage and current:

$$P = VI = I^2R = V^2/R \tag{B.2}$$

FIGURE B.2   Symbol for a resistor.

The generated heat must be dissipated by the resistor, or it will overheat and fail. Different kinds of physical resistors used in electronic circuits range in power dissipation capacity from 100mW to 10W.

An important point to note about the relationship between voltage and current for a resistor is that it is not time dependent. The voltage at any given instant depends only on the current at that instant, and *vice versa*. This stems from the fact that resistors do not store energy, they simply transform it.

B.1.3   CAPACITORS

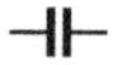

A capacitor (Figure B.3) is a component that does store energy, rather than dissipating it as heat. As the symbol suggests, a capacitor conceptually consists of two plates, separated by an insulator. Current, consisting of a flow of charge, transports charge away from one plate and onto the other, creating an electric field between the plates. This results in a voltage difference across the plates, and hence between the two terminals of the capacitor. The relationship between the voltage and the stored charge is

$$Q = CV \tag{B.3}$$

where the constant C is the *capacitance*, measured in Farads. Since we are usually interested in current, not charge storage, we can take the time derivative of this equation, recognizing that I, Q and V are functions of time:

$$I = \frac{dQ}{dt} = C\frac{dV}{dt} \tag{B.4}$$

This equation shows that the behavior of a capacitor is dynamic, that is, time dependent. For example, we cannot determine the current flowing through the terminals of a capacitor simply by measuring the instantaneous voltage across the terminals; we need to know how fast the voltage is changing. The time-dependent behavior is the main cause of delay in digital circuits. However, the fact that a capacitor stores energy also allows it to be used as a storage element, giving us time-dependent behavior where needed in digital systems.

Practical capacitors come in a range of capacitance, from a few picoFarads ($1pF = 10^{-12}F$) to thousands of microFarads ($1\mu F = 10^{-6}F$). Capacitances in digital circuits are usually toward the smaller end of the range, either in capacitors that are deliberately included in circuits, or in stray capacitances that arise between adjacent conductors in the circuit.

B.1.4   INDUCTORS

An inductor (Figure B.4) is another component that stores energy, but in the form of a magnetic field. As the symbol suggests, one form of inductor

FIGURE B.3   Symbol for a capacitor.

FIGURE B.4   Symbol for an inductor.

FIGURE B.5 Magnetic fields around current-carrying conductors.

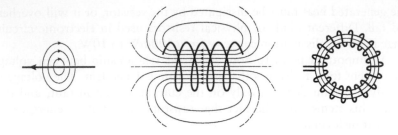

is a helical coil of wire. Even a straight wire conducting current creates a magnetic field, as shown at the left of Figure B.5. Coiling the wire concentrates the field, as shown in the center of the figure, and winding the coil around a toroidal core with magnetic permeability ideally confines the field to the core, as shown at the right of the figure. (In a real core, there will be some stray leakage field.)

If the current through the inductor is constant, the magnetic field does not change, and the voltage across the inductor is zero. When the current changes, the magnetic field changes. This changing magnetic field induces a voltage difference across the inductor, which tends to oppose the current change. The relationship between the voltage and current change is

$$V = L\frac{dI}{dt} \qquad (B.5)$$

where V and I are functions of time, and L is a constant called the *inductance*, measured in Henries. Thus, the behavior of an inductor, like that of a capacitor, is dynamic. Practical inductors used in electronic circuits have inductance in the range of a few microHenries ($1\mu H = 10^{-6}H$) to a few milliHenries ($1mH = 10^{-3}H$). However, most of the inductors we encounter in digital circuits are parasitic, rather than being components deliberately included in a design. We discuss the effects of parasitic inductance on digital circuits in Chapter 6.

B.1.5   MOSFETS

Nearly all digital circuits these days are made from *metal-oxide semiconductor field-effect transistors* (MOSFETs) fabricated on the surface of a crystalline silicon wafer. The manufacturing process adds atoms of impurities in selected areas of the silicon, and builds thin layers of material on the surface. The resulting structure for an *n-channel* MOSFET, along with its circuit symbol, is shown in Figure B.6. The pure silicon material forms a regular crystal lattice, with four bonds between neighboring atoms, since silicon is a Group IV element. The n regions are areas that have been

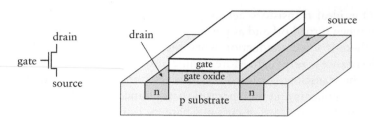

FIGURE B.6 Symbol for a MOSFET (left), and structure of an n-channel MOSFET (right).

doped (infused) with atoms of an element from Group V of the periodic table. These atoms have one more valence electron than the silicon atoms, so n-type material has a surplus of electrons. The p substrate is an area that has been doped with atoms of an element from Group III of the periodic table, which have one less valence electron than silicon. Hence, p-type material has "holes" corresponding to missing lattice bonds. The gate oxide is a thin insulating layer made of silicon dioxide, and the gate is a conducting layer made of polycrystalline silicon (polysilicon). Conducting contacts with the n-type source and drain terminals are usually made with metal, typically copper in modern circuits.

When the gate terminal is at the same voltage as the substrate, no current can flow between the source and drain terminals. However, when a positive voltage is applied to the gate, electrons are attracted toward it. If the gate voltage is greater than a threshold voltage, sufficient electrons are attracted to form an n-type channel immediately under the gate oxide. This channel can conduct current between the source and drain terminals. Note that no current flows between the gate and the source, drain, or substrate, due to the insulation of the gate oxide.

We can also manufacture *p-channel* MOSFETs on silicon wafers. The symbol is shown in Figure B.7. The structure is similar to that for n-channel MOSFETs, but the substrate is made of n-type material and the source and drain regions are p-type. In a p-channel MOSFET, the threshold voltage is negative with respect to the substrate. Applying a negative voltage to the gate repels electrons. We can think of this conversely as attracting holes to the area under gate oxide. When there are enough holes, a p-type channel is formed, allowing current to flow.

In complementary MOS (CMOS) digital circuits, we use both n-channel and p-channel MOSFETs. We arrange for the substrates of the n-channel MOSFETs to be connected to the ground voltage ($V_{SS} = 0\,V$) and the substrates of the p-channel MOSFETs to be connected to the positive power-supply voltage (V_{DD}). The threshold voltages for both are then between V_{SS} and V_{DD}. A gate voltage near V_{SS} causes an n-channel transistor to turn off (not to conduct current between source and drain) and a p-channel transistor to turn on (to conduct current between source and drain). Conversely, a gate voltage near V_{DD} causes an n-channel transistor to turn on and a p-channel transistor to turn off.

FIGURE B.7 Symbol for a p-channel MOSFET.

Our ideal transistors act as perfect insulators between source and drain when turned off and as perfect conductors when turned on. In practice, there is some conduction when the transistor is turned off, though the resistance is very high. More significantly, there is a nonzero resistance when the transistor is turned on. This resistance combines with that of other components and with parasitic capacitance and inductance to cause nonideal effects that we discuss in Chapters 1 and 6.

Another important property of MOSFETs is the gate capacitance. The gate and the substrate are two conductive plates separated by a thin layer of insulator. This is the structure described for a capacitor. The gate capacitance has some adverse effects, described in Chapter 1, but is is also central to construction of certain kinds of memory components described in Chapter 5.

B.1.6   DIODES

In Section B.1.5, we described the use of n-type and p-type regions in silicon to form MOSFETs. If we manufacture a silicon device with an n region immediately adjacent to a p region, we form a diode, illustrated in Figure B.8. For an ideal diode, if the voltage at the anode is positive with respect to the cathode, the diode is *forward biased* and acts like a conductor. On the other hand, if the voltage at the anode is negative with respect to the cathode, the diode is *reverse biased* and acts like an insulator.

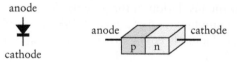

FIGURE B.8   Symbol for a diode (left), and its silicon structure (right).

In practice, the behavior of a diode is more complex than this two-state ideal model suggests. The actual current/voltage relationship is closer to that shown in Figure B.9. As the forward voltage increases, the current increases exponentially. For the ranges of voltages seen in digital circuits, we can approximate this behavior with a threshold voltage of around 0.7 V. Below the threshold, the current flow is small, approximating an insulator. Above the threshold, the current flow is large, approximating a conductor.

The two-state behavior of a diode arises from the electrical effects that occur in the narrow junction region between the p-type and n-type materials. In a silicon diode doped with the usual Group III and Group V elements used in digital IC manufacture (elements such as boron or aluminium in Group III and phosphorus in Group V), the power consumed in the diode is dissipated as heat. However, if the silicon is doped with other

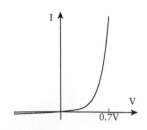

FIGURE B.9   Current/voltage characteristics of a diode.

combinations of elements, some of the energy is emitted in the form of photons. Such diodes are called *light-emitting diodes* (LEDs). The wavelengths of light emitted range from infrared through the visible spectrum to blue, depending on the materials used and the construction of the diode. For some LEDs, the substrate material is not silicon, but combinations of Group III and V elements manufactured to form crystalline lattices. In LEDs, the threshold voltage is typically larger than 0.7 V, ranging up to three or four volts for some high-energy short-wavelength LEDS.

B.1.7   BIPOLAR TRANSISTORS

In Section B.1.6, we saw that we can form a diode at the junction between n-type and p-type materials in silicon. If we sandwich three layers, we form a *bipolar transistor*. Figure B.10 shows an NPN transistor, in which a layer of p-type material is sandwiched between layers of n-type material.

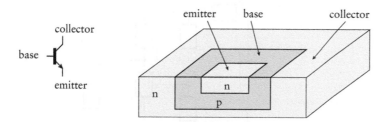

FIGURE B.10   Symbol for an NPN bipolar transistor (left), and its silicon structure (right).

The three terminals are called the *base*, *emitter* and *collector*. In an ideal NPN transistor, the current flowing between the emitter and collector terminals is proportional to the current flowing into the base, typically by a factor of 100 or so. This property allows a transistor to be used as an amplifier. However, for digital applications, we usually use transistors in an operating region called *saturation*. In that case, absence of base current turns the transistor off (no emitter-collector current), and sufficient base current turns the transistor on (the emitter-collector path conducts current). This is similar in operation to a MOSFET, but it is current that controls the device, not voltage.

We can also make a PNP transistor by sandwiching a layer of n-type material between layers of p-type material. The symbol for a PNP transistor is shown in Figure B.11. In a PNP transistor, the emitter-collector current is controlled by the current flowing out of the base terminal. Thus, it is complementary to an NPN transistor.

The main advantage of bipolar transistors over MOSFETs in digital applications is that, when individually packaged, they are somewhat more robust, being less susceptible to damage from electrostatic discharge.

FIGURE B.11   Symbol for a PNP bipolar transistor.

Hence, they are often used in connecting external input/output components to digital systems.

B.2   CIRCUITS

Now that we have identified the components that we use in digital systems, we describe the effects of connecting them together with wires. An ideal wire is a perfect conductor. We can consider it to be a two-terminal component with equal voltages at the terminals. However, it is generally simpler just to think of the wires in a circuit as joining terminals of other components, making their voltages equal and linking their current flows.

We define a *circuit* to be an interconnection of the terminals of a number of components. Figure B.12 shows an example of a circuit comprising several two- and three-terminal components. We call each point at which

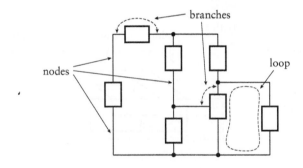

FIGURE B.12 A circuit consisting of interconnected components, with nodes, branches and loops.

two or more terminals are connected a *node*. Note that the terminals don't have to be connected at the same physical point. We think of a wire joining several terminals as a single node. We call each connection through a component from one node to another a *branch*. For a two-terminal component, the component itself forms a single branch. For a three-terminal component, there are three branches, one between each pair of terminals. Finally, we call a closed path through branches of a circuit a *loop*.

B.2.1   KIRCHHOFF'S LAWS

We can analyze a circuit to determine the current flowing through each branch (the *branch current*) and the voltage difference across each branch (the *branch voltage*). If all the branch currents and voltages are constant, the behavior of the circuit is static. Usually, however, the branch currents and voltages are time dependent, in which case the behavior is dynamic. In this section, we will identify two laws of circuit behavior that allow us to build systems of equations relating branch currents and voltages. We can then solve the equations to determine the circuit behavior. If the behavior is static, the equations will be algebraic. If the behavior is dynamic, we may have to solve differential equations.

The first law, *Kirchhoff's current law* (KCL), states that the current flowing out of any node and the current flowing into the node must be the same; that is, the sum of the currents must be zero. The intuition behind this law is that charge does not accumulate at nodes of a circuit.

The second law, *Kirchhoff's voltage law* (KVL), states that the sum of the branch voltages around any closed path in a circuit must be zero. The intuition here is that, if we start at any given node in the path, we determine the voltage at some other node by accumulating the branch voltages between the starting node and the other node. When we get back to the starting node, we should have accumulated the same voltage that we started with, since a node is an equipotential region.

B.2.2   SERIES AND PARALLEL R, C, AND L

Suppose we connect a voltage source and two resistors, R_1 and R_2, in the simple circuit shown in Figure B.13. We say the resistors are connected *in series*. Since the current flowing into and out of each node is zero, the same current I flows through both resistors. Since the accumulated voltage around the circuit loop is zero, the sum of the voltages across the resistors must be V. Recalling Ohm's law from our discussion of resistor behavior in Section B.1.2, we can write the following equation describing the circuit:

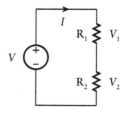

FIGURE B.13 A circuit with two resistors in series.

$$V = V_1 + V_2 = IR_1 + IR_2$$
$$= I(R_1 + R_2) \qquad (B.6)$$

This indicates that series combination of resistors behaves in the same way as a single resistor with resistance equal to the sum of the two individual resistances. The analysis can be extended to multiple resistors in series: the combination is equivalent to a single resistor whose resistance is the sum of the individual resistances:

$$R = R_1 + R_2 + R_3 + \cdots \qquad (B.7)$$

Since the sum of the voltages across the two resistors is V, the voltage at the node between the resistors is a fraction of V given by

$$V_2 = V \frac{R_2}{R_1 + R_2} \qquad (B.8)$$

For this reason, we sometimes call a series connection of resistors a *voltage divider*.

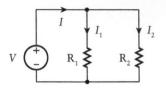

FIGURE B.14 A circuit with two resistors in parallel.

We can perform a similar analysis for a circuit comprising two resistors connected *in parallel*, as shown in Figure B.14. In this case, the voltage across each resistor is V, since each resistor is part of a loop containing the resistor and the voltage source. Since the net current into each node is zero, the current I is split between I_1 and I_2. Thus, we can write the equation:

$$I = I_1 + I_2 = \frac{V}{R_1} + \frac{V}{R_2}$$

$$= V\left(\frac{R_1 + R_2}{R_1 R_2}\right) \tag{B.9}$$

In other words, the parallel combination of resistors behaves in the same way as a single resistor with resistance given by:

$$R = \frac{R_1 R_2}{R_1 + R_2} \quad \text{or} \quad \frac{1}{R} = \frac{1}{R_1} + \frac{1}{R_2} \tag{B.10}$$

We can generalize this to a parallel connection of multiple resistors, which has an equivalent resistance given by

$$\frac{1}{R} = \frac{1}{R_1} + \frac{1}{R_2} + \frac{1}{R_3} + \cdots \tag{B.11}$$

Having performed this analysis for circuits with resistors in series and in parallel, we can perform similar analyses with capacitors and inductors. In the case of capacitors connected in series, we note that the same current flows through all capacitors, and so the same charge Q (being the integral of current) accumulates on each capacitor:

$$V = \frac{Q}{C} = V_1 + V_2 + V_3 + \cdots = \frac{Q}{C_1} + \frac{Q}{C_2} + \frac{Q}{C_3} + \cdots \tag{B.12}$$

Dividing by Q gives the equivalent capacitance for capacitors connected in series:

$$\frac{1}{C} = \frac{1}{C_1} + \frac{1}{C_2} + \frac{1}{C_3} + \cdots \tag{B.13}$$

For capacitors connected in parallel, the voltage across each capacitor is V, and the current is split between the branches. Integrating the currents gives the total charge as the sum of the individual capacitors' charges:

$$Q = VC = Q_1 + Q_2 + Q_3 + \cdots = VC_1 + VC_2 + VC_3 + \cdots \tag{B.14}$$

Dividing by V gives the equivalent capacitance for capacitors connected in parallel:

$$C = C_1 + C_2 + C_3 + \cdots \qquad \text{(B.15)}$$

Similarly, for inductors connected in series, the equivalent inductance is

$$L = L_1 + L_2 + L_3 + \cdots \qquad \text{(B.16)}$$

and for inductors connected in parallel, the equivalent inductance is

$$\frac{1}{L} = \frac{1}{L_1} + \frac{1}{L_2} + \frac{1}{L_3} + \cdots \qquad \text{(B.17)}$$

B.2.3   RC CIRCUITS

A number of nonideal effects in digital circuits arise from combinations of resistors and capacitors as circuit components. The components are usually not designed into the circuits but arise as properties of the transistors, wires, and packaging of real components. We will start with analysis of a simple circuit consisting of a resistor and capacitor connected in series, as shown in Figure B.15, driven by a switched voltage source between a low voltage (0V) and a high voltage (V). Assume initially, at time $t = 0$, that there is no charge accumulated on the capacitor, so the voltage V_C is 0V, and that the switch is in the 0V state. KVL and KCL tell us that V_R is also 0V and that the current I is zero.

Now suppose that the switch changes at $t = 0$ to the nonzero voltage V. Since there is no charge on the capacitor, V_C is still 0V, so the full voltage V appears across the resistor. This causes current to flow, starting to charge the capacitor. As charge accumulates over time, V_C increases and V_R decreases. To determine the voltage over time at the node between the resistor and the capacitor, we note that

$$V = V_R + V_C = IR + V_C \qquad \text{(B.18)}$$

and that

$$I = C \frac{dV_C}{dt} \qquad \text{(B.19)}$$

Substituting and rearranging gives the first-order differential equation:

$$RC \frac{dV_C}{dt} + V_C - V = 0 \qquad \text{(B.20)}$$

The solution of this equation gives the following formula for V_C as a function of time:

$$V_C(t) = V\left(1 - e^{-t/RC}\right) \qquad \text{(B.21)}$$

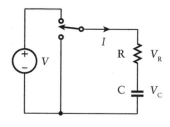

FIGURE B.15   A circuit with a resistor and a capacitor in series.

Figure B.16 shows a graph of the voltage plotted against time with $V = 5.0$ and $RC = 0.001$. The voltage approaches V asymptotically, reaching approximately 63% of the value after a time interval given by RC. We call this interval the *time constant* of the circuit.

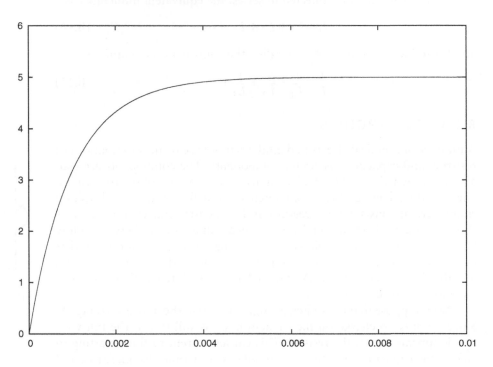

FIGURE B.16 Graph of capacitor voltage as it charges through a resistor from a voltage source.

We simplified the analysis above by assuming an initial voltage of 0V. In general, if the initial voltage across the capacitor is V_0, and the driving voltage changes from V_0 to V, the voltage on the capacitor is given by the function:

$$V_C(t) = V + (V_0 - V)\, e^{-t/RC} \tag{B.22}$$

Thus, if the driving voltage switches periodically between 0V and V, the voltage on the capacitor is as shown in Figure B.17.

B.2.4   RLC CIRCUITS

While we consider ideal wires to be pure conductors, real wires, especially long ones, have nonzero inductance. This, combined with parasitic capacitance and resistance in various components, leads to a number of the signal integrity issues that we discussed in Chapter 6. We can understand

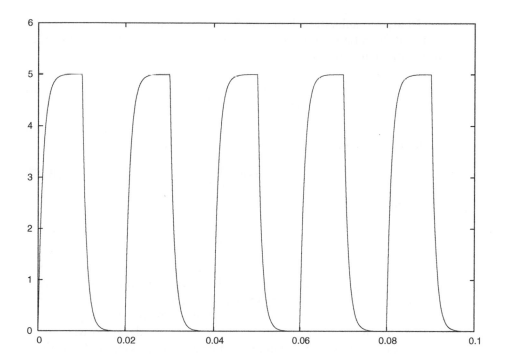

FIGURE B.17   Graph of capacitor voltage with the driving voltage switching periodically.

these effects by considering a series connection of a resistor, a capacitor and an inductor, as shown in Figure B.18. In practice, the resistor might be the effective on resistance of a transistor, the capacitor might be the gate capacitance of a transistor connected to a component input, and the inductor might be the series inductance of the wire connecting them.

We start our analysis of the RLC circuit by considering KCL at the nodes on either side of the resistor. For the node on the right, the sum of the currents through the capacitor and resistor is zero, giving the equation:

$$C\frac{dV_C}{dt} + \frac{V_C - V_L}{R} = 0 \qquad (B.23)$$

We can rearrange this to express V_L in terms of V_C:

$$V_L = RC\frac{dV_C}{dt} + V_C \qquad (B.24)$$

For the node on the left, the sum of the currents through the inductor and resistor is zero. We can determine the current through the inductor by integrating Equation B.5, giving the equation:

$$\frac{V_C - V_L}{R} + \frac{1}{L}\int_{-\infty}^{t} V_L d\tilde{t} = 0 \qquad (B.25)$$

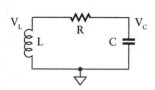

FIGURE B.18   A series RLC circuit.

Next, we substitute Equation B.24 into Equation B.25, divide by C, and differentiate with respect to t to yield the following second-order differential equation for V_C:

$$\frac{d^2V_C}{dt^2} + \frac{R}{L}\frac{dV_C}{dt} + \frac{1}{LC}V_C = 0 \tag{B.26}$$

This form of equation describes a resonant harmonic system with fundamental frequency:

$$\omega_0 = \frac{1}{\sqrt{LC}} \tag{B.27}$$

and damping factor:

$$\alpha = \frac{1}{2}\frac{R}{L} \tag{B.28}$$

If $\alpha < \omega_0$, we say the circuit is underdamped. From a nonzero initial state, the voltage V_C oscillates sinusoidally, with the amplitude decaying exponentially. If $\alpha = \omega_0$, we say the circuit is critically damped. The voltage V_C decays toward zero in the minimal time possible. Finally, if $\alpha > \omega_0$, we say the circuit is overdamped. The voltage V_C decays toward zero, but settles at a slower rate than in a critically damped circuit. These three cases are illustrated in Figure B.19.

FIGURE B.19   The behavior of underdamped, critically damped and overdamped RLC circuits.

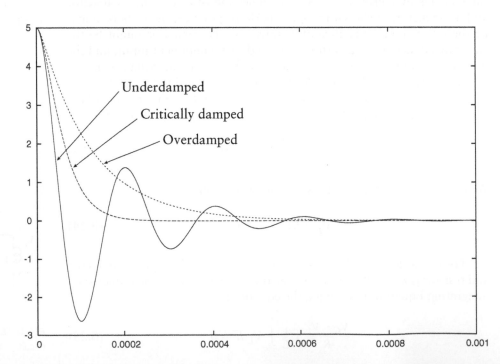

In practical systems, we are interested in the RLC circuit's response to changes in a driving voltage. The driven circuit is illustrated in Figure B.20. The voltage source represents an output from one part of a system driving the input to another part. The node at which V_C is measured represents the input. We can perform a similar analysis for this circuit, giving the following differential equation for V_C:

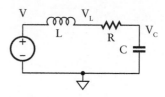

FIGURE B.20 An RLC circuit driven by a voltage source.

$$\frac{d^2 V_C}{dt^2} + \frac{R}{L}\frac{dV_C}{dt} + \frac{1}{LC}V_C = \frac{1}{LC}V \qquad (B.29)$$

The behavior of this circuit depends on the function V of time. For analyzing digital circuits, we consider V to be a step function, changing from zero to some given voltage level, or *vice versa*. The behavior in response to a step from a positive voltage to 0V is as shown in Figure B.19, since in that case $V = 0$. This indicates that the voltage seen on the capacitor may undershoot and "ring" before settling at its final level. Similarly, the behavior in response to a step from 0V to a positive voltage involves possible overshoot and ringing before settling. Increasing the resistance can help to dampen the ringing, at the expense of a slower transition to the final voltage level. Decreasing parasitic capacitance and inductance can help by increasing the fundamental frequency, making the circuit respond faster.

B.3   FURTHER READING

Foundations of Analog and Digital Electronic Circuits, Anant Agarwal and Jeffrey H. Lang, Morgan Kaufmann Publishers, 2005.

CMOS VLSI Design: A Circuits and Systems Perspective, 3rd Edition, Neil H. E. Weste and David Harris, Addison-Wesley, 2005.

APPENDIX C

VHDL FOR SYNTHESIS

Throughout this book, we have used VHDL, both for description of a digital circuit under design and for verification testbenches. Our design methodology involves use of synthesis tools to transform VHDL design descriptions expressed at the register-transfer level of abstraction into implementations at the gate level. The style of VHDL that we write for RTL design descriptions differs from that for testbenches, since not all aspects of testbenches can sensibly be implemented in hardware. Most RTL synthesis tools only accept VHDL descriptions that are written in a subset of the language that can sensibly be implemented in hardware. Testbench code, on the other hand, is more like code in general purpose programming languages, and can use the full suite of language features.

In this Appendix, we describe a subset of VHDL that is accepted by most RTL synthesis tools. We have used this subset in examples throughout the book. Most tools will accept a greater subset of the language than that described here. However, if we write code using such language features accepted by one tool, the code may not be portable for use with other tools.

C.1 DATA TYPES AND OPERATIONS

In this book, we have consistently used the standard type std_logic for single-bit signals and the standard type std_logic_vector for multibit signals. These types are accepted, and indeed preferred, for synthesis. We can perform operations on std_logic and std_logic_vector values, such as logical operations (and, or, not, etc.), shift operations (shift_left and shift_right), concatenation (&), indexing and slicing. These operations can be implemented by synthesis tools.

For numeric operations, we can take two approaches. The first is to use the types unsigned and signed and the related operations from the numeric_std package. This gives us fine control over the number of bits used to represent values. We just declare signals with the required index values. The second approach is to use the built-in types natural and integer. In this case, we rely on the synthesis tool to work out the appropriate number of bits to represent the values. We can give the tool a hint by defining a range of values allowed for each signal, for example:

```
signal num1 : integer range –50 to +50;
signal num2 : natural range 0 to 359;
```

A synthesis tool can infer a 7-bit 2s-complement representation for num1 and a 9-bit unsigned binary representation for num2. However, the number of bits used for operators and intermediate results in arithmetic expressions may vary between tools.

In examples in this book, we have also used the fixed-point types (ufixed and sfixed) and related operations defined in the fixed_pkg package. At the time of writing, not all synthesis tools accept use of these types and operations, since the packages have only recently been standardized. However, we should expect synthesis support for their use to become more widespread in the future.

We have also described the floating-point types and operations defined in the float_pkg package. These types and operations are less likely to be supported for synthesis, since the hardware required to implement them is much more complex than that for integer and fixed-point types. The hardware is typically sequential, and varies in internal organization depending on the application requirements. While synthesis may be feasible for some specialized applications, for other applications, we would expect to implement floating-point operations using library components or components created by generator tools.

C.2 COMBINATIONAL FUNCTIONS

Combinational circuits, discussed in Chapter 2, use the values of one or more inputs to determine the value of each output. VHDL allows us to express such behavior using *concurrent assignment statements* within an architecture body. Each such assignment statement assigns to a target signal that represents an output of a combinational logic block.

A simple assignment is of the form:

```
target  <=  expression;
```

in which the target is a signal name, and the expression represents the function performed by the combinational circuit. The expression includes the names of input signals, and combines their values using the operators that we mentioned in Section C.1. The expression must not include any reference to the target signal, since that would imply a feedback loop, and the circuit would not be combinational. Moreover, we must not imply feedback indirectly with multiple assignment statements, such as the following:

```
s1 <= a and b and s2;
s2 <= c or s1;
```

VHDL also allows us to write a number of combinational assignments as part of larger compound assignment statements. There are two forms. The first, a *selected assignment,* chooses between simple assignments based on the value of a controlling expression. The general form is

```
with select-expression select
    target <= expression-1 when choice-1,
              expression-2 when choice-2,
              ...
              expression-n when choice-n;
```

A synthesis tool would infer a multiplexer, as shown in Figure C.1. The select inputs of the multiplexer are connected to the output of the combinational logic inferred from the select expression. Each of the expressions before the when keywords would be synthesized to combinational logic connected to the particular data input of the multiplexer identified by the corresponding choice value. The choice values are expressions, but they must not involve any inputs. Usually, they are just literal values. Since all possible values of the select expression must be accounted for in the choices, the last choice can be the special keyword others, representing all values not mentioned previously in the selected assignment. We often use

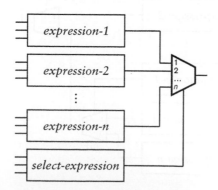

FIGURE C.1   Hardware inferred for a selected assignment.

an others choice when the select expression yields a std_logic_vector value, since we need to account for 'X' and 'Z' elements in the value.

The second form of compound assignment statement is a conditional assignment, which chooses among alternative simple assignments based on the outcomes of successive condition tests. The general form is

```
target <= expression-1 when condition-1 else
          expression-2 when condition-2 else
          ...
          expression-n;
```

Each of the expressions and conditions implies combinational logic. The outputs of the expression logic are connected to decision logic driven by the condition logic, such as that shown in Figure C.2. Since the conditions are tested one by one until a true condition is found, the decision logic is priority based, with conditions appearing earlier in the conditional assignment having priority over those appearing later. As a consequence, the propagation delay for the inferred logic may be as long as the sum of propagation delays of the inferred decision component. Of course, a synthesis tool may optimize the circuit, and may be able to implement the assignment as a single multiplexer if the conditions are mutually exclusive.

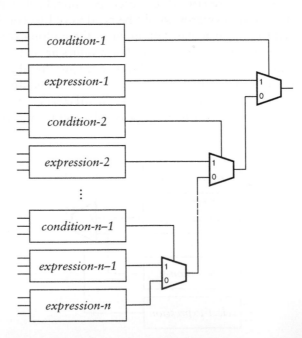

FIGURE C.2   Hardware inferred for a conditional assignment.

We can identify two special cases of the conditional assignment. The first is the following:

```
target <= '1' when condition else '0';
```

This special case arises when the target signal is of type std_logic, but the condition expression results in a Boolean (true or false) value. Typical examples involve conditions that include tests for equality or inequality of values. Since we cannot assign a Boolean value to a std_logic signal, we must use the form shown. A synthesis tool would infer combinational logic for the condition and connect its output to the target signal.

The second special case is the following:

```
target <= expression when condition else 'Z';
```

In this case, a synthesis tool would infer a tristate driver, as shown in Figure C.3, with its input connected to logic inferred from the expression, and its enable input connected to logic inferred from the condition. In the case of the target being a std_logic_vector value, the high-impedance value 'Z' is replaced by a vector of high-impedance values, such as "ZZZZZZZZ".

In some cases, we might find it more convenient to express combinational logic using a process statement rather than a collection of concurrent assignment statements. Typical cases are combinational logic blocks (such as those in finite-state machines) with multiple outputs all controlled by the same or similar conditions, or where the expressions are complex and are more readily understood by being broken into smaller subexpressions. For a synthesis tool to be able to infer combinational logic from a process, we must follow a number of rules. First, the format of the process must be

```
process-name: process (sensitivity-list) is
    declarations;
begin
    statements;
end process process-name;
```

The process name should be chosen to indicate the intended purpose of the combinational block. The sensitivity list is a list of all of the input

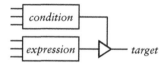

FIGURE C.3   Hardware inferred for an assignment with 'Z' as an alternative.

signals to the combinational block. This ensures that whenever any of the input signals changes value, the process determines new values for the output signals. If any signal is read in the process but is not in the sensitivity list, the synthesis tool would typically issue a warning message.

The declarations between the process header and the begin keyword can include declarations of constants, types, subtypes and variables. The last of these can be used for intermediate values in computing the outputs of the combinational logic. For example, instead of writing

```
s1 <= (c * a) + shift_left(b, 4) + offset1;
s2 <= (c * a) + shift_left(b, 4) + offset2;
```

we might declare a variable for the common part of the expression and use it as follows:

```
variable base : unsigned(17 downto 0);
...
base := (c * a) + shift_left(b, 4);
s1 <= base + offset1;
s2 <= base + offset2;
```

The statements within a process representing combinational logic should all be simple assignments or control-flow statements (if statements and case statements). All of the expressions and conditions in control-flow statements should imply combination logic. In particular, they should not use rising_edge or falling_edge, since those expressions are used to imply sequential logic. If we do use control-flow statements, we must ensure that each signal and variable is assigned in all possible paths through the process. Otherwise, there is some combination of input values for which no new output value is determined, implying storage for the previous value. The process would not exhibit combinational behavior in that case; instead, it would imply a level-sensitive latch.

The circuit inferred by a synthesis tool has a propagation delay determined by the gates in the circuit. The delay depends on the particular technology library used and the circuit's ultimate placement and routing. Any delay that we specify in a signal assignment statement has no effect on the synthesized circuit. Thus, we should generally not write assignments such as:

```
s <= a + b after 300 ps;
```

A synthesis tool would ignore the delay in the assignment, or perhaps issue a warning. We usually only write assignments with delays in test-bench models for stimulus generation.

C.3 SEQUENTIAL CIRCUITS

As we discussed in Chapter 4, most sequential systems use edge-triggered clocked timing. For RTL synthesis, we describe the storage elements involved in such systems (the flip-flops and registers) using processes based on a small number of templates. Whether a process represents a flip-flop or a register depends on whether the output signal assigned by the process is a single bit or a vector of bits, respectively.

The simplest form of flip-flop or register updates its output on every clock cycle. We use the following process template to express such a register:

```
process-name: process (clock) is
begin
  if rising_edge(clock) then
    target <= input;
  end if;
end process process-name;
```

In this and other templates for registers, the process name should be chosen to indicate the intended purpose of the register. A synthesis tool infers a register that updates the target signal with the value of the input signal on every rising edge of the clock signal. For a falling-edge-triggered register, we substitute falling_edge for rising_edge in the above process.

More frequently, we use registers with control signals. If the control signals are synchronous, that is, they take effect only on the active clock edge, we test for them in an inner if statement nested within the outer if statement. An example is the following process representing a register with synchronous reset and clock enable:

```
process-name: process (clock) is
begin
  if rising_edge(clock) then
    if reset = '1' then
      target <= "0000";
    elsif clock_en = '1' then
      target <= input;
```

(continued)

```
      end if;
    end if;
  end process process-name;
```

If, on the other hand, the control signals are asynchronous, we test for them before testing for the active clock edge. An example is the following process representing a flip-flop with asynchronous negative-logic reset and asynchronous positive-logic preset:

```
process-name: process (clock, reset, preset) is
begin
  if reset = '0' then
    target <= '0';
  if preset = '1' then
    target <= '1';
  elsif rising_edge(clock) then
    target <= input;
  end if;
end process process-name;
```

Note that we must include the asynchronous control signal names in the sensitivity list in such cases. We can combine synchronous and asynchronous control in a single register. This is illustrated by the following example, with asynchronous reset and synchronous clock enable:

```
process-name: process (clock, reset) is
begin
  if reset = '1' then
    target <= "00000000";
  elsif rising_edge(clock) then
    if clock_en = '1' then
      target <= input;
    end if;
  end if;
end process process-name;
```

RTL designs typically involve combinational logic connected to the inputs of registers. While we can use separate concurrent assignments and processes to model the logic and the registers, respectively, it is often clearer to combine the two. The registers are modeled by processes according to the templates we described above, and the combinational logic is modeled by expressions on the right-hand sides of the assignments within the processes. For example, the following process represents a combinational multiplier connected to the input of a register with synchronous control signals.

```
process-name: process (clock) is
begin
  if rising_edge(clock) then
    if reset = '1' then
      target <= "0000";
    elsif clock_en = '1' then
      target <= a * b;
    end if;
  end if;
end process process-name;
```

We can extend this principle to represent multiplexers and other more involved combinational circuits, using if statements and case statements within processes representing registers. For example, the following represents a register whose input comes from a multiplexer:

```
process-name: process (clock) is
begin
  if rising_edge(clock) then
    if reset = '1' then
      target <= "0000";
    elsif clock_en = '1' then
      case select_expr is
        when choice_1 =>
          target <= expression_1;
        when choice_2 =>
          target <= expression_2;
        ...
        when others =>
          target <= ...;
      end case;
    end if;
  end if;
end process process-name;
```

Two special cases of registers with combinational logic are counters and shift registers, each of which includes a combinational function of its output to determine its input. A counter simply increments or decrements its output, for example:

```
process-name: process (clock) is
begin
  if rising_edge(clock) then
```

(continued)

```
      if reset = '1' then
        target <= "0000";
      elsif count_en = '1' then
        target <= target + 1;
      end if;
    end if;
  end process process-name;
```

A shift register forms its input from a shifted version of its output, to which is concatenated an input bit:

```
process-name: process (clock) is
begin
  if rising_edge(clock) then
    if reset = '1' then
      target <= "0000";
    elsif shift_en = '1' then
      target <= data_in & target(3 downto 1);
    end if;
  end if;
end process process-name;
```

If an RTL design includes several registers that are all clocked together and all have the same control signals, we can combine them into a single process, for example:

```
process-name: process (clock) is
begin
  if rising_edge(clock) then
    if reset = '1' then
      target1 <= "0000";
      target2 <= "00000000";
    elsif clock_en = '1' then
      target1 <= expression1;
      target2 <= expression2;
    end if;
  end if;
end process process-name;
```

A synthesis tool would infer the required number of registers, with inputs for each taken from the corresponding inferred combinational logic. The combined process is a more succinct model than separate processes.

C.3.1 FINITE-STATE MACHINES

We introduced finite-state machines (FSMs) in Chapter 4, and described their use in the control sections of sequential digital systems. We use an enumeration type to specify the set of states for the FSM, for example:

```
type state_type is (state1, state2, state3, state4);
signal current_state, next_state : state_type;
```

Each of the values in the enumeration type represents one of the FSM states. A synthesis tool would typically select a binary encoding for the state values to optimize the hardware that it generates.

An FSM is usually implemented using a register to store the current state and combination logic for the next-state and output functions. We can write three separate processes for these elements, using the templates described previously, as follows:

```
state_reg : process (clock) is
begin
  if rising_edge(clock) then
    if reset = '1' then
      current_state <= initial-state;
    else
      current_state <= next_state;
    end if;
  end if;
end process state_reg;

next_state_logic : process (current_state,
                              input-1, input-2, ...) is
begin
  case current_state is
    when state1 =>
      if condition-1 then
        next_state <= state-value;
      elsif condition-2 then
        next_state <= state-value;
      ...
      else
        next_state <= state-value;
      end if;
    when state2 =>
      ...
```

(continued)

```
      end case;
  end process next_state_logic;

  output_logic : process (current_state,
                                input-1, input-2, ...) is
  begin
    case current_state is
      when state1 =>
        moore-output-1 <= value; moore-output-2 <= value; ...
        if condition-1 then
          mealy-output-1 <= value; mealy-output-2 <= value; ...
        elsif condition-2 then
          mealy-output-1 <= value; mealy-output-2 <= value; ...
        ...
        else
          mealy-output-1 <= value; mealy-output-2 <= value; ...
        end if;
      when state2 =>
        ...
    end case;
  end process output_logic;
```

Often, the processes representing the two combinational logic blocks may be combined, especially if the conditions governing the Mealy-style outputs are the same as those governing the state transitions. The combined process would be of the form:

```
  fsm_logic : process (current_state, input-1, input-2, ...) is
  begin
    case current_state is
      when state1 =>
        if condition-1 then
          next_state <= state-value;
          mealy-output-1 <= value; mealy-output-2 <= value; ...
        elsif condition-2 then
          next_state <= state-value;
          mealy-output-1 <= value; mealy-output-2 <= value; ...
        ...
        else
          next_state <= state-value;
          mealy-output-1 <= value; mealy-output-2 <= value; ...
        end if;
        moore-output-1 <= value; moore-output-2 <= value; ...
      when state2 =>
        ...
    end case;
  end process fsm_logic;
```

In some designs, the FSM outputs are Moore-style and are activated only during one or two states. In such cases, the output function can be expressed using simple concurrent assignments rather than using a process, for example:

```
moore-output-1 <= '1' when current_state = state1 else '0';
moore-output-2 <= '1' when current_state = state3
                          or current_state = state4 else
                  '0';
```

C.4 MEMORIES

RTL synthesis support for memories is not as well established as that for combinational and sequential circuits. Most ASIC designs would use memories generated using a memory component generator or provided as library cells by an IP vendor, since memory circuits can be highly customized in ASICs. On the other hand, synthesis tools can take advantage of the memory blocks built into most larger FPGAs, and so support inference of memories from VHDL code targeted at an FPGA implementation. We should consult the documentation for the tool we plan to use for an FPGA design to determine how it expects a memory to be described. In this section, we outline templates that should be compatible with most FPGA synthesis tools.

We represent the data stored in a memory using a signal of an array type. We need to declare the type, specifying the range of address values and the type of the data stored in each location, for example:

```
type memory_type is array (0 to 2**18-1)
                     of std_logic_vector(15 downto 0);
signal data_ram : memory_type;
```

This describes storage for a memory with 2^{18} locations, each of 16 bits. The memory is then described using a process that implements the read and write operations. For example, to describe a flow-through SSRAM, we can use a process of the form:

```
process-name : process (clock) is
begin
  if rising_edge(clock) then
```

(continued)

```
        if enable = '1' then
          if write = '1' then
            data_ram(to_integer(address)) <= data-in;
            data-out <= data-in;
          else
            data-out <= data_ram(to_integer(address));
          end if;
        end if;
    end if;
end process process-name;
```

Describing read-only memories (ROMs) is substantially simpler, since only read operations are required. Instead of using a signal for the data storage, we can use a constant initialized to the array of data values, for example:

```
type rom_type is array (0 to 128) of unsigned(11 downto 0);
constant data_rom : rom_type :=
  (X"000", X"021", X"1B3", X"7C0", ...);
```

Alternatively, a synthesis tool might provide a way of initializing a signal representing ROM storage to values saved in a file. This can be useful for ROMs containing instructions for an embedded processor.

The read operation for a ROM can be as simple as an assignment, such as the following:

```
data-out <= data_rom(to_integer(address));
```

This represents a combinational ROM, in which the output is a function of the address. Alternatively, if the ROM is to be implemented using a synchronous block memory in an FPGA, we can use a process of the following form to describe the ROM:

```
process-name : process (clock) is
begin
  if rising_edge(clock) then
    if enable = '1' then
      data-out <= data_rom(to_integer(address));
    end if;
  end if;
end process process-name;
```

APPENDIX D

THE GUMNUT
MICROCONTROLLER CORE

This Appendix is a complete reference for the Gumnut microcontroller core introduced in Chapter 7. We provide details of the instruction set and the bus interface for connecting memory and I/O controllers. Documentation for an assembler and its assembly language are included on the companion website.

D.1 THE GUMNUT INSTRUCTION SET

The Gumnut has an instruction memory of up to 4096 18-bit instructions (using 12-bit addresses) and a data memory of 256 bytes (using 8-bit addresses). When the CPU is reset, it clears the PC to 0 and starts executing instructions. Within the CPU, there are eight general purpose registers, named r0 through r7, that can hold data to be operated upon by instructions. Register r0 is special, in that it is hard-wired to have the value 0, and any updates to it are ignored. The CPU also has two single-bit condition-code registers called Z (zero) and C (carry). They are set to 1 or cleared to 0 depending on the result of certain instructions, and can be tested to decide among alternative courses of action in the program.

The complete Gumnut instruction set is summarized in Table D.1. For the arithmetic and logical instructions, *op2* is either a second register (*rs2*) or an immediate value (*immed*). Details of each instruction, including its instruction word encoding, are provided in this section.

D.1.1 ARITHMETIC AND LOGICAL INSTRUCTIONS

▶ add *rd, rs, rs2*

17 16 15 14	13 12 11	10 9 8	7 6 5	4 3	2 1 0
1 1 1 0	*rd*	*rs*	*rs2*		0 0 0

Add values in registers *rs* and *rs2*, and place result in register *rd*. Set C to carry out of the addition. Set Z to 1 if the result is 0, or to 0 otherwise.

INSTRUCTION	DESCRIPTION
Arithmetic and logical instructions	
add *rd*, *rs*, *op2*	Add *rs* and *op2*, result in *rd*
addc *rd*, *rs*, *op2*	Add *rs* and *op2* with carry, result in *rd*
sub *rd*, *rs*, *op2*	Subtract *op2* from *rs*, result in *rd*
subc *rd*, *rs*, *op2*	Subtract *op2* from *rs* with carry, result in *rd*
and *rd*, *rs*, *op2*	Logical AND of *rs* and *op2*, result in *rd*
or *rd*, *rs*, *op2*	Logical OR of *rs* and *op2*, result in *rd*
xor *rd*, *rs*, *op2*	Logical XOR of *rs* and *op2*, result in *rd*
mask *rd*, *rs*, *op2*	Logical AND of *rs* and NOT *op2*, result in *rd*
Shift instructions	
shl *rd*, *rs*, *count*	Shift *rs* value left *count* places, result in *rd*
shr *rd*, *rs*, *count*	Shift *rs* value right *count* places, result in *rd*
rol *rd*, *rs*, *count*	Rotate *rs* value left *count* places, result in *rd*
ror *rd*, *rs*, *count*	Rotate *rs* value right *count* places, result in *rd*
Memory and I/O instructions	
ldm *rd*, (*rs*) ± *offset*	Load to *rd* from memory
stm *rd*, (*rs*) ± *offset*	Store to memory from *rd*
inp *rd*, (*rs*) ± *offset*	Input to *rd* from input controller register
out *rd*, (*rs*) ± *offset*	Output to output controller register from *rd*
Branch instructions	
bz ±*disp*	Branch if Z is set
bnz ±*disp*	Branch if Z is not set
bc ±*disp*	Branch if C is set
bnc ±*disp*	Branch if C is not set
Jump instructions	
jmp *addr*	Jump to *addr*
jsb *addr*	Jump to subroutine at *addr*

TABLE D.1   The Gumnut instruction set.

(continued)

INSTRUCTION	DESCRIPTION
Miscellaneous instructions	
ret	Return from subroutine
reti	Return from interrupt
enai	Enable interrupts
disi	Disable interrupts
wait	Wait for interrupts
stby	Enter low-power standby mode

TABLE D.1 (*continued*)
The Gumnut instruction set.

▶ add *rd, rs, immed*

17 16 15 14	13 12 11	10 9 8	7 6 5 4 3 2 1 0
0 0 0 0	*rd*	*rs*	*immed*

Add value in register *rs* and the immediate operand *immed*, and place result in register *rd*. Set C to carry out of the addition. Set Z to 1 if the result is 0, or to 0 otherwise.

▶ addc *rd, rs, rs2*

17 16 15 14	13 12 11	10 9 8	7 6 5	4 3	2 1 0
1 1 1 0	*rd*	*rs*	*rs2*		0 0 1

Add values in registers *rs* and *rs2* and C, and place result in register *rd*. Set C to carry out of the addition. Set Z to 1 if the result is 0, or to 0 otherwise.

▶ addc *rd, rs, immed*

17 16 15 14	13 12 11	10 9 8	7 6 5 4 3 2 1 0
0 0 0 1	*rd*	*rs*	*immed*

Add value in register *rs*, the immediate operand *immed*, and C, and place result in register *rd*. Set C to carry out of the addition. Set Z to 1 if the result is 0, or to 0 otherwise.

▶ sub *rd, rs, rs2*

17 16 15 14	13 12 11	10 9 8	7 6 5	4 3	2 1 0
1 1 1 0	*rd*	*rs*	*rs2*		0 1 0

Subtract value in register *rs2* from value in register *rs*, and place result in register *rd*. Set C to borrow out of the subtraction. Set Z to 1 if the result is 0, or to 0 otherwise.

▶ sub *rd, rs, immed*

17 16 15	14 13 12	11 10 9	8 7 6	5 4 3 2 1 0
0 0 1 0	rd	rs		immed

Subtract immediate operand *immed* from value in register *rs*, and place result in register *rd*. Set C to borrow out of the subtraction. Set Z to 1 if the result is 0, or to 0 otherwise.

▶ subc *rd, rs, rs2*

17 16 15 14	13 12 11	10 9 8	7 6 5	4	3 2 1 0
1 1 1 0	rd	rs	rs2		0 1 1

Subtract value in register *rs2* and C from value in register *rs*, and place result in register *rd*. Set C to borrow out of the subtraction. Set Z to 1 if the result is 0, or to 0 otherwise.

▶ subc *rd, rs, immed*

17 16 15	14 13 12	11 10 9	8 7 6	5 4 3 2 1 0
0 0 1 1	rd	rs		immed

Subtract immediate operand *immed* and C from value in register *rs*, and place result in register *rd*. Set C to borrow out of the subtraction. Set Z to 1 if the result is 0, or to 0 otherwise.

▶ and *rd, rs, rs2*

17 16 15 14	13 12 11	10 9 8	7 6 5	4	3 2 1 0
1 1 1 0	rd	rs	rs2		1 0 0

Form logical AND of values in registers *rs* and *rs2*, and place result in register *rd*. Set C to 0. Set Z to 1 if the result is 0, or to 0 otherwise.

▶ and *rd, rs, immed*

17 16 15	14 13 12	11 10 9	8 7 6	5 4 3 2 1 0
0 1 0 0	rd	rs		immed

Form logical AND of value in register *rs* and the immediate operand *immed*, and place result in register *rd*. Set C to 0. Set Z to 1 if the result is 0, or to 0 otherwise.

▶ or *rd, rs, rs2*

17 16 15 14	13 12 11	10 9 8	7 6 5	4	3 2 1 0
1 1 1 0	rd	rs	rs2		1 0 1

Form logical inclusive-OR of values in registers *rs* and *rs2*, and place result in register *rd*. Set C to 0. Set Z to 1 if the result is 0, or to 0 otherwise.

▶ or *rd, rs, immed*

17 16 15	14 13 12	11 10 9	8 7 6	5 4 3 2 1 0
0 1 0 1	rd	rs		immed

Form logical inclusive-OR of value in register *rs* and the immediate operand *immed*, and place result in register *rd*. Set C to 0. Set Z to 1 if the result is 0, or to 0 otherwise.

▶ xor *rd, rs, rs2*

17 16 15 14	13 12 11	10 9 8	7 6 5	4 3	2 1 0
1 1 1 0	*rd*	*rs*	*rs2*		1 1 0

Form logical exclusive-OR of values in registers *rs* and *rs2*, and place result in register *rd*. Set C to 0. Set Z to 1 if the result is 0, or to 0 otherwise.

▶ xor *rd, rs, immed*

17	16 15 14	13 12 11	10 9 8	7 6 5 4 3 2 1 0
0	1 1 0	*rd*	*rs*	*immed*

Form logical exclusive-OR of value in register *rs* and the immediate operand *immed*, and place result in register *rd*. Set C to 0. Set Z to 1 if the result is 0, or to 0 otherwise.

▶ mask *rd, rs, rs2*

17 16 15 14	13 12 11	10 9 8	7 6 5	4 3	2 1 0
1 1 1 0	*rd*	*rs*	*rs2*		1 1 1

Form logical AND of value in register *rs* and logical NOT of value in *rs2*, and place result in register *rd*. Set C to 0. Set Z to 1 if the result is 0, or to 0 otherwise.

▶ mask *rd, rs, immed*

17	16 15 14	13 12 11	10 9 8	7 6 5 4 3 2 1 0
0	1 1 1	*rd*	*rs*	*immed*

Form logical AND of value in register *rs* and logical NOT of the immediate operand *immed*, and place result in register *rd*. Set C to 0. Set Z to 1 if the result is 0, or to 0 otherwise.

D.1.2 SHIFT INSTRUCTIONS

▶ shl *rd, rs, count*

17 16 15	14 13 12	11 10 9	8 7 6	5 4 3	2	1 0
1 1 0	*rd*	*rs*	*count*			0 0

Perform a logical shift left of the value in register *rs* by *count* places, and place result in register *rd*. Set C to the value of the last bit shifted past the left-hand end of the byte. Set Z to 1 if the result is 0, or to 0 otherwise.

▶ shr *rd, rs, count*

17 16 15	14 13 12	11 10 9	8 7 6	5 4 3	2	1 0
1 1 0	*rd*	*rs*	*count*			0 1

Perform a logical shift right of the value in register *rs* by *count* places, and place result in register *rd*. Set C to the value of the last bit shifted past the right-hand end of the byte. Set Z to 1 if the result is 0, or to 0 otherwise.

▶ rol *rd, rs, count*

17 16 15	14 13 12	11 10 9	8 7 6 5 4	3 2	1 0
1 1 0	rd	rs	count		1 0

Rotate the value in register *rs* left by *count* places, and place result in register *rd*. Set C to the value of the last bit rotated past the left-hand end of the byte. Set Z to 1 if the result is 0, or to 0 otherwise.

▶ ror *rd, rs, count*

17 16 15	14 13 12	11 10 9	8 7 6 5 4	3 2	1 0
1 1 0	rd	rs	count		1 1

Rotate the value in register *rs* right by *count* places, and place result in register *rd*. Set C to the value of the last bit rotated past the right-hand end of the byte. Set Z to 1 if the result is 0, or to 0 otherwise.

D.1.3   MEMORY AND I/O INSTRUCTIONS

▶ ldm *rd*, (*rs*) ± *offset*

17 16 15 14 13	12 11 10	9 8 7	6 5 4 3 2 1 0
1 0 0 0	rd	rs	offset

Load into register *rd* from the memory location whose address is the sum of the value in register *rs* and the value *offset*. C and Z are unaffected.

▶ stm *rd*, (*rs*) ± *offset*

17 16 15 14 13	12 11 10	9 8 7	6 5 4 3 2 1 0
1 0 0 1	rd	rs	offset

Store the value in register *rd* to the memory location whose address is the sum of the value in register *rs* and the value *offset*. C and Z are unaffected.

▶ inp *rd*, (*rs*) ± *offset*

17 16 15 14 13	12 11 10	9 8 7	6 5 4 3 2 1 0
1 0 1 0	rd	rs	offset

Input into register *rd* from the I/O controller register whose address is the sum of the value in register *rs* and the value *offset*. C and Z are unaffected.

▶ out *rd*, (*rs*) ± *offset*

17 16 15 14 13	12 11 10	9 8 7	6 5 4 3 2 1 0
1 0 1 1	rd	rs	offset

Output the value in register *rd* to the I/O controller register whose address is the sum of the value in register *rs* and the value *offset*. C and Z are unaffected.

D.1.4   BRANCH INSTRUCTIONS

▶ bz ±*disp*

17 16 15 14 13 12	11 10	9 8 7 6 5 4 3 2 1 0
1 1 1 1 0	0 0	*disp*

If Z is 1, branch by adding the value *disp* to the PC; otherwise, the PC is unaffected. C and Z are unaffected.

▶ bnz ±*disp*

17 16 15 14 13 12	11 10	9 8 7 6 5 4 3 2 1 0
1 1 1 1 0	0 1	*disp*

If Z is 0, branch by adding the value *disp* to the PC; otherwise, the PC is unaffected. C and Z are unaffected.

▶ bc ±*disp*

17 16 15 14 13 12	11 10	9 8 7 6 5 4 3 2 1 0
1 1 1 1 0	1 0	*disp*

If C is 1, branch by adding the value *disp* to the PC; otherwise, the PC is unaffected. C and Z are unaffected.

▶ bnc ±*disp*

17 16 15 14 13 12	11 10	9 8 7 6 5 4 3 2 1 0
1 1 1 1 0	1 1	*disp*

If C is 0, branch by adding the value *disp* to the PC; otherwise, the PC is unaffected. C and Z are unaffected.

D.1.5   JUMP INSTRUCTIONS

▶ jmp *addr*

17 16 15 14 13	12	11 10 9 8 7 6 5 4 3 2 1 0
1 1 1 1 0	0	*addr*

Jump by setting the PC to the value *addr*. C and Z are unaffected.

▶ jsb *addr*

17 16 15 14 13	12	11 10 9 8 7 6 5 4 3 2 1 0
1 1 1 1 0	1	*addr*

Jump to a subroutine by first pushing the value of the PC onto the return-address stack, and then setting the PC to the value *addr*. The return-address stack is eight entries deep. If the stack is full before execution of the instruction, the bottom entry is lost. C and Z are unaffected.

D.1.6 MISCELLANEOUS INSTRUCTIONS

▶ ret

17 16 15 14 13 12	11 10 9	8 7 6 5 4 3 2 1 0
1 1 1 1 1 0	0 0 0	

Return from a subroutine by setting the PC to the value in the top entry of the return-address stack and then popping the stack. If the stack is empty before execution of the instruction, the value copied to the PC is undefined.

▶ reti

17 16 15 14 13 12	11 10 9	8 7 6 5 4 3 2 1 0
1 1 1 1 1 0	0 0 1	

Return from an interrupt handler by setting the PC, C and Z to the values saved when the interrupt was acknowledged and by re-enabling interrupts. If this instruction is executed other than in an interrupt handler, the values written to PC, C and Z are undefined.

▶ enai

17 16 15 14 13 12	11 10 9	8 7 6 5 4 3 2 1 0
1 1 1 1 1 0	0 1 0	

Enable response to interrupts.

▶ disi

17 16 15 14 13 12	11 10 9	8 7 6 5 4 3 2 1 0
1 1 1 1 1 0	0 1 1	

Disable response to interrupts.

▶ wait

17 16 15 14 13 12	11 10 9	8 7 6 5 4 3 2 1 0
1 1 1 1 1 0	1 0 0	

Suspend execution and wait until an interrupt occurs.

▶ stby

17 16 15 14 13 12	11 10 9	8 7 6 5 4 3 2 1 0
1 1 1 1 1 0	1 0 1	

Enter low-power standby mode until an interrupt occurs.

D.2   THE GUMNUT BUS INTERFACE

The Gumnut microcontroller core uses Wishbone buses to connect to the instruction memory, the data memory, and the I/O controller ports, as shown in Figure D.1. Each of the buses uses classic single-read and single-write bus cycles, as described in the Wishbone bus specification included for reference on the companion website. The companion website also contains VHDL behavioral and synthesizable RTL models for the Gumnut, as well as models for an instruction memory and a data memory. These models can be used for embedded system designs implemented in FPGAs or ASICs. The website also contains a simple assembler, called gasm, together with documentation and sample assembly-language programs.

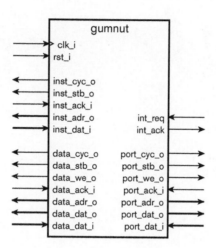

FIGURE D.1   Wishbone bus connections on the Gumnut core.

INDEX

Printed and bound by CPI Group (UK) Ltd, Croydon, CR0 4YY

03/10/2024

01040319-0001